Volume III: Chapters 14–17

Complete Solutions Guide to Accompany

Calculus
Fourth Edition

Larson/Hostetler/Edwards

Dianna L. Zook
The Pennsylvania State University

D. C. Heath and Company

Lexington, Massachusetts   Toronto

Preface

This solutions guide is a supplement to *Calculus with Analytic Geometry, Fourth Edition,* by Roland E. Larson, Robert P. Hostetler, and Bruce H. Edwards. Solutions to every exercise in the text are given with all essential algebraic steps included. There are three volumes in the complete set of solutions guides. Volume I contains Chapters 1–7, Volume II contains Chapters 8–13, and Volume III contains Chapters 14–17. Also available is a one-volume *Study and Solutions Guide to Accompany Calculus, Fourth Edition,* written by David E. Heyd, which contains worked-out solutions to *selected* representative exercises from the text.

I have made every effort to see that the solutions are correct. However, I would appreciate hearing about any errors or other suggestions for improvement.

I would like to thank several people who helped in the production of this guide: David E. Heyd, who assisted the authors of the text; Linda M. Bollinger, who was in charge of the production and typing of the guide; Randall R. Hammond, who produced the art; Kathy M. Evanoff and Paula M. Sibeto, who double-checked my solutions; and Jill D. Larson, who helped prepare the final copy of the guide. I would also like to thank the students in my mathematics classes. Finally, I would like to thank my husband, Ed Schlindwein, for his support during the many months I have worked on this project.

Dianna L. Zook
The Pennsylvania State University
Erie, Pennsylvania 16563

CONTENTS

CHAPTER 14
Functions of Several Variables

Section 14.1 Introduction to Functions of Several Variables

1. $f(x, y) = \dfrac{x}{y}$

  (a) $f(3, 2) = \dfrac{3}{2}$

  (b) $f(-1, 4) = -\dfrac{1}{4}$

  (c) $f(30, 5) = \dfrac{30}{5} = 6$

  (d) $f(5, y) = \dfrac{5}{y}$

  (e) $f(x, 2) = \dfrac{x}{2}$

  (f) $f(5, t) = \dfrac{5}{t}$

2. $f(x, y) = 4 - x^2 - 4y^2$

  (a) $f(0, 0) = 4$

  (b) $f(0, 1) = 4 - 0 - 4 = 0$

  (c) $f(2, 3) = 4 - 4 - 36 = -36$

  (d) $f(1, y) = 4 - 1 - 4y^2 = 3 - 4y^2$

  (e) $f(x, 0) = 4 - x^2 - 0 = 4 - x^2$

  (f) $f(t, 1) = 4 - t^2 - 4 = -t^2$

3. $f(x, y) = xe^y$

  (a) $f(5, 0) = 5e^0 = 5$

  (b) $f(3, 2) = 3e^2$

  (c) $f(2, -1) = 2e^{-1} = \dfrac{2}{e}$

  (d) $f(5, y) = 5e^y$

  (e) $f(x, 2) = xe^2$

  (f) $f(t, t) = te^t$

4. $g(x, y) = \ln|x + y|$

  (a) $g(2, 3) = \ln|2 + 3| = \ln 5$

  (b) $g(5, 6) = \ln|5 + 6| = \ln 11$

  (c) $g(e, 0) = \ln|e + 0| = 1$

  (d) $g(0, 1) = \ln|0 + 1| = 0$

  (e) $g(2, -3) = \ln|2 - 3| = \ln 1 = 0$

  (f) $g(e, e) = \ln|e + e| = \ln 2e$

$$= \ln 2 + \ln e = (\ln 2) + 1$$

5. $h(x, y, z) = \dfrac{xy}{z}$

  (a) $h(2, 3, 9) = \dfrac{(2)(3)}{9} = \dfrac{2}{3}$

  (b) $h(1, 0, 1) = \dfrac{(1)(0)}{1} = 0$

6. $f(x, y, z) = \sqrt{x + y + z}$

  (a) $f(0, 5, 4) = \sqrt{0 + 5 + 4} = 3$

  (b) $f(6, 8, -3) = \sqrt{6 + 8 - 3} = \sqrt{11}$

7. $f(x, y) = x \sin y$

  (a) $f\left(2, \dfrac{\pi}{4}\right) = 2 \sin \dfrac{\pi}{4} = \sqrt{2}$

  (b) $f(3, 1) = 3 \sin 1$

8. $V(r, h) = \pi r^2 h$

  (a) $V(3, 10) = \pi(3)^2(10) = 90\pi$

  (b) $V(5, 2) = \pi(5)^2(2) = 50\pi$

9. $f(x, y) = \displaystyle\int_x^y (2t - 3)\, dt$

  (a) $f(0, 4) = \displaystyle\int_0^4 (2t - 3)\, dt = \Big[t^2 - 3t\Big]_0^4 = 4$

  (b) $f(1, 4) = \displaystyle\int_1^4 (2t - 3)\, dt = \Big[t^2 - 3t\Big]_1^4 = 6$

10. $g(x, y) = \int_x^y \frac{1}{t} \, dt$

(a) $g(4, 1) = \int_4^1 \frac{1}{t} \, dt = \ln |t| \Big]_4^1 = -\ln 4$

(b) $g(6, 3) = \int_6^3 \frac{1}{t} \, dt = \ln |t| \Big]_6^3 = \ln 3 - \ln 6 = \ln\left(\frac{1}{2}\right)$

11. $f(x, y) = \sqrt{4 - x^2 - y^2}$
Domain: $4 - x^2 - y^2 \geq 0$
$$x^2 + y^2 \leq 4$$
$$\{(x, y) : x^2 + y^2 \leq 4\}$$
Range: $0 \leq z \leq 2$

12. $f(x, y) = \sqrt{4 - x^2 - 4y^2}$
Domain: $4 - x^2 - 4y^2 \geq 0$
$$x^2 + 4y^2 \leq 4$$
$$\frac{x^2}{4} + \frac{y^2}{1} \leq 1$$
$$\left\{(x, y) : \frac{x^2}{4} + \frac{y^2}{1} \leq 1\right\}$$
Range: $0 \leq z \leq 2$

13. $f(x, y) = \arcsin(x + y)$
Domain: $\{(x, y) : -1 \leq x + y \leq 1\}$
Range: $-\frac{\pi}{2} \leq z \leq \frac{\pi}{2}$

14. $f(x, y) = \arccos \frac{y}{x}$
Domain: $\left\{(x, y) : -1 \leq \frac{y}{x} \leq 1\right\}$
Range: $0 \leq z \leq \pi$

15. $z = \frac{x + y}{xy}$
Domain: $\{(x, y) : x \neq 0, \ y \neq 0\}$
Range: all real numbers

16. $z = \frac{xy}{x - y}$
Domain: $\{(x, y) : x \neq y\}$
Range: all real numbers

17. $f(x, y) = \ln(4 - x - y)$
Domain: $4 - x - y > 0$
$$x + y < 4$$
$$\{(x, y) : \ y < -x + 4\}$$
Range: all real numbers

18. $f(x, y) = \ln(4 - xy)$
Domain: $4 - xy > 0$
$$xy < 4$$
$$\{(x, y) : \ xy < 4\}$$
Range: all real numbers

19. $f(x, y) = e^{x/y}$
Domain: $\{(x, y) : \ y \neq 0\}$
Range: $z > 0$

20. $f(x, y) = x^2 + y^2$
Domain: $\{(x, y) : \ x$ any real number, y any real number$\}$
Range: $z \geq 0$

21. $g(x, y) = \frac{1}{xy}$
Domain: $\{(x, y) : x \neq 0, \ y \neq 0\}$
Range: all real numbers except zero

22. $g(x, y) = x\sqrt{y}$
Domain: $\{(x, y) : y \geq 0\}$
Range: all real numbers

23. $z = e^{1-x^2-y^2}$

Level curves:
$$c = e^{1-x^2-y^2}$$
$$\ln c = 1 - x^2 - y^2$$
$$x^2 + y^2 = 1 - \ln c$$

Circles centered at $(0, 0)$
Matches (c)

24. $z = e^{1-x^2+y^2}$

Level curves:
$$c = e^{1-x^2+y^2}$$
$$\ln c = 1 - x^2 + y^2$$
$$x^2 - y^2 = 1 - \ln c$$

Hyperbolas centered at $(0, 0)$
Matches (d)

25. $z = |y|^{1+|x|}$

Level curves:
$$c = |y|^{1+|x|}$$
$$c^{1/(1+|x|)} = |y|$$
$$\pm c^{1/(1+|x|)} = y$$

Matches (f)

26. $z = -\dfrac{x}{y^2 - x}$

Level curves:
$$c = -\frac{x}{y^2 - x}$$
$$cy^2 - cx = -x$$
$$cy^2 = x(c - 1)$$
$$\frac{c}{c-1}y^2 = x$$

Parabolas with vertex $(0, 0)$
Matches (b)

27. $z = \ln|y - x^2|$

Level curves:
$$c = \ln|y - x^2|$$
$$\pm e^c = y - x^2$$
$$y = x^2 \pm e^c$$

Parabolas
Matches (e)

28. $z = \cos\left(\dfrac{x^2 + 2y^2}{4}\right)$

Level curves:
$$c = \cos\left(\frac{x^2 + 2y^2}{4}\right)$$
$$\cos^{-1} c = \frac{x^2 + 2y^2}{4}$$
$$x^2 + 2y^2 = 4\cos^{-1} c$$

Ellipses
Matches (a)

29. $z = 4 - x^2 - y^2$

Paraboloid
Domain: entire xy-plane
Range: $z \leq 4$

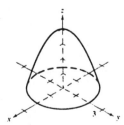

30. $z = \sqrt{x^2 + y^2}$

Cone
Domain of f: entire xy-plane
Range: $z \geq 0$

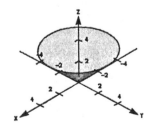

31. $f(x, y) = y^2$

Since the variable x is missing, the surface is a cylinder with rulings parallel to the x-axis. The generating curve is $z = y^2$. The domain is the entire xy-plane and the range is $z \geq 0$.

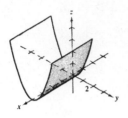

32. $z = y^2 - x^2 + 1$

Hyperbolic paraboloid

Domain: entire xy-plane

Range: $-\infty < z < \infty$

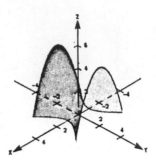

33. $f(x, y) = 6 - 2x - 3y$

Plane

Domain: entire xy-plane

Range: $-\infty < z < \infty$

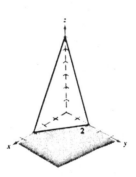

34. $f(x, y) = \dfrac{1}{12}\sqrt{144 - 16x^2 - 9y^2}$

Semi-ellipsoid

Domain: set of all points lying on or inside

the ellipse $\dfrac{x^2}{9} + \dfrac{y^2}{16} = 1$

Range: $0 \leq z \leq 1$

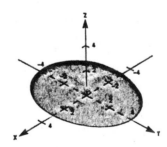

35. $f(x, y) = e^{-x}$

Since the variable y is missing, the surface is a cylinder with rulings parallel to the y-axis. The generating curve is $z = e^{-x}$. The domain is the entire xy-plane and the range is $z > 0$.

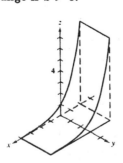

36. $f(x, y) = \begin{cases} xy, & x \geq 0, \ y \geq 0 \\ 0, & \text{elsewhere} \end{cases}$

Domain of f: entire xy-plane

Range: $z \geq 0$

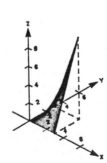

37. $f(x, y) = x^2 + y^2$

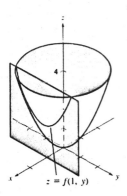

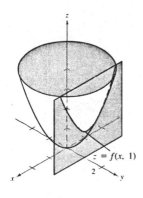

38. $z = f(x, y) = xy$

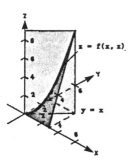

39. $f(x, y) = \sqrt{25 - x^2 - y^2}$

The level curves are of the form $c = \sqrt{25 - x^2 - y^2}$, $x^2 + y^2 = 25 - c^2$. Thus, the level curves are circles of radius 5 or less, centered at the origin.

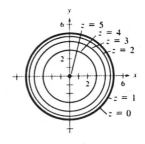

40. $f(x, y) = x^2 + y^2$

The level curves are of the form $x^2 + y^2 = c$. Thus, the level curves are circles centered at the origin.

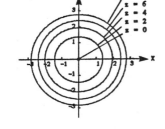

41. $f(x, y) = xy$

The level curves are hyperbolas of the form $xy = c$.

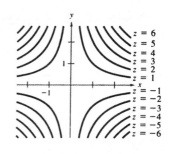

42. $f(x, y) = 6 - 2x - 3y$

The level curves are of the form

$6 - 2x - 3y = c$ or $2x + 3y = 6 - c$.

Thus, the level curves are straight lines with a slope of $-\frac{2}{3}$.

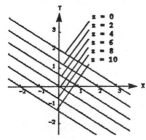

43. $f(x, y) = \dfrac{x}{x^2 + y^2}$

The level curves are of the form

$$c = \frac{x}{x^2 + y^2}$$

$$x^2 - \frac{x}{c} + y^2 = 0$$

$$\left(x - \frac{1}{2c}\right)^2 + y^2 = \left(\frac{1}{2c}\right)^2.$$

Thus, the level curves are circles passing through the origin and centered at $(1/2c, 0)$.

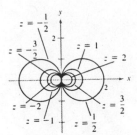

44. $f(x, y) = \arctan\left(\dfrac{y}{x}\right)$

The level curves are of the form

$$c = \arctan\left(\frac{y}{x}\right)$$

$$\tan(c) = \frac{y}{x}$$

$$y = (\tan c)x.$$

Thus, the level curves are lines through the origin.

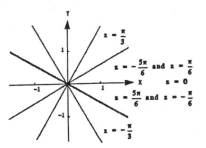

45. $f(x, y) = \ln(x - y)$

The level curves are of the form

$$c = \ln(x - y)$$

$$e^c = x - y$$

$$y = x - e^c.$$

Thus, the level curves are parallel lines of slope 1 passing through the fourth quadrant.

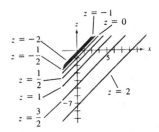

46. $f(x, y) = \dfrac{x + y}{x - y}$

The level curves are of the form

$$c = \frac{x + y}{x - y}$$

$$y = \left(\frac{c - 1}{c + 1}\right)x.$$

Thus, the level curves are lines passing through the origin.

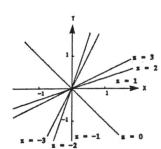

47. $f(x, y) = e^{xy}$

The level curves are of the form

$c = e^{xy}, \quad \ln(c) = xy.$

Thus, the level curves are hyperbolas centered at the origin with the x- and y-axes as asymptotes.

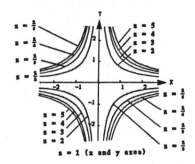

48. $f(x, y) = \cos(x + y)$

The level curves are of the form

$c = \cos(x + y), \quad \arccos(c) = x + y.$

Thus, the level curves are lines of slope -1.

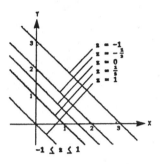

49. $f(x, y, z) = x - 2y + 3z$

$c = 6$

$6 = x - 2y + 3z$

Plane

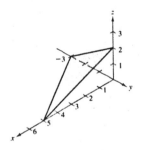

50. $f(x, y, z) = 4x + y + 2z$

$c = 4$

$4 = 4x + y + 2z$

Plane

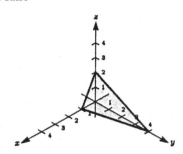

51. $f(x, y, z) = x^2 + y^2 + z^2$

$c = 9$

$9 = x^2 + y^2 + z^2$

Sphere

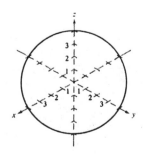

52. $f(x, y, z) = x^2 + y^2 - z$

$c = 1$

$1 = x^2 + y^2 - z$

Elliptic paraboloid

Vertex: $(0, 0, -1)$

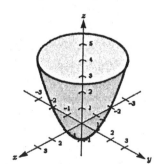

53. $f(x, y, z) = 4x^2 + 4y^2 - z^2$

$c = 0$

$0 = 4x^2 + 4y^2 - z^2$

Elliptic cone

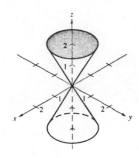

54. $f(x, y, z) = \sin x - z$

$c = 0$

$0 = \sin x - z$ or $z = \sin x$

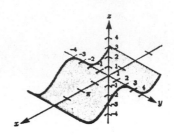

55. $N(d, L) = \left(\dfrac{d-4}{4}\right)^2 L$

(a) $N(22, 12) = \left(\dfrac{22-4}{4}\right)^2 (12) = 243$ board-feet

(b) $N(30, 12) = \left(\dfrac{30-4}{4}\right)^2 (12) = 507$ board-feet

56. $A(r, t) = 1000e^{rt}$

Rate	Number of years			
	5	10	15	20
0.08	$1491.82	$2225.54	$3320.12	$4953.03
0.10	$1648.72	$2718.28	$4481.69	$7389.06
0.12	$1822.12	$3320.12	$6049.65	$11,023.18
0.14	$2013.75	$4055.20	$8166.17	$16,444.65

57. $W(x, y) = \dfrac{1}{x-y}, \quad y < x$

(a) $W(15, 10) = \dfrac{1}{15-10} = \dfrac{1}{5}$ hr = 12 min

(b) $W(12, 9) = \dfrac{1}{12-9} = \dfrac{1}{3}$ hr = 20 min

(c) $W(12, 6) = \dfrac{1}{12-6} = \dfrac{1}{6}$ hr = 10 min

(d) $W(4, 2) = \dfrac{1}{4-2} = \dfrac{1}{2}$ hr = 30 min

58. $f(x, y) = 100x^{0.6}y^{0.4}$

$\quad f(2x, 2y) = 100(2x)^{0.6}(2y)^{0.4}$

$\qquad = 100(2)^{0.6}x^{0.6}(2)^{0.4}y^{0.4}$

$\qquad = 100(2)^{0.6}(2)^{0.4}x^{0.6}y^{0.4}$

$\qquad = 2[100x^{0.6}y^{0.4}]$

$\qquad = 2f(x, y)$

59. $T = 600 - 0.75x^2 - 0.75y^2$

The level curves are of the form

$c = 600 - 0.75x^2 - 0.75y^2$

$x^2 + y^2 = \dfrac{600 - c}{0.75}$

Thus, the level curves are circles centered at the origin.

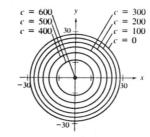

60. $PV = kT$, $\quad 20(2600) = k(40)$

(a) $k = \dfrac{20(2600)}{40} = 1300$

(b) $P = \dfrac{kT}{V} = 1300\left(\dfrac{T}{V}\right)$

The level curves are of the form

$c = 1300\left(\dfrac{T}{V}\right)$

$V = \dfrac{1300}{c}T$

Thus, the level curves are lines through the origin with slope $\dfrac{1300}{c}$.

61. $C = 0.75xy + \quad 2(0.40)xz \quad + 2(0.40)yz$

$\qquad$ base $\;+\;$ front & back $\;+\;$ two ends

$\quad = 0.75xy + 0.80(xz + yz)$

62. $V = \pi r^2 l + \dfrac{4}{3}\pi r^3$

$\qquad = \dfrac{\pi r^2}{3}(3l + 4r)$

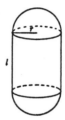

63. (a) Highest pressure at C

(b) Lowest pressure at A

(c) Highest wind velocity at B

64. Southwest

65. $f(x,\ y) = 4x^2 + y^2$

Level curves: $c = 4x^2 + y^2$

Ellipses

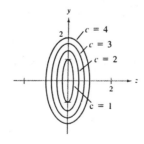

66. $f(x,\ y) = \dfrac{y^2}{7} - \dfrac{x^2}{5}$

Level curves: $c = \dfrac{y^2}{7} - \dfrac{x^2}{5}$

Hyperbolas

Section 14.2 Limits and Continuity

1. $\lim\limits_{(x,y)\to(a,b)} [f(x,\ y) - g(x,\ y)] = \lim\limits_{(x,y)\to(a,b)} f(x,\ y) - \lim\limits_{(x,y)\to(a,b)} g(x,\ y) = 5 - 3 = 2$

2. $\lim\limits_{(x,y)\to(a,b)} \left[\dfrac{4f(x,\ y)}{g(x,\ y)}\right] = \dfrac{4\left[\lim\limits_{(x,y)\to(a,b)} f(x,\ y)\right]}{\lim\limits_{(x,y)\to(a,b)} g(x,\ y)} = \dfrac{4(5)}{3} = \dfrac{20}{3}$

3. $\lim\limits_{(x,y)\to(a,b)} [f(x,\ y)g(x,\ y)] = \left[\lim\limits_{(x,y)\to(a,b)} f(x,\ y)\right]\left[\lim\limits_{(x,y)\to(a,b)} g(x,\ y)\right] = 5(3) = 15$

4. $\lim\limits_{(x,y)\to(a,b)} \left[\dfrac{f(x,\ y) - g(x,\ y)}{f(x,\ y)}\right] = \dfrac{\lim\limits_{(x,y)\to(a,b)} f(x,y) - \lim\limits_{(x,y)\to(a,b)} g(x,y)}{\lim\limits_{(x,y)\to(a,b)} f(x,y)} = \dfrac{5-3}{5} = \dfrac{2}{5}$

5. $\lim\limits_{(x,y)\to(2,1)} (x + 3y^2) = 2 + 3(1)^2 = 5$
Continuous

6. $\lim\limits_{(x,y)\to(0,0)} (5x + 3xy + y + 1) = 0 + 0 + 0 + 1 = 1$
Continuous

7. $\lim\limits_{(x,y)\to(2,4)} \dfrac{x+y}{x-y} = \dfrac{2+4}{2-4} = -3$
Continuous for $x \neq y$

8. $\lim\limits_{(x,y)\to(1,1)} \dfrac{x}{\sqrt{x+y}} = \dfrac{1}{\sqrt{1+1}} = \dfrac{\sqrt{2}}{2}$
Continuous for $x + y > 0$

9. $\lim\limits_{(x,y)\to(0,1)} \dfrac{\arcsin(x/y)}{1 + xy} = \arcsin 0 = 0$
Continuous for $xy \neq -1, \quad y \neq 0$

10. $\lim\limits_{(x,y)\to(\pi/4,2)} y\sin(xy) = 2\sin\dfrac{\pi}{2} = 2$
Continuous

11. $\lim\limits_{(x,y)\to(0,0)} e^{xy} = e^0 = 1$
Continuous

12. $\lim\limits_{(x,y)\to(1,1)} \dfrac{xy}{x^2 + y^2} = \dfrac{1}{2}$
Continuous except at $(0,\ 0)$

13. $\lim\limits_{(x,y,z)\to(1,2,5)} \sqrt{x+y+z} = \sqrt{8} = 2\sqrt{2}$
Continuous for $x + y + z \geq 0$

14. $\lim\limits_{(x,y,z)\to(2,0,1)} xe^{yz} = 2e^0 = 2$
Continuous

15. $f(x,\ y) = \dfrac{-xy^2}{x^2 + y^4}$

Along the line $x = y^2$: $\lim\limits_{(x,y)\to(0,0)} \left(\dfrac{-xy^2}{x^2 + y^4}\right) = \lim\limits_{(x,y)\to(0,0)} \left(\dfrac{-y^4}{2y^4}\right) = -\dfrac{1}{2}$

Along the line $x = -y^2$: $\lim\limits_{(x,y)\to(0,0)} \left(\dfrac{-xy^2}{x^2 + y^4}\right) = \lim\limits_{(x,y)\to(0,0)} \left(\dfrac{y^4}{2y^4}\right) = \dfrac{1}{2}$

Therefore, the limit does not exist.

16. $f(x,\ y) = \dfrac{x^2}{(x^2 + 1)(y^2 + 1)}$

$\lim\limits_{(x,y)\to(0,0)} \left(\dfrac{x^2}{(x^2 + 1)(y^2 + 1)}\right) = \dfrac{0}{(0+1)(0+1)} = 0$

17. $\displaystyle\lim_{(x,y)\to(0,0)} \frac{\sin(x^2 + y^2)}{x^2 + y^2} = \lim_{r\to 0}\frac{\sin r^2}{r^2} = \lim_{r\to 0}\frac{2r\cos r^2}{2r} = \lim_{r\to 0}\cos r^2 = 1$

Continuous except at $(0, 0)$

(Use polar coordinates and L'Hôpital's Rule.)

18. $\displaystyle\lim_{(x,y)\to(0,0)} e^{xy} = 1$

Continuous

19. $\displaystyle\lim_{(x,y)\to(0,0)} \frac{xy}{x^2 + y^2}$

Along $y = x$: $\displaystyle\lim_{(x,y)\to(0,0)} \frac{xy}{x^2 + y^2} = \lim_{x\to 0}\frac{x^2}{2x^2} = \lim_{x\to 0}\frac{1}{2} = \frac{1}{2}$

Along $y = 0$: $\displaystyle\lim_{(x,y)\to(0,0)} \frac{xy}{x^2 + y^2} = \lim_{x\to 0}\frac{0}{x^2} = 0$

Therefore, the limit does not exist. Continuous except at $(0, 0)$

20. $\displaystyle\lim_{(x,y)\to(0,0)} \left[1 - \frac{\cos(x^2 + y^2)}{x^2 + y^2}\right] = -\infty$

Continuous except at $(0, 0)$

21. $\displaystyle\lim_{(x,y)\to(0,0)} (\sin x + \cos y) = 1$

Continuous

22. $\displaystyle\lim_{(x,y)\to(0,0)} \frac{y}{x^2 + y^2}$

Along $y = 0$: $\displaystyle\lim_{(x,y)\to(0,0)} \frac{y}{x^2 + y^2} = \frac{0}{x^2} = 0$

Along $y = x$: $\displaystyle\lim_{(x,y)\to(0,0)} \frac{y}{x^2 + y^2} = \lim_{x\to 0}\frac{x}{2x^2} = \lim_{x\to 0}\frac{1}{2x} = \infty$

Therefore, the limit does not exist. Continuous except at $(0, 0)$

23. $\displaystyle\lim_{(x,y)\to(0,0)} \ln(x^2 + y^2) = \ln(0) = -\infty$

Continuous except at $(0, 0)$

24. $\displaystyle\lim_{(x,y)\to(0,0)} \frac{2x - y^2}{2x^2 + y}$

Along $y = 0$: $\displaystyle\lim_{(x,y)\to(0,0)} \frac{2x - y^2}{2x^2 + y} = \lim_{x\to 0}\frac{2x}{2x^2} = \lim_{x\to 0}\frac{1}{x} = \infty$

Along $x = 0$: $\displaystyle\lim_{(x,y)\to(0,0)} \frac{2x - y^2}{2x^2 + y} = \lim_{y\to 0}\frac{-y^2}{y} = \lim_{y\to 0}(-y) = 0$

Therefore, the limit does not exist. Discontinous at $y = -2x^2$

25. $\displaystyle\lim_{(x,y)\to(0,0)} \frac{\sin(x^2 + y^2)}{x^2 + y^2} = \lim_{r\to 0}\frac{\sin r^2}{r^2} = \lim_{r\to 0}\frac{2r\cos r^2}{2r} = \lim_{r\to 0}\cos r^2 = 1$

26. $\lim\limits_{(x,y)\to(0,0)} \dfrac{xy^2}{x^2+y^2} = \lim\limits_{r\to0} \dfrac{(r\cos\theta)(r^2\sin^2\theta)}{r^2} = \lim\limits_{r\to0} (r\cos\theta\sin^2\theta) = 0$

27. $\lim\limits_{(x,y)\to(0,0)} \dfrac{x^3+y^3}{x^2+y^2} = \lim\limits_{r\to0} \dfrac{r^3(\cos^3\theta+\sin^3\theta)}{r^2} = 0$

28. $\lim\limits_{(x,y)\to(0,0)} \dfrac{x^2y^2}{x^2+y^2} = \lim\limits_{r\to0} \dfrac{r^4\cos^2\theta\sin^2\theta}{r^2} = \lim\limits_{r\to0} r^2\cos^2\theta\sin^2\theta = 0$

29. $f(x,\ y,\ z) = \dfrac{1}{\sqrt{x^2+y^2+z^2}}$

Continuous except at $(0,\ 0,\ 0)$

30. $f(x,\ y,\ z) = \dfrac{z}{x^2+y^2-4}$

Continuous for $x^2+y^2\neq4$

31. $f(x,\ y,\ z) = \dfrac{\sin z}{e^x+e^y}$

Continuous

32. $f(x,\ y,\ z) = xy\sin z$

Continuous

33. $\quad\quad f(t) = t^2$

$g(x,\ y) = 3x - 2y$

$f(g(x,\ y)) = f(3x - 2y)$

$\quad\quad = (3x-2y)^2 = 9x^2 - 12xy + 4y^2$

Continuous

34. $\quad\quad f(t) = \dfrac{1}{t}$

$g(x,\ y) = x^2 + y^2$

$f(g(x,\ y)) = f(x^2+y^2) = \dfrac{1}{x^2+y^2}$

Continuous except at $(0,\ 0)$

35. $\quad\quad f(t) = \dfrac{1}{t}$

$g(x,\ y) = 3x - 2y$

$f(g(x,\ y)) = f(3x-2y) = \dfrac{1}{3x-2y}$

Continuous for $y\neq\dfrac{3x}{2}$

36. $\quad\quad f(t) = \dfrac{1}{4-t}$

$g(x,\ y) = x^2 + y^2$

$f(g(x,\ y)) = f(x^2+y^2) = \dfrac{1}{4-x^2-y^2}$

Continuous for $x^2+y^2\neq4$

37. $f(x,\ y) = x^2 - 4y$

(a) $\lim\limits_{\Delta x\to0} \dfrac{f(x+\Delta x,\ y) - f(x,\ y)}{\Delta x} = \lim\limits_{\Delta x\to0} \dfrac{[(x+\Delta x)^2 - 4y] - (x^2-4y)}{\Delta x} = \lim\limits_{\Delta x\to0} \dfrac{2x\Delta x - (\Delta x)^2}{\Delta x}$

$\quad\quad = \lim\limits_{\Delta x\to0} (2x - \Delta x) = 2x$

(b) $\lim\limits_{\Delta y\to0} \dfrac{f(x,\ y+\Delta y) - f(x,\ y)}{\Delta y} = \lim\limits_{\Delta y\to0} \dfrac{[x^2 - 4(y+\Delta y)] - (x^2-4y)}{\Delta y} = \lim\limits_{\Delta y\to0} \dfrac{-4\Delta y}{\Delta y}$

$\quad\quad = \lim\limits_{\Delta y\to0} (-4) = -4$

38. $f(x,\ y) = x^2 + y^2$

(a) $\lim\limits_{\Delta x\to0} \dfrac{f(x+\Delta x,\ y) - f(x,\ y)}{\Delta x} = \lim\limits_{\Delta x\to0} \dfrac{[(x+\Delta x)^2 + y^2] - (x^2+y^2)}{\Delta x} = \lim\limits_{\Delta x\to0} \dfrac{2x\Delta x + (\Delta x)^2}{\Delta x}$

$\quad\quad = \lim\limits_{\Delta x\to0} (2x + \Delta x) = 2x$

(b) $\lim\limits_{\Delta y\to0} \dfrac{f(x,\ y+\Delta y) - f(x,\ y)}{\Delta y} = \lim\limits_{\Delta y\to0} \dfrac{[x^2 + (y+\Delta y)^2] - (x^2+y^2)}{\Delta y} = \lim\limits_{\Delta y\to0} \dfrac{2y\Delta y + (\Delta y)^2}{\Delta y}$

$\quad\quad = \lim\limits_{\Delta y\to0} (2y + \Delta y) = 2y$

39. $f(x, y) = 2x + xy - 3y$

(a) $\displaystyle \lim_{\Delta x \to 0} \frac{f(x + \Delta x, y) - f(x, y)}{\Delta x} = \lim_{\Delta x \to 0} \frac{[2(x + \Delta x) + (x + \Delta x)y - 3y] - (2x + xy - 3y)}{\Delta x}$

$$= \lim_{\Delta x \to 0} \frac{2\Delta x + \Delta xy}{\Delta x}$$

$$= \lim_{\Delta x \to 0} (2 + y) = 2 + y$$

(b) $\displaystyle \lim_{\Delta y \to 0} \frac{f(x, y + \Delta y) - f(x, y)}{\Delta y} = \lim_{\Delta y \to 0} \frac{[2x + x(y + \Delta y) - 3(y + \Delta y)] - (2x + xy - 3y)}{\Delta y}$

$$= \lim_{\Delta y \to 0} \frac{x\Delta y - 3\Delta y}{\Delta y}$$

$$= \lim_{\Delta y \to 0} (x - 3) = x - 3$$

40. $f(x, y) = \sqrt{y}(y + 1)$

(a) $\displaystyle \lim_{\Delta x \to 0} \frac{f(x + \Delta x, y) - f(x, y)}{\Delta x} = \lim_{\Delta x \to 0} \frac{\sqrt{y}(y + 1) - \sqrt{y}(y + 1)}{\Delta x} = 0$

(b) $\displaystyle \lim_{\Delta y \to 0} \frac{f(x, y + \Delta y) - f(x, y)}{\Delta y} = \lim_{\Delta y \to 0} \frac{(y + \Delta y)^{3/2} + (y + \Delta y)^{1/2} - (y^{3/2} + y^{1/2})}{\Delta y}$

$$= \lim_{\Delta y \to 0} \frac{(y + \Delta y)^{3/2} - y^{3/2}}{\Delta y} + \lim_{\Delta y \to 0} \frac{(y + \Delta y)^{1/2} - y^{1/2}}{\Delta y}$$

$$= \frac{3}{2}y^{1/2} + \frac{1}{2}y^{-1/2} \quad \text{(L'Hôpital's Rule)}$$

$$= \frac{3y + 1}{2\sqrt{y}}$$

41. $f(x, y) = \dfrac{0.2x^2 y}{x^4 + 4y^2}$

(a) Along $y = x$: $\displaystyle \lim_{(x,y) \to (0,0)} \frac{0.2x^2 y}{x^4 + 4y^2} = \lim_{x \to 0} \frac{0.2x^3}{x^4 + 4x^2} = \lim_{x \to 0} \frac{0.2x}{x^2 + 4} = 0$

(b) Along $y = x^2$: $\displaystyle \lim_{(x,y) \to (0,0)} \frac{0.2x^2 y}{x^4 + 4y^2} = \lim_{x \to 0} \frac{0.2x^4}{x^4 + 4x^4} = \frac{0.2}{5} = 0.04$

Therefore, the limit does not exist.

42. $f(x, y) = 2x + 3y \sin \dfrac{1}{x}$

(a) Along $y = 0$: $\displaystyle \lim_{(x,y) \to (0,0)} \left(2x + 3y \sin \frac{1}{x}\right) = \lim_{x \to 0} (2x) = 0$

(b) Along $y = x$: $\displaystyle \lim_{(x,y) \to (0,0)} \left(2x + 3y \sin \frac{1}{x}\right) = \lim_{x \to 0} \left[x\left(2 + 3 \sin \frac{1}{x}\right)\right] = 0$

43. Since $\lim\limits_{(x,y)\to(a,b)} f(x,\,y) = L_1$, then for $\epsilon/2 > 0$, there corresponds $\delta_1 > 0$ such that $|f(x,\,y) - L_1| < \epsilon/2$ whenever $0 < \sqrt{(x-x_0)^2 + (y-y_0)^2} < \delta_1$. Since $\lim\limits_{(x,y)\to(a,b)} g(x,\,y) = L_2$, then for $\epsilon/2 > 0$, there corresponds $\delta_2 > 0$ such that $|g(x,\,y) - L_2| < \epsilon/2$ whenever $0 < \sqrt{(x-x_0)^2 + (y-y_0)^2} < \delta_2$. Let δ be the larger of δ_1 and δ_2. By the triangle inequality,

$$|f(x,\,y) + g(x,\,y) - (L_1 + L_2)| = |(f(x,\,y) - L_1) + (g(x,\,y) - L_2)|$$

$$\leq |f(x,\,y) - L_1| + |g(x,\,y) - L_2|$$

$$< \frac{\epsilon}{2} + \frac{\epsilon}{2} = \epsilon.$$

Therefore, $\lim\limits_{(x,y)\to(a,b)} [f(x,\,y) + g(x,\,y)] = L_1 + L_2$.

44. Given that $f(x,\,y)$ is continuous, then $\lim\limits_{(x,y)\to(a,b)} f(x,\,y) = f(a,\,b) < 0$, which means that for each $\epsilon > 0$, there corresponds a $\delta > 0$ such that $|f(x,\,y) - f(a,\,b)| < \epsilon$ whenever $0 < \sqrt{(x-x_0)^2 + (y-y_0)^2} < \delta$. Let $\epsilon = \dfrac{|f(a,\,b)|}{2}$, then $f(x,\,y) < 0$ for every point in the corresponding δ neighborhood since

$$|f(x,\,y) - f(a,\,b)| < \frac{|f(a,\,b)|}{2} \Rightarrow -\frac{|f(a,\,b)|}{2} < f(x,\,y) - f(a,\,b) < \frac{|f(a,\,b)|}{2}$$

$$\Rightarrow \frac{3}{2}f(a,\,b) < f(x,\,y) < \frac{1}{2}f(a,\,b) < 0.$$

Section 14.3   Partial Derivatives

1. $f(x,\,y) = 2x - 3y + 5$
  $f_x(x,\,y) = 2$
  $f_y(x,\,y) = -3$

2. $f(x,\,y) = x^2 - 3y^2 + 7$
  $f_x(x,\,y) = 2x$
  $f_y(x,\,y) = -6y$

3. $f(x,\,y) = xy$
  $f_x(x,\,y) = y$
  $f_y(x,\,y) = x$

4. $f(x,\,y) = \dfrac{x}{y}$
  $f_x(x,\,y) = \dfrac{1}{y}$
  $f_y(x,\,y) = -\dfrac{x}{y^2}$

5. $z = x\sqrt{y}$
  $\dfrac{\partial z}{\partial x} = \sqrt{y}$
  $\dfrac{\partial z}{\partial y} = \dfrac{x}{2\sqrt{y}}$

6. $z = x^2 - 3xy + y^2$
  $\dfrac{\partial z}{\partial x} = 2x - 3y$
  $\dfrac{\partial z}{\partial y} = -3x + 2y$

7. $z = x^2 e^{2y}$
  $\dfrac{\partial z}{\partial x} = 2xe^{2y}$
  $\dfrac{\partial z}{\partial y} = 2x^2 e^{2y}$

8. $z = xe^{x/y}$
  $\dfrac{\partial z}{\partial x} = \dfrac{x}{y}e^{x/y} + e^{x/y} = e^{x/y}\left(\dfrac{x}{y} + 1\right)$
  $\dfrac{\partial z}{\partial y} = xe^{x/y}\left(-\dfrac{x}{y^2}\right) = -\dfrac{x^2}{y^2}e^{x/y}$

9. $z = \ln(x^2 + y^2)$

$$\frac{\partial z}{\partial x} = \frac{2x}{x^2 + y^2}$$

$$\frac{\partial z}{\partial y} = \frac{2y}{x^2 + y^2}$$

10. $z = \ln \sqrt{xy} = \frac{1}{2}\ln(xy)$

$$\frac{\partial z}{\partial x} = \frac{1}{2}\frac{y}{xy} = \frac{1}{2x}$$

$$\frac{\partial z}{\partial y} = \frac{1}{2}\frac{x}{xy} = \frac{1}{2y}$$

11. $z = \ln\left(\frac{x+y}{x-y}\right) = \ln(x+y) - \ln(x-y)$

$$\frac{\partial z}{\partial x} = \frac{1}{x+y} - \frac{1}{x-y} = -\frac{2y}{x^2-y^2}$$

$$\frac{\partial z}{\partial y} = \frac{1}{x+y} + \frac{1}{x-y} = \frac{2x}{x^2-y^2}$$

12. $z = \frac{x^2}{2y} + \frac{4y^2}{x}$

$$\frac{\partial z}{\partial x} = \frac{2x}{2y} - \frac{4y^2}{x^2} = \frac{x^3 - 4y^3}{x^2 y}$$

$$\frac{\partial z}{\partial y} = -\frac{x^2}{2y^2} + \frac{8y}{x} = \frac{-x^3 + 16y^3}{2xy^2}$$

13. $h(x, y) = e^{-(x^2+y^2)}$

$$h_x(x, y) = -2xe^{-(x^2+y^2)}$$

$$h_y(x, y) = -2ye^{-(x^2+y^2)}$$

14. $g(x, y) = \ln \sqrt{x^2 + y^2} = \frac{1}{2}\ln(x^2 + y^2)$

$$g_x(x, y) = \frac{1}{2}\frac{2x}{x^2 + y^2} = \frac{x}{x^2 + y^2}$$

$$g_y(x, y) = \frac{1}{2}\frac{2y}{x^2 + y^2} = \frac{y}{x^2 + y^2}$$

15. $f(x, y) = \sqrt{x^2 + y^2}$

$$f_x(x, y) = \frac{1}{2}(x^2 + y^2)^{-1/2}(2x) = \frac{x}{\sqrt{x^2 + y^2}}$$

$$f_y(x, y) = \frac{1}{2}(x^2 + y^2)^{-1/2}(2y) = \frac{y}{\sqrt{x^2 + y^2}}$$

16. $f(x, y) = \frac{xy}{x^2 + y^2}$

$$f_x(x, y) = \frac{(x^2 + y^2)(y) - (xy)(2x)}{(x^2 + y^2)^2} = \frac{y^3 - x^2 y}{(x^2 + y^2)^2}$$

$$f_y(x, y) = \frac{(x^2 + y^2)(x) - (xy)(2y)}{(x^2 + y^2)^2} = \frac{x^3 - xy^2}{(x^2 + y^2)^2}$$

17. $z = \sin(2x - y)$

$$\frac{\partial z}{\partial x} = 2\cos(2x - y)$$

$$\frac{\partial z}{\partial y} = -\cos(2x - y)$$

18. $z = \sin 3x \cos 3y$

$$\frac{\partial z}{\partial x} = 3\cos 3x \cos 3y$$

$$\frac{\partial z}{\partial y} = -3\sin 3x \sin 3y$$

19. $z = e^y \sin xy$

$$\frac{\partial z}{\partial x} = ye^y \cos xy$$

$$\frac{\partial z}{\partial y} = e^y \sin xy + xe^y \cos xy$$

$$= e^y(x \cos xy + \sin xy)$$

20. $z = \cos(x^2 + y^2)$

$$\frac{\partial z}{\partial x} = -2x \sin(x^2 + y^2)$$

$$\frac{\partial z}{\partial y} = -2y \sin(x^2 + y^2)$$

21. $f(x, y) = \int_x^y (t^2 - 1)\, dt = \left[\frac{t^3}{3} - t\right]_x^y = \left(\frac{y^3}{3} - y\right) - \left(\frac{x^3}{3} - x\right)$

$$f_x(x, y) = -x^2 + 1 = 1 - x^2$$

$$f_y(x, y) = y^2 - 1$$

22. $f(x, y) = \int_x^y (2t + 1)\, dt + \int_y^x (2t - 1)\, dt = \int_x^y (2t + 1)\, dt - \int_x^y (2t - 1)\, dt = \int_x^y 2\, dt = 2t\Big]_x^y = 2y - 2x$

$$f_x(x, y) = -2$$

$$f_y(x, y) = 2$$

23. $f(x, y) = \arctan \dfrac{y}{x}$

$f_x(x, y) = \dfrac{1}{1 + (y^2/x^2)}\left(-\dfrac{y}{x^2}\right) = \dfrac{-y}{x^2 + y^2}$

At $(2, -2)$, $f_x(2, -2) = \dfrac{1}{4}$.

$f_y(x, y) = \dfrac{1}{1 + (y^2/x^2)}\left(\dfrac{1}{x}\right) = \dfrac{x}{x^2 + y^2}$

At $(2, -2)$, $f_y(2, -2) = \dfrac{1}{4}$.

24. $f(x, y) = \arcsin xy$

$f_x(x, y) = \dfrac{y}{\sqrt{1 - (xy)^2}}$

At $(1, 0)$, $f_x(1, 0) = 0$.

$f_y(x, y) = \dfrac{x}{\sqrt{1 - (xy)^2}}$

At $(1, 0)$, $f_y(1, 0) = 1$.

25. $f(x, y) = \dfrac{xy}{x - y}$

$f_x(x, y) = \dfrac{y(x - y) - xy}{(x - y)^2} = \dfrac{-y^2}{(x - y)^2}$

At $(2, -2)$, $f_x(2, -2) = -\dfrac{1}{4}$.

$f_y(x, y) = \dfrac{x(x - y) + xy}{(x - y)^2} = \dfrac{x^2}{(x - y)^2}$

At $(2, -2)$, $f_y(2, -2) = \dfrac{1}{4}$.

26. $f(x, y) = \dfrac{4xy}{\sqrt{x^2 + y^2}}$

$f_x(x, y) = \dfrac{4y^3}{(x^2 + y^2)^{3/2}}$

At $(1, 0)$, $f_x(1, 0) = 0$.

$f_y(x, y) = \dfrac{4x^3}{(x^2 + y^2)^{3/2}}$

At $(1, 0)$, $f_y(1, 0) = 4$.

27. $w = \sqrt{x^2 + y^2 + z^2}$

$\dfrac{\partial w}{\partial x} = \dfrac{x}{\sqrt{x^2 + y^2 + z^2}}$

$\dfrac{\partial w}{\partial y} = \dfrac{y}{\sqrt{x^2 + y^2 + z^2}}$

$\dfrac{\partial w}{\partial z} = \dfrac{z}{\sqrt{x^2 + y^2 + z^2}}$

28. $w = \dfrac{xy}{x + y + z}$

$\dfrac{\partial w}{\partial x} = \dfrac{y(x + y + z) - xy}{(x + y + z)^2} = \dfrac{y(y + z)}{(x + y + z)^2}$

$\dfrac{\partial w}{\partial y} = \dfrac{x(x + y + z) - xy}{(x + y + z)^2} = \dfrac{x(x + z)}{(x + y + z)^2}$

$\dfrac{\partial w}{\partial z} = \dfrac{-xy}{(x + y + z)^2}$

29. $F(x, y, z) = \ln \sqrt{x^2 + y^2 + z^2}$

$\qquad = \dfrac{1}{2} \ln(x^2 + y^2 + z^2)$

$F_x(x, y, z) = \dfrac{x}{x^2 + y^2 + z^2}$

$F_y(x, y, z) = \dfrac{y}{x^2 + y^2 + z^2}$

$F_z(x, y, z) = \dfrac{z}{x^2 + y^2 + z^2}$

30. $G(x, y, z) = \dfrac{1}{\sqrt{1 - x^2 - y^2 - z^2}}$

$G_x(x, y, z) = \dfrac{x}{(1 - x^2 - y^2 - z^2)^{3/2}}$

$G_y(x, y, z) = \dfrac{y}{(1 - x^2 - y^2 - z^2)^{3/2}}$

$G_z(x, y, z) = \dfrac{z}{(1 - x^2 - y^2 - z^2)^{3/2}}$

31. $H(x, y, z) = \sin(x + 2y + 3z)$

$H_x(x, y, z) = \cos(x + 2y + 3z)$

$H_y(x, y, z) = 2\cos(x + 2y + 3z)$

$H_z(x, y, z) = 3\cos(x + 2y + 3z)$

32. $f(x, y, z) = 3x^2y - 5xyz + 10yz^2$

$f_x(x, y, z) = 6xy - 5yz$

$f_y(x, y, z) = 3x^2 - 5xz + 10z^2$

$f_z(x, y, z) = -5xy + 20yz$

33.
$$z = x^2 - 2xy + 3y^2$$
$$\frac{\partial z}{\partial x} = 2x - 2y$$
$$\frac{\partial^2 z}{\partial x^2} = 2$$
$$\frac{\partial^2 z}{\partial y \partial x} = -2$$
$$\frac{\partial z}{\partial y} = -2x + 6y$$
$$\frac{\partial^2 z}{\partial y^2} = 6$$
$$\frac{\partial^2 z}{\partial x \partial y} = -2$$

34.
$$z = x^4 - 3x^2 y^2 + y^4$$
$$\frac{\partial z}{\partial x} = 4x^3 - 6xy^2$$
$$\frac{\partial^2 z}{\partial x^2} = 12x^2 - 6y^2$$
$$\frac{\partial^2 z}{\partial y \partial x} = -12xy$$
$$\frac{\partial z}{\partial y} = -6x^2 y + 4y^3$$
$$\frac{\partial^2 z}{\partial y^2} = -6x^2 + 12y^2$$
$$\frac{\partial^2 z}{\partial x \partial y} = -12xy$$

35.
$$z = e^x \tan y$$
$$\frac{\partial z}{\partial x} = e^x \tan y$$
$$\frac{\partial^2 z}{\partial x^2} = e^x \tan y$$
$$\frac{\partial^2 z}{\partial y \partial x} = e^x \sec^2 y$$
$$\frac{\partial z}{\partial y} = e^x \sec^2 y$$
$$\frac{\partial^2 z}{\partial y^2} = 2e^x \sec^2 y \tan y$$
$$\frac{\partial^2 z}{\partial x \partial y} = e^x \sec^2 y$$

36.
$$z = 2e^{xy^2}$$
$$\frac{\partial z}{\partial x} = 2y^2 e^{xy^2}$$
$$\frac{\partial^2 z}{\partial x^2} = 2y^4 e^{xy^2}$$
$$\frac{\partial^2 z}{\partial y \partial x} = 2y^2(2xy)e^{xy^2} + 4ye^{xy^2} = 4ye^{xy^2}(xy^2 + 1)$$
$$\frac{\partial z}{\partial y} = 2(2x)e^{xy^2} = 4xye^{xy^2}$$
$$\frac{\partial^2 z}{\partial y^2} = 4xy(2xy)e^{xy^2} + 4xe^{xy^2} = 4xe^{xy^2}(2xy^2 + 1)$$
$$\frac{\partial^2 z}{\partial x \partial y} = 4xy(y^2)e^{xy^2} + e^{xy^2}(4y) = 4ye^{xy^2}(xy^2 + 1)$$

37.
$$z = \arctan \frac{y}{x}$$
$$\frac{\partial z}{\partial x} = \frac{1}{1 + (y^2/x^2)}\left(-\frac{y}{x^2}\right) = \frac{-y}{x^2 + y^2}$$
$$\frac{\partial^2 z}{\partial x^2} = \frac{2xy}{(x^2 + y^2)^2}$$
$$\frac{\partial^2 z}{\partial y \partial x} = \frac{-(x^2 + y^2) + y(2y)}{(x^2 + y^2)^2} = \frac{y^2 - x^2}{(x^2 + y^2)^2}$$
$$\frac{\partial z}{\partial y} = \frac{1}{1 + (y^2/x^2)}\left(\frac{1}{x}\right) = \frac{x}{x^2 + y^2}$$
$$\frac{\partial^2 z}{\partial y^2} = \frac{-2xy}{(x^2 + y^2)^2}$$
$$\frac{\partial^2 z}{\partial x \partial y} = \frac{(x^2 + y^2) - x(2x)}{(x^2 + y^2)^2} = \frac{y^2 - x^2}{(x^2 + y^2)^2}$$

38.
$$z = \sin(x - 2y)$$
$$\frac{\partial z}{\partial x} = \cos(x - 2y)$$
$$\frac{\partial^2 z}{\partial x^2} = -\sin(x - 2y)$$
$$\frac{\partial^2 z}{\partial y \partial x} = 2\sin(x - 2y)$$
$$\frac{\partial z}{\partial y} = -2\cos(x - 2y)$$
$$\frac{\partial^2 z}{\partial y^2} = -4\sin(x - 2y)$$
$$\frac{\partial^2 z}{\partial x \partial y} = 2\sin(x - 2y)$$

39. $z = \sqrt{x^2 + y^2}$

$$\frac{\partial z}{\partial x} = \frac{x}{\sqrt{x^2 + y^2}}$$

$$\frac{\partial^2 z}{\partial x^2} = \frac{y^2}{(x^2 + y^2)^{3/2}}$$

$$\frac{\partial^2 z}{\partial y \partial x} = \frac{-xy}{(x^2 + y^2)^{3/2}}$$

$$\frac{\partial z}{\partial y} = \frac{y}{\sqrt{x^2 + y^2}}$$

$$\frac{\partial^2 z}{\partial y^2} = \frac{x^2}{(x^2 + y^2)^{3/2}}$$

$$\frac{\partial^2 z}{\partial x \partial y} = \frac{-xy}{(x^2 + y^2)^{3/2}}$$

40. $z = \dfrac{xy}{x - y}$

$$\frac{\partial z}{\partial x} = \frac{y(x - y) - xy}{(x - y)^2} = \frac{-y^2}{(x - y)^2}$$

$$\frac{\partial^2 z}{\partial x^2} = \frac{2y^2}{(x - y)^3}$$

$$\frac{\partial^2 z}{\partial y \partial x} = \frac{(x - y)^2(-2y) + y^2(2)(x - y)(-1)}{(x - y)^4}$$

$$= \frac{-2xy}{(x - y)^3}$$

$$\frac{\partial z}{\partial y} = \frac{x(x - y) + xy}{(x - y)^2} = \frac{x^2}{(x - y)^2}$$

$$\frac{\partial^2 z}{\partial y^2} = \frac{2x^2}{(x - y)^3}$$

$$\frac{\partial^2 z}{\partial x \partial y} = \frac{(x - y)^2(2x) - x^2(2)(x - y)}{(x - y)^4}$$

$$= \frac{-2xy}{(x - y)^3}$$

41. $z = x^3 + 3x^2 y$

$$\frac{\partial z}{\partial x} = 3x^2 + 6xy$$

$$\frac{\partial^2 z}{\partial y \partial x} = 6x$$

$$\frac{\partial z}{\partial y} = 3x^2$$

$$\frac{\partial^2 z}{\partial x \partial y} = 6x$$

Therefore, $\dfrac{\partial^2 z}{\partial y \partial x} = \dfrac{\partial^2 z}{\partial x \partial y}$.

42. $z = \ln(x - y)$

$$\frac{\partial z}{\partial x} = \frac{1}{x - y}$$

$$\frac{\partial^2 z}{\partial y \partial x} = \frac{1}{(x - y)^2}$$

$$\frac{\partial z}{\partial y} = \frac{-1}{x - y}$$

$$\frac{\partial^2 z}{\partial x \partial y} = \frac{1}{(x - y)^2}$$

Therefore, $\dfrac{\partial^2 z}{\partial y \partial x} = \dfrac{\partial^2 z}{\partial x \partial y}$.

43. $z = x \sec y$

$$\frac{\partial z}{\partial x} = \sec y$$

$$\frac{\partial^2 z}{\partial y \partial x} = \sec y \tan y$$

$$\frac{\partial z}{\partial y} = x \sec y \tan y$$

$$\frac{\partial^2 z}{\partial x \partial y} = \sec y \tan y$$

Therefore, $\dfrac{\partial^2 z}{\partial y \partial x} = \dfrac{\partial^2 z}{\partial x \partial y}$.

44. $z = \sqrt{9 - x^2 - y^2}$

$$\frac{\partial z}{\partial x} = \frac{-x}{\sqrt{9 - x^2 - y^2}}$$

$$\frac{\partial^2 z}{\partial y \partial x} = \frac{-xy}{(9 - x^2 - y^2)^{3/2}}$$

$$\frac{\partial z}{\partial y} = \frac{-y}{\sqrt{9 - x^2 - y^2}}$$

$$\frac{\partial^2 z}{\partial x \partial y} = \frac{-xy}{(9 - x^2 - y^2)^{3/2}}$$

Therefore, $\dfrac{\partial^2 z}{\partial y \partial x} = \dfrac{\partial^2 z}{\partial x \partial y}$.

45.
$$z = xe^{-y^2}$$
$$\frac{\partial z}{\partial x} = e^{-y^2}$$
$$\frac{\partial^2 z}{\partial y \partial x} = -2ye^{-y^2}$$
$$\frac{\partial z}{\partial y} = -2xye^{-y^2}$$
$$\frac{\partial^2 z}{\partial x \partial y} = -2ye^{-y^2}$$
Therefore, $\dfrac{\partial^2 z}{\partial y \partial x} = \dfrac{\partial^2 z}{\partial x \partial y}$.

46.
$$z = xe^y + ye^x$$
$$\frac{\partial z}{\partial x} = e^y + ye^x$$
$$\frac{\partial^2 z}{\partial y \partial x} = e^y + e^x$$
$$\frac{\partial z}{\partial y} = xe^y + e^x$$
$$\frac{\partial^2 z}{\partial x \partial y} = e^y + e^x$$
Therefore, $\dfrac{\partial^2 z}{\partial y \partial x} = \dfrac{\partial^2 z}{\partial x \partial y}$.

47.
$$f(x, y, z) = xyz$$
$$f_x(x, y, z) = yz$$
$$f_y(x, y, z) = xz$$
$$f_{yy}(x, y, z) = 0$$
$$f_{xy}(x, y, z) = z$$
$$f_{yx}(x, y, z) = z$$
$$f_{yyx}(x, y, z) = 0$$
$$f_{xyy}(x, y, z) = 0$$
$$f_{yxy}(x, y, z) = 0$$
Therefore, $f_{xyy} = f_{yxy} = f_{yyx} = 0$.

48.
$$f(x, y, z) = x^2 - 3xy + 4yz + z^3$$
$$f_x(x, y, z) = 2x - 3y$$
$$f_y(x, y, z) = -3x + 4z$$
$$f_{yy}(x, y, z) = 0$$
$$f_{xy}(x, y, z) = -3$$
$$f_{yx}(x, y, z) = -3$$
$$f_{yyx}(x, y, z) = 0$$
$$f_{xyy}(x, y, z) = 0$$
$$f_{yxy}(x, y, z) = 0$$
Therefore, $f_{xyy} = f_{yxy} = f_{yyx} = 0$.

49.
$$f(x, y, z) = e^{-x} \sin yz$$
$$f_x(x, y, z) = -e^{-x} \sin yz$$
$$f_y(x, y, z) = ze^{-x} \cos yz$$
$$f_{yy}(x, y, z) = -z^2 e^{-x} \sin yz$$
$$f_{xy}(x, y, z) = -ze^{-x} \cos yz$$
$$f_{yx}(x, y, z) = -ze^{-x} \cos yz$$
$$f_{yyx}(x, y, z) = z^2 e^{-x} \sin yz$$
$$f_{xyy}(x, y, z) = z^2 e^{-x} \sin yz$$
$$f_{yxy}(x, y, z) = z^2 e^{-x} \sin yz$$
Therefore, $f_{xyy} = f_{yxy} = f_{yyx}$.

50.
$$f(x, y, z) = \frac{x}{y+z}$$
$$f_x(x, y, z) = \frac{1}{y+z}$$
$$f_y(x, y, z) = -\frac{x}{(y+z)^2}$$
$$f_{yy}(x, y, z) = \frac{2x}{(y+z)^3}$$
$$f_{xy}(x, y, z) = -\frac{1}{(y+z)^2}$$
$$f_{yx}(x, y, z) = -\frac{1}{(y+z)^2}$$
$$f_{yyx}(x, y, z) = \frac{2}{(y+z)^3}$$
$$f_{xyy}(x, y, z) = \frac{2}{(y+z)^3}$$
$$f_{yxy}(x, y, z) = \frac{2}{(y+z)^3}$$
Therefore, $f_{xyy} = f_{yxy} = f_{yyx}$.

51. $z = 5xy$

$$\frac{\partial z}{\partial x} = 5y$$

$$\frac{\partial^2 z}{\partial x^2} = 0$$

$$\frac{\partial z}{\partial y} = 5x$$

$$\frac{\partial^2 z}{\partial y^2} = 0$$

Therefore, $\dfrac{\partial^2 z}{\partial x^2} + \dfrac{\partial^2 z}{\partial y^2} = 0 + 0 = 0.$

52. $z = \sin x \left(\dfrac{e^y - e^{-y}}{2} \right)$

$$\frac{\partial z}{\partial x} = \cos x \left(\frac{e^y - e^{-y}}{2} \right)$$

$$\frac{\partial^2 z}{\partial x^2} = -\sin x \left(\frac{e^y - e^{-y}}{2} \right)$$

$$\frac{\partial z}{\partial y} = \sin x \left(\frac{e^y + e^{-y}}{2} \right)$$

$$\frac{\partial^2 z}{\partial y^2} = \sin x \left(\frac{e^y - e^{-y}}{2} \right)$$

Therefore, $\dfrac{\partial^2 z}{\partial x^2} + \dfrac{\partial^2 z}{\partial y^2} = -\sin x \left(\dfrac{e^y - e^{-y}}{2} \right) + \sin x \left(\dfrac{e^y - e^{-y}}{2} \right) = 0.$

53. $z = e^x \sin y$

$$\frac{\partial z}{\partial x} = e^x \sin y$$

$$\frac{\partial^2 z}{\partial x^2} = e^x \sin y$$

$$\frac{\partial z}{\partial y} = e^x \cos y$$

$$\frac{\partial^2 z}{\partial y^2} = -e^x \sin y$$

Therefore, $\dfrac{\partial^2 z}{\partial x^2} + \dfrac{\partial^2 z}{\partial y^2} = e^x \sin y - e^x \sin y = 0.$

54. $z = \arctan \dfrac{y}{x}$

From Exercise 37, we have

$$\frac{\partial^2 z}{\partial x^2} + \frac{\partial^2 z}{\partial y^2} = \frac{2xy}{(x^2 + y^2)^2} + \frac{-2xy}{(x^2 + y^2)^2} = 0.$$

55. $z = \sin(x - ct)$

$$\frac{\partial z}{\partial t} = -c \cos(x - ct)$$

$$\frac{\partial^2 z}{\partial t^2} = -c^2 \sin(x - ct)$$

$$\frac{\partial z}{\partial x} = \cos(x - ct)$$

$$\frac{\partial^2 z}{\partial x^2} = -\sin(x - ct)$$

Therefore, $\dfrac{\partial^2 z}{\partial t^2} = c^2 \dfrac{\partial^2 z}{\partial x^2}.$

56. $z = \sin(wct) \sin(wx)$

$$\frac{\partial z}{\partial t} = wc \cos(wct) \sin(wx)$$

$$\frac{\partial^2 z}{\partial t^2} = -w^2 c^2 \sin(wct) \sin(wx)$$

$$\frac{\partial z}{\partial x} = w \sin(wct) \cos(wx)$$

$$\frac{\partial^2 z}{\partial x^2} = -w^2 \sin(wct) \sin(wx)$$

Therefore, $\dfrac{\partial^2 z}{\partial t^2} = c^2 \dfrac{\partial^2 z}{\partial x^2}.$

57. $z = e^{-t} \cos \dfrac{x}{c}$

$$\frac{\partial z}{\partial t} = -e^{-t} \cos \frac{x}{c}$$

$$\frac{\partial z}{\partial x} = -\frac{1}{c} e^{-t} \sin \frac{x}{c}$$

$$\frac{\partial^2 z}{\partial x^2} = -\frac{1}{c^2} e^{-t} \cos \frac{x}{c}$$

Therefore, $\dfrac{\partial z}{\partial t} = c^2 \dfrac{\partial^2 z}{\partial x^2}$.

58. $z = e^{-t} \sin \dfrac{x}{c}$

$$\frac{\partial z}{\partial t} = -e^{-t} \sin \frac{x}{c}$$

$$\frac{\partial z}{\partial x} = \frac{1}{c} e^{-t} \cos \frac{x}{c}$$

$$\frac{\partial^2 z}{\partial x^2} = -\frac{1}{c^2} e^{-t} \sin \frac{x}{c}$$

Therefore, $\dfrac{\partial z}{\partial t} = c^2 \dfrac{\partial^2 z}{\partial x^2}$.

59. $f(x, y) = 2x + 3y$

$$\frac{\partial f}{\partial x} = \lim_{\Delta x \to 0} \frac{f(x + \Delta x, y) - f(x, y)}{\Delta x} = \lim_{\Delta x \to 0} \frac{2(x + \Delta x) + 3y - 2x - 3y}{\Delta x} = \lim_{\Delta x \to 0} \frac{2\Delta x}{\Delta x} = 2$$

$$\frac{\partial f}{\partial y} = \lim_{\Delta y \to 0} \frac{f(x, y + \Delta y) - f(x, y)}{\Delta y} = \lim_{\Delta y \to 0} \frac{2x + 3(y + \Delta y) - 2x - 3y}{\Delta y} = \lim_{\Delta y \to 0} \frac{3\Delta y}{\Delta y} = 3$$

60. $f(x, y) = \dfrac{1}{x + y}$

$$\frac{\partial f}{\partial x} = \lim_{\Delta x \to 0} \frac{f(x + \Delta x, y) - f(x, y)}{\Delta x} = \lim_{\Delta x \to 0} \frac{\dfrac{1}{x + \Delta x + y} - \dfrac{1}{x + y}}{\Delta x} = \lim_{\Delta x \to 0} \frac{-1}{(x + \Delta x + y)(x + y)} = \frac{-1}{(x + y)^2}$$

$$\frac{\partial f}{\partial y} = \lim_{\Delta y \to 0} \frac{f(x, y + \Delta y) - f(x, y)}{\Delta y} = \lim_{\Delta y \to 0} \frac{\dfrac{1}{x + y + \Delta y} - \dfrac{1}{x + y}}{\Delta y} = \lim_{\Delta y \to 0} \frac{-1}{(x + y + \Delta y)(x + y)} = \frac{-1}{(x + y)^2}$$

61. $f(x, y) = \sqrt{x + y}$

$$\frac{\partial f}{\partial x} = \lim_{\Delta x \to 0} \frac{f(x + \Delta x, y) - f(x, y)}{\Delta x} = \lim_{\Delta x \to 0} \frac{\sqrt{x + \Delta x + y} - \sqrt{x + y}}{\Delta x}$$

$$= \lim_{\Delta x \to 0} \frac{(\sqrt{x + \Delta x + y} - \sqrt{x + y})(\sqrt{x + \Delta x + y} + \sqrt{x + y})}{\Delta x(\sqrt{x + \Delta x + y} + \sqrt{x + y})}$$

$$= \lim_{\Delta x \to 0} \frac{1}{\sqrt{x + \Delta x + y} + \sqrt{x + y}} = \frac{1}{2\sqrt{x + y}}$$

$$\frac{\partial f}{\partial y} = \lim_{\Delta y \to 0} \frac{f(x, y + \Delta y) - f(x, y)}{\Delta y} = \lim_{\Delta y \to 0} \frac{\sqrt{x + y + \Delta y} - \sqrt{x + y}}{\Delta y}$$

$$= \lim_{\Delta y \to 0} \frac{(\sqrt{x + y + \Delta y} - \sqrt{x + y})(\sqrt{x + y + \Delta y} + \sqrt{x + y})}{\Delta y(\sqrt{x + y + \Delta y} + \sqrt{x + y})}$$

$$= \lim_{\Delta y \to 0} \frac{1}{\sqrt{x + y + \Delta y} + \sqrt{x + y}} = \frac{1}{2\sqrt{x + y}}$$

62. $f(x, y) = x^2 - 2xy + y^2 = (x - y)^2$

$$\frac{\partial f}{\partial x} = \lim_{\Delta x \to 0} \frac{f(x + \Delta x, y) - f(x, y)}{\Delta x}$$

$$= \lim_{\Delta x \to 0} \frac{(x + \Delta x)^2 - 2(x + \Delta x)y + y^2 - x^2 + 2xy - y^2}{\Delta x} = \lim_{\Delta x \to 0} (2x + \Delta x - 2y) = 2(x - y)$$

$$\frac{\partial f}{\partial y} = \lim_{\Delta y \to 0} \frac{f(x, y + \Delta y) - f(x, y)}{\Delta y}$$

$$= \lim_{\Delta y \to 0} \frac{x^2 - 2x(y + \Delta y) + (y + \Delta y)^2 - x^2 + 2xy - y^2}{\Delta y} = \lim_{\Delta y \to 0} (-2x + 2y + \Delta y) = 2(y - x)$$

63. $z = \sqrt{49 - x^2 - y^2}, \quad x = 2, \quad (2, 3, 6)$

Intersecting curve: $z = \sqrt{45 - y^2}$

$\dfrac{\partial z}{\partial y} = \dfrac{-y}{\sqrt{45 - y^2}}$

At $(2, 3, 6)$:

$\dfrac{\partial z}{\partial y} = \dfrac{-3}{\sqrt{45 - 9}} = -\dfrac{1}{2}$

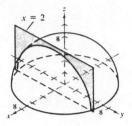

64. $z = x^2 + 4y^2, \quad y = 1, \quad (2, 1, 8)$

Intersecting curve: $z = x^2 + 4$

$\dfrac{\partial z}{\partial x} = 2x$

At $(2, 1, 8)$: $\dfrac{\partial z}{\partial x} = 2(2) = 4$

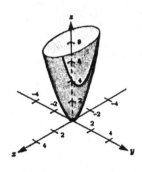

65. $z = 9x^2 - y^2, \quad y = 3, \quad (1, 3, 0)$

Intersecting curve: $z = 9x^2 - 9$

$\dfrac{\partial z}{\partial x} = 18x$

At $(1, 3, 0)$: $\dfrac{\partial z}{\partial x} = 18(1) = 18$

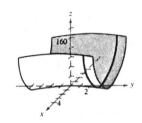

66. $z = 9x^2 - y^2, \quad x = 1, \quad (1, 3, 0)$

Intersecting curve: $z = 9 - y^2$

$\dfrac{\partial z}{\partial y} = -2y$

At $(1, 3, 0)$: $\dfrac{\partial z}{\partial y} = -2(3) = -6$

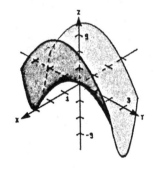

67. $C = 32\sqrt{xy} + 175x + 205y + 1050$

$\dfrac{\partial C}{\partial x} = 16\sqrt{(y/x)} + 175$

$\dfrac{\partial C}{\partial x}\bigg]_{(80,20)} = 16\sqrt{(1/4)} + 175 = 183$

$\dfrac{\partial C}{\partial y} = 16\sqrt{(x/y)} + 205$

$\dfrac{\partial C}{\partial y}\bigg]_{(80,20)} = 16\sqrt{4} + 205 = 237$

68. $f(x, y) = 100x^{0.6}y^{0.4}$

(a) $\dfrac{\partial f}{\partial x} = 60x^{-0.4}y^{0.4} = 60\left(\dfrac{y}{x}\right)^{0.4} = 60\left(\dfrac{500}{1000}\right)^{0.4} \approx 45.47$

(b) $\dfrac{\partial f}{\partial y} = 40x^{0.6}y^{-0.6} = 40\left(\dfrac{x}{y}\right)^{0.6} = 40\left(\dfrac{1000}{500}\right)^{0.6} \approx 60.63$

69. An increase in price will cause a decrease in demand.

70. $R = \dfrac{1}{32}v_0{}^2 \sin 2\theta$

$\dfrac{\partial R}{\partial v_0} = \dfrac{1}{16}v_0 \sin 2\theta$

$\dfrac{\partial R}{\partial v_0}\bigg]_{(2000,5)} = \dfrac{2000}{16}\sin(10°) \approx 21.7$

$\dfrac{\partial R}{\partial \theta} = \dfrac{1}{16}v_0{}^2 \cos 2\theta$

$\dfrac{\partial R}{\partial \theta}\bigg]_{(2000,5)} = \dfrac{(2000)^2}{16}\cos(10°) \approx 246,201.9$

71. $T = 500 - 0.6x^2 - 1.5y^2$

$\dfrac{\partial T}{\partial x} = -1.2x = -2.4°/\text{ft}$

$\dfrac{\partial T}{\partial y} = -3y = -9°/\text{ft}$

72. $PV = kT$

(a) $P = \dfrac{kT}{V}, \quad \dfrac{\partial P}{\partial T} = \dfrac{k}{V}$

(b) $V = \dfrac{kT}{P}, \quad \dfrac{\partial V}{\partial P} = -\dfrac{kT}{P^2}$

73. $f(x, y) = \begin{cases} \dfrac{xy(x^2 - y^2)}{x^2 + y^2}, & (x, y) \neq (0, 0) \\ 0, & (x, y) = (0, 0) \end{cases}$

(a) $f_x(x, y) = \dfrac{(x^2 + y^2)(3x^2y - y^3) - (x^3y - xy^3)(2x)}{(x^2 + y^2)^2} = \dfrac{y(x^4 + 4x^2y^2 - y^4)}{(x^2 + y^2)^2}$

$f_y(x, y) = \dfrac{(x^2 + y^2)(x^3 - 3xy^2) - (x^3y - xy^3)(2y)}{(x^2 + y^2)^2} = \dfrac{x(x^4 - 4x^2y^2 - y^4)}{(x^2 + y^2)^2}$

(b) $f_x(0, 0) = \lim\limits_{\Delta x \to 0} \dfrac{f(\Delta x, 0) - f(0, 0)}{\Delta x} = \lim\limits_{\Delta x \to 0} \dfrac{0/[(\Delta x)^2] - 0}{\Delta x} = 0$

$f_y(0, 0) = \lim\limits_{\Delta y \to 0} \dfrac{f(0, \Delta y) - f(0, 0)}{\Delta y} = \lim\limits_{\Delta y \to 0} \dfrac{0/[(\Delta y)^2] - 0}{\Delta y} = 0$

(c) $f_{xy}(0, 0) = \dfrac{\partial}{\partial y}\left(\dfrac{\partial f}{\partial x}\right)\bigg|_{(0,0)}$

$= \lim\limits_{\Delta y \to 0} \dfrac{f_x(0, \Delta y) - f_x(0, 0)}{\Delta y} = \lim\limits_{\Delta y \to 0} \dfrac{\Delta y(-(\Delta y)^4)}{((\Delta y)^2)^2(\Delta y)} = \lim\limits_{\Delta y \to 0} 1 = -1$

$f_{yx}(0, 0) = \dfrac{\partial}{\partial x}\left(\dfrac{\partial f}{\partial y}\right)\bigg|_{(0,0)}$

$= \lim\limits_{\Delta x \to 0} \dfrac{f_y(\Delta x, 0) - f_y(0, 0)}{\Delta x} = \lim\limits_{\Delta x \to 0} \dfrac{\Delta x((\Delta x)^4)}{((\Delta x)^2)^2(\Delta x)} = \lim\limits_{\Delta x \to 0} 1 = 1$

(d) f, f_x, f_y, f_{xy}, and f_{yx} are continuous.

Section 14.4   Differentials

1. $z = 3x^2y^3$

$dz = 6xy^3\,dx + 9x^2y^2\,dy$

2. $z = \dfrac{x^2}{y}$

$dz = \dfrac{2x}{y}\,dx - \dfrac{x^2}{y^2}\,dy$

3. $z = \dfrac{-1}{x^2 + y^2}$

$dz = \dfrac{2x}{(x^2 + y^2)^2}\,dx + \dfrac{2y}{(x^2 + y^2)^2}\,dy = \dfrac{2}{(x^2 + y^2)^2}(x\,dx + y\,dy)$

4. $z = e^x \sin y$

$dz = (e^x \sin y)\,dx + (e^x \cos y)\,dy$

5. $z = x \cos y - y \cos x$

$dz = (\cos y + y \sin x)\,dx + (-x \sin y - \cos x)\,dy = (\cos y + y \sin x)\,dx - (x \sin y + \cos x)\,dy$

6. $z = \left(\dfrac{1}{2}\right)\left(e^{x^2+y^2} - e^{-x^2-y^2}\right)$

$dz = 2x\left(\dfrac{e^{x^2+y^2} + e^{-x^2-y^2}}{2}\right)dx + 2y\left(\dfrac{e^{x^2+y^2} + e^{-x^2-y^2}}{2}\right)dy = \left(e^{x^2+y^2} + e^{-x^2-y^2}\right)(x\,dx + y\,dy)$

7. $w = 2z^3 y \sin x$

$dw = 2z^3 y \cos x\,dx + 2z^3 \sin x\,dy + 6z^2 y \sin x\,dz$

8. $w = e^x \cos y + z$

$dw = e^x \cos y\,dx - e^x \sin y\,dy + dz$

9. $w = \dfrac{x + y}{z - 2y}$

$dw = \dfrac{1}{z - 2y}\,dx + \dfrac{z + 2x}{(z - 2y)^2}\,dy - \dfrac{x + y}{(z - 2y)^2}\,dz$

10. $w = x^2 yz^2 + \sin yz$

$dw = 2xyz^2\,dx + (x^2 z^2 + x \cos yz)\,dy + (2x^2 yz + y \cos yz)\,dz$

11. (a) $f(1,\ 2) = 4$

$f(1.05,\ 2.1) = 3.4875$

$\Delta z = f(1.05,\ 2.1) - f(1,\ 2) = -0.5125$

   (b) $dz = -2x\,dx - 2y\,dy$

$= -2(0.05) - 4(0.1) = -0.5$

12. (a) $f(1,\ 2) = \sqrt{5} \approx 2.2361$

$f(1.05,\ 2.1) = \sqrt{5.5125} \approx 2.3479$

$\Delta z = 0.11180$

   (b) $dz = \dfrac{x}{\sqrt{x^2 + y^2}}\,dx + \dfrac{y}{\sqrt{x^2 + y^2}}\,dy$

$= \dfrac{x\,dx + y\,dy}{\sqrt{x^2 + y^2}} = \dfrac{0.05 + 2(0.1)}{\sqrt{5}} \approx 0.11180$

13. (a) $f(1,\ 2) = \sin 2$

$f(1.05,\ 2.1) = 1.05 \sin 2.1$

$\Delta z = f(1.05,\ 2.1) - f(1,\ 2) \approx -0.00293$

   (b) $dz = \sin y\,dx + x \cos y\,dy$

$= (\sin 2)(0.05) + (\cos 2)(0.1) \approx 0.00385$

14. (a) $f(1,\ 2) = 2$

$f(1.05,\ 2.1) = 2.205$

$\Delta z = 0.205$

   (b) $dz = y\,dx + x\,dy$

$= 2(0.05) + (0.1) = 0.2$

15. (a) $f(1, \ 2) = -5$

$f(1.05, \ 2.1) = -5.25$

$\Delta z = -0.25$

(b) $dz = 3 \, dx - 4 \, dy$

$= 3(0.05) - 4(0.1) = -0.25$

16. (a) $f(1, \ 2) = \dfrac{1}{2} = 0.5$

$f(1.05, \ 2.1) = \dfrac{1.05}{2.1} = 0.5$

$\Delta z = 0$

(b) $dz = \dfrac{1}{y} \, dx - \dfrac{x}{y^2} \, dy$

$= \dfrac{1}{2}(0.05) - \dfrac{1}{4}(0.1) = 0$

17. $f(x, \ y) = x^2 - 2x + y$

$f_x(x, \ y) = 2x - 2 = 2(x - 1)$

$f_y(x, \ y) = 1$

$\epsilon_1 = f_x(x_1, \ y) - f_x(x, \ y)$ where $x \le x_1 \le x + \Delta x$

$= 2(x_1 - 1) - 2(x - 1) = 2(x_1 - x) \le 2\Delta x$

As $(\Delta x, \ \Delta y) \Rightarrow (0, \ 0), \quad x_1 \Rightarrow x$ and $\epsilon_1 \Rightarrow 0$.

$\epsilon_2 = f_y(x + \Delta x, \ y_1) - f_y(x, \ y)$ where $y \le y_1 \le y + \Delta y$

$= 1 - 1 = 0$

As $(\Delta x, \ \Delta y) \Rightarrow (0, \ 0), \quad \epsilon_2 \Rightarrow 0$.

18. $f(x, \ y) = x^2 + y^2$

$f_x(x, \ y) = 2x$

$f_y(x, \ y) = 2y$

$\epsilon_1 = f_x(x_1, \ y) - f_x(x, \ y)$ where $x \le x_1 \le x + \Delta x$

$= 2x_1 - 2x = 2(x_1 - x)$

As $(\Delta x, \ \Delta y) \Rightarrow (0, \ 0), \quad x_1 \Rightarrow x$ and $\epsilon_1 \Rightarrow 0$.

$\epsilon_2 = f_y(x + \Delta x, \ y_1) - f_y(x, \ y)$ where $y \le y_1 \le y + \Delta y$

$= 2y_1 - 2y = 2(y_1 - y)$

As $(\Delta x, \ \Delta y) \Rightarrow (0, \ 0), \quad y_1 \Rightarrow y$ and $\epsilon_2 \Rightarrow 0$.

19. $f(x, \ y) = x^2 y$

$f_x(x, \ y) = 2xy$

$f_y(x, \ y) = x^2$

$\epsilon_1 = f_x(x_1, \ y) - f_x(x, \ y)$ where $x \le x_1 \le x + \Delta x$

$= 2x_1 y - 2xy = 2y(x_1 - x)$

As $(\Delta x, \ \Delta y) \Rightarrow (0, \ 0), \quad x_1 \Rightarrow x$ and $\epsilon_1 \Rightarrow 0$.

$\epsilon_2 = f_y(x + \Delta x, \ y_1) - f_y(x, \ y)$ where $y \le y_1 \le y + \Delta y$

$= (x + \Delta x)^2 - x^2 = 2x(\Delta x) + (\Delta x)^2$

As $(\Delta x, \ \Delta y) \Rightarrow (0, \ 0), \quad \epsilon_2 \Rightarrow 0$.

20. $f(x, y) = 5x - 10y + y^3$

$f_x(x, y) = 5$

$f_y(x, y) = -10 + 3y^2$

$\quad \epsilon_1 = f_x(x_1, y) - f_x(x, y)$ where $x \leq x_1 \leq x + \Delta x$

$\quad\quad = 5 - 5 = 0$

As $(\Delta x, \Delta y) \Rightarrow (0, 0)$, $\epsilon_1 = 0$.

$\epsilon_2 = f_y(x + \Delta x, y_1) - f_y(x, y)$ where $y \leq y_1 \leq y + \Delta y$

$\quad = (-10 + 3y_1^2) - (-10 + 3y^2) = 3(y_1^2 - y^2)$

As $(\Delta x, \Delta y) \Rightarrow (0, 0)$, $y_1 \Rightarrow y$ and $\epsilon_2 \Rightarrow 0$.

21. $dV = (2\pi rh)\, dr + (\pi r^2)\, dh$

$\dfrac{dV}{V} = 2\dfrac{dr}{r} + \dfrac{dh}{h} = 2(0.04) + (0.02) = 0.10 = 10\%$

22. $a = \dfrac{v^2}{r}$

$da = \dfrac{2v}{r}\, dv - \dfrac{v^2}{r^2}\, dr$

$\dfrac{da}{a} = 2\dfrac{dv}{v} - \dfrac{dr}{r} = 2(0.02) - (-0.01) = 0.05 = 5\%$

Note: The maximum error will occur when dv and dr differ in signs.

23. $P = \dfrac{E^2}{R}$

$dP = \dfrac{2E}{R}\, dE - \dfrac{E^2}{R^2}\, dR$

$\dfrac{dP}{P} = 2\dfrac{dE}{E} - \dfrac{dR}{R} = 2(0.02) - (-0.03) = 0.07 = 7\%$

24. $\dfrac{1}{R} = \dfrac{1}{R_1} + \dfrac{1}{R_2}$

$R = \dfrac{R_1 R_2}{R_1 + R_2}$

$dR_1 = \Delta R_1 = 0.5$

$dR_2 = \Delta R_2 = -2$

$\Delta R \approx dR = \dfrac{\partial R}{\partial R_1}\, dR_1 + \dfrac{\partial R}{\partial R_2}\, dR_2 = \dfrac{R_2^2}{(R_1 + R_2)^2}\Delta R_1 + \dfrac{R_1^2}{(R_1 + R_2)^2}\Delta R_2$

When $R_1 = 10$ and $R_2 = 15$, we have $\Delta R \approx \dfrac{15^2}{(10 + 15)^2}(0.5) + \dfrac{10^2}{(10 + 15)^2}(-2) = -0.14$ ohm.

25. $T = 2\pi\sqrt{\dfrac{L}{g}}$

$dg = \Delta g = 32.24 - 32.09 = 0.15$

$dL = \Delta L = 2.48 - 2.5 = -0.02$

$\Delta T \approx dT = \dfrac{\partial T}{\partial g}\, dg + \dfrac{\partial T}{\partial L}\, dL = -\dfrac{\pi}{g}\sqrt{\dfrac{L}{g}}\Delta g + \dfrac{\pi}{\sqrt{Lg}}\Delta L$

When $g = 32.09$ and $L = 2.5$, we have $\Delta T \approx -\dfrac{\pi}{32.09}\sqrt{\dfrac{2.5}{32.09}}(0.15) + \dfrac{\pi}{\sqrt{(2.5)(32.09)}}(-0.02) \approx -0.0111$ sec.

26. $A = \dfrac{1}{2}ab\sin C$

$$dA = \dfrac{1}{2}[(b\sin C)\,da + (a\sin C)\,db + (ab\cos C)\,dC]$$

$$\dfrac{dA}{A} = \dfrac{da}{a} + \dfrac{db}{b} + (\cot C)\,dC = \dfrac{1/16}{3} + \dfrac{1/16}{4} + (1)(0.02) \approx 0.05646 = 5.646\%$$

27. $L = 0.00021\left(\ln\dfrac{2h}{r} - 0.75\right)$

$$dL = 0.00021\left[\dfrac{dh}{h} - \dfrac{dr}{r}\right] = 0.00021\left[\dfrac{(\pm 1/100)}{100} - \dfrac{(\pm 1/16)}{2}\right] \approx (\pm 6.5) \times 10^{-6}$$

$$L = 0.00021(\ln 100 - 0.75) \approx 8.096 \times 10^{-4} \pm dL = 8.096 \times 10^{-4} \pm 6.5 \times 10^{-6} \text{ micro-henrys}$$

28. $V = \dfrac{\pi r^2 h}{3}$

$r = 3$

$h = 6$

$$dV = \dfrac{2\pi rh}{3}\,dr + \dfrac{\pi r^2}{3}\,dh = \dfrac{\pi r}{3}(2h\,dr + r\,dh)$$

Δr	Δh	dV	ΔV	$\Delta V - dV$
0.1	0.1	4.7124	4.8391	0.1267
0.1	−0.1	2.8274	2.8264	−0.0010
0.001	0.002	0.0566	0.0566	0.0000
−0.0001	0.0002	−0.0019	−0.0019	0.0000

29. $f(x,\,y) = \begin{cases} \dfrac{3x^2 y}{x^4 + y^2}, & (x,\,y) \neq (0,\,0) \\ 0, & (x,\,y) = (0,\,0) \end{cases}$

(a) $f_x(0,\,0) = \displaystyle\lim_{\Delta x \to 0} \dfrac{f(\Delta x,\,0) - f(0,\,0)}{\Delta x} = \lim_{\Delta x \to 0} \dfrac{0/[(\Delta x)^4] - 0}{\Delta x} = 0$

$f_y(0,\,0) = \displaystyle\lim_{\Delta y \to 0} \dfrac{f(0,\,\Delta y) - f(0,\,0)}{\Delta y} = \lim_{\Delta y \to 0} \dfrac{0/[(\Delta y)^2] - 0}{\Delta y} = 0$

Thus, the partial derivatives exist at $(0,\,0)$.

(b)   Along the line $y = x$: $\displaystyle\lim_{(x,y)\to(0,0)} f(x,\,y) = \lim_{x \to 0} \dfrac{3x^3}{x^4 + x^2} = \lim_{x \to 0} \dfrac{3x}{x^2 + 1} = 0$

Along the curve $y = x^2$: $\displaystyle\lim_{(x,y)\to(0,0)} f(x,\,y) = \dfrac{3x^4}{2x^4} = \dfrac{3}{2}$

f is not continuous at $(0,\,0)$. Therefore, f is not differentiable at $(0,\,0)$.

30. $f(x, y) = \begin{cases} \dfrac{2x^2 y^2}{x^4 + y^4}, & (x, y) \neq (0, 0) \\ 0, & (x, y) = (0, 0) \end{cases}$

(a) $f_x(0, 0) = \lim\limits_{\Delta x \to 0} \dfrac{f(\Delta x, 0) - f(0, 0)}{\Delta x} = \lim\limits_{\Delta x \to 0} \dfrac{0 - 0}{\Delta x} = 0$

$f_y(0, 0) = \lim\limits_{\Delta y \to 0} \dfrac{f(0, \Delta y) - f(0, 0)}{\Delta y} = \lim\limits_{\Delta y \to 0} \dfrac{0 - 0}{\Delta y} = 0$

Thus, the partial derivatives exist at $(0, 0)$.

(b) Along the line $y = x$: $\lim\limits_{(x,y) \to (0,0)} f(x, y) = \lim\limits_{x \to 0} \dfrac{2x^4}{2x^4} = 1$

     Along the curve $y = x^2$: $\lim\limits_{(x,y) \to (0,0)} f(x, y) = \lim\limits_{x \to 0} \dfrac{2x^6}{x^4 + x^8} = \lim\limits_{x \to 0} \dfrac{2x^2}{1 + x^4} = 0$

f is not continuous at $(0, 0)$. Therefore, f is not differentiable at $(0, 0)$.

Section 14.5   Chain Rules for Functions of Several Variables

1. $w = x^2 + y^2$

     $x = e^t$

     $y = e^{-t}$

     $\dfrac{dw}{dt} = 2xe^t + 2y(-e^{-t}) = 2(e^{2t} - e^{-2t})$

2. $w = \sqrt{x^2 + y^2}$

     $x = \sin t$

     $y = e^t$

     $\dfrac{dw}{dt} = \dfrac{x}{\sqrt{x^2 + y^2}} \cos t + \dfrac{y}{\sqrt{x^2 + y^2}} e^t = \dfrac{x \cos t + y e^t}{\sqrt{x^2 + y^2}} = \dfrac{\sin t \cos t + e^{2t}}{\sqrt{\sin^2 t + e^{2t}}}$

3. $w = x \sec y$

     $x = e^t$

     $y = \pi - t$

     $\dfrac{dw}{dt} = (\sec y)(e^t) + (x \sec y \tan y)(-1) = e^t \sec(\pi - t)[1 - \tan(\pi - t)]$

4. $w = \ln \dfrac{y}{x}$

     $x = \cos t$

     $y = \sin t$

     $\dfrac{dw}{dt} = \left(\dfrac{-1}{x}\right)(-\sin t) + \left(\dfrac{1}{y}\right)(\cos t) = \tan t + \cot t$

5. $w = x^2 + y^2 + z^2$

$x = e^t \cos t$

$y = e^t \sin t$

$z = e^t$

$\dfrac{dw}{dt} = 2x(-e^t \sin t + e^t \cos t) + 2y(e^t \cos t + e^t \sin t) + 2ze^t = 4e^{2t}$

6. $w = xy \cos z$

$x = t$

$y = t^2$

$z = \arccos t$

$\dfrac{dw}{dt} = (y \cos z)(1) + (x \cos z)(2t) + (-xy \sin z)\left(-\dfrac{1}{\sqrt{1-t^2}}\right) = t^3 + 2t^3 + t^3 = 4t^3$

7. $w = x^2 + y^2$

$x = s + t$

$y = s - t$

$\dfrac{\partial w}{\partial s} = 2x + 2y = 2(x + y) = 4s$

$\dfrac{\partial w}{\partial t} = 2x + 2y(-1) = 2(x - y) = 4t$

When $s = 2$ and $t = -1$,

$\qquad \dfrac{\partial w}{\partial s} = 8$ and $\dfrac{\partial w}{\partial t} = -4.$

8. $w = y^3 - 3x^2 y$

$x = e^s$

$y = e^t$

$\dfrac{\partial w}{\partial s} = -6xy(e^s) + (3y^2 - 3x^2)(0)$

$\qquad = -6e^{2s+t}$

$\dfrac{\partial w}{\partial t} = -6xy(0) + (3y^2 - 3x^2)(e^t)$

$\qquad = 3e^t(e^{2t} - e^{2s})$

When $s = 0$ and $t = 1$,

$\qquad \dfrac{\partial w}{\partial s} = -6e$ and $\dfrac{\partial w}{\partial t} = 3e(e^2 - 1).$

9. $w = x^2 - y^2$

$x = s \cos t$

$y = s \sin t$

$\dfrac{\partial w}{\partial s} = 2x \cos t - 2y \sin t = 2s \cos 2t$

$\dfrac{\partial w}{\partial t} = 2x(-s \sin t) - 2y(s \cos t) = -2s^2 \sin 2t$

When $s = 3$ and $t = \dfrac{\pi}{4}$, $\dfrac{\partial w}{\partial s} = 0$ and $\dfrac{\partial w}{\partial t} = -18.$

10. $w = \sin(2x + 3y)$

$x = s + t$

$y = s - t$

$\dfrac{\partial w}{\partial s} = 2 \cos(2x + 3y) + 3 \cos(2x + 3y) = 5 \cos(2x + 3y) = 5 \cos(5s - t)$

$\dfrac{\partial w}{\partial t} = 2 \cos(2x + 3y) - 3 \cos(2x + 3y) = -\cos(2x + 3y) = -\cos(5s - t)$

When $s = 0$ and $t = \dfrac{\pi}{2}$, $\dfrac{\partial w}{\partial s} = 0$ and $\dfrac{\partial w}{\partial t} = 0.$

11. $w = xy, \quad x = 2\sin t, \quad y = \cos t$

(a) $\dfrac{dw}{dt} = 2y\cos t + x(-\sin t) = 2y\cos t - x\sin t = 2(\cos^2 t - \sin^2 t) = 2\cos 2t$

(b) $w = 2\sin t\cos t = \sin 2t, \quad \dfrac{dw}{dt} = 2\cos 2t$

12. $w = \cos(x - y), \quad x = t^2, \quad y = 1$

(a) $\dfrac{dw}{dt} = -\sin(x - y)(2t) + \sin(x - y)(0) = -2t\sin(x - y) = -2t\sin(t^2 - 1)$

(b) $w = \cos(t^2 - 1), \quad \dfrac{dw}{dt} = -2t\sin(t^2 - 1)$

13. $w = xy + xz + yz, \quad x = t - 1, \quad y = t^2 - 1, \quad z = t$

(a) $\dfrac{dw}{dt} = (y + z) + (x + z)(2t) + (x + y) = (t^2 - 1 + t) + (t - 1 + t)(2t) + (t - 1 + t^2 - 1) = 3(2t^2 - 1)$

(b) $\quad w = (t - 1)(t^2 - 1) + (t - 1)t + (t^2 - 1)t$

$\dfrac{dw}{dt} = 2t(t - 1) + (t^2 - 1) + 2t - 1 + 3t^2 - 1 = 3(2t^2 - 1)$

14. $w = xyz, \quad x = t^2, \quad y = 2t, \quad z = e^{-t}$

(a) $\dfrac{dw}{dt} = yz(2t) + xz(2) + (xy)(-e^{-t})$

$\quad = (2t)(e^{-t})(2t) + (t^2)(e^{-t})(2) + (t^2)(2t)(-e^{-t}) = 2t^2 e^{-t}(2 + 1 - t) = 2t^2 e^{-t}(3 - t)$

(b) $\quad w = (t^2)(2t)(e^{-t}) = 2t^3 e^{-t}$

$\dfrac{dw}{dt} = (2t^3)(-e^{-t}) + (e^{-t})(6t^2) = 2t^2 e^{-t}(-t + 3)$

15. $w = x^2 - 2xy + y^2, \quad x = r + \theta, \quad y = r - \theta$

(a) $\dfrac{\partial w}{\partial r} = (2x - 2y)(1) + (-2x + 2y)(1) = 0$

$\dfrac{\partial w}{\partial \theta} = (2x - 2y)(1) + (-2x + 2y)(-1) = 4x - 4y = 4(x - y) = 4[(r + \theta) - (r - \theta)] = 8\theta$

(b) $\quad w = (r + \theta)^2 - 2(r + \theta)(r - \theta) + (r - \theta)^2 = (r^2 + 2r\theta + \theta^2) - 2(r^2 - \theta^2) + (r^2 - 2r\theta + \theta^2) = 4\theta^2$

$\dfrac{\partial w}{\partial r} = 0$

$\dfrac{\partial w}{\partial \theta} = 8\theta$

16. $w = \sqrt{4 - 2x^2 - 2y^2}, \quad x = r\cos\theta, \quad y = r\sin\theta$

(a) $\dfrac{\partial w}{\partial r} = \dfrac{-2x}{\sqrt{4 - 2x^2 - 2y^2}}\cos\theta + \dfrac{-2y}{\sqrt{4 - 2x^2 - 2y^2}}\sin\theta = \dfrac{-2r\cos^2\theta - 2r\sin^2\theta}{\sqrt{4 - 2r^2}} = \dfrac{-2r}{\sqrt{4 - 2r^2}}$

$\dfrac{\partial w}{\partial \theta} = \dfrac{-2x}{\sqrt{4 - 2x^2 - 2y^2}}(-r\sin\theta) + \dfrac{-2y}{\sqrt{4 - 2x^2 - 2y^2}}(r\cos\theta) = \dfrac{2r\sin\theta\cos\theta - 2r\sin\theta\cos\theta}{\sqrt{4 - 2r^2}} = 0$

(b) $\quad w = \sqrt{4 - 2r^2\cos^2\theta - 2r^2\sin^2\theta} = \sqrt{4 - 2r^2}$

$\dfrac{\partial w}{\partial r} = \dfrac{-2r}{\sqrt{4 - 2r^2}}$

$\dfrac{\partial w}{\partial \theta} = 0$

17. (a) $\dfrac{\partial w}{\partial r} = \dfrac{-y}{x^2 + y^2}\cos\theta + \dfrac{x}{x^2 + y^2}\sin\theta = \dfrac{-r\sin\theta\cos\theta}{r^2} + \dfrac{r\cos\theta\sin\theta}{r^2} = 0$

$\dfrac{\partial w}{\partial\theta} = \dfrac{-y}{x^2 + y^2}(-r\sin\theta) + \dfrac{x}{x^2 + y^2}(r\cos\theta) = \dfrac{-(r\sin\theta)(-r\sin\theta)}{r^2} + \dfrac{(r\cos\theta)(r\cos\theta)}{r^2} = 1$

(b) $\quad w = \arctan\dfrac{r\sin\theta}{r\cos\theta} = \arctan(\tan\theta) = \theta$

$\dfrac{\partial w}{\partial r} = 0$

$\dfrac{\partial w}{\partial\theta} = 1$

18. $w = \dfrac{yx}{z}, \quad x = r + \theta, \quad y = r - \theta, \quad z = \theta^2$

(a) $\dfrac{\partial w}{\partial r} = \dfrac{y}{z} + \dfrac{x}{z} = \dfrac{r - \theta}{\theta^2} + \dfrac{r + \theta}{\theta^2} = \dfrac{2r}{\theta^2}$

(b) $\dfrac{\partial w}{\partial\theta} = \dfrac{y}{z} - \dfrac{x}{z} - \dfrac{xy}{z^2}(2\theta) = \dfrac{r - \theta}{\theta^2} - \dfrac{r + \theta}{\theta^2} - \dfrac{(r + \theta)(r - \theta)}{\theta^4}(2\theta) = \dfrac{-2r^2}{\theta^3}$

19. $F(x,\, y,\, z) = x^2 + y^2 + z^2 - 25$

$\qquad F_x = 2x$

$\qquad F_y = 2y$

$\qquad F_z = 2z$

$\qquad \dfrac{\partial z}{\partial x} = -\dfrac{F_x}{F_z} = -\dfrac{x}{z}$

$\qquad \dfrac{\partial z}{\partial y} = -\dfrac{F_y}{F_z} = -\dfrac{y}{z}$

20. $F(x,\, y,\, z) = xz + yz + xy$

$\qquad F_x = z + y$

$\qquad F_y = z + x$

$\qquad F_z = x + y$

$\qquad \dfrac{\partial z}{\partial x} = -\dfrac{F_x}{F_z} = -\dfrac{y + z}{x + y}$

$\qquad \dfrac{\partial z}{\partial y} = -\dfrac{F_y}{F_z} = -\dfrac{x + z}{x + y}$

21. $F(x,\, y,\, z) = \tan(x + y) + \tan(y + z) - 1$

$\qquad F_x = \sec^2(x + y)$

$\qquad F_y = \sec^2(x + y) + \sec^2(y + z)$

$\qquad F_z = \sec^2(y + z)$

$\qquad \dfrac{\partial z}{\partial x} = -\dfrac{F_x}{F_z} = -\dfrac{\sec^2(x + y)}{\sec^2(y + z)}$

$\qquad \dfrac{\partial z}{\partial y} = -\dfrac{F_y}{F_z} = -\dfrac{\sec^2(x + y) + \sec^2(y + z)}{\sec^2(y + z)} = -\left(\dfrac{\sec^2(x + y)}{\sec^2(y + z)} + 1\right)$

22. $F(x,\, y,\, z) = e^x\sin(y + z) - z$

$\qquad F_x = e^x\sin(y + z)$

$\qquad F_y = e^x\cos(y + z)$

$\qquad F_z = e^x\cos(y + z) - 1$

$\qquad \dfrac{\partial z}{\partial x} = -\dfrac{F_x}{F_z} = \dfrac{e^x\sin(y + z)}{1 - e^x\cos(y + z)}$

$\qquad \dfrac{\partial z}{\partial y} = -\dfrac{F_y}{F_z} = \dfrac{e^x\cos(y + z)}{1 - e^x\cos(y + z)}$

23. $x^2 + 2yz + z^2 - 1 = 0$

(i) $2x + 2y\dfrac{\partial z}{\partial x} + 2z\dfrac{\partial z}{\partial x} = 0$ implies $\dfrac{\partial z}{\partial x} = -\dfrac{x}{y + z}$.

(ii) $2y\dfrac{\partial z}{\partial y} + 2z + 2z\dfrac{\partial z}{\partial y} = 0$ implies $\dfrac{\partial z}{\partial y} = -\dfrac{z}{y + z}$.

Differentiate (i) with respect to x.

$$2 + 2y\dfrac{\partial^2 z}{\partial x^2} + 2z\dfrac{\partial^2 z}{\partial x^2} + 2\left(\dfrac{\partial z}{\partial x}\right)^2 = 0$$

$$\dfrac{\partial^2 z}{\partial x^2} = \dfrac{-1 - (\partial z/\partial x)^2}{y + z} = -\dfrac{(y + z)^2 + x^2}{(y + z)^3}$$

Differentiate (i) with respect to y.

$$2y\dfrac{\partial^2 z}{\partial y\partial x} + 2\dfrac{\partial z}{\partial x} + 2z\dfrac{\partial^2 z}{\partial y\partial x} + 2\left(\dfrac{\partial z}{\partial x}\right)\left(\dfrac{\partial z}{\partial y}\right) = 0$$

$$\dfrac{\partial^2 z}{\partial y\partial z} = -\dfrac{\partial z/\partial x + (\partial z/\partial x)(\partial z/\partial y)}{y + z} = \dfrac{x(y + z) - xz}{(y + z)^3} = \dfrac{xy}{(x + y)^3} = \dfrac{\partial^2 z}{\partial x\partial y}$$

Differentiate (ii) with respect to y.

$$2y\dfrac{\partial^2 z}{\partial y^2} + 2\dfrac{\partial z}{\partial y} + 2\dfrac{\partial z}{\partial y} + 2z\dfrac{\partial^2 z}{\partial y^2} + 2\left(\dfrac{\partial z}{\partial y}\right)^2 = 0$$

$$\dfrac{\partial^2 z}{\partial y^2} = \dfrac{-4(\partial z/\partial y) - 2(\partial z/\partial y)^2}{2y + 2z} = \dfrac{2z(y + z) - z}{(y + z)^3} = \dfrac{z(2y + z)}{(y + z)^3}$$

24. $x + \sin(y + z) = 0$

(i) $1 + \dfrac{\partial z}{\partial x}\cos(y + z) = 0$ implies $\dfrac{\partial z}{\partial x} = -\dfrac{1}{\cos(y + z)} = -\sec(y + z)$.

(ii) $\left(1 + \dfrac{\partial z}{\partial y}\right)\cos(y + z) = 0$ implies $\dfrac{\partial z}{\partial y} = -1$.

Differentiate (i) with respect to x.

$$\dfrac{\partial^2 z}{\partial x^2}\cos(y + z) - \left(\dfrac{\partial z}{\partial x}\right)^2\sin(y + z) = 0$$

$$\dfrac{\partial^2 z}{\partial x^2} = \dfrac{\sin(y + z)\sec^2(y + z)}{\cos(y + z)} = \tan(y + z)\sec^2(y + z)$$

Differentiate (ii) with respect to x.

$$\dfrac{\partial^2 z}{\partial x\partial y}\cos(y + z) = 0 \text{ implies } \dfrac{\partial^2 z}{\partial x\partial y} = 0 = \dfrac{\partial^2 z}{\partial y\partial x}.$$

Differentiate (ii) with respect to y.

$$\dfrac{\partial^2 z}{\partial y^2}\cos(y + z) - \dfrac{\partial z}{\partial y}\left(1 + \dfrac{\partial z}{\partial y}\right)\sin(y + z) = 0$$

$$\dfrac{\partial^2 z}{\partial y^2} = 0$$

25. $F(x, y, z, w) = xyz + xzw - yzw + w^2 - 5$

$$F_x = yz + zw$$

$$F_y = xz - zw$$

$$F_z = xy + xw - yw$$

$$F_w = xz - yz + 2w$$

$$\frac{\partial w}{\partial x} = -\frac{F_x}{F_w} = -\frac{z(y + w)}{xz - yz + 2w}$$

$$\frac{\partial w}{\partial y} = -\frac{F_y}{F_w} = -\frac{z(x - w)}{xz - yz + 2w}$$

$$\frac{\partial w}{\partial z} = -\frac{F_z}{F_w} = \frac{xy + xw - yw}{xz - yz + 2w}$$

26. $F(x, y, z, w) = x^2 + y^2 + z^2 + 6xw - 8w^2 - 5$

$$F_x = 2x + 6w$$

$$F_y = 2y$$

$$F_z = 2z$$

$$F_w = 6x - 16w$$

$$\frac{\partial w}{\partial x} = -\frac{F_x}{F_w} = -\frac{x + 3w}{3x - 8w}$$

$$\frac{\partial w}{\partial y} = -\frac{F_y}{F_w} = -\frac{y}{3x - 8w}$$

$$\frac{\partial w}{\partial z} = -\frac{F_z}{F_w} = -\frac{z}{3x - 8w}$$

27. (a) $V = xyz$

$$\frac{dV}{dt} = yz\frac{dx}{dt} + xz\frac{dy}{dt} + xy\frac{dz}{dt} = 6(4)(3) + 10(4)(2) + 10(6)\frac{1}{2} = 182 \text{ ft}^3/\text{min}$$

(b) $S = 2xy + 2xz + 2yz$

$$\frac{dS}{dt} = (2y + 2z)\frac{\partial x}{\partial t} + (2x + 2z)\frac{\partial y}{\partial t} + (2x + 2y)\frac{\partial z}{\partial t}$$

$$= [2(6) + 2(4)](3) + [2(10) + 2(4)](2) + [2(10) + 2(6)]\left(\frac{1}{2}\right) = 132 \text{ ft}^2/\text{min}$$

28. (a) $V = \pi r^2 h$

$$\frac{dV}{dt} = \pi\left(2rh\frac{dr}{dt} + r^2\frac{dh}{dt}\right) = \pi r\left(2h\frac{dr}{dt} + r\frac{dh}{dt}\right) = \pi(12)[2(36)(6) + 12(-4)] = 4608\pi \text{ in}^3/\text{min}$$

(b) $S = 2\pi r(r + h)$

$$\frac{dS}{dt} = 2\pi\left[(2r + h)\frac{dr}{dt} + r\frac{dh}{dt}\right] = 2\pi[(24 + 36)(6) + 12(-4)] = 624\pi \text{ in}^2/\text{min}$$

29. (a) $V = \frac{1}{3}\pi r^2 h$

$$\frac{dV}{dt} = \frac{1}{3}\pi\left(2rh\frac{dr}{dt} + r^2\frac{dh}{dt}\right) = \frac{1}{3}\pi[2(12)(36)(6) + (12)^2(-4)] = 1536\pi \text{ in}^3/\text{min}$$

(b) $S = \pi r\sqrt{r^2 + h^2} + \pi r^2$

$$\frac{dS}{dt} = \pi\left[\left(\sqrt{r^2 + h^2} + \frac{r^2}{\sqrt{r^2 + h^2}} + 2r\right)\frac{dr}{dt} + \frac{rh}{\sqrt{r^2 + h^2}}\frac{dh}{dt}\right]$$

$$= \pi\left[\left(\sqrt{12^2 + 36^2} + \frac{144}{\sqrt{12^2 + 36^2}} + 2(12)\right)(6) + \frac{36(12)}{\sqrt{12^2 + 36^2}}(-4)\right]$$

$$= \pi\left[\left(12\sqrt{10} + \frac{12}{\sqrt{10}}\right)(6) + 144 + \frac{36}{\sqrt{10}}(-4)\right]$$

$$= \frac{648\pi}{\sqrt{10}} + 144\pi \text{ in}^2/\text{min}$$

30. (a) $V = \dfrac{\pi}{3}(r^2 + rR + R^2)h$

$\dfrac{dV}{dt} = \dfrac{\pi}{3}\left[(2r + R)h\dfrac{dr}{dt} + (r + 2R)h\dfrac{dR}{dt} + (r^2 + rR + R^2)\dfrac{dh}{dt}\right]$

$\qquad = \dfrac{\pi}{3}\left[[2(15) + 25](10)(4) + [15 + 2(25)](10)(4) + [(15)^2 + (15)(25) + (25)^2](12)\right]$

$\qquad = \dfrac{\pi}{3}(19500) = 6500\pi \ \text{cm}^3/\text{min}$

(b) $S = \pi(R + r)\sqrt{(R - r)^2 + h^2}$

$\dfrac{dS}{dt} = \pi\left[\left[\sqrt{(R - r)^2 + h^2} - (R + r)\dfrac{(R - r)}{\sqrt{(R - r)^2 + h^2}}\right]\dfrac{dr}{dt}\right.$

$\qquad\qquad + \left[\sqrt{(R - r)^2 + h^2} + (R + r)\dfrac{(R - r)}{\sqrt{(R - r)^2 + h^2}}\right]\dfrac{dR}{dt}$

$\qquad\qquad \left. +(R + r)\dfrac{h}{\sqrt{(R - r)^2 + h^2}}\dfrac{dh}{dt}\right]$

$\qquad = \pi\left[\left[\sqrt{(25 - 15)^2 + 10^2} - (25 + 15)\dfrac{25 - 15}{\sqrt{(25 - 15)^2 + 10^2}}\right](4)\right.$

$\qquad\qquad + \left[\sqrt{(25 - 15)^2 + 10^2} + (25 + 15)\dfrac{25 - 10}{\sqrt{(25 - 15)^2 + 10^2}}\right](4)$

$\qquad\qquad \left. +(25 + 15)\dfrac{10}{\sqrt{(25 - 15)^2 + 10^2}}(12)\right]$

$\qquad = 320\sqrt{2}\,\pi\text{cm}^2/\text{min}$

31. $I = \dfrac{1}{2}m(r_1{}^2 + r_2{}^2)$

$\dfrac{dI}{dt} = \dfrac{1}{2}m\left[2r_1\dfrac{dr_1}{dt} + 2r_2\dfrac{dr_2}{dt}\right]$

$\qquad = m[(6)(2) + (8)(2)] = 28m$

32. $pV = RT$

$T = \dfrac{1}{R}(pV)$

$\dfrac{dT}{dt} = \dfrac{1}{R}\left[V\dfrac{dp}{dt} + p\dfrac{dV}{dt}\right]$

33. $f(x,\ y) = x^3 - 3xy^2 + y^3$

$f(tx,\ ty) = (tx)^3 - 3(tx)(ty)^2 + (ty)^3 = t^3(x^3 - 3xy^2 + y^3) = t^3 f(x,\ y)$

Degree: 3

$xf_x(x,\ y) + yf_y(x,\ y) = x(3x^2 - 3y^2) + y(-6xy + 3y^2) = 3x^2 - 9xy^2 + 3y^3 = 3f(x,\ y)$

34. $f(x,\ y) = \dfrac{xy}{\sqrt{x^2 + y^2}}$

$f(tx,\ ty) = \dfrac{(tx)(ty)}{\sqrt{(tx)^2 + (ty)^2}} = t\left(\dfrac{xy}{\sqrt{x^2 + y^2}}\right) = tf(x,\ y)$

Degree: 1

$xf_x(x,\ y) + yf_y(x,\ y) = x\left(\dfrac{y^3}{(x^2 + y^2)^{3/2}}\right) + y\left(\dfrac{x^3}{(x^2 + y^2)^{3/2}}\right) = \dfrac{xy}{\sqrt{x^2 + y^2}} = 1f(x,\ y)$

35. $f(x, y) = e^{x/y}$

$f(tx, ty) = e^{tx/ty} = e^{x/y} = f(x, y)$

Degree: 0

$xf_x(x, y) + yf_y(x, y) = x\left(\dfrac{1}{y}e^{x/y}\right) + y\left(-\dfrac{x}{y^2}e^{x/y}\right) = 0$

36. $f(x, y) = 2x^3 - 3xy^2$

$f(tx, ty) = 2(tx)^3 - 3(tx)(ty)^2 = t^3 f(x, y)$

Degree: 3

$xf_x(x, y) + yf_y(x, y) = x(6x^2 - 3y^2) + y(-6xy) = 6x^3 - 9xy^2 = 3f(x, y)$

37. $f(x, y) = \dfrac{x^2}{\sqrt{x^2 + y^2}}$

$f(tx, ty) = \dfrac{(tx)^2}{\sqrt{(tx)^2 + (ty)^2}} = t\left(\dfrac{x^2}{\sqrt{x^2 + y^2}}\right) = tf(x, y)$

Degree: 1

$xf_x(x, y) + yf_y(x, y) = x\left[\dfrac{x^3 + 2xy^2}{(x^2 + y^2)^{3/2}}\right] + y\left[\dfrac{-x^2 y}{(x^2 + y^2)^{3/2}}\right] = \dfrac{x^4 + x^2 y^2}{(x^2 + y^2)^{3/2}} = \dfrac{x^2(x^2 + y^2)}{(x^2 + y^2)^{3/2}}$

$$= \dfrac{x^2}{\sqrt{x^2 + y^2}} = f(x, y)$$

38. $g(t) = f(xt, yt) = t^n f(x, y)$

Let $u = xt,\quad v = yt,$ then

$$g'(t) = \frac{\partial f}{\partial u} \cdot \frac{du}{dt} + \frac{\partial f}{\partial v} \cdot \frac{dv}{dt} = \frac{\partial f}{\partial u}x + \frac{\partial f}{\partial v}y \quad \text{and} \quad g'(t) = nt^{n-1}f(x, y).$$

Now, let $t = 1$ and we have $u = x,\quad v = y.$ Thus,

$$\frac{\partial f}{\partial x}x + \frac{\partial f}{\partial y}y = nf(x, y).$$

39.
$$w = f(x, y)$$
$$x = u - v$$
$$y = v - u$$
$$\frac{\partial w}{\partial u} = \frac{\partial w}{\partial x} - \frac{\partial w}{\partial y}$$
$$\frac{\partial w}{\partial v} = -\frac{\partial w}{\partial x} + \frac{\partial w}{\partial y}$$
$$\frac{\partial w}{\partial u} + \frac{\partial w}{\partial v} = 0$$

40.
$$w = (x - y)\sin(y - x)$$
$$\frac{\partial w}{\partial x} = -(x - y)\cos(y - x) + \sin(y - x)$$
$$\frac{\partial w}{\partial y} = (x - y)\cos(y - x) - \sin(y - x)$$
$$\frac{\partial w}{\partial x} + \frac{\partial w}{\partial y} = 0$$

41. $w = f(x, y), \quad x = r\cos\theta, \quad y = r\sin\theta$

$$\frac{\partial w}{\partial r} = \frac{\partial w}{\partial x}\cos\theta + \frac{\partial w}{\partial y}\sin\theta$$

$$\frac{\partial w}{\partial \theta} = \frac{\partial w}{\partial x}(-r\sin\theta) + \frac{\partial w}{\partial y}(r\cos\theta)$$

(a)
$$r\cos\theta\frac{\partial w}{\partial r} = \frac{\partial w}{\partial x}r\cos^2\theta + \frac{\partial w}{\partial y}r\sin\theta\cos\theta$$

$$-\sin\theta\frac{\partial w}{\partial \theta} = \frac{\partial w}{\partial x}(r\sin^2\theta) - \frac{\partial w}{\partial x}r\sin\theta\cos\theta$$

$$r\cos\theta\frac{\partial w}{\partial r} - \sin\theta\frac{\partial w}{\partial \theta} = \frac{\partial w}{\partial x}(r\cos^2\theta + r\sin^2\theta)$$

$$r\frac{\partial w}{\partial x} = \frac{\partial w}{\partial r}(r\cos\theta) - \frac{\partial w}{\partial \theta}\sin\theta$$

$$\frac{\partial w}{\partial x} = \frac{\partial w}{\partial r}\cos\theta - \frac{\partial w}{\partial \theta}\frac{\sin\theta}{r}$$

$$r\sin\theta\frac{\partial w}{\partial r} = \frac{\partial w}{\partial x}r\sin\theta\cos\theta + \frac{\partial w}{\partial y}r\sin^2\theta$$

$$\cos\theta\frac{\partial w}{\partial \theta} = \frac{\partial w}{\partial x}(-r\sin\theta\cos\theta) + \frac{\partial w}{\partial y}(r\cos^2\theta)$$

$$r\sin\theta\frac{\partial w}{\partial r} + \cos\theta\frac{\partial w}{\partial \theta} = \frac{\partial w}{\partial y}(r\sin^2\theta + r\cos^2\theta)$$

$$r\frac{\partial w}{\partial y} = \frac{\partial w}{\partial r}r\sin\theta + \frac{\partial w}{\partial \theta}\cos\theta$$

$$\frac{\partial w}{\partial y} = \frac{\partial w}{\partial y}\sin\theta + \frac{\partial w}{\partial \theta}\frac{\cos\theta}{r}$$

(b) $\left(\dfrac{\partial w}{\partial r}\right)^2 + \dfrac{1}{r^2}\left(\dfrac{\partial w}{\partial \theta}\right)^2 = \left(\dfrac{\partial w}{\partial x}\right)^2\cos^2\theta + 2\dfrac{\partial w}{\partial x}\dfrac{\partial w}{\partial y}\sin\theta\cos\theta + \left(\dfrac{\partial w}{\partial y}\right)^2\sin^2\theta + \left(\dfrac{\partial w}{\partial x}\right)^2\sin^2\theta$

$$- 2\frac{\partial w}{\partial x}\frac{\partial w}{\partial y}\sin\theta\cos\theta + \left(\frac{\partial w}{\partial y}\right)^2\cos^2\theta = \left(\frac{\partial w}{\partial x}\right)^2 + \left(\frac{\partial w}{\partial y}\right)^2$$

42. $w = \arctan\dfrac{y}{x}, \quad x = r\cos\theta, \quad y = r\sin\theta$

$$= \arctan\left(\frac{r\sin\theta}{r\cos\theta}\right) = \arctan(\tan\theta) = \theta \text{ for } -\frac{\pi}{2} < \theta < \frac{\pi}{2}$$

$$\frac{\partial w}{\partial x} = \frac{-y}{x^2 + y^2}, \quad \frac{\partial w}{\partial y} = \frac{x}{x^2 + y^2}, \quad \frac{\partial w}{\partial r} = 0, \quad \frac{\partial w}{\partial \theta} = 1$$

$$\left(\frac{\partial w}{\partial x}\right)^2 + \left(\frac{\partial w}{\partial y}\right)^2 = \frac{y^2}{(x^2+y^2)^2} + \frac{x^2}{(x^2+y^2)^2} = \frac{1}{x^2+y^2} = \frac{1}{r^2}$$

$$\left(\frac{\partial w}{\partial r}\right)^2 + \left(\frac{1}{r^2}\right)\left(\frac{\partial w}{\partial \theta}\right)^2 = 0 + \frac{1}{r^2}(1) = \frac{1}{r^2}$$

Therefore, $\left(\dfrac{\partial w}{\partial x}\right)^2 + \left(\dfrac{\partial w}{\partial y}\right)^2 = \left(\dfrac{\partial w}{\partial r}\right)^2 + \dfrac{1}{r^2}\left(\dfrac{\partial w}{\partial \theta}\right)^2.$

43. Given $\dfrac{\partial u}{\partial x} = \dfrac{\partial v}{\partial y}$ and $\dfrac{\partial u}{\partial y} = -\dfrac{\partial v}{\partial x}$, $\;x = r\cos\theta$ and $y = r\sin\theta$.

$$\frac{\partial u}{\partial r} = \frac{\partial u}{\partial x}\cos\theta + \frac{\partial u}{\partial y}\sin\theta = \frac{\partial v}{\partial y}\cos\theta - \frac{\partial v}{\partial x}\sin\theta$$

$$\frac{\partial v}{\partial \theta} = \frac{\partial v}{\partial x}(-r\sin\theta) + \frac{\partial v}{\partial y}(r\cos\theta) = r\left[\frac{\partial v}{\partial y}\cos\theta - \frac{\partial v}{\partial x}\sin\theta\right]$$

Therefore, $\dfrac{\partial u}{\partial r} = \dfrac{1}{r}\dfrac{\partial v}{\partial \theta}$.

$$\frac{\partial v}{\partial r} = \frac{\partial v}{\partial x}\cos\theta + \frac{\partial v}{\partial y}\sin\theta = -\frac{\partial u}{\partial y}\cos\theta + \frac{\partial u}{\partial x}\sin\theta$$

$$\frac{\partial u}{\partial \theta} = \frac{\partial u}{\partial x}(-r\sin\theta) + \frac{\partial u}{\partial y}(r\cos\theta) = -r\left[-\frac{\partial u}{\partial y}\cos\theta + \frac{\partial u}{\partial x}\sin\theta\right]$$

Therefore, $\dfrac{\partial v}{\partial r} = -\dfrac{1}{r}\dfrac{\partial u}{\partial \theta}$.

44. $u = \ln\sqrt{x^2 + y^2}, \quad v = \arctan\dfrac{y}{x}$

$$\frac{\partial u}{\partial r} = \frac{x}{x^2 + y^2}\cos\theta + \frac{y}{x^2 + y^2}\sin\theta = \frac{r\cos^2\theta + r\sin^2\theta}{r^2} = \frac{1}{r}$$

$$\frac{\partial v}{\partial \theta} = \frac{-y}{x^2 + y^2}(-r\sin\theta) + \frac{x}{x^2 + y^2}(r\cos\theta) = \frac{r^2\sin^2\theta + r^2\cos^2\theta}{r^2} = 1$$

Thus, $\dfrac{\partial u}{\partial r} = \dfrac{1}{r}\dfrac{\partial v}{\partial \theta}$.

$$\frac{\partial v}{\partial r} = \frac{-y}{x^2 + y^2}\cos\theta + \frac{x}{x^2 + y^2}\sin\theta = \frac{-r\sin\theta\cos\theta + r\sin\theta\cos\theta}{r^2} = 0$$

$$\frac{\partial u}{\partial \theta} = \frac{x}{x^2 + y^2}(-r\sin\theta) + \frac{y}{x^2 + y^2}(r\cos\theta) = \frac{-r^2\sin\theta\cos\theta + r^2\sin\theta\cos\theta}{r^2} = 0$$

Thus, $\dfrac{\partial v}{\partial r} = -\dfrac{1}{r}\dfrac{\partial u}{\partial \theta}$.

45. $u(x,\ t) = \dfrac{1}{2}[f(x - ct) + f(x + ct)]$

Let $r = x - ct$ and $s = x + ct$. Then $u(r,\ s) = \dfrac{1}{2}[f(r) + f(s)]$.

$$\frac{\partial u}{\partial t} = \frac{\partial u}{\partial r}\frac{dr}{dt} + \frac{\partial u}{\partial s}\frac{\theta s}{\theta t} = \frac{1}{2}\frac{df}{dr}(-c) + \frac{1}{2}\frac{df}{ds}(c)$$

$$\frac{\partial^2 u}{\partial t^2} = \frac{1}{2}\frac{d^2 f}{dr^2}(-c)^2 + \frac{1}{2}\frac{d^2 f}{ds^2}(c)^2 = \frac{c^2}{2}\left[\frac{d^2 f}{dr^2} + \frac{d^2 f}{ds^2}\right]$$

$$\frac{\partial u}{\partial x} = \frac{\partial u}{\partial r}\frac{\partial r}{\partial x} + \frac{\partial u}{\partial s}\frac{\partial s}{\partial x} = \frac{1}{2}\frac{df}{dr}(1) + \frac{1}{2}\frac{df}{ds}(1)$$

$$\frac{\partial^2 u}{\partial x^2} = \frac{1}{2}\frac{d^2 f}{dr^2}(1)^2 + \frac{1}{2}\frac{d^2 f}{ds^2}(1)^2 = \frac{1}{2}\left[\frac{d^2 f}{dr^2} + \frac{d^2 f}{ds^2}\right]$$

Thus, $\dfrac{\partial^2 u}{\partial t^2} = c^2\dfrac{\partial^2 u}{\partial x^2}$.

Section 14.6 Directional Derivatives and Gradients

1. $f(x, y) = 3x - 4xy + 5y$

$$v = \frac{1}{2}(i + \sqrt{3}\,j)$$

$\nabla f(x, y) = (3 - 4y)i + (-4x + 5)j$

$\nabla f(1, 2) = -5i + j$

$$u = \frac{v}{\|v\|} = \frac{1}{2}i + \frac{\sqrt{3}}{2}j$$

$$D_u f(1, 2) = \nabla f(1, 2) \cdot u = \frac{1}{2}(-5 + \sqrt{3})$$

2. $f(x, y) = x^2 - y^2$

$$v = \frac{\sqrt{2}}{2}(i + j)$$

$\nabla f(x, y) = 2xi - 2yj$

$\nabla f(4, 3) = 8i - 6j$

$$u = \frac{v}{\|v\|} = \frac{\sqrt{2}}{2}i + \frac{\sqrt{2}}{2}j$$

$$D_u f(4, 3) = \nabla f(4, 3) \cdot u = \sqrt{2}$$

3. $f(x, y) = xy$

$$v = i + j$$

$\nabla f(x, y) = yi + xj$

$\nabla f(2, 3) = 3i + 2j$

$$u = \frac{v}{\|v\|} = \frac{\sqrt{2}}{2}i + \frac{\sqrt{2}}{2}j$$

$$D_u f(2, 3) = \nabla f(2, 3) \cdot u = \frac{5\sqrt{2}}{2}$$

4. $f(x, y) = \dfrac{x}{y}$

$$v = -j$$

$$\nabla f(x, y) = \frac{1}{y}i - \frac{x}{y^2}j$$

$\nabla f(1, 1) = i - j$

$$u = \frac{v}{\|v\|} = -j$$

$$D_u f(1, 1) = \nabla f(1, 1) \cdot u = 1$$

5. $g(x, y) = \sqrt{x^2 + y^2}$

$$v = 3i - 4j$$

$$\nabla g = \frac{x}{\sqrt{x^2 + y^2}}i + \frac{y}{\sqrt{x^2 + y^2}}j$$

$$\nabla g(3, 4) = \frac{3}{5}i + \frac{4}{5}j$$

$$u = \frac{v}{\|v\|} = \frac{3}{5}i - \frac{4}{5}j$$

$$D_u g(3, 4) = \nabla g(3, 4) \cdot u = -\frac{7}{25}$$

6. $g(x, y) = \arcsin xy$

$$v = i + 5j$$

$$\nabla g(x, y) = \frac{y}{\sqrt{1 - (xy)^2}}i + \frac{x}{\sqrt{1 - (xy)^2}}j$$

$\nabla g(1, 0) = j$

$$u = \frac{v}{\|v\|} = \frac{1}{\sqrt{26}}i + \frac{5}{\sqrt{26}}j$$

$$D_u g(1, 0) = \nabla g(1, 0) \cdot u = \frac{5\sqrt{26}}{26}$$

7. $h(x, y) = e^x \sin y$

$$v = -i$$

$\nabla h = e^x \sin y\,i + e^x \cos y\,j$

$$h\left(1, \frac{\pi}{2}\right) = ei$$

$$u = \frac{v}{\|v\|} = -i$$

$$D_u h\left(1, \frac{\pi}{2}\right) = \nabla h\left(1, \frac{\pi}{2}\right) \cdot u = -e$$

8. $h(x, y) = e^{-(x^2+y^2)}$

$$v = i + j$$

$\nabla h = -2xe^{-(x^2+y^2)}i - 2ye^{-(x^2+y^2)}j$

$\nabla h(0, 0) = 0$

$$D_u h(0, 0) = \nabla h(0, 0) \cdot u = 0$$

9. $f(x, y, z) = xy + yz + xz$

$$\mathbf{v} = 2\mathbf{i} + \mathbf{j} - \mathbf{k}$$

$$\nabla f(x, y, z) = (y + z)\mathbf{i} + (x + z)\mathbf{j} + (x + y)\mathbf{k}$$

$$\nabla f(1, 1, 1) = 2\mathbf{i} + 2\mathbf{j} + 2\mathbf{k}$$

$$\mathbf{u} = \frac{\mathbf{v}}{\|\mathbf{v}\|} = \frac{\sqrt{6}}{3}\mathbf{i} + \frac{\sqrt{6}}{6}\mathbf{j} - \frac{\sqrt{6}}{6}\mathbf{k}$$

$$D_{\mathbf{u}}f(1, 1, 1) = \nabla f(1, 1, 1) \cdot \mathbf{u} = \frac{2\sqrt{6}}{3}$$

10. $f(x, y, z) = x^2 + y^2 + z^2$

$$\mathbf{v} = \mathbf{i} - 2\mathbf{j} + 3\mathbf{k}$$

$$\nabla f = 2x\mathbf{i} + 2y\mathbf{j} + 2z\mathbf{k}$$

$$\nabla f(1, 2, -1) = 2\mathbf{i} + 4\mathbf{j} - 2\mathbf{k}$$

$$\mathbf{u} = \frac{\mathbf{v}}{\|\mathbf{v}\|} = \frac{1}{\sqrt{14}}\mathbf{i} - \frac{2}{\sqrt{14}}\mathbf{j} + \frac{3}{\sqrt{14}}\mathbf{k}$$

$$D_{\mathbf{u}}f(1, 2, -1) = \nabla f(1, 2, -1) \cdot \mathbf{u} = -\frac{6}{7}\sqrt{14}$$

11. $h(x, y, z) = x\arctan yz$

$$\mathbf{v} = \langle 1, 2, -1 \rangle$$

$$\nabla h(x, y, z) = \arctan yz\,\mathbf{i} + \frac{xz}{1 + (yz)^2}\mathbf{j} + \frac{xy}{1 + (yz)^2}\mathbf{k}$$

$$\nabla h(4, 1, 1) = \frac{\pi}{4}\mathbf{i} + 2\mathbf{j} + 2\mathbf{k}$$

$$\mathbf{u} = \frac{\mathbf{v}}{\|\mathbf{v}\|} = \left\langle \frac{1}{\sqrt{6}}, \frac{2}{\sqrt{6}}, -\frac{1}{\sqrt{6}} \right\rangle$$

$$D_{\mathbf{u}}h(4, 1, 1) = \nabla h(4, 1, 1) \cdot \mathbf{u} = \frac{\pi + 8}{4\sqrt{6}} = \frac{(\pi + 8)\sqrt{6}}{24}$$

12. $h(x, y, z) = xyz$

$$\mathbf{v} = \langle 2, 1, 2 \rangle$$

$$\nabla h = yz\mathbf{i} + xz\mathbf{j} + xy\mathbf{k}$$

$$\nabla h(2, 1, 1) = \mathbf{i} + 2\mathbf{j} + 2\mathbf{k}$$

$$\mathbf{u} = \frac{\mathbf{v}}{\|\mathbf{v}\|} = \frac{2}{3}\mathbf{i} + \frac{1}{3}\mathbf{j} + \frac{2}{3}\mathbf{k}$$

$$D_{\mathbf{u}}h(2, 1, 1) = \nabla h(2, 1, 1) \cdot \mathbf{u} = \frac{8}{3}$$

13. $f(x, y) = x^2 + y^2$

$$\mathbf{u} = \frac{1}{\sqrt{2}}\mathbf{i} + \frac{1}{\sqrt{2}}\mathbf{j}$$

$$\nabla f = 2x\mathbf{i} + 2y\mathbf{j}$$

$$D_{\mathbf{u}}f = \nabla f \cdot \mathbf{u} = \frac{2}{\sqrt{2}}x + \frac{2}{\sqrt{2}}y = \sqrt{2}(x + y)$$

14. $f(x, y) = \dfrac{y}{x + y}$

$$\mathbf{u} = \frac{\sqrt{3}}{2}\mathbf{i} - \frac{1}{2}\mathbf{j}$$

$$\nabla f = -\frac{y}{(x + y)^2}\mathbf{i} + \frac{x}{(x + y)^2}\mathbf{j}$$

$$D_{\mathbf{u}}f = \nabla f \cdot \mathbf{u} = -\frac{\sqrt{3}\,y}{2(x + y)^2} - \frac{x}{2(x + y)^2}$$

$$= -\frac{1}{2(x + y)^2}\left(\sqrt{3}\,y + x\right)$$

15. $f(x, y) = \sin(2x - y)$

$$\mathbf{u} = \frac{1}{2}\mathbf{i} - \frac{\sqrt{3}}{2}\mathbf{j}$$

$$\nabla f = 2\cos(2x - y)\mathbf{i} - \cos(2x - y)\mathbf{j}$$

$$D_{\mathbf{u}}f = \nabla f \cdot \mathbf{u} = \cos(2x - y) + \frac{\sqrt{3}}{2}\cos(2x - y)$$

$$= \left(\frac{2 + \sqrt{3}}{2}\right)\cos(2x - y)$$

16. $g(x, y) = xe^y$

$$\mathbf{u} = -\frac{1}{2}\mathbf{i} + \frac{\sqrt{3}}{2}\mathbf{j}$$

$$\nabla g = e^y\mathbf{i} + xe^y\mathbf{j}$$

$$D_{\mathbf{u}}g = -\frac{1}{2}e^y + \frac{\sqrt{3}}{2}xe^y = \frac{e^y}{2}\left(\sqrt{3}\,x - 1\right)$$

17. $f(x, y) = x^2 + 4y^2$

$\mathbf{v} = -2\mathbf{i} - 2\mathbf{j}$

$\nabla f = 2x\mathbf{i} + 8y\mathbf{j}$

$\mathbf{u} = \dfrac{\mathbf{v}}{\|\mathbf{v}\|} = -\dfrac{1}{\sqrt{2}}\mathbf{i} - \dfrac{1}{\sqrt{2}}\mathbf{j}$

$D_{\mathbf{u}}f = -\dfrac{2}{\sqrt{2}}x - \dfrac{8}{\sqrt{2}}y = -\sqrt{2}(x + 4y)$

At $P = (3, 1)$, $D_{\mathbf{u}}f = -7\sqrt{2}$.

18. $f(x, y) = \cos(x + y)$

$\mathbf{v} = \dfrac{\pi}{2}\mathbf{i} - \pi\mathbf{j}$

$\nabla f = -\sin(x + y)\mathbf{i} - \sin(x + y)\mathbf{j}$

$\mathbf{u} = \dfrac{\mathbf{v}}{\|\mathbf{v}\|} = \dfrac{1}{\sqrt{5}}\mathbf{i} - \dfrac{2}{\sqrt{5}}\mathbf{j}$

$D_{\mathbf{u}}f = -\dfrac{1}{\sqrt{5}}\sin(x + y) + \dfrac{2}{\sqrt{5}}\sin(x + y)$

$= \dfrac{1}{\sqrt{5}}\sin(x + y) = \dfrac{\sqrt{5}}{5}\sin(x + y)$

At $(0, \pi)$, $D_{\mathbf{u}}f = 0$.

19. $h(x, y, z) = \ln(x + y + z)$

$\mathbf{v} = 3\mathbf{i} + 3\mathbf{j} + \mathbf{k}$

$\nabla h = \dfrac{1}{x + y + z}(\mathbf{i} + \mathbf{j} + \mathbf{k})$

At $(1, 0, 0)$, $\nabla h = \mathbf{i} + \mathbf{j} + \mathbf{k}$.

$\mathbf{u} = \dfrac{\mathbf{v}}{\|\mathbf{v}\|} = \dfrac{1}{\sqrt{19}}(3\mathbf{i} + 3\mathbf{j} + \mathbf{k})$

$D_{\mathbf{u}}h = \nabla h \cdot \mathbf{u} = \dfrac{7}{\sqrt{19}} = \dfrac{7\sqrt{19}}{19}$

20. $g(x, y, z) = xye^z$

$\mathbf{v} = -2\mathbf{i} - 4\mathbf{j}$

$\nabla g = ye^z\mathbf{i} + xe^z\mathbf{j} + xye^z\mathbf{k}$

At $(1, 0, 0)$, $\nabla g = \mathbf{j}$.

$\mathbf{u} = \dfrac{\mathbf{v}}{\|\mathbf{v}\|} = -\dfrac{1}{\sqrt{5}}\mathbf{i} - \dfrac{2}{\sqrt{5}}\mathbf{j}$

$D_{\mathbf{u}}g = \nabla g \cdot \mathbf{u} = -\dfrac{2}{\sqrt{5}} = -\dfrac{2\sqrt{5}}{5}$

21. $f(x, y) = x^2 - 3xy + y^2$

$\nabla f(x, y) = (2x - 3y)\mathbf{i} + (2y - 3x)\mathbf{j}$

$\nabla f(4, 2) = 2\mathbf{i} - 8\mathbf{j}$

$\|\nabla f(4, 2)\| = \sqrt{68} = 2\sqrt{17}$

22. $f(x, y) = y\sqrt{x}$

$\nabla f(x, y) = \dfrac{y}{2\sqrt{x}}\mathbf{i} + \sqrt{x}\mathbf{j}$

$\nabla f(4, 2) = \dfrac{1}{2}\mathbf{i} + 2\mathbf{j}$

$\|\nabla f(4, 2)\| = \dfrac{\sqrt{17}}{2}$

23. $h(x, y) = x\tan y$

$\nabla h(x, y) = \tan y\mathbf{i} + x\sec^2 y\mathbf{j}$

$\nabla h\left(2, \dfrac{\pi}{4}\right) = \mathbf{i} + 4\mathbf{j}$

$\left\|\nabla h\left(2, \dfrac{\pi}{4}\right)\right\| = \sqrt{17}$

24. $h(x, y) = y\cos(x - y)$

$\nabla h(x, y) = -y\sin(x - y)\mathbf{i} + [\cos(x - y) + y\sin(x - y)]\mathbf{j}$

$\nabla h\left(0, \dfrac{\pi}{3}\right) = \dfrac{\sqrt{3}\,\pi}{6}\mathbf{i} + \left(\dfrac{3 - \sqrt{3}\,\pi}{6}\right)\mathbf{j}$

$\left\|\nabla h\left(0, \dfrac{\pi}{3}\right)\right\| = \sqrt{\dfrac{3\pi^2}{36} + \dfrac{9 - 6\sqrt{3}\,\pi + 3\pi^2}{36}} = \dfrac{\sqrt{3(2\pi^2 - 2\sqrt{3}\,\pi + 3)}}{6}$

25. $g(x, y) = \ln \sqrt[3]{x^2 + y^2} = \dfrac{1}{3}\ln(x^2 + y^2)$

$\nabla g(x, y) = \dfrac{1}{3}\left[\dfrac{2x}{x^2 + y^2}\mathbf{i} + \dfrac{2y}{x^2 + y^2}\mathbf{j}\right]$

$\nabla g(1, 2) = \dfrac{1}{3}\left(\dfrac{2}{5}\mathbf{i} + \dfrac{4}{5}\mathbf{j}\right) = \dfrac{2}{15}(\mathbf{i} + 2\mathbf{j})$

$\|\nabla g(1, 2)\| = \dfrac{2\sqrt{5}}{15}$

26. $g(x, y) = ye^{-x^2}$

$\nabla g(x, y) = -2xye^{-x^2}\mathbf{i} + e^{-x^2}\mathbf{j}$

$\nabla g(0, 5) = \mathbf{j}$

$\|\nabla g(0, 5)\| = 1$

27. $f(x, y, z) = \sqrt{x^2 + y^2 + z^2}$

$\nabla f(x, y, z) = \dfrac{1}{\sqrt{x^2 + y^2 + z^2}}(x\mathbf{i} + y\mathbf{j} + z\mathbf{k})$

$\nabla f(1, 4, 2) = \dfrac{1}{\sqrt{21}}(\mathbf{i} + 4\mathbf{j} + 2\mathbf{k})$

$\|\nabla f(1, 4, 2)\| = 1$

28. $f(x, y, z) = xe^{yz}$

$\nabla f(x, y, z) = e^{yz}\mathbf{i} + xze^{yz}\mathbf{j} + xye^{yz}\mathbf{k}$

$\nabla f(2, 0, -4) = \mathbf{i} - 8\mathbf{j}$

$\|\nabla f(2, 0, -4)\| = \sqrt{65}$

29. $w = \dfrac{1}{\sqrt{1 - x^2 - y^2 - z^2}}$

$\nabla w = \dfrac{1}{(\sqrt{1 - x^2 - y^2 - z^2})^3}(x\mathbf{i} + y\mathbf{j} + z\mathbf{k})$

$\nabla w(0, 0, 0) = \mathbf{0}$

$\|\nabla w(0, 0, 0)\| = 0$

30. $w = xy^2z^2$

$\nabla w = y^2z^2\mathbf{i} + 2xyz^2\mathbf{j} + 2xy^2z\mathbf{k}$

$\nabla w(2, 1, 1) = \mathbf{i} + 4\mathbf{j} + 4\mathbf{k}$

$\|\nabla w(2, 1, 1)\| = \sqrt{33}$

31. $f(x, y) = 3 - \dfrac{x}{3} - \dfrac{y}{2}$

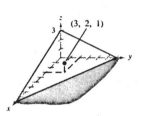

For Exercises 32–38,

$f(x, y) = 3 - \dfrac{x}{3} - \dfrac{y}{2}$ and

$D_\theta f(x, y) = -\left(\dfrac{1}{3}\right)\cos\theta - \left(\dfrac{1}{2}\right)\sin\theta.$

32. (a) $D_{\pi/4}f(3, 2) = -\left(\dfrac{1}{3}\right)\dfrac{\sqrt{2}}{2} - \left(\dfrac{1}{2}\right)\dfrac{\sqrt{2}}{2} = -\dfrac{5\sqrt{2}}{12}$

(b) $D_{2\pi/3}f(3, 2) = -\left(\dfrac{1}{3}\right)\left(-\dfrac{1}{2}\right) - \left(\dfrac{1}{2}\right)\dfrac{\sqrt{3}}{2} = \dfrac{2 - 3\sqrt{3}}{12}$

33. (a) $D_{4\pi/3}f(3, 2) = -\left(\dfrac{1}{3}\right)\left(-\dfrac{1}{2}\right) - \left(\dfrac{1}{2}\right)\left(-\dfrac{\sqrt{3}}{2}\right) = \dfrac{2 + 3\sqrt{3}}{12}$

(b) $D_{-\pi/6}f(3, 2) = -\left(\dfrac{1}{3}\right)\left(\dfrac{\sqrt{3}}{2}\right) - \left(\dfrac{1}{2}\right)\left(-\dfrac{1}{2}\right) = \dfrac{3 - 2\sqrt{3}}{12}$

34. (a) $\mathbf{u} = \left(\dfrac{1}{\sqrt{2}}\right)(\mathbf{i} + \mathbf{j})$

$D_{\mathbf{u}}f = \nabla f \cdot \mathbf{u}$

$= -\left(\dfrac{1}{3}\right)\dfrac{1}{\sqrt{2}} - \left(\dfrac{1}{2}\right)\dfrac{1}{\sqrt{2}} = -\dfrac{5\sqrt{2}}{12}$

(b) $\mathbf{v} = -3\mathbf{i} - 4\mathbf{j}$

$\|\mathbf{v}\| = \sqrt{9 + 16} = 5$

$\mathbf{u} = -\dfrac{3}{5}\mathbf{i} - \dfrac{4}{5}\mathbf{j}$

$D_{\mathbf{u}}f = \nabla f \cdot \mathbf{u} = \dfrac{1}{5} + \dfrac{2}{5} = \dfrac{3}{5}$

35. (a) $\mathbf{v} = -3\mathbf{i} + 4\mathbf{j}$

$\|\mathbf{v}\| = \sqrt{9 + 16} = 5$

$\mathbf{u} = -\dfrac{3}{5}\mathbf{i} + \dfrac{4}{5}\mathbf{j}$

$D_{\mathbf{u}}f = \nabla f \cdot \mathbf{u} = \dfrac{1}{5} - \dfrac{2}{5} = -\dfrac{1}{5}$

(b) $\mathbf{v} = \mathbf{i} + 3\mathbf{j}$

$\|\mathbf{v}\| = \sqrt{10}$

$\mathbf{u} = \dfrac{1}{\sqrt{10}}\mathbf{i} + \dfrac{3}{\sqrt{10}}\mathbf{j}$

$D_{\mathbf{u}}f = \nabla f \cdot \mathbf{u} = \dfrac{-11}{6\sqrt{10}} = -\dfrac{11\sqrt{10}}{60}$

36. $\nabla f = -\left(\dfrac{1}{3}\right)\mathbf{i} - \left(\dfrac{1}{2}\right)\mathbf{j}$

37. $\|\nabla f\| = \sqrt{\dfrac{1}{9} + \dfrac{1}{4}} = \dfrac{1}{6}\sqrt{13}$

38. $\nabla f = -\dfrac{1}{3}\mathbf{i} - \dfrac{1}{2}\mathbf{j}$

$\dfrac{\nabla f}{\|\nabla f\|} = \dfrac{1}{\sqrt{13}}(-2\mathbf{i} - 3\mathbf{j})$

Therefore,

$\mathbf{u} = \dfrac{1}{\sqrt{13}}(3\mathbf{i} - 2\mathbf{j})$ and $D_{\mathbf{u}}f(3,\ 2) = \nabla f \cdot \mathbf{u} = 0$.

∇f is the direction of greatest rate of change of f. Hence, in a direction orthogonal to ∇f, the rate of change of f is 0.

39. $f(x,\ y) = 9 - x^2 - y^2$

For Exercises 40–42,

$f(x,\ y) = 9 - x^2 - y^2$ and

$D_\theta f(x,\ y) = -2x\cos\theta - 2y\sin\theta = -2(x\cos\theta + y\sin\theta)$.

40. (a) $D_{-\pi/4}f(1,\ 2) = -2\left(\dfrac{\sqrt{2}}{2} - \sqrt{2}\right) = \sqrt{2}$

(b) $D_{\pi/3}f(1,\ 2) = -2\left(\dfrac{1}{2} + \sqrt{3}\right) = -(1 + 2\sqrt{3})$

41. $\nabla f(1,\ 2) = -2\mathbf{i} - 4\mathbf{j}$

$\|\nabla f(1,\ 2)\| = \sqrt{4 + 16} = \sqrt{20} = 2\sqrt{5}$

42. $\nabla f(1,\ 2) = -2\mathbf{i} - 4\mathbf{j}$

$\dfrac{\nabla f(1,\ 2)}{\|\nabla f(1,\ 2)\|} = \dfrac{1}{\sqrt{5}}(-\mathbf{i} - 2\mathbf{j})$

Therefore, $\mathbf{u} = \dfrac{1}{\sqrt{5}}(-2\mathbf{i} + \mathbf{j})$ and

$D_{\mathbf{u}}f(1,\ 2) = \nabla f(1,\ 2) \cdot \mathbf{u} = 0$.

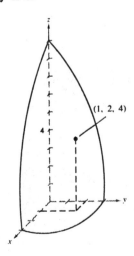

(1, 2, 4)

43. $f(x, y) = x^2 + y^2$

$c = 25, \quad P = (3, 4)$

$\nabla f(x, y) = 2x\mathbf{i} + 2y\mathbf{j}$

$x^2 + y^2 = 25$

$\nabla f(3, 4) = 6\mathbf{i} + 8\mathbf{j}$

44. $f(x, y) = 6 - 2x - 3y$

$c = 6, \quad P = (0, 0)$

$\nabla f(x, y) = -2\mathbf{i} - 3\mathbf{j}$

$6 - 2x - 3y = 6$

$0 = 2x + 3y$

$\nabla f(0, 0) = -2\mathbf{i} - 3\mathbf{j}$

45. $f(x, y) = \dfrac{x}{x^2 + y^2}$

$c = \dfrac{1}{2}, \quad P = (1, 1)$

$\nabla f(x, y) = \dfrac{y^2 - x^2}{(x^2 + y^2)^2}\mathbf{i} - \dfrac{2xy}{(x^2 + y^2)^2}\mathbf{j}$

$\dfrac{x}{x^2 + y^2} = \dfrac{1}{2}$

$x^2 + y^2 - 2x = 0$

$\nabla f(1, 1) = -\dfrac{1}{2}\mathbf{j}$

46. $f(x, y) = xy$

$c = -3, \quad P = (-1, 3)$

$\nabla f(x, y) = y\mathbf{i} + x\mathbf{j}$

$xy = -3$

$\nabla f(-1, 3) = 3\mathbf{i} - \mathbf{j}$

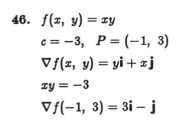

47. $4x^2 - y = 6$

$f(x, y) = 4x^2 - y$

$\nabla f(x, y) = 8x\mathbf{i} - \mathbf{j}$

$\nabla f(2, 10) = 16\mathbf{i} - \mathbf{j}$

$\dfrac{\nabla f(2, 10)}{\|\nabla f(2, 10)\|} = \dfrac{1}{\sqrt{257}}(16\mathbf{i} - \mathbf{j}) = \dfrac{\sqrt{257}}{257}(16\mathbf{i} - \mathbf{j})$

48. $3x^2 - 2y^2 = 1$

$f(x, y) = 3x^2 - 2y^2$

$\nabla f(x, y) = 6x\mathbf{i} - 4y\mathbf{j}$

$\nabla f(1, 1) = 6\mathbf{i} - 4\mathbf{j}$

$\dfrac{\nabla f(1, 1)}{\|\nabla f(1, 1)\|} = \dfrac{1}{\sqrt{13}}(3\mathbf{i} - 2\mathbf{j}) = \dfrac{\sqrt{13}}{13}(3\mathbf{i} - 2\mathbf{j})$

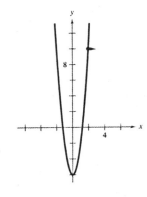

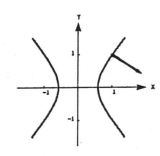

49. $9x^2 + 4y^2 = 40$

$f(x, y) = 9x^2 + 4y^2$

$\nabla f(x, y) = 18x\mathbf{i} + 8y\mathbf{j}$

$\nabla f(2, -1) = 36\mathbf{i} - 8\mathbf{j}$

$\dfrac{\nabla f(2, -1)}{\|\nabla f(2, -1)\|} = \dfrac{1}{\sqrt{85}}(9\mathbf{i} - 2\mathbf{j}) = \dfrac{\sqrt{85}}{85}(9\mathbf{i} - 2\mathbf{j})$

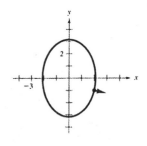

50. $xe^y - y = 5$

$$f(x, y) = xe^y - y$$

$$\nabla f(x, y) = e^y \mathbf{i} + (xe^y - 1)\mathbf{j}$$

$$\nabla f(5, 0) = \mathbf{i} + 4\mathbf{j}$$

$$\frac{\nabla f(5, 0)}{\|\nabla f(5, 0)\|} = \frac{1}{\sqrt{17}}(\mathbf{i} + 4\mathbf{j}) = \frac{\sqrt{17}}{17}(\mathbf{i} + 4\mathbf{j})$$

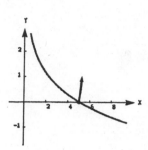

51. $T = \dfrac{x}{x^2 + y^2}$

$$\nabla T = \frac{y^2 - x^2}{(x^2 + y^2)^2}\mathbf{i} - \frac{2xy}{(x^2 + y^2)^2}\mathbf{j}$$

$$\nabla T(3, 4) = \frac{7}{625}\mathbf{i} - \frac{24}{625}\mathbf{j} = \frac{1}{625}(7\mathbf{i} - 24\mathbf{j})$$

52. $h(x, y) = 4000 - 0.001x^2 - 0.004y^2$

$$\nabla h = -0.002x\mathbf{i} - 0.008y\mathbf{j}$$

$$\nabla h(500, 300) = -\mathbf{i} - 2.4\mathbf{j}$$

OR $5\nabla h = -(5\mathbf{i} + 12\mathbf{j})$

53. $T(x, y) = 400 - 2x^2 - y^2,$ $P = (10, 10)$

$$\frac{dx}{dt} = -4x \qquad\qquad \frac{dy}{dt} = -2y$$

$$x(t) = C_1 e^{-4t} \qquad\quad y(t) = C_2 e^{-2t}$$

$$10 = x(0) = C_1 \qquad 10 = y(0) = C_2$$

$$x(t) = 10e^{-4t} \qquad\quad y(t) = 10e^{-2t}$$

$$x = \frac{y^2}{10} \qquad\qquad\quad y^2(t) = 100e^{-4x}$$

$$y^2 = 10x$$

54. $T(x, y) = 50 - x^2 - 2y^2,$ $P = (0, 0)$

$$\frac{dx}{dt} = -2x \qquad\qquad \frac{dy}{dt} = -4y$$

$$x(t) = C_1 e^{-2t} \qquad\quad y(t) = C_2 e^{-4t}$$

$$0 = x(0) = C_1 \qquad 0 = y(0) = C_2$$

$$x(t) = 0 \qquad\qquad\quad y(t) = 0$$

$$x = y$$

55. $\dfrac{x^2}{a^2} + \dfrac{y^2}{b^2} = 1$

$$f(x, y) = d_1 + d_2 = 2a$$

$$\nabla f(x, y) = \mathbf{0}$$

Therefore, $\mathbf{T} \cdot \nabla f(x, y) = 0$. $f(x, y)$ is the plane $z = 2a$ which is parallel to the xy-plane, and $D_\mathbf{u} f(x, y) = 0$ for all $\mathbf{u}$.

Section 14.7   Tangent Planes and Normal Lines

1. $F(x, y, z) = x + y + z - 4$

$$\nabla F = \mathbf{i} + \mathbf{j} + \mathbf{k}$$

$$\mathbf{n} = \frac{\nabla F}{\|\nabla F\|} = \frac{1}{\sqrt{3}}(\mathbf{i} + \mathbf{j} + \mathbf{k}) = \frac{\sqrt{3}}{3}(\mathbf{i} + \mathbf{j} + \mathbf{k})$$

2. $F(x, y, z) = x^2 + y^2 + z^2 - 11$

$\nabla F(x, y, z) = 2x\mathbf{i} + 2y\mathbf{j} + 2z\mathbf{k}$

$\nabla F(3, 1, 1) = 6\mathbf{i} + 2\mathbf{j} + 2\mathbf{k}$

$\mathbf{n} = \dfrac{\nabla F}{\|\nabla F\|} = \dfrac{1}{\sqrt{44}}(6\mathbf{i} + 2\mathbf{j} + 2\mathbf{k}) = \dfrac{1}{\sqrt{11}}(3\mathbf{i} + \mathbf{j} + \mathbf{k}) = \dfrac{\sqrt{11}}{11}(3\mathbf{i} + \mathbf{j} + \mathbf{k})$

3. $F(x, y, z) = \sqrt{x^2 + y^2} - z$

$\nabla F(x, y, z) = \dfrac{x}{\sqrt{x^2 + y^2}}\mathbf{i} + \dfrac{y}{\sqrt{x^2 + y^2}}\mathbf{j} - \mathbf{k}$

$\nabla F(3, 4, 5) = \dfrac{3}{5}\mathbf{i} + \dfrac{4}{5}\mathbf{j} - \mathbf{k}$

$\mathbf{n} = \dfrac{\nabla F}{\|\nabla F\|} = \dfrac{5}{5\sqrt{2}}\left(\dfrac{3}{5}\mathbf{i} + \dfrac{4}{5}\mathbf{j} - \mathbf{k}\right) = \dfrac{1}{5\sqrt{2}}(3\mathbf{i} + 4\mathbf{j} - 5\mathbf{k}) = \dfrac{\sqrt{2}}{10}(3\mathbf{i} + 4\mathbf{j} - 5\mathbf{k})$

4. $F(x, y, z) = x^3 - z$

$\nabla F(x, y, z) = 3x^2\mathbf{i} - \mathbf{k}$

$\nabla F(2, 1, 8) = 12\mathbf{i} - \mathbf{k}$

$\mathbf{n} = \dfrac{\nabla F}{\|\nabla F\|} = \dfrac{1}{\sqrt{145}}(12\mathbf{i} - \mathbf{k}) = \dfrac{\sqrt{145}}{145}(12\mathbf{i} - \mathbf{k})$

5. $F(x, y, z) = x^2y^4 - z$

$\nabla F(x, y, z) = 2xy^4\mathbf{i} + 4x^2y^3\mathbf{j} - \mathbf{k}$

$\nabla F(1, 2, 16) = 32\mathbf{i} + 32\mathbf{j} - \mathbf{k}$

$\mathbf{n} = \dfrac{\nabla F}{\|\nabla F\|} = \dfrac{1}{\sqrt{2049}}(32\mathbf{i} + 32\mathbf{j} - \mathbf{k}) = \dfrac{\sqrt{2049}}{2049}(32\mathbf{i} + 32\mathbf{j} - \mathbf{k})$

6. $F(x, y, z) = x^2 + 3y + z^3 - 9$

$\nabla F(x, y, z) = 2x\mathbf{i} + 3\mathbf{j} + 3z^2\mathbf{k}$

$\nabla F(2, -1, 2) = 4\mathbf{i} + 3\mathbf{j} + 12\mathbf{k}$

$\mathbf{n} = \dfrac{\nabla F}{\|\nabla F\|} = \dfrac{1}{13}(4\mathbf{i} + 3\mathbf{j} + 12\mathbf{k})$

7. $F(x, y, z) = -x\sin y + z - 4$

$\nabla F(x, y, z) = -\sin y\mathbf{i} - x\cos y\mathbf{j} + \mathbf{k}$

$\nabla F\left(6, \dfrac{\pi}{6}, 7\right) = -\dfrac{1}{2}\mathbf{i} - 3\sqrt{3}\,\mathbf{j} + \mathbf{k}$

$\mathbf{n} = \dfrac{\nabla F}{\|\nabla F\|} = \dfrac{2}{\sqrt{113}}\left(-\dfrac{1}{2}\mathbf{i} - 3\sqrt{3}\,\mathbf{j} + \mathbf{k}\right) = \dfrac{1}{\sqrt{113}}(-\mathbf{i} - 6\sqrt{3}\,\mathbf{j} + 2\mathbf{k}) = \dfrac{\sqrt{113}}{113}(-\mathbf{i} - 6\sqrt{3}\,\mathbf{j} + 2\mathbf{k})$

8. $F(x, y, z) = ze^{x^2 - y^2} - 3$

$\nabla F(x, y, z) = 2xze^{x^2 - y^2}\mathbf{i} - 2yze^{x^2 - y^2}\mathbf{j} + e^{x^2 - y^2}\mathbf{k}$

$\nabla F(2, 2, 3) = 12\mathbf{i} - 12\mathbf{j} + \mathbf{k}$

$\mathbf{n} = \dfrac{\nabla F}{\|\nabla F\|} = \dfrac{1}{17}(12\mathbf{i} - 12\mathbf{j} + \mathbf{k})$

9. $F(x, y, z) = \ln\left(\dfrac{x}{y-z}\right) = \ln x - \ln(y-z)$

$\nabla F(x, y, z) = \dfrac{1}{x}\mathbf{i} - \dfrac{1}{y-z}\mathbf{j} + \dfrac{1}{y-z}\mathbf{k}$

$\nabla F(1, 4, 3) = \mathbf{i} - \mathbf{j} + \mathbf{k}$

$\mathbf{n} = \dfrac{\nabla F}{\|\nabla F\|} = \dfrac{1}{\sqrt{3}}(\mathbf{i} - \mathbf{j} + \mathbf{k}) = \dfrac{\sqrt{3}}{3}(\mathbf{i} - \mathbf{j} + \mathbf{k})$

10. $\qquad F(x, y, z) = \sin(x - y) - z - 2$

$\qquad\quad \nabla F(x, y, z) = \cos(x - y)\mathbf{i} - \cos(x - y)\mathbf{j} - \mathbf{k}$

$\nabla F\left(\dfrac{\pi}{3}, \dfrac{\pi}{6}, -\dfrac{3}{2}\right) = \dfrac{\sqrt{3}}{2}\mathbf{i} - \dfrac{\sqrt{3}}{2}\mathbf{j} - \mathbf{k}$

$\qquad \mathbf{n} = \dfrac{\nabla F}{\|\nabla F\|} = \dfrac{2}{\sqrt{10}}\left(\dfrac{\sqrt{3}}{2}\mathbf{i} - \dfrac{\sqrt{3}}{2}\mathbf{j} - \mathbf{k}\right) = \dfrac{1}{\sqrt{10}}(\sqrt{3}\,\mathbf{i} - \sqrt{3}\,\mathbf{j} - 2\mathbf{k}) = \dfrac{\sqrt{10}}{10}(\sqrt{3}\,\mathbf{i} - \sqrt{3}\,\mathbf{j} - 2\mathbf{k})$

11. $\quad f(x, y) = 25 - x^2 - y^2, \quad (3, 1, 15)$

$\quad F(x, y, z) = 25 - x^2 - y^2 - z$

$F_x(x, y, z) = -2x$	$F_y(x, y, z) = -2y$	$F_z(x, y, z) = -1$
$F_x(3, 1, 15) = -6$	$F_y(3, 1, 15) = -2$	$F_z(3, 1, 15) = -1$

$\quad -6(x - 3) - 2(y - 1) - (z - 15) = 0$

$\quad 0 = 6x + 2y + z - 35$

$\quad 6x + 2y + z = 35$

12. $\quad f(x, y) = \sqrt{x^2 + y^2}, \quad (3, 4, 5)$

$\quad F(x, y, z) = \sqrt{x^2 + y^2} - z$

$F_x(x, y, z) = \dfrac{x}{\sqrt{x^2 + y^2}}$	$F_y(x, y, z) = \dfrac{y}{\sqrt{x^2 + y^2}}$	$F_z(x, y, z) = -1$
$F_x(3, 4, 5) = \dfrac{3}{5}$	$F_y(3, 4, 5) = \dfrac{4}{5}$	$F_z(3, 4, 5) = -1$

$\quad \dfrac{3}{5}(x - 3) + \dfrac{4}{5}(y - 4) - (z - 5) = 0$

$\quad 3(x - 3) + 4(y - 4) - 5(z - 5) = 0$

$\quad 3x + 4y - 5z = 0$

13. $\quad f(x, y) = \dfrac{y}{x}, \quad (1, 2, 2)$

$\quad F(x, y, z) = \dfrac{y}{x} - z$

$F_x(x, y, z) = -\dfrac{y}{x^2}$	$F_y(x, y, z) = \dfrac{1}{x}$	$F_z(x, y, z) = -1$
$F_x(1, 2, 2) = -2$	$F_y(1, 2, 2) = 1$	$F_z(1, 2, 2) = -1$

$\quad -2(x - 1) + (y - 2) - (z - 2) = 0$

$\quad -2x + y - z + 2 = 0$

$\quad 2x - y + z = 2$

14. $f(x, y) = 2 - \frac{2}{3}x - y,\quad (3, -1, 1)$

$F(x, y, z) = 2 - \frac{2}{3}x - y - z$

$F_x(x, y, z) = -\frac{2}{3},\quad F_y(x, y, z) = -1,\quad F_z(x, y, z) = -1$

$-\frac{2}{3}(x - 3) - (y + 1) - (z - 1) = 0$

$-\frac{2}{3}x - y - z + 2 = 0$

$2x + 3y + 3z = 6$

15. $g(x, y) = x^2 - y^2,\quad (5, 4, 9)$

$G(x, y, z) = x^2 - y^2 - z$

$G_x(x, y, z) = 2x \qquad G_y(x, y, z) = -2y \qquad G_z(x, y, z) = -1$

$G_x(5, 4, 9) = 10 \qquad G_y(5, 4, 9) = -8 \qquad G_z(5, 4, 9) = -1$

$10(x - 5) - 8(y - 4) - (z - 9) = 0$

$10x - 8y - z = 9$

16. $g(x, y) = \arctan \dfrac{y}{x},\quad (1, 0, 0)$

$G(x, y, z) = \arctan \dfrac{y}{x} - z$

$G_x(x, y, z) = \dfrac{-(y/x^2)}{1 + (y^2/x^2)} = \dfrac{-y}{x^2 + y^2} \qquad G_y(x, y, z) = \dfrac{1/x}{1 + (y^2/x^2)} = \dfrac{x}{x^2 + y^2} \qquad G_z(x, y, z) = -1$

$G_x(1, 0, 0) = 0 \qquad\qquad G_y(1, 0, 0) = 1 \qquad\qquad G_z(1, 0, 0) = -1$

$y - z = 0$

17. $z = e^x(\sin y + 1),\quad \left(0, \dfrac{\pi}{2}, 2\right)$

$F(x, y, z) = e^x(\sin y + 1) - z$

$F_x(x, y, z) = e^x(\sin y + 1) \qquad F_y(x, y, z) = e^x \cos y \qquad F_z(x, y, z) = -1$

$F_x\left(0, \dfrac{\pi}{2}, 2\right) = 2 \qquad F_y\left(0, \dfrac{\pi}{2}, 2\right) = 0 \qquad F_z\left(0, \dfrac{\pi}{2}, 2\right) = -1$

$2x - z = -2$

18. $z = x^3 - 3xy + y^3,\quad (1, 2, 3)$

$F(x, y, z) = x^3 - 3xy + y^3 - z$

$F_x(x, y, z) = 3x^2 - 3y \qquad F_y(x, y, z) = -3x + 3y^2 \qquad F_z(x, y, z) = -1$

$F_x(1, 2, 3) = -3 \qquad F_y(1, 2, 3) = 9 \qquad F_z(1, 2, 3) = -1$

$-3(x - 1) + 9(y - 2) - (z - 3) = 0$

$-3x + 9y - z - 12 = 0$

$3x - 9y + z = -12$

19. $h(x, y) = \ln \sqrt{x^2 + y^2}$, $(3, 4, \ln 5)$

$H(x, y, z) = \ln \sqrt{x^2 + y^2} - z = \frac{1}{2} \ln(x^2 + y^2) - z$

$$H_x(x, y, z) = \frac{x}{x^2 + y^2} \qquad H_y(x, y, z) = \frac{y}{x^2 + y^2} \qquad H_z(x, y, z) = -1$$

$$H_x(3, 4, \ln 5) = \frac{3}{25} \qquad H_y(3, 4, \ln 5) = \frac{4}{25} \qquad H_z(3, 4, \ln 5) = -1$$

$\frac{3}{25}(x - 3) + \frac{4}{25}(y - 4) - (z - \ln 5) = 0$

$3(x - 3) + 4(y - 4) - 25(z - \ln 5) = 0$

$3x + 4y - 25z = 25(1 - \ln 5)$

20. $h(x, y) = \cos y$, $\left(5, \frac{\pi}{4}, \frac{\sqrt{2}}{2}\right)$

$H(x, y, z) = \cos y - z$

$$H_x(x, y, z) = 0 \qquad H_y(x, y, z) = -\sin y \qquad H_z(x, y, z) = -1$$

$$H_x\left(5, \frac{\pi}{4}, \frac{\sqrt{2}}{2}\right) = 0 \qquad H_y\left(5, \frac{\pi}{4}, \frac{\sqrt{2}}{2}\right) = -\frac{\sqrt{2}}{2} \qquad H_z\left(5, \frac{\pi}{4}, \frac{\sqrt{2}}{2}\right) = -1$$

$-\frac{\sqrt{2}}{2}\left(y - \frac{\pi}{4}\right) - \left(z - \frac{\sqrt{2}}{2}\right) = 0$

$-\frac{\sqrt{2}}{2}y - z + \frac{\sqrt{2}\,\pi}{8} + \frac{\sqrt{2}}{2} = 0$

$4\sqrt{2}\,y + 8z = \sqrt{2}(\pi + 4)$

21. $x^2 + 4y^2 + z^2 = 36$, $(2, -2, 4)$

$F(x, y, z) = x^2 + 4y^2 + z^2 - 36$

$$F_x(x, y, z) = 2x \qquad F_y(x, y, z) = 8y \qquad F_z(x, y, z) = 2z$$

$$F_x(2, -2, 4) = 4 \qquad F_y(2, -2, 4) = -16 \qquad F_z(2, -2, 4) = 8$$

$4(x - 2) - 16(y + 2) + 8(z - 4) = 0$

$(x - 2) - 4(y + 2) + 2(z - 4) = 0$

$x - 4y + 2z = 18$

22. $x^2 + 2z^2 = y^2$, $(1, 3, -2)$

$F(x, y, z) = x^2 - y^2 + 2z^2$

$$F_x(x, y, z) = 2x \qquad F_y(x, y, z) = -2y \qquad F_z(x, y, z) = 4z$$

$$F_x(1, 3, -2) = 2 \qquad F_y(1, 3, -2) = -6 \qquad F_z(1, 3, -2) = -8$$

$2(x - 1) - 6(y - 3) - 8(z + 2) = 0$

$(x - 1) - 3(y - 3) - 4(z + 2) = 0$

$x - 3y - 4z = 0$

23. $xy^2 + 3x - z^2 = 4$, $(2, 1, -2)$

$F(x, y, z) = xy^2 + 3x - z^2 - 4$

$F_x(x, y, z) = y^2 + 3$ $F_y(x, y, z) = 2xy$ $F_z(x, y, z) = -2z$

$F_x(2, 1, -2) = 4$ $F_y(2, 1, -2) = 4$ $F_z(2, 1, -2) = 4$

$4(x - 2) + 4(y - 1) + 4(z + 2) = 0$

$x + y + z = 1$

24. $y = x(2z - 1)$, $(4, 4, 1)$

$F(x, y, z) = x(2z - 1) - y$

$F_x(x, y, z) = 2z - 1$ $F_y(x, y, z) = -1$ $F_z(x, y, z) = 2x$

$F_x(4, 4, 1) = 1$ $F_y(4, 4, 1) = -1$ $F_z(4, 4, 1) = 8$

$(x - 4) - (y - 4) + 8(z - 1) = 0$

$x - y + 8z = 8$

25. $x^2 + y^2 + z = 9$, $(1, 2, 4)$

$F(x, y, z) = x^2 + y^2 + z - 9$

$F_x(x, y, z) = 2x$ $F_y(x, y, z) = 2y$ $F_z(x, y, z) = 1$

$F_x(1, 2, 4) = 2$ $F_y(1, 2, 4) = 4$ $F_z(1, 2, 4) = 1$

Plane: $2(x - 1) + 4(y - 2) + (z - 4) = 0$, $2x + 4y + z = 14$

Line: $\dfrac{x - 1}{2} = \dfrac{y - 2}{4} = \dfrac{z - 4}{1}$

26. $x^2 + y^2 + z^2 = 9$, $(1, 2, 2)$

$F(x, y, z) = x^2 + y^2 + z^2 - 9$

$F_x(x, y, z) = 2x$ $F_y(x, y, z) = 2y$ $F_z(x, y, z) = 2z$

$F_x(1, 2, 2) = 2$ $F_y(1, 2, 2) = 4$ $F_z(1, 2, 2) = 4$

Direction numbers: 1, 2, 2

Plane: $(x - 1) + 2(y - 2) + 2(z - 2) = 0$, $x + 2y + 2z = 9$

Line: $\dfrac{x - 1}{1} = \dfrac{y - 2}{2} = \dfrac{z - 2}{2}$

27. $xy - z = 0$, $(-2, -3, 6)$

$F(x, y, z) = xy - z$

$F_x(x, y, z) = y$ $F_y(x, y, z) = x$ $F_z(x, y, z) = -1$

$F_x(-2, -3, 6) = -3$ $F_y(-2, -3, 6) = -2$ $F_z(-2, -3, 6) = -1$

Direction numbers: 3, 2, 1

Plane: $3(x + 2) + 2(x + 3) + (z - 6) = 0$, $3x + 2y + z = -6$

Line: $\dfrac{x + 2}{3} = \dfrac{y + 3}{2} = \dfrac{z - 6}{1}$

28. $x^2 + y^2 - z^2 = 0$, $(5, 12, 13)$

$F(x, y, z) = x^2 + y^2 - z^2$

$F_x(x, y, z) = 2x \qquad F_y(x, y, z) = 2y \qquad F_z(x, y, z) = -2z$

$F_x(5, 12, 13) = 10 \qquad F_y(5, 12, 13) = 24 \qquad F_z(5, 12, 13) = -26$

Direction numbers: $5, 12, -13$

Plane: $5(x - 5) + 12(y - 12) - 13(z - 13) = 0$, $\quad 5x + 12y - 13z = 0$

Line: $\dfrac{x - 5}{5} = \dfrac{y - 12}{12} = \dfrac{z - 13}{-13}$

29. $z = \arctan \dfrac{y}{x}$, $\quad \left(1, 1, \dfrac{\pi}{4}\right)$

$F(x, y, z) = \arctan \dfrac{y}{x} - z$

$F_x(x, y, z) = \dfrac{-y}{x^2 + y^2} \qquad F_y(x, y, z) = \dfrac{x}{x^2 + y^2} \qquad F_z(x, y, z) = -1$

$F_x\left(1, 1, \dfrac{\pi}{4}\right) = -\dfrac{1}{2} \qquad F_y\left(1, 1, \dfrac{\pi}{4}\right) = \dfrac{1}{2} \qquad F_z\left(1, 1, \dfrac{\pi}{4}\right) = -1$

Direction numbers: $1, -1, 2$

Plane: $(x - 1) - (y - 1) + 2\left(z - \dfrac{\pi}{4}\right) = 0$, $\quad x - y + 2z = \dfrac{\pi}{2}$

Line: $\dfrac{x - 1}{1} = \dfrac{y - 1}{-1} = \dfrac{z - (\pi/4)}{2}$

30. $xyz = 10$, $\quad (1, 2, 5)$

$F(x, y, z) = xyz - 10$

$F_x(x, y, z) = yz \qquad F_y(x, y, z) = xz \qquad F_z(x, y, z) = xy$

$F_x(1, 2, 5) = 10 \qquad F_y(1, 2, 5) = 5 \qquad F_z(1, 2, 5) = 2$

Direction numbers: $10, 5, 2$

Plane: $10(x - 1) + 5(y - 2) + 2(z - 5) = 0$, $\quad 10x + 5y + 2z = 30$

Line: $\dfrac{x - 1}{10} = \dfrac{y - 2}{5} = \dfrac{z - 5}{2}$

31. $F(x, y, z) = 3x^2 + 2y^2 - z - 15$, $(2, 2, 5)$

$\nabla F(x, y, z) = 6x\mathbf{i} + 4y\mathbf{j} - \mathbf{k}$

$\nabla F(2, 2, 5) = 12\mathbf{i} + 8\mathbf{j} - \mathbf{k}$

$\cos \theta = \dfrac{|\nabla F(2, 2, 5) \cdot \mathbf{k}|}{\|\nabla F(2, 2, 5)\|} = \dfrac{1}{\sqrt{209}}$

$\theta = \arccos\left(\dfrac{1}{\sqrt{209}}\right) \approx 86°$

32. $F(x, y, z) = xy - z^2$, $(2, 2, 2)$

$\nabla F(x, y, z) = y\mathbf{i} + x\mathbf{j} - 2z\mathbf{k}$

$\nabla F(2, 2, 2) = 2\mathbf{i} + 2\mathbf{j} - 4\mathbf{k}$

$\cos \theta = \dfrac{|\nabla F(2, 2, 2) \cdot \mathbf{k}|}{\|\nabla F(2, 2, 2)\|} = \dfrac{4}{\sqrt{24}} = \dfrac{2}{\sqrt{6}}$

$\theta = \arccos \dfrac{2}{\sqrt{6}} \approx 35.26°$

33. $F(x, y, z) = x^2 - y^2 + z$, $(1, 2, 3)$

$\nabla F(x, y, z) = 2x\mathbf{i} - 2y\mathbf{j} + \mathbf{k}$

$\nabla F(1, 2, 3) = 2\mathbf{i} - 4\mathbf{j} + \mathbf{k}$

$\cos\theta = \dfrac{|\nabla F(1, 2, 3) \cdot \mathbf{k}|}{\|\nabla F(1, 2, 3)\|} = \dfrac{1}{\sqrt{21}}$

$\theta = \arccos\dfrac{1}{\sqrt{21}} \approx 77.40°$

34. $F(x, y, z) = x^2 + y^2 - 5$, $(2, 1, 3)$

$\nabla F(x, y, z) = 2x\mathbf{i} + 2y\mathbf{j}$

$\nabla F(2, 1, 3) = 4\mathbf{i} + 2\mathbf{j}$

$\cos\theta = \dfrac{|\nabla F(2, 1, 3) \cdot \mathbf{k}|}{\|\nabla F(2, 1, 3)\|} = 0$

$\theta = \arccos 0 = 90°$

35. $F(x, y, z) = x^2 + y^2 - 5$ $G(x, y, z) = x - z$

$\nabla F(x, y, z) = 2x\mathbf{i} + 2y\mathbf{j}$ $\nabla G(x, y, z) = \mathbf{i} - \mathbf{k}$

$\nabla F(2, 1, 2) = 4\mathbf{i} + 2\mathbf{j}$ $\nabla G(2, 1, 2) = \mathbf{i} - \mathbf{k}$

(a) $\nabla F \times \nabla G = \begin{vmatrix} \mathbf{i} & \mathbf{j} & \mathbf{k} \\ 4 & 2 & 0 \\ 1 & 0 & -1 \end{vmatrix} = -2\mathbf{i} + 4\mathbf{j} - 2\mathbf{k} = -2(\mathbf{i} - 2\mathbf{j} + \mathbf{k})$

Direction numbers: $1, -2, 1$, $\dfrac{x-2}{1} = \dfrac{y-1}{-2} = \dfrac{z-2}{1}$

(b) $\cos\theta = \dfrac{|\nabla F \cdot \nabla G|}{\|\nabla F\| \, \|\nabla G\|} = \dfrac{4}{\sqrt{20}\sqrt{2}} = \dfrac{2}{\sqrt{10}} = \dfrac{\sqrt{10}}{5}$

Not orthogonal

36. $F(x, y, z) = x^2 + y^2 - z$ $G(x, y, z) = 4 - y - z$

$\nabla F(x, y, z) = 2x\mathbf{i} + 2y\mathbf{j} - \mathbf{k}$ $\nabla G(x, y, z) = -\mathbf{j} - \mathbf{k}$

$\nabla F(2, -1, 5) = 4\mathbf{i} - 2\mathbf{j} - \mathbf{k}$ $\nabla G(2, -1, 5) = -\mathbf{j} - \mathbf{k}$

(a) $\nabla F \times \nabla G = \begin{vmatrix} \mathbf{i} & \mathbf{j} & \mathbf{k} \\ 4 & -2 & -1 \\ 0 & -1 & -1 \end{vmatrix} = \mathbf{i} + 4\mathbf{j} - 4\mathbf{k}$

Direction numbers: $1, 4, -4$, $\dfrac{x-2}{1} = \dfrac{y+1}{4} = \dfrac{z-5}{-4}$

(b) $\cos\theta = \dfrac{|\nabla F \cdot \nabla G|}{\|\nabla F\| \, \|\nabla G\|} \dfrac{3}{\sqrt{21}\sqrt{2}} = \dfrac{3}{\sqrt{42}} = \dfrac{\sqrt{42}}{14}$

Not orthogonal

37. $F(x, y, z) = x^2 + z^2 - 25$ $G(x, y, z) = y^2 + z^2 - 25$

$\nabla F = 2x\mathbf{i} + 2z\mathbf{k}$ $\nabla G = 2y\mathbf{j} + 2z\mathbf{k}$

$\nabla F(3, 3, 4) = 6\mathbf{i} + 8\mathbf{k}$ $\nabla G(3, 3, 4) = 6\mathbf{j} + 8\mathbf{k}$

(a) $\nabla F \times \nabla G = \begin{vmatrix} \mathbf{i} & \mathbf{j} & \mathbf{k} \\ 6 & 0 & 8 \\ 0 & 6 & 8 \end{vmatrix} = -48\mathbf{i} - 48\mathbf{j} + 36\mathbf{k}$

Direction numbers: $4, 4, -3$, $\dfrac{x-3}{4} = \dfrac{y-3}{4} = \dfrac{z-4}{-3}$

(b) $\cos\theta = \dfrac{|\nabla F \cdot \nabla G|}{\|\nabla F\| \, \|\nabla G\|} = \dfrac{64}{(10)(10)} = \dfrac{16}{25}$

Not orthogonal

38. $F(x, y, z) = \sqrt{x^2 + y^2} - z$ $G(x, y, z) = 2x + y + 2z - 20$

$\nabla F(x, y, z) = \dfrac{x}{\sqrt{x^2 + y^2}}\mathbf{i} + \dfrac{y}{\sqrt{x^2 + y^2}}\mathbf{j} - \mathbf{k}$ $\nabla G(x, y, z) = 2\mathbf{i} + \mathbf{j} + 2\mathbf{k}$

$\nabla G(3, 4, 5) = 2\mathbf{i} + \mathbf{j} + 2\mathbf{k}$

$\nabla F(3, 4, 5) = \dfrac{3}{5}\mathbf{i} + \dfrac{4}{5}\mathbf{j} - \mathbf{k}$

(a) $\nabla F \times \nabla G = \begin{vmatrix} \mathbf{i} & \mathbf{j} & \mathbf{k} \\ \frac{3}{5} & \frac{4}{5} & -1 \\ 2 & 1 & 2 \end{vmatrix} = \dfrac{13}{5}\mathbf{i} - \dfrac{16}{5}\mathbf{j} - \mathbf{k} = \dfrac{1}{5}(13\mathbf{i} - 16\mathbf{j} - 5\mathbf{k})$

Direction numbers: 13, -16, -5, $\dfrac{x - 3}{13} = \dfrac{y - 4}{-16} = \dfrac{z - 5}{-5}$

(b) $\cos\theta = \dfrac{|\nabla F \cdot \nabla G|}{\|\nabla F\|\,\|\nabla G\|} = 0$

Orthogonal

39. $F(x, y, z) = x^2 + y^2 + z^2 - 6$ $G(x, y, z) = x - y - z$

$\nabla F(x, y, z) = 2x\mathbf{i} + 2y\mathbf{j} + 2z\mathbf{k}$ $\nabla G(x, y, z) = \mathbf{i} - \mathbf{j} - \mathbf{k}$

$\nabla F(2, 1, 1) = 4\mathbf{i} + 2\mathbf{j} + 2\mathbf{k}$ $\nabla G(2, 1, 1) = \mathbf{i} - \mathbf{j} - \mathbf{k}$

(a) $\nabla F \times \nabla G = \begin{vmatrix} \mathbf{i} & \mathbf{j} & \mathbf{k} \\ 4 & 2 & 2 \\ 1 & -1 & -1 \end{vmatrix} = 6\mathbf{j} - 6\mathbf{k} = 6(\mathbf{j} - \mathbf{k})$

Direction numbers: 0, 1, -1, $x = 2$, $\dfrac{y - 1}{1} = \dfrac{z - 1}{-1}$

(b) $\cos\theta = \dfrac{|\nabla F \cdot \nabla G|}{\|\nabla F\|\,\|\nabla G\|} = 0$

Orthogonal

40. $F(x, y, z) = x^2 + y^2 - z$ $G(x, y, z) = x + y + 6z - 33$

$\nabla F(x, y, z) = 2x\mathbf{i} + 2y\mathbf{j} - \mathbf{k}$ $\nabla G(x, y, z) = \mathbf{i} + \mathbf{j} + 6\mathbf{k}$

$\nabla F(1, 2, 5) = 2\mathbf{i} + 4\mathbf{j} - \mathbf{k}$ $\nabla G(1, 2, 5) = \mathbf{i} + \mathbf{j} + 6\mathbf{k}$

(a) $\nabla F \times \nabla G = \begin{vmatrix} \mathbf{i} & \mathbf{j} & \mathbf{k} \\ 2 & 4 & -1 \\ 1 & 1 & 6 \end{vmatrix} = 25\mathbf{i} - 13\mathbf{j} - 2\mathbf{k}$

Direction numbers: 25, -13, -2, $\dfrac{x - 1}{25} = \dfrac{y - 2}{-13} = \dfrac{z - 5}{-2}$

(b) $\cos\theta = \dfrac{|\nabla F \cdot \nabla G|}{\|\nabla F\|\,\|\nabla G\|} = 0$

Orthogonal

41. $F(x, y, z) = 3 - x^2 - y^2 + 6y - z$

$\nabla F(x, y, z) = -2x\mathbf{i} + (-2y + 6)\mathbf{j} - \mathbf{k}$

$-2x = 0, \quad x = 0$

$-2y + 6 = 0, \quad y = 3$

$z = 3 - 0^2 - 3^2 + 6(3) = 12$

$(0, 3, 12)$

42. $F(x, y, z) = 3x^2 + 2y^2 - 3x + 4y - z - 5$

$\nabla F(x, y, z) = (6x - 3)\mathbf{i} + (4y + 4)\mathbf{j} - \mathbf{k}$

$6x - 3 = 0, \quad x = \frac{1}{2}$

$4y + 4 = 0, \quad y = -1$

$z = 3\left(\frac{1}{2}\right)^2 + 2(-1)^2 - 3\left(\frac{1}{2}\right) + 4(-1) - 5 = -\frac{31}{4}$

$\left(\frac{1}{2}, -1, -\frac{31}{4}\right)$

43. $T(x, y, z) = 400 - 2x^2 - y^2 - 4z^2, \quad (4, 3, 10)$

$\dfrac{dx}{dt} = -4x \qquad \dfrac{dy}{dt} = -2y \qquad \dfrac{dz}{dt} = -8z$

$x(t) = C_1 e^{-4t} \qquad y(t) = C_2 e^{-2t} \qquad z(t) = C_3 e^{-8t}$

$x(0) = C_1 = 4 \qquad y(0) = C_2 = 3 \qquad z(0) = C_3 = 10$

$x = 4e^{-4t} \qquad y = 3e^{-2t} \qquad z = 10e^{-8t}$

44. $T(x, y, z) = 100 - 3x - y - z^2, \quad (2, 2, 5)$

$\dfrac{dx}{dt} = -3 \qquad \dfrac{dy}{dt} = -1 \qquad \dfrac{dz}{dt} = -2z$

$x(t) = -3t + C_1 \qquad y(t) = -t + C_2 \qquad z(t) = C_3 e^{-2t}$

$x(0) = C_1 = 2 \qquad y(0) = C_2 = 2 \qquad z(0) = C_3 = 5$

$x = -3t + 2 \qquad y = -t + 2 \qquad z = 5e^{-2t}$

45. $F(x, y, z) = \dfrac{x^2}{a^2} + \dfrac{y^2}{b^2} + \dfrac{z^2}{c^2} - 1$

$F_x(x, y, z) = \dfrac{2x}{a^2}$

$F_y(x, y, z) = \dfrac{2y}{b^2}$

$F_z(x, y, z) = \dfrac{2z}{c^2}$

Plane: $\dfrac{2x_0}{a^2}(x - x_0) + \dfrac{2y_0}{b^2}(y - y_0) + \dfrac{2z_0}{c^2}(z - z_0) = 0$

$\dfrac{x_0 x}{a^2} + \dfrac{y_0 y}{b^2} + \dfrac{z_0 z}{c^2} = \dfrac{x_0^2}{a^2} + \dfrac{y_0^2}{b^2} + \dfrac{z_0^2}{c^2} = 1$

46. $F(x, y, z) = \dfrac{x^2}{a^2} + \dfrac{y^2}{b^2} - \dfrac{z^2}{c^2} - 1$

$F_x(x, y, z) = \dfrac{2x}{a^2}$

$F_y(x, y, z) = \dfrac{2y}{b^2}$

$F_z(x, y, z) = \dfrac{-2z}{c^2}$

Plane: $\dfrac{2x_0}{a^2}(x - x_0) + \dfrac{2y_0}{b^2}(y - y_0) - \dfrac{2z_0}{c^2}(z - z_0) = 0$

$\dfrac{x_0 x}{a^2} + \dfrac{y_0 y}{b^2} - \dfrac{z_0 z}{c^2} = \dfrac{x_0^2}{a^2} + \dfrac{y_0^2}{b^2} - \dfrac{z_0^2}{c^2} = 1$

47. $F(x, y, z) = a^2x^2 + b^2y^2 - z^2$

$F_x(x, y, z) = 2a^2x$

$F_y(x, y, z) = 2b^2y$

$F_z(x, y, z) = -2z$

Plane: $2a^2x_0(x - x_0) + 2b^2y_0(y - y_0) - 2z_0(z - z_0) = 0$

$a^2x_0x + b^2y_0y - z_0z = a^2x_0^2 + b^2y_0^2 - z_0^2 = 0$

Therefore, the plane passes through the origin.

48. $F(x, y, z) = (x - h)^2 + (y - k)^2 + (z - l)^2 - r^2$

$F_x = 2(x - h)$

$F_y = 2(y - k)$

$F_z = 2(z - l)$

Line: $\dfrac{x - x_0}{x_0 - h} = \dfrac{y - y_0}{y_0 - k} = \dfrac{z - z_0}{z_0 - l}$

At the center (h, k, l): $-1 = -1 = -1$

Therefore, the line passes through the center.

49. Given $w = F(x, y, z)$ where F is differentiable at (x_0, y_0, z_0) and $\nabla F(x_0, y_0, z_0) \neq \mathbf{0}$, the level surface of F at (x_0, y_0, z_0) is of the form $F(x, y, z) = C$ for some constant C. Let $G(x, y, z) = F(x, y, z) - C = 0$. Then $\nabla G(x_0, y_0, z_0) = \nabla F(x_0, y_0, z_0)$ where $\nabla G(x_0, y_0, z_0)$ is normal to $F(x_0, y_0, z_0) - C = 0$. Therefore, $\nabla F(x_0, y_0, z_0)$ is normal to $F(x_0, y_0, z_0) = C$.

50. Given $z = f(x, y)$, then

$F(x, y, z) = f(x, y) - z = 0$

$\nabla F(x_0, y_0, z_0) = f_x(x_0, y_0)\mathbf{i} + f_y(x_0, y_0)\mathbf{j} - \mathbf{k}$

$$\cos\theta = \frac{|\nabla F(x_0, y_0, z_0) \cdot \mathbf{k}|}{\|\nabla F(x_0, y_0, z_0)\|\,\|\mathbf{k}\|}$$

$$= \frac{|-1|}{\sqrt{[f_x(x_0, y_0)]^2 + [f_y(x_0, y_0)]^2 + (-1)^2}} = \frac{1}{\sqrt{[f_x(x_0, y_0)]^2 + [f_y(x_0, y_0)]^2 + 1}}$$

Section 14.8 Extrema of Functions of Two Variables

1. $f(x, y) = 2x^2 + 2xy + y^2 + 2x - 3$

$\left.\begin{array}{l} f_x = 4x + 2y + 2 = 0 \\[4pt] f_y = 2x + 2y = 0 \end{array}\right\}$ Solving simultaneously yields $x = -1$ and $y = 1$.

$f_{xx} = 4, \quad f_{yy} = 2, \quad f_{xy} = 2$

At the critical point $(-1, 1)$, $f_{xx} > 0$ and $f_{xx}f_{yy} - (f_{xy})^2 > 0$. Therefore, $(-1, 1, -4)$ is a relative minimum.

2. $f(x, y) = -x^2 - 5y^2 + 8x - 10y - 13$

$\left.\begin{array}{l} f_x = -2x + 8 = 0 \\ f_y = -10y - 10 = 0 \end{array}\right\}$ Solving simultaneously yields $x = 4$ and $y = -1$.

$f_{xx} = -2, \quad f_{yy} = -10, \quad f_{xy} = 0$

At the critical point $(4, -1)$, $f_{xx} < 0$ and $f_{xx}f_{yy} - (f_{xy})^2 > 0$. Therefore, $(4, -1, 8)$ is a relative maximum.

3. $f(x, y) = -5x^2 + 4xy - y^2 + 16x + 10$

$\left.\begin{array}{l} f_x = -10x + 4y + 16 = 0 \\ f_y = 4x - 2y = 0 \end{array}\right\}$ Solving simultaneously yields $x = 8$ and $y = 16$.

$f_{xx} = -10, \quad f_{yy} = -2, \quad f_{xy} = 4$

At the critical point $(8, 16)$, $f_{xx} < 0$ and $f_{xx}f_{yy} - (f_{xy})^2 > 0$. Therefore, $(8, 16, 74)$ is a relative maximum.

4. $f(x, y) = x^2 + 6xy + 10y^2 - 4y + 4$

$\left.\begin{array}{l} f_x = 2x + 6y = 0 \\ f_y = 6x + 20y - 4 = 0 \end{array}\right\}$ Solving simultaneously yields $x = -6$ and $y = 2$.

$f_{xx} = 2, \quad f_{yy} = 20, \quad f_{xy} = 6$

At the critical point $(-6, 2)$, $f_{xx} > 0$ and $f_{xx}f_{yy} - (f_{xy})^2 > 0$. Therefore, $(-6, 2, 0)$ is a relative minimum.

5. $f(x, y) = 2x^2 + 3y^2 - 4x - 12y + 13$

$f_x = 4x - 4 = 4(x - 1) = 0$ when $x = 1$.

$f_y = 6y - 12 = 6(y - 2) = 0$ when $y = 2$.

$f_{xx} = 4, \quad f_{yy} = 6, \quad f_{xy} = 0$

At the critical point $(1, 2)$, $f_{xx} > 0$ and $f_{xx}f_{yy} - (f_{xy})^2 > 0$. Therefore, $(1, 2, -1)$ is a relative minimum.

6. $f(x, y) = 5 + 3x - 4y - 3x^2 - 2y^2$

$f_x = 3 - 6x = 3(1 - 2x) = 0$ when $x = \frac{1}{2}$.

$f_y = -4 - 4y = -4(1 + y) = 0$ when $y = -1$.

$f_{xx} = -6, \quad f_{yy} = -4, \quad f_{xy} = 0$

At the critical point $\left(\frac{1}{2}, -1\right)$, $f_{xx} < 0$ and $f_{xx}f_{yy} - (f_{xy})^2 > 0$. Therefore, $\left(\frac{1}{2}, -1, \frac{31}{4}\right)$ is a relative maximum.

7. $h(x, y) = x^2 - y^2 - 2x - 4y - 4$

$h_x = 2x - 2 = 2(x - 1) = 0$ when $x = 1$.

$h_y = -2y - 4 = -2(y + 2) = 0$ when $y = -2$.

$h_{xx} = 2, \quad h_{yy} = -2, \quad h_{xy} = 0$

At the critical point $(1, -2)$, $h_{xx}h_{yy} - (h_{xy})^2 < 0$. Therefore, $(1, -2, -1)$ is a saddle point.

8. $h(x, y) = x^2 - 3xy - y^2$

$\left.\begin{array}{l} h_x = 2x - 3y = 0 \\ h_y = -3x - 2y = 0 \end{array}\right\}$ Solving simultaneously yields $x = 0$ and $y = 0$.

$h_{xx} = 2, \quad h_{yy} = -2, \quad h_{xy} = -3$

At the critical point $(0, 0)$, $h_{xx}h_{yy} - (h_{xy})^2 < 0$. Therefore, $(0, 0, 0)$ is a saddle point.

9. $g(x, y) = xy$

$\left.\begin{array}{l} g_x = y \\ g_y = x \end{array}\right\}$ $x = 0$ and $y = 0$

$g_{xx} = 0$, $g_{yy} = 0$, $g_{xy} = 1$

At the critical point $(0, 0)$, $g_{xx}g_{yy} - (g_{xy})^2 < 0$. Therefore, $(0, 0, 0)$ is a saddle point.

10. $g(x, y) = 120x + 120y - xy - x^2 - y^2$

$\left.\begin{array}{l} g_x = 120 - y - 2x = 0 \\ g_y = 120 - x - 2y = 0 \end{array}\right\}$ Solving simultaneously yields $x = 40$ and $y = 40$.

$g_{xx} = -2$, $g_{yy} = -2$, $g_{xy} = -1$

At the critical point $(40, 40)$, $g_{xx} < 0$ and $g_{xx}g_{yy} - (g_{xy})^2 > 0$. Therefore, $(40, 40, 4800)$ is a relative maximum.

11. $f(x, y) = x^3 - 3xy + y^3$

$\left.\begin{array}{l} f_x = 3(x^2 - y) = 0 \\ f_y = 3(-x + y^2) = 0 \end{array}\right\}$ Solving by substitution yields two critical points $(0, 0)$ and $(1, 1)$.

$f_{xx} = 6x$, $f_{yy} = 6y$, $f_{xy} = -3$

At the critical point $(0, 0)$, $f_{xx}f_{yy} - (f_{xy})^2 < 0$. Therefore, $(0, 0, 0)$ is a saddle point. At the critical point $(1, 1)$, $f_{xx} = 6 > 0$ and $f_{xx}f_{yy} - (f_{xy})^2 > 0$. Therefore, $(1, 1, -1)$ is a relative minimum.

12. $f(x, y) = 4xy - x^4 - y^4$

$\left.\begin{array}{l} f_x = 4y - 4x^3 = 0 \\ f_y = 4x - 4y^3 = 0 \end{array}\right\}$ Solving by substitution yields three critical points $(0, 0)$, $(1, 1)$, $(-1, -1)$.

$f_{xx} = -12x^2$, $f_{yy} = -12y^2$, $f_{xy} = 4$

At the critical point $(0, 0)$, $f_{xx}f_{yy} - (f_{xy})^2 < 0$. Therefore, $(0, 0, 0)$ is a saddle point. At the critical point $(1, 1)$, $f_{xx} < 0$ and $f_{xx}f_{yy} - (f_{xy})^2 > 0$. Therefore, $(1, 1, 2)$ is a relative maximum. At the critical point $(-1, -1)$, $f_{xx} < 0$ and $f_{xx}f_{yy} - (f_{xy})^2 > 0$. Therefore, $(-1, -1, 2)$ is a relative maximum.

13. $f(x, y) = \dfrac{-4x}{x^2 + y^2 + 1}$

$\left.\begin{array}{l} f_x = \dfrac{4(x^2 - y^2 - 1)}{(x^2 + y^2 + 1)^2} \\[3mm] f_y = \dfrac{8xy}{(x^2 + y^2 + 1)^2} \end{array}\right\}$ Critical points: $(1, 0)$ and $(-1, 0)$

$f_{xx} = \dfrac{8x(-x^2 + 3y^2 + 3)}{(x^2 + y^2 + 1)^3}$

$f_{yy} = \dfrac{8x(x^2 - 3y^2 + 1)}{(x^2 + y^2 + 1)^3}$

$f_{xy} = \dfrac{-8y(3x^2 - y^2 - 1)}{(x^2 + y^2 + 1)^3}$

At $(1, 0, -2)$, $f_{xx} > 0$ and $f_{xx}f_{yy} - (f_{xy})^2 > 0$; relative minimum.
At $(-1, 0, 2)$, $f_{xx} < 0$ and $f_{xx}f_{yy} - (f_{xy})^2 > 0$; relative maximum.

14. $f(x, y) = y^3 - 3yx^2 - 3y^2 - 3x^2 + 1$

$\left. \begin{array}{l} f_x = -6xy - 6x = -6x(y+1) = 0 \\ f_y = 3y^2 - 3x^2 - 6y = 0 \end{array} \right\}$ Solving simultaneously yields the critical points $(0, 0), (0, 2), (\sqrt{3}, -1),$ and $(-\sqrt{3}, -1).$

$f_{xx} = -6(y+1), \quad f_{yy} = 6(y-1), \quad f_{xy} = -6x$

At the critical point $(0, 0), \quad f_{xx} = -6 < 0$ and $f_{xx}f_{yy} - (f_{xy})^2 = 36 > 0.$ Therefore, $(0, 0, 1)$ is a relative maximum. At $(0, 2), (\sqrt{3}, -1),$ and $(-\sqrt{3}, -1), \quad f_{xx}f_{yy} - (f_{xy})^2 < 0.$ Therefore, $(0, 2, -3), (\sqrt{3}, -1, -3),$ and $(-\sqrt{3}, -1, -3)$ are saddle points.

15. $f(x, y) = (x^2 + 4y^2)e^{1-x^2-y^2}$

$\left. \begin{array}{l} f_x = 2x(1 - x^2 - 4y^2)e^{1-x^2-y^2} = 0 \\ f_y = 2y(4 - x^2 - 4y^2)e^{1-x^2-y^2} = 0 \end{array} \right\}$ Solving yields the critical points $(0, 0), (1, 0), (-1, 0), (0, 1),$ and $(0, -1).$

$f_{xx} = 2(2x^4 + 8x^2y^2 - 5x^2 - 4y^2 + 1)e^{1-x^2-y^2}$

$f_{yy} = 2(8y^4 + 2x^2y^2 - x^2 - 20y^2 + 4)e^{1-x^2-y^2}$

$f_{xy} = 4y(x^3 + 4xy^2 - 5x)e^{1-x^2-y^2}$

At $(0, 0), \quad f_{xx} > 0$ and $f_{xx}f_{yy} - (f_{xy})^2 > 0.$ Therefore, $(0, 0, 0)$ is a relative minimum.
At $(\pm 1, 0), \quad f_{xx}f_{yy} - (f_{xy})^2 < 0.$ Therefore, $(\pm 1, 0, 1)$ are saddle points.
At $(0, \pm 1), \quad f_{xx} < 0$ and $f_{xx}f_{yy} - (f_{xy})^2 > 0.$ Therefore, $(0, \pm 1, 4)$ are relative maxima.

16. $f(x, y) = e^{-x} \sin y$

$\left. \begin{array}{l} f_x = -e^{-x} \sin y = 0 \\ f_y = e^{-x} \cos y = 0 \end{array} \right\}$ Since $e^{-x} > 0$ for all x and $\sin y$ and $\cos y$ are never both zero for a given value of y, there are no critical points.

17. $f(x, y) = \left(\frac{1}{2} - x^2 + y^2\right)e^{1-x^2-y^2}$

$\left. \begin{array}{l} f_x = (2x^3 - 2xy^2 - 3x)e^{1-x^2-y^2} = 0 \\ f_y = (2x^2y - 2y^3 + y)e^{1-x^2-y^2} = 0 \end{array} \right\}$ Solving yields the critical points $(0, 0), \left(0, \pm\frac{\sqrt{2}}{2}\right), \left(\pm\frac{\sqrt{6}}{2}, 0\right).$

$f_{xx} = (-4x^4 + 4x^2y^2 + 12x^2 - 2y^2 - 3)e^{1-x^2-y^2}$

$f_{yy} = (4y^4 - 4x^2y^2 + 2x^2 - 8y^2 + 1)e^{1-x^2-y^2}$

$f_{xy} = (-4x^3y + 4xy^3 + 2xy)e^{1-x^2-y^2}$

At the critical point $(0, 0), \quad f_{xx}f_{yy} - (f_{xy})^2 < 0.$ Therefore, $(0, 0, e/2)$ is a saddle point. At the critical points $(0, \pm\sqrt{2}/2), \quad f_{xx} < 0$ and $f_{xx}f_{yy} - (f_{xy})^2 > 0.$ Therefore, $(0, \pm\sqrt{2}/2, \sqrt{e})$ are relative maxima. At the critical points $(\pm\sqrt{6}/2, 0), \quad f_{xx} > 0$ and $f_{xx}f_{yy} - (f_{xy})^2 > 0.$ Therefore, $(\pm\sqrt{6}/2, 0, -\sqrt{e}/e)$ are relative minima.

18. $f(x, y) = e^{-(x^2+y^2)}$

$\left. \begin{array}{l} f_x = -2xe^{-(x^2+y^2)} = 0 \\ f_y = -2ye^{-(x^2+y^2)} = 0 \end{array} \right\}$ Solving yields $x = y = 0.$

$f_{xx} = 2(2x^2 - 1)e^{-(x^2+y^2)}$

$f_{yy} = 2(2y^2 - 1)e^{-(x^2+y^2)}$

$f_{xy} = 4xye^{-(x^2+y^2)}$

At the critical point $(0, 0), \quad f_{xx} < 0$ and $f_{xx}f_{yy} - (f_{xy})^2 > 0.$ Therefore, $(0, 0, 1)$ is a relative maximum.

19. $f(x, y) = x^2 + xy$

$R = \{(x, y) : |x| \leq 2, \ |y| \leq 1\}$

$\left.\begin{array}{l} f_x = 2x + y = 0 \\ f_y = x = 0 \end{array}\right\}$ Solving yields $x = y = 0$ or $y = -2x$.

$f(2, 1) = 6, \quad f(-2, -1) = 6, \quad f(0, 0) = 0$

$f\left(\frac{1}{2}, -1\right) = -\frac{1}{4}, \quad f\left(-\frac{1}{2}, 1\right) = -\frac{1}{4}$

Absolute maxima: $(2, 1, 6)$ and $(-2, -1, 6)$

Absolute minima: $\left(\frac{1}{2}, -1, -\frac{1}{4}\right)$ and $\left(-\frac{1}{2}, 1, -\frac{1}{4}\right)$

20. $f(x, y) = x^2 + 2xy + y^2$

$R = \{(x, y) : |x| \leq 2, \ |y| \leq 1\}$

$\left.\begin{array}{l} f_x = 2x + 2y = 0 \\ f_y = 2x + 2y = 0 \end{array}\right\}$ Solving yields the line $y = -x$.

$f(x, -x) = 0, \quad f(2, 1) = 9, \quad f(-2, -1) = 9$

Absolute maxima: $(2, 1, 9)$ and $(-2, -1, 9)$

Absolute minima: $(x, -x, 0)$ where $|x| \leq 1$

21. $f(x, y) = x^2 + 2xy + y^2$

$R = \{(x, y) : x^2 + y^2 \leq 8\}$

$\left.\begin{array}{l} f_x = 2x + 2y = 0 \\ f_y = 2x + 2y = 0 \end{array}\right\}$ Solving yields $y = -x$.

$f(x, -x), \ = 0, \quad f(2, 2) = 16, \quad f(-2, -2) = 16$

Absolute maxima: $(2, 2, 16)$ and $(-2, -2, 16)$

Absolute minima: $(x, -x, 0)$ where $|x| \leq 2$.

22. $f(x, y) = x^2 - 4xy$

$R = \{(x, y) : 0 \leq x \leq 4, \ 0 \leq y \leq \sqrt{x}\}$

$\left.\begin{array}{l} f_x = 2x - 4y = 0 \\ f_y = -4x = 0 \end{array}\right\}$ Solving yields $x = y = 0$.

$f(0, 0) = 0, \quad f(4, 0) = 16, \quad f(4, 2) = -16$

Absolute maxima: $(4, 0, 16)$

Absolute minima: $(4, 2, -16)$

23. $f(x, y) = x^3 + y^3$

$\left.\begin{array}{l} f_x = 3x^2 = 0 \\ f_y = 3y^2 = 0 \end{array}\right\}$ Solving yields $x = y = 0$.

$f_{xx} = 6x, \quad f_{yy} = 6y, \quad f_{xy} = 0$

At $(0, 0)$, $f_{xx}f_{yy} - (f_{xy})^2 = 0$ and the test fails. $(0, 0, 0)$ is a saddle point.

24. $f(x, y) = x^3 + y^3 - 3x^2 + 6y^2 + 3x + 12y + 7$

$\left.\begin{array}{l} f_x = 3x^2 - 6x + 3 = 0 \\ f_y = 3y^2 + 12y + 12 = 0 \end{array}\right\}$ Solving yields $x = 1$ and $y = -2$.

$f_{xx} = 6x - 6, \quad f_{yy} = 6y + 12, \quad f_{xy} = 0$

At $(1, -2)$, $f_{xx}f_{yy} - (f_{xy})^2 = 0$ and the test fails. $(1, -2, 0)$ is a saddle point.

25. $f(x, y) = (x - 1)^2(y + 4)^2$

$\left.\begin{array}{l} f_x = 2(x - 1)(y + 4)^2 = 0 \\ f_y = 2(x - 1)^2(y + 4) = 0 \end{array}\right\}$ Solving yields the critical points $(1, a)$ and $(b, -4)$.

$f_{xx} = 2(y + 4)^2$

$f_{yy} = 2(x - 1)^2$

$f_{xy} = 4(x - 1)(y + 4)$

At both $(1, a)$ and $(b, -4)$, $f_{xx}f_{yy} - (f_{xy})^2 = 0$ and the test fails.

Absolute minima: $(1, a, 0)$ and $(b, -4, 0)$

26. $f(x, y) = \sqrt{(x-1)^2 + (y+2)^2}$

$f_x = \dfrac{x-1}{\sqrt{(x-1)^2 + (y+2)^2}} = 0$

$f_y = \dfrac{y+2}{\sqrt{(x-1)^2 + (y+2)^2}} = 0$ $\quad\Biggr\}$ Solving yields $x = 1$ and $y = -2$.

$f_{xx} = \dfrac{(y+2)^2}{[(x-1)^2 + (y+2)^2]^{3/2}}$

$f_{yy} = \dfrac{(x-1)^2}{[(x-1)^2 + (y+2)^2]^{3/2}}$

$f_{xy} = \dfrac{(x-1)(y+2)}{[(x-1)^2 + (y+2)^2]^{3/2}}$

At $(1, -2)$, $f_{xx}f_{yy} - (f_{xy})^2$ is undefined and the test fails.

Absolute minimum: $(1, -2, 0)$

27. $f(x, y) = x^{2/3} + y^{2/3}$

$f_x = \dfrac{2}{3\sqrt[3]{x}}$

$f_y = \dfrac{2}{3\sqrt[3]{y}}$ $\quad\Biggr\}$ f_x and f_y are undefined at $x = 0$, $y = 0$. The critical point is $(0, 0)$.

$f_{xx} = -\dfrac{2}{9x\sqrt[3]{x}}$, $\quad f_{yy} = -\dfrac{2}{9y\sqrt[3]{y}}$, $\quad f_{xy} = 0$

At $(0, 0)$, $f_{xx}f_{yy} - (f_{xy})^2$ is undefined and the test fails.

Absolute minimum: $(0, 0, 0)$

28. $f(x, y) = (x^2 + y^2)^{2/3}$

$f_x = \dfrac{4x}{3(x^2 + y^2)^{1/3}}$

$f_y = \dfrac{4y}{3(x^2 + y^2)^{1/3}}$ $\quad\Biggr\}$ f_x and f_y are undefined at $x = 0$, $y = 0$. The critical point is $(0, 0)$.

$f_{xx} = \dfrac{4(x^2 + 3y^2)}{9(x^2 + y^2)^{4/3}}$

$f_{yy} = \dfrac{4(3x^2 + y^2)}{9(x^2 + y^2)^{4/3}}$

$f_{xy} = \dfrac{-8xy}{9(x^2 + y^2)^{4/3}}$

At $(0, 0)$, $f_{xx}f_{yy} - (f_{xy})^2$ is undefined and the test fails.

Absolute minimum: $(0, 0, 0)$

29. $f(x, y, z) = x^2 + (y-3)^2 + (z+1)^2$

$f_x = 2x = 0$

$f_y = 2(y-3) = 0$ $\quad\Biggr\}$ Solving yields the critical point $(0, 3, -1)$.

$f_z = 2(z+1) = 0$

30. $f(x, y, z) = 4 - [x(y-1)(z+2)]^2$

$\left.\begin{array}{l} f_x = -2x(y-1)^2(z+2)^2 = 0 \\ f_y = -2x^2(y-1)(z+2)^2 = 0 \\ f_z = -2x^2(y-1)^2(z+2) = 0 \end{array}\right\}$ Solving yields the critical points $(0, a, b)$, $(c, 1, d)$, $(e, f, -2)$.

31. $f(x, y, z) = x^2 + y^2 + 2xz - 4yz + 10z$

$\left.\begin{array}{l} f_x = 2x + 2z = 0 \\ f_y = 2y - 4z = 0 \\ f_z = 2x - 4y + 10 = 0 \end{array}\right\}$ Solving yields the critical point $(-1, 2, 1)$.

32. $f(x, y, z) = x^2 - y^2 + yz - x^2z$

$\left.\begin{array}{l} f_x = 2x - 2xz = 2x(1-z) = 0 \\ f_y = -2y + z = 0 \\ f_z = y - x^2 = 0 \end{array}\right\}$ Solving yields the critical points $(0, 0, 0)$, $\left(\pm\dfrac{\sqrt{2}}{2}, \dfrac{1}{2}, 1\right)$.

33. $f(x, y, z) = (x-1)^2(1-z) + y(z-y)$

$\left.\begin{array}{l} f_x = 2(x-1)(1-z) = 0 \\ f_y = z - 2y = 0 \\ f_z = -(x-1)^2 + y = 0 \end{array}\right\}$ Solving yields the critical points $(1, 0, 0)$, $\left(1 \pm \dfrac{\sqrt{2}}{2}, \dfrac{1}{2}, 1\right)$.

34. $f(x, y, z) = (x+z-3)^2 + y^2 - z^2 + 2z(5-2y)$

$\left.\begin{array}{l} f_x = 2(x+z-3) = 0 \\ f_y = 2y - 4z = 0 \\ f_z = 2(x+z-3) - 2z + 10 - 4y = 0 \end{array}\right\}$ Solving yields the critical point $(2, 2, 1)$.

35. Given that f is a differentiable function such that $\nabla f(x_0, y_0) = \mathbf{0}$, the $f_x(x_0, y_0) = 0$ and $f_y(x_0, y_0) = 0$. Therefore, the tangent plane is $-(z - z_0) = 0$ or $z = z_0 = f(x_0, y_0)$ which is horizontal.

Section 14.9 Applications of Extrema of Functions of Two Variables

1. A point on the plane is given by $(x,\ y,\ 12 - 2x - 3y)$. The square of the distance from the origin to this point is given by

$$S = x^2 + y^2 + (12 - 2x - 3y)^2$$
$$S_x = 2x + 2(12 - 2x - 3y)(-2)$$
$$S_y = 2y + 2(12 - 2x - 3y)(-3).$$

From the equations $S_x = 0$ and $S_y = 0$, we obtain the system

$$5x + 6y = 24$$
$$3x + 5y = 18.$$

Solving simultaneously, we have $x = \frac{12}{7}$, $y = \frac{18}{7}$, $z = 12 - \frac{24}{7} - \frac{54}{7} = \frac{6}{7}$. Therefore, the distance from the origin to $\left(\frac{12}{7},\ \frac{18}{7},\ \frac{6}{7}\right)$ is

$$\sqrt{\left(\tfrac{12}{7}\right)^2 + \left(\tfrac{18}{7}\right)^2 + \left(\tfrac{6}{7}\right)^2} = \frac{6\sqrt{14}}{7}.$$

2. A point on the plane is given by $(x,\ y,\ 12 - 2x - 3y)$. The square of the distance from $(1, 2, 3)$ to a point on the plane is given by

$$S = (x - 1)^2 + (y - 2)^2 + (9 - 2x - 3y)^2$$
$$S_x = 2(x - 1) + 2(9 - 2x - 3y)(-2)$$
$$S_y = 2(y - 2) + 2(9 - 2x - 3y)(-3).$$

From the equations $S_x = 0$ and $S_y = 0$, we obtain the system

$$5x + 6y = 19$$
$$6x + 10y = 29.$$

Solving simultaneously, we have $x = \frac{16}{14}$, $y = \frac{31}{14}$, $z = \frac{43}{14}$ and the distance is

$$\sqrt{\left(\tfrac{16}{14} - 1\right)^2 + \left(\tfrac{31}{14} - 2\right)^2 + \left(\tfrac{43}{14} - 3\right)^2} = \frac{1}{\sqrt{14}}.$$

3. A point on the paraboloid is given by $(x,\ y,\ x^2 + y^2)$. The square of the distance from $(5, 5, 0)$ to a point on the paraboloid is given by

$$S = (x - 5)^2 + (y - 5)^2 + (x^2 + y^2)^2$$
$$S_x = 2(x - 5) + 4x(x^2 + y^2) = 0$$
$$S_y = 2(y - 5) + 4y(x^2 + y^2) = 0.$$

From the equations $S_x = 0$ and $S_y = 0$, we obtain the system

$$2x^3 + 2xy^2 + x - 5 = 0$$
$$2y^3 + 2x^2y + y - 5 = 0.$$

Solving, we have $x = 1$, $y = 1$, $z = 2$ and the distance is

$$\sqrt{(1 - 5)^2 + (1 - 5)^2 + (2 - 0)^2} = 6.$$

4. A point on the paraboloid is given by $(x, y, x^2 + y^2)$. The square of the distance from $(5, 0, 0)$ to a point on the paraboloid is given by

$$S = (x - 5)^2 + y^2 + (x^2 + y^2)^2$$
$$S_x = 2(x - 5) + 4x(x^2 + y^2) = 0$$
$$S_y = 2y + 4y(x^2 + y^2) = 0.$$

From the equations $S_x = 0$ and $S_y = 0$, we obtain the system

$$2x^3 + 2xy^2 + x - 5 = 0$$
$$2y^3 + 2x^2y + y = 0.$$

Solving, we have $x \approx 1.235$, $y = 0$, $z \approx 1.525$ and the distance is

$$\sqrt{(1.235 - 5)^2 + (1.525)^2} \approx 4.06.$$

5. Let x, y and z be the numbers. Since $x + y + z = 30$, $z = 30 - x - y$.

$$P = xyz = 30xy - x^2y - xy^2$$
$$P_x = 30y - 2xy - y^2 = y(30 - 2x - y) = 0 \left.\right\} \begin{array}{l} 2x + y = 30 \\ x + 2y = 30 \end{array}$$
$$P_y = 30x - x^2 - 2xy = x(30 - x - 2y) = 0$$

Solving simultaneously yields $x = 10$, $y = 10$, and $z = 10$.

6. Since $x + y + z = 32$, $z = 32 - x - y$. Therefore,

$$P = xy^2z = 32xy^2 - x^2y^2 - xy^3$$
$$P_x = 32y^2 - 2xy^2 - y^3 = y^2(32 - 2x - y) = 0$$
$$P_y = 64xy - 2x^2y - 3xy^2 = y(64x - 2x^2 - 3xy) = 0.$$

Ignoring the solution $y = 0$ and substituting $y = 32 - 2x$ into $P_y = 0$, we have

$$64x - 2x^2 - 3x(32 - 2x) = 0$$
$$4x(x - 8) = 0.$$

Therefore, $x = 8$, $y = 16$, and $z = 8$.

7. Let x, y, and z be the numbers and let $S = x^2 + y^2 + z^2$. Since $x + y + z = 30$, we have

$$S = x^2 + y^2 + (30 - x - y)^2$$
$$S_x = 2x + 2(30 - x - y)(-1) = 0 \left.\right\} \begin{array}{l} 2x + y = 30 \\ x + 2y = 30. \end{array}$$
$$S_y = 2y + 2(30 - x - y)(-1) = 0$$

Solving simultaneously yields $x = 10$, $y = 10$, and $z = 10$.

8. Let x, y, and z be the numbers and let $S = x^2 + y^2 + z^2$. Since $x + y + z = 1$, we have

$$S = x^2 + y^2 + (1 - x - y)^2$$
$$S_x = 2x - 2(1 - x - y) = 0 \left.\right\} \begin{array}{l} 2x + y = 1 \\ x + 2y = 1. \end{array}$$
$$S_y = 2y - 2(1 - x - y) = 0$$

Solving simultaneously yields $x = \frac{1}{3}$, $y = \frac{1}{3}$, and $z = \frac{1}{3}$.

9. Let x, y, and z be the length, width, and height, respectively. Then the sum of the length and girth is given by $x + (2y + 2z) = 108$ or $x = 108 - 2y - 2z$. The volume is given by

$$V = xyz = 108zy - 2zy^2 - 2yz^2$$

$$V_y = 108z - 4yz - 2z^2 = z(108 - 4y - 2z) = 0$$

$$V_z = 108y - 2y^2 - 4yz = y(108 - 2y - 4z) = 0.$$

Solving the system $4y + 2z = 108$ and $2y + 4z = 108$, we obtain the solution $x = 36$ inches, $y = 18$ inches, and $z = 18$ inches.

10. Let x, y, and z be the length, width, and height, respectively. Then the sum of the two perimeters of the two cross sections is given by $(2x + 2z) + (2y + 2z) = 108$ or $x = 54 - y - 2z$. The volume is given by

$$V = xyz = 54yz - y^2z - 2yz^2$$

$$V_y = 54z - 2yz - 2z^2 = z(54 - 2y - 2z) = 0$$

$$V_z = 54y - y^2 - 4yz = y(54 - y - 4z) = 0.$$

Solving the system $2y + 2z = 54$ and $y + 4z = 54$, we obtain the solution $x = 18$ inches, $y = 18$ inches, and $z = 9$ inches.

11. The distance from P to Q is $\sqrt{x^2 + 4}$. The distance from Q to R is $\sqrt{(y - x)^2 + 1}$. The distance from R to S is $10 - y$.

$$C = 3k\sqrt{x^2 + 4} + 2k\sqrt{(y - x)^2 + 1} + k(10 - y)$$

$$C_x = 3k\left(\frac{x}{\sqrt{x^2 + 4}}\right) + 2k\left(\frac{-(y - x)}{\sqrt{(y - x)^2 + 1}}\right) = 0$$

$$C_y = 2k\left(\frac{y - x}{\sqrt{(y - x)^2 + 1}}\right) - k = 0 \Rightarrow \frac{y - x}{\sqrt{(y - x)^2 + 1}} = \frac{1}{2}$$

$$3k\left(\frac{x}{\sqrt{x^2 + 4}}\right) + 2k\left(-\frac{1}{2}\right) = 0$$

$$\frac{x}{\sqrt{x^2 + 4}} = \frac{1}{3}$$

$$3x = \sqrt{x^2 + 4}$$

$$9x^2 = x^2 + 4$$

$$x^2 = \frac{1}{2}$$

$$x = \frac{\sqrt{2}}{2}$$

$$2(y - x) = \sqrt{(y - x)^2 + 1}$$

$$4(y - x)^2 = (y - x)^2 + 1$$

$$(y - x)^2 = \frac{1}{3}$$

$$y = \frac{1}{\sqrt{3}} - \frac{1}{\sqrt{2}} = \frac{2\sqrt{3} + 3\sqrt{2}}{6}$$

Therefore, $x = \dfrac{\sqrt{2}}{2} \approx 0.707$ mile and $y = \dfrac{2\sqrt{3} + 3\sqrt{2}}{6} \approx 1.284$ miles.

12. Let x, y, and z be the length, width, and height, respectively. Then $C_0 = 1.5xy + 2yz + 2xz$ and $z = \dfrac{C_0 - 1.5xy}{2(x+y)}$.

The volume is given by

$$V = xyz = \frac{C_0 xy - 1.5x^2 y^2}{2(x+y)}$$

$$V_x = \frac{y^2(2C_0 - 3x^2 - 6xy)}{4(x+y)^2}$$

$$V_y = \frac{x^2(2C_0 - 3y^2 - 6xy)}{4(x+y)^2}.$$

In solving the system $V_x = 0$ and $V_y = 0$, we note by the symmetry of the equations that $y = x$. Substituting $y = x$ into $V_x = 0$ yields

$$\frac{x^2(2C_0 - 9x^2)}{16x^2} = 0, \quad 2C_0 = 9x^2, \quad x = \frac{1}{3}\sqrt{2C_0}, \quad y = \frac{1}{3}\sqrt{2C_0}, \text{ and } z = \frac{1}{4}\sqrt{2C_0}.$$

13. Let $a + b + c = k$. Then

$$V = \frac{4\pi abc}{3} = \frac{4}{3}\pi ab(k - a - b) = \frac{4}{3}\pi(kab - a^2 b - ab^2)$$

$$\left.\begin{array}{l} V_a = \dfrac{4\pi}{3}(kb - 2ab - b^2) = 0 \\[2mm] V_b = \dfrac{4\pi}{3}(ka - a^2 - 2ab) = 0 \end{array}\right\} \quad \begin{array}{l} kb - 2ab - b^2 = 0 \\[2mm] ka - a^2 - 2ab = 0 \end{array}$$

Solving this system simultaneously yields $a = b$ and substitution yields $b = k/3$. Therefore, the solution is $a = b = c = k/3$.

14. Consider the sphere given by $x^2 + y^2 + z^2 = r^2$ and let a vertex of the rectangular box be $(x, y, \sqrt{r^2 - x^2 - y^2}\,)$. Then the volume is given by

$$V = (2x)(2y)(2\sqrt{r^2 - x^2 - y^2}\,) = 8xy\sqrt{r^2 - x^2 - y^2}$$

$$V_x = 8\left(xy\frac{-x}{\sqrt{r^2 - x^2 - y^2}} + y\sqrt{r^2 - x^2 - y^2}\right) = \frac{8y}{\sqrt{r^2 - x^2 - y^2}}(r^2 - 2x^2 - y^2) = 0$$

$$V_y = 8\left(xy\frac{-y}{\sqrt{r^2 - x^2 - y^2}} + x\sqrt{r^2 - x^2 - y^2}\right) = \frac{8x}{\sqrt{r^2 - x^2 - y^2}}(r^2 - x^2 - 2y^2) = 0.$$

Solving the system

$$2x^2 + y^2 = r^2$$
$$x^2 + 2y^2 = r^2$$

yields the solution $x = y = z = \dfrac{r}{\sqrt{3}}$.

15. Let x, y, and z be the length, width, and height, respectively and let V_0 be the given volume. Then $V_0 = xyz$ and $z = V_0/xy$. The surface area is

$$S = 2xy + 2yz + 2xz = 2\left(xy + \frac{V_0}{x} + \frac{V_0}{y}\right)$$

$$\left.\begin{array}{l} S_x = 2\left(y - \dfrac{V_0}{x^2}\right) = 0 \\[2mm] S_y = 2\left(x - \dfrac{V_0}{y^2}\right) = 0 \end{array}\right\} \quad \begin{array}{l} x^2 y - V_0 = 0 \\[2mm] xy^2 - V_0 = 0 \end{array}$$

Solving simultaneously yields $x = \sqrt[3]{V_0}$, $y = \sqrt[3]{V_0}$, and $z = \sqrt[3]{V_0}$.

16. $A = \frac{1}{2}[(10 - 2x) + (10 - 2x) + 2x\cos\theta]x\sin\theta$

$\qquad = 10x\sin\theta - 2x^2\sin\theta + x^2\sin\theta\cos\theta$

$\dfrac{\partial A}{\partial x} = 10\sin\theta - 4x\sin\theta + 2x\sin\theta\cos\theta = 0$

$\dfrac{\partial A}{\partial \theta} = 10x\cos\theta - 2x^2\cos\theta + x^2(2\cos^2\theta - 1) = 0$

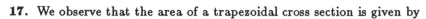

From $\partial A/\partial x = 0$, we have

$\qquad 5 - 2x + x\cos\theta = 0$

$$\cos\theta = \frac{2x - 5}{x}.$$

Substituting this expression for $\cos\theta$ in $\partial A/\partial\theta = 0$, we have

$$10x\left(\frac{2x - 5}{x}\right) - 2x^2\left(\frac{2x - 5}{x}\right) + x^2\left[2\left(\frac{2x - 5}{x}\right)^2 - 1\right] = 0$$

which simplifies to $x(3x - 10) = 0 \Rightarrow x = 0$ or $10/3$. Using $x = 10/3$, we have $\cos\theta = \dfrac{2(10/3) - 5}{10/3} = \dfrac{1}{2}$. Therefore, $\theta = 60°$, $x = 10/3$ inches.

17. We observe that the area of a trapezoidal cross section is given by

$$A = h\left[\frac{(w - 2r) + [(w - 2r) + 2x]}{2}\right] = (w - 2r + x)h$$

where $x = r\cos\theta$ and $h = r\sin\theta$. Substituting these expressions for x and h, we have

$$A(r,\ \theta) = (w - 2r + r\cos\theta)(r\sin\theta) = wr\sin\theta - 2r^2\sin\theta + r^2\sin\theta\cos\theta$$

Now $A_r(r,\ \theta) = w\sin\theta - 4r\sin\theta + 2r\sin\theta\cos\theta = \sin\theta(w - 4r + 2r\cos\theta) = 0 \Rightarrow w = r(4 - 2\cos\theta)$

$\qquad A_\theta(r,\ \theta) = wr\cos\theta - 2r^2\cos\theta + r^2\cos 2\theta = 0.$

Substituting the expression for w from $A_r(r,\ \theta) = 0$ into the equation $A_\theta(r,\ \theta) = 0$, we have

$\qquad r^2(4 - 2\cos\theta)\cos\theta - 2r^2\cos\theta + r^2(2\cos^2\theta - 1) = 0$

$$r^2(2\cos\theta - 1) = 0 \quad\text{or}\quad \cos\theta = \frac{1}{2}.$$

Therefore, the first partial derivatives are zero when $\theta = \pi/3$ and $r = w/3$. (Ignore the solution $r = \theta = 0$.) Thus, the trapezoid of maximum area occurs when each edge of width $w/3$ is turned up 60° from the horizontal.

18. $R(x_1,\ x_2) = -5x_1^2 - 8x_2^2 - 2x_1x_2 + 42x_1 + 102x_2$

$\qquad R_{x_1} = -10x_1 - 2x_2 + 42 = 0, \quad 5x_1 + x_2 = 21$

$\qquad R_{x_2} = -16x_2 - 2x_1 + 102 = 0, \quad x_1 + 8x_2 = 51$

Solving this system yields $x_1 = 3$ and $x_2 = 6$.

$R_{x_1x_1} = -10$

$R_{x_1x_2} = -2$

$R_{x_2x_2} = -16$

$R_{x_1x_1} < 0$ and $R_{x_1x_1}R_{x_2x_2} - (R_{x_1x_2})^2 > 0$

Thus, revenue is maximized when $x_1 = 3$ and $x_2 = 6$.

19. $R(p_1, \; p_2) = 500p_1 + 800p_2 + 1.5p_1p_2 - 1.5p_1{}^2 - p_2{}^2$

$\qquad R_{p_1} = 500 + 1.5p_2 - 3p_1 = 0, \quad -3p_1 + 1.5p_2 = -500$

$\qquad R_{p_2} = 800 + 1.5p_1 - 2p_2 = 0, \quad 3p_1 - 4.0p_2 = -1600$

Solving this system yields $-2.5p_2 = -2100$ which implies that $p_2 = \$840$ and $p_1 = \$586.67$.

20. $P(x_1, \; x_2) = 15(x_1 + x_2) - C_1 - C_2$

$\qquad\qquad = 15x_1 + 15x_2 - (0.02x_1{}^2 + 4x_1 + 500) - (0.05x_2{}^2 + 4x_2 + 275)$

$\qquad\qquad = -0.02x_1{}^2 - 0.05x_2{}^2 + 11x_1 + 11x_2 - 775$

$\qquad P_{x_1} = -0.04x_1 + 11 = 0, \quad x_1 = 275$

$\qquad P_{x_2} = -0.10x_2 + 11 = 0, \quad x_2 = 110$

$\quad P_{x_1 x_1} = -0.04$

$\quad P_{x_1 x_2} = 0$

$\quad P_{x_2 x_2} = -0.10$

$P_{x_1 x_1} < 0$ and $P_{x_1 x_1} P_{x_2 x_2} - (P_{x_1 x_2})^2 > 0$

Therefore, profit is maximized when $x_1 = 275$ and $x_2 = 110$.

21. $S(a, \; b, \; c) = \displaystyle\sum_{i=1}^{n} (y_i - ax_i{}^2 - bx_i - c)^2$

$\dfrac{\partial S}{\partial a} = \displaystyle\sum_{i=1}^{n} -2x_i{}^2(y_i - ax_i{}^2 - bx_i - c) = 0$

$\dfrac{\partial S}{\partial b} = \displaystyle\sum_{i=1}^{n} -2x_i(y_i - ax_i{}^2 - bx_i - c) = 0$

$\dfrac{\partial S}{\partial c} = -2\displaystyle\sum_{i=1}^{n} (y_i - ax_i{}^2 - bx_i - c) = 0$

$a\displaystyle\sum_{i=1}^{n} x_i{}^4 + b\sum_{i=1}^{n} x_i{}^3 + c\sum_{i=1}^{n} x_i{}^2 = \sum_{i=1}^{n} x_i{}^2 y_i$

$a\displaystyle\sum_{i=1}^{n} x_i{}^3 + b\sum_{i=1}^{n} x_i{}^2 + c\sum_{i=1}^{n} x_i = \sum_{i=1}^{n} x_i y_i$

$a\displaystyle\sum_{i=1}^{n} x_i{}^2 + b\sum_{i=1}^{n} x_i + cn = \sum_{i=1}^{n} y_i$

22. $S(a, \; b) = \displaystyle\sum_{i=1}^{n} (ax_i + b - y_i)^2$

$\dfrac{\partial S}{\partial a} = \displaystyle\sum_{i=1}^{n} 2x_i(ax_i + b - y_i) = 0$

$\dfrac{\partial S}{\partial b} = 2\displaystyle\sum_{i=1}^{n} (ax_i + b - y_i) = 0$

$\displaystyle\sum_{i=1}^{n} ax_i{}^2 + b\sum_{i=1}^{n} x_i - \sum_{i=1}^{n} x_i y_i = 0$

$\left(\displaystyle\sum_{i=1}^{n} x_i\right) a + \sum_{i=1}^{n} b - \sum_{i=1}^{n} y_i = 0$

$\left(\displaystyle\sum_{i=1}^{n} x_i\right) b + \left(\sum_{i=1}^{n} x_i{}^2\right) a = \sum_{i=1}^{n} x_i y_i$

$nb + \left(\displaystyle\sum_{i=1}^{n} x_i\right) a = \sum_{i=1}^{n} y_i$

23. (a)

x	y	xy	x^2
-2	0	0	4
0	1	0	0
2	3	6	4
$\sum x_i = 0$	$\sum y_i = 4$	$\sum x_i y_i = 6$	$\sum x_i^2 = 8$

$$a = \frac{3(6) - 0(4)}{3(8) - 0^2} = \frac{3}{4}, \quad b = \frac{1}{3}\left[4 - \frac{3}{4}(0)\right] = \frac{4}{3}, \quad y = \frac{3}{4}x + \frac{4}{3}$$

(b) $S = \left(-\dfrac{3}{2} + \dfrac{4}{3} - 0\right)^2 + \left(\dfrac{4}{3} - 1\right)^2 + \left(\dfrac{3}{2} + \dfrac{4}{3} - 3\right)^2 = \dfrac{1}{6}$

24. (a)

x	y	xy	x^2
-3	0	0	9
-1	1	-1	1
1	1	1	1
3	2	6	9
$\sum x_i = 0$	$\sum y_i = 4$	$\sum x_i y_i = 6$	$\sum x_i^2 = 20$

$$a = \frac{4(6) - 0(4)}{4(20) - (0)^2} = \frac{3}{10}, \quad b = \frac{1}{4}\left[4 - \frac{3}{10}(0)\right] = 1, \quad y = \frac{3}{10}x + 1$$

(b) $S = \left(\dfrac{1}{10} - 0\right)^2 + \left(\dfrac{7}{10} - 1\right)^2 + \left(\dfrac{13}{10} - 1\right)^2 + \left(\dfrac{19}{10} - 2\right)^2 = \dfrac{1}{5}$

25. (a)

x	y	xy	x^2
0	4	0	0
1	3	3	1
1	1	1	1
2	0	0	4
$\sum x_i = 4$	$\sum y_i = 8$	$\sum x_i y_i = 4$	$\sum x_i^2 = 6$

$$a = \frac{4(4) - 4(8)}{4(6) - 4^2} = -2, \quad b = \frac{1}{4}[8 + 2(4)] = 4, \quad y = -2x + 4$$

(b) $S = (4 - 4)^2 + (2 - 3)^2 + (2 - 1)^2 + (0 - 0)^2 = 2$

26. (a)

x	y	xy	x^2
3	0	0	9
1	0	0	1
2	0	0	4
3	1	3	9
4	1	4	16
4	2	8	16
5	2	10	25
6	2	12	36
$\sum x_i = 28$	$\sum y_i = 8$	$\sum x_i y_i = 37$	$\sum x_i^2 = 116$

$$a = \frac{8(37) - (28)(8)}{8(116) - (28)^2} = \frac{72}{144} = \frac{1}{2}, \quad b = \frac{1}{8}\left[8 - \frac{1}{2}(28)\right] = -\frac{3}{4}, \quad y = \frac{1}{2}x - \frac{3}{4}$$

(b) $S = \left(\frac{3}{4} - 0\right)^2 + \left(-\frac{1}{4} - 0\right)^2 + \left(\frac{1}{4} - 0\right)^2 + \left(\frac{3}{4} - 1\right)^2 + \left(\frac{5}{4} - 1\right)^2 + \left(\frac{5}{4} - 2\right)^2$

$$+ \left(\frac{7}{4} - 2\right)^2 + \left(\frac{9}{4} - 2\right)^2 = \frac{3}{2}$$

27. $(0, 0)$, $(1, 1)$, $(3, 4)$, $(4, 2)$, $(5, 5)$

$$\sum x_i = 13, \qquad \sum y_i = 12,$$
$$\sum x_i y_i = 46, \qquad \sum x_i^2 = 51$$
$$a = \frac{5(46) - 13(12)}{5(51) - (13)^2} = \frac{74}{86} = \frac{37}{43}$$
$$b = \frac{1}{5}\left[12 - \frac{37}{43}(13)\right] = \frac{7}{43}$$
$$y = \frac{37}{43}x + \frac{7}{43}$$

28. $(1, 0)$, $(3, 3)$, $(5, 6)$

$$\sum x_i = 9, \qquad \sum y_i = 9,$$
$$\sum x_i y_i = 39, \qquad \sum x_i^2 = 35$$
$$a = \frac{3(39) - 9(9)}{3(35) - (9)^2} = \frac{36}{24} = \frac{3}{2}$$
$$b = \frac{1}{3}\left[9 - \frac{3}{2}(9)\right] = -\frac{9}{6} = -\frac{3}{2}$$
$$y = \frac{3}{2}x - \frac{3}{2}$$

29. $(0, 6)$, $(4, 3)$, $(5, 0)$, $(8, -4)$, $(10, -5)$

$$\sum x_i = 27, \qquad \sum y_i = 0,$$
$$\sum x_i y_i = -70, \qquad \sum x_i^2 = 205$$
$$a = \frac{5(-70) - (27)(0)}{5(205) - (27)^2} = \frac{-350}{296} = -\frac{175}{148}$$
$$b = \frac{1}{5}\left[0 - \left(-\frac{175}{148}\right)(27)\right] = \frac{945}{148}$$
$$y = -\frac{175}{148}x + \frac{945}{148}$$

30. $(5, 2)$, $(0, 0)$, $(2, 1)$, $(7, 4)$, $(10, 6)$, $(12, 6)$

$$\sum x_i = 36, \qquad \sum y_i = 19,$$
$$\sum x_i y_i = 172, \qquad \sum x_i^2 = 322$$
$$a = \frac{6(172) - (36)(19)}{6(322) - (36)^2} = \frac{348}{636} = \frac{29}{53}$$
$$b = \frac{1}{6}\left[19 - \frac{29}{53}(36)\right] = -\frac{37}{318}$$
$$y = \frac{29}{53}x - \frac{37}{318}$$

31. $(-2, 0)$, $(-1, 0)$, $(0, 1)$, $(1, 2)$, $(2, 5)$

$$\sum x_i = 0$$

$$\sum y_i = 8$$

$$\sum x_i{}^2 = 10$$

$$\sum x_i{}^3 = 0$$

$$\sum x_i{}^4 = 34$$

$$\sum x_i y_i = 12$$

$$\sum x_i{}^2 y_i = 22$$

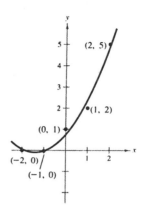

$34a + 10c = 22$, $10b = 12$, $10a + 5c = 8$

$a = \frac{3}{7}$, $b = \frac{6}{5}$, $c = \frac{26}{35}$, $y = \frac{3}{7}x^2 + \frac{6}{5}x + \frac{26}{35}$

32. $(-4, 5)$, $(-2, 6)$, $(2, 6)$, $(4, 2)$

$$\sum x_i = 0$$

$$\sum y_i = 19$$

$$\sum x_i{}^2 = 40$$

$$\sum x_i{}^3 = 0$$

$$\sum x_i{}^4 = 544$$

$$\sum x_i y_i = -12$$

$$\sum x_i{}^2 y_i = 160$$

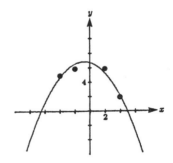

$544a + 40c = 160$, $40b = -12$, $40a + 4c = 19$

$a = -\frac{5}{24}$, $b = -\frac{3}{10}$, $c = \frac{41}{6}$, $y = -\frac{5}{24}x^2 - \frac{3}{10}x + \frac{41}{6}$

33. $(0, 0)$, $(2, 2)$, $(3, 6)$, $(4, 12)$

$$\sum x_i = 9$$

$$\sum y_i = 20$$

$$\sum x_i{}^2 = 29$$

$$\sum x_i{}^3 = 99$$

$$\sum x_i{}^4 = 353$$

$$\sum x_i y_i = 70$$

$$\sum x_i{}^2 y_i = 254$$

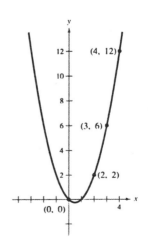

$353a + 99b + 29c = 254$

$99a + 29b + 9c = 70$

$29a + 9b + 4c = 20$

$a = 1$, $b = -1$, $c = 0$, $y = x^2 - x$

34. $(0, 10)$, $(1, 9)$, $(2, 6)$, $(3, 0)$

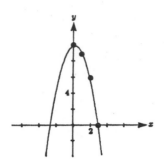

$$\sum x_i = 6$$

$$\sum y_i = 25$$

$$\sum x_i{}^2 = 14$$

$$\sum x_i{}^3 = 36$$

$$\sum x_i{}^4 = 98$$

$$\sum x_i y_i = 21$$

$$\sum x_i{}^2 y_i = 33$$

$98a + 36b + 14c = 33$

$36a + 14b + 6c = 21$

$14a + 6b + 4c = 25$

$a = -\frac{5}{4}$, $b = \frac{9}{20}$, $c = \frac{199}{20}$, $y = -\frac{5}{4}x^2 + \frac{9}{20}x + \frac{199}{20}$

35. (a) $(1.00, 450)$, $(1.25, 375)$, $(1.50, 330)$

$$\sum x_i = 3.75, \quad \sum y_i = 1155, \quad \sum x_i{}^2 = 4.8125, \quad \sum x_i y_i = 1413.75$$

$$a = \frac{3(1413.75) - (3.75)(1155)}{3(4.8125) - (3.75)^2} = -240$$

$$b = \frac{1}{3}[1155 - (-240)(3.75)] = 685$$

$$y = -240x + 685$$

(b) When $x = 1.40$, $y = -240(1.40) + 685 = 349$.

36. $(2.67, 135.2)$, $(1.35, 118.5)$, $(3.93, 167.3)$, $(5.14, 197.6)$, $(7.43, 204.7)$

$$\sum x_i = 20.52, \quad \sum y_i = 823.3, \quad \sum x_i y_i = 3715.033, \quad \sum x_i{}^2 = 106.0208$$

$$a = \frac{5(3715.033) - (20.52)(823.3)}{5(106.0208) - (20.52)^2} \approx 15.4177$$

$$b = \frac{1}{5}[823.3 - 15.4177(20.52)] \approx 101.3857$$

$$y = 15.4177x + 101.3857$$

37. $(1.0, 32)$, $(1.5, 41)$, $(2.0, 48)$, $(2.5, 53)$

$$\sum x_i = 7, \quad \sum y_i = 174, \quad \sum x_i y_i = 322, \quad \sum x_i{}^2 = 13.5$$

$a = 14$, $b = 19$, $y = 14x + 19$

When $x = 1.6$, $y = 41.4$ bushels per acre.

38. (a) $(-15, 3)$, $(-5, 3.7)$, $(0, 4.1)$, $(5, 4.5)$, $(10, 4.8)$

$$\sum x_i = -5, \qquad \sum y_i = 20.1, \qquad \sum x_i^2 = 375, \qquad \sum x_i^3 = -2375,$$

$$\sum x_i^4 = 61875, \qquad \sum x_iy_i = 7, \qquad \sum x_i^2y_i = 1360$$

$$61875a - 2375b + 375c = 1360$$

$$-2375a + 375b - 5c = 7$$

$$375a - 5b + 5c = 20.1$$

$$y = 0.000011316x^2 + 0.073304413x + 4.092455714$$

(b) When $x = 15$, $y \approx 5.2$ billion.

39. $(0, 0)$, $(2, 15)$, $(4, 30)$, $(6, 50)$, $(8, 65)$, $(10, 70)$

$$\sum x_i = 30, \qquad \sum y_i = 230, \qquad \sum x_i^2 = 220, \qquad \sum x_i^3 = 1800,$$

$$\sum x_i^4 = 15664, \qquad \sum x_iy_i = 1670, \qquad \sum x_i^2y_i = 13500$$

$$15664a + 1800b + 220c = 13500$$

$$1800a + 220b + 30c = 1670$$

$$220a + 30b + 6c = 230$$

$$y = -\tfrac{25}{112}x^2 + \tfrac{541}{56}x - \tfrac{25}{14}$$

Section 14.10   Lagrange Multipliers

1. Maximize $f(x, y) = xy$ subject to $x + y - 10 = 0$.

$$F(x, y, \lambda) = xy + \lambda(x + y - 10)$$

$$\left.\begin{array}{l} F_x = y + \lambda = 0 \\ F_y = x + \lambda = 0 \end{array}\right\} \; x = y$$

$$F_\lambda = x + y - 10 = 0$$

$$2x = 10 \Rightarrow x = y = 5$$

$$f(5, 5) = 25$$

2. Maximize $f(x, y) = xy$ subject to $2x + y - 4 = 0$.

$$F(x, y, \lambda) = xy + \lambda(2x + y - 4)$$

$$F_x = y + 2\lambda = 0, \qquad y = -2\lambda$$

$$F_y = x + \lambda = 0, \qquad x = -\lambda$$

$$F_\lambda = 2x + y - 4 = 0$$

$$-2\lambda - 2\lambda = 4 \Rightarrow \lambda = -1, \quad x = 1, \; y = 2$$

$$f(1, 2) = 2$$

3. Minimize $f(x, y) = x^2 + y^2$ subject to $x + y - 4 = 0$.

$$F(x, y, \lambda) = x^2 + y^2 + \lambda(x + y - 4)$$

$$\left.\begin{array}{l} F_x = 2x + \lambda = 0 \\ F_y = 2y + \lambda = 0 \end{array}\right\} \; x = y$$

$$F_\lambda = x + y - 4 = 0$$

$$2x = 4 \Rightarrow x = y = 2$$

$$f(2, 2) = 8$$

4. Minimize $f(x, y) = x^2 + y^2$ subject to $2x - 4y + 5 = 0$.

$$F(x, y, \lambda) = x^2 + y^2 + \lambda(2x - 4y + 5)$$

$$F_x = 2x + 2\lambda = 0, \qquad x = -\lambda$$

$$F_y = 2y - 4\lambda = 0, \qquad y = 2\lambda$$

$$F_\lambda = 2x - 4y + 5 = 0, \qquad -10\lambda = -5$$

$$\lambda = \tfrac{1}{2} \Rightarrow x = -\tfrac{1}{2}, \; y = 1$$

$$f\left(-\tfrac{1}{2}, 1\right) = \tfrac{5}{4}$$

5. Maximize $f(x, y) = x^2 - y^2$ subject to $y - x^2 = 0$.

$F(x, y, \lambda) = x^2 - y^2 + \lambda(y - x^2)$

$$F_x = 2x - 2x\lambda = 0, \quad x = 0 \text{ or } \lambda = 1$$

$$F_y = -2y + \lambda = 0, \quad y = \frac{\lambda}{2} = \frac{1}{2}$$

$$F_\lambda = y - x^2 = 0$$

$y = x^2 \Rightarrow x = \sqrt{1/2} = \dfrac{\sqrt{2}}{2}$ or $y = 0$

$f\left(\dfrac{\sqrt{2}}{2}, \dfrac{1}{2}\right) = \dfrac{1}{4}$

Note: $f(0, 0) = 0$ does not give a maximum value.

6. Maximize $f(x, y) = x^2 - y^2$ subject to $x - 2y + 6 = 0$.

$F(x, y, \lambda) = x^2 - y^2 + \lambda(x - 2y + 6)$

$$F_x = 2x + \lambda = 0, \quad x = -\frac{\lambda}{2}$$

$$F_y = -2y - 2\lambda = 0, \quad y = -\lambda$$

$$F_\lambda = x - 2y + 6 = 0, \quad \frac{3\lambda}{2} = -6$$

$\lambda = -4 \Rightarrow x = 2, \quad y = 4$

$f(2, 4) = -12$

7. Maximize $f(x, y) = 2x + 2xy + y$ subject to $2x + y - 100 = 0$.

$F(x, y, \lambda) = 2x + 2xy + y + \lambda(2x + y - 100)$

$$\left.\begin{array}{ll} F_x = 2 + 2y + 2\lambda = 0, & \lambda = -1 - y \\ F_y = 2x + \lambda + 1 = 0, & 2x - y = 0 \\ F_\lambda = 2x + y - 100 = 0, & 2x + y = 100 \end{array}\right\} \quad 4x = 100 \Rightarrow x = 25, \quad y = 50$$

$f(25, 50) = 2600$

8. Maximize $f(x, y) = 3x + y + 10$ subject to $x^2 y - 6 = 0$.

$F(x, y, \lambda) = 3x + y + 10 + \lambda(x^2 y - 6)$

$$\left.\begin{array}{ll} F_x = 3 + 2xy\lambda = 0, & \lambda = -\dfrac{3}{2xy} \\ \\ F_y = 1 + x^2\lambda = 0, & \lambda = -\dfrac{1}{x^2} \end{array}\right\} \quad 3x^2 = 2xy, \quad x = 0 \text{ or } y = \dfrac{3x}{2}$$

$$F_\lambda = x^2 y - 6 = 0$$

$x^2\left(\dfrac{3x}{2}\right) - 6 = 0, \quad x^3 = 4 \Rightarrow x = \sqrt[3]{4}, \quad y = \dfrac{3\sqrt[3]{4}}{2}$

$f\left(\sqrt[3]{4}, \dfrac{3\sqrt[3]{4}}{2}\right) = \dfrac{9\sqrt[3]{4} + 20}{2}$

9. Maximize $f(x, y) = \sqrt{6 - x^2 - y^2}$ subject to $x + y - 2 = 0$.

$F(x, y, \lambda) = 6 - x^2 - y^2 + \lambda(x + y - 2)$

$$F_x = -2x + \lambda = 0, \quad 2x = \lambda$$

$$F_y = -2y + \lambda = 0, \quad 2y = \lambda$$

$$F_\lambda = x + y - 2 = 0, \quad 2x = 2$$

$x = y = 1$

$f(1, 1) = 2$

Note: $f(x, y)$ is maximum when $g(x, y) = 6 - x^2 - y^2$ is maximum.

10. Minimize $f(x, y) = \sqrt{x^2 + y^2}$ subject to $2x + 4y - 15 = 0$.

$F(x, y, \lambda) = x^2 + y^2 + \lambda(2x + 4y - 15)$

$$\left. \begin{aligned} F_x &= 2x + 2\lambda = 0, \quad \lambda = -x \\ F_y &= 2y + 4\lambda = 0, \quad \lambda = -\frac{1}{2}y \end{aligned} \right\} \; y = 2x$$

$F_\lambda = 2x + 4y - 15 = 0$

$10x = 15 \Rightarrow x = \dfrac{3}{2}, \quad y = 3$

$f\left(\dfrac{3}{2}, 3\right) = \dfrac{3\sqrt{5}}{4}$

Note: $f(x, y)$ is minimum when $g(x, y) = x^2 + y^2$ is minimum.

11. Maximize $f(x, y) = e^{xy}$ subject to $x^2 + y^2 - 8 = 0$.

$F(x, y, \lambda) = e^{xy} + \lambda(x^2 + y^2 - 8)$

$$\left. \begin{aligned} F_x &= ye^{xy} + 2x\lambda = 0, \quad e^{xy} = -2x\lambda \\ F_y &= xe^{xy} + 2y\lambda = 0, \quad e^{xy} = -2y\lambda \end{aligned} \right\} \; x = y$$

$F_\lambda = x^2 + y^2 - 8 = 0$

$2x^2 = 8 \Rightarrow x = y = 2$

$f(2, 2) = e^4$

12. Minimize $f(x, y) = 2x + y$ subject to $xy - 32 = 0$.

$F(x, y, \lambda) = 2x + y + \lambda(xy - 32)$

$$\left. \begin{aligned} F_x &= 2 + y\lambda = 0, \quad y = -\frac{2}{\lambda} \\ F_y &= 1 + x\lambda = 0, \quad x = -\frac{1}{\lambda} \end{aligned} \right\} \; y = 2x$$

$F_\lambda = xy - 32 = 0$

$2x^2 = 32 \Rightarrow x = 4, \quad y = 8$

$f(4, 8) = 16$

13. Minimize $f(x, y, z) = x^2 + y^2 + z^2$ subject to $x + y + z - 6 = 0$.

$F(x, y, z, \lambda) = x^2 + y^2 + z^2 + \lambda(x + y + z - 6)$

$$\left. \begin{aligned} F_x &= 2x + \lambda = 0 \\ F_y &= 2y + \lambda = 0 \\ F_z &= 2z + \lambda = 0 \end{aligned} \right\} \; x = y = z$$

$F_\lambda = x + y + z - 6 = 0$

$3x = 6 \Rightarrow x = y = z = 2$

$f(2, 2, 2) = 12$

14. Maximize $f(x, y, z) = xyz$ subject to $x + y + z - 6 = 0$.

$F(x, y, z, \lambda) = xyz + \lambda(x + y + z - 6)$

$$\left.\begin{array}{l} F_x = yz + \lambda = 0 \\ F_y = xz + \lambda = 0 \\ F_z = xy + \lambda = 0 \end{array}\right\} \; x = y = z$$

$F_\lambda = x + y + z - 6 = 0$

$3x = 6 \Rightarrow x = y = z = 2$

$f(2, 2, 2) = 8$

15. Minimize $f(x, y, z) = x^2 + y^2 + z^2$ subject to $x + y + z - 1 = 0$.

$F(x, y, z, \lambda) = x^2 + y^2 + z^2 + \lambda(x + y + z - 1)$

$$\left.\begin{array}{l} F_x = 2x + \lambda = 0 \\ F_y = 2y + \lambda = 0 \\ F_z = 2z + \lambda = 0 \end{array}\right\} \; x = y = z$$

$F_\lambda = x + y + z - 1 = 0$

$3x = 1 \Rightarrow x = y = z = \frac{1}{3}$

$f\left(\frac{1}{3}, \frac{1}{3}, \frac{1}{3}\right) = \frac{3}{9} = \frac{1}{3}$

16. Minimize $f(x, y) = x^2 - 8x + y^2 - 12y + 48$ subject to $x + y = 8$.

$F(x, y, \lambda) = x^2 - 8x + y^2 - 12y + 48 + \lambda(x + y - 8)$

$$\left.\begin{array}{l} F_x = 2x - 8 + \lambda = 0 \\ F_y = 2y - 12 + \lambda = 0 \end{array}\right\} \; 2x - 8 = 2y - 12 \text{ or } x = y - 2$$

$F_\lambda = x + y - 8 = 0$

Substituting into $F_\lambda = 0$ yields $y - 2 + y - 8 = 0$ or $y = 5$ and $x = 3$.

$f(3, 5) = -2$

17. Maximize $f(x, y, z) = xyz$ subject to $x + y + z = 32$ and $x - y + z = 0$.

$F(x, y, z, \lambda, \mu) = xyz + \lambda(x + y + z - 32) + \mu(x - y + z)$

$$\left.\begin{array}{l} F_x = yz + \lambda + \mu = 0 \\ F_y = xz + \lambda - \mu = 0 \\ F_z = xy + \lambda + \mu = 0 \end{array}\right\} \; yz = xy \Rightarrow x = z$$

$$\left.\begin{array}{l} F_\lambda = x + y + z - 32 = 0 \\ F_\mu = x - y + z = 0 \end{array}\right\} \; 2y = 32, \quad y = 16, \quad x = \frac{y}{2} = 8 = z$$

$f(8, 16, 8) = 8(16)(8) = 1024$

18. Miminize $f(x, y, z) = x^2 + y^2 + z^2$ subject to $x + 2z = 4$ and $x + y = 8$.

$F(x, y, z, \lambda, \mu) = x^2 + y^2 + z^2 + \lambda(x + 2z - 4) + \mu(x + y - 8)$

$$F_x = 2x + \lambda + \mu = 0,$$

$$\left. \begin{array}{ll} F_y = 2y + \mu = 0, & \mu = -2y \\ F_z = 2z - 2\lambda = 0, & \lambda = -z \end{array} \right\} \; 2x - 2y - z = 0$$

$$F_\lambda = x + 2z - 4 = 0, \quad x + 2z = 4$$

$$F_\mu = x + y - 8 = 0, \quad x + y = 8$$

$x = y = 4, \quad z = 0$

$f(4, 4, 0) = 4^2 + 4^2 + 0^2 = 32$

19. Maximize $f(x, y, z) = xy + yz$ subject to $x + 2y = 6$ and $x - 3z = 0$.

$F(x, y, z, \lambda, \mu) = xy + yz + \lambda(x + 2y - 6) + \mu(x - 3z)$

(1) $F_x = y + \lambda + \mu = 0$ (2) $F_y = x + z + 2\lambda = 0$

(3) $F_z = y - 3\mu = 0$ (4) $F_\lambda = x + 2y - 6 = 0$

(5) $F_\mu = x - 3z = 0$

(5) $\Rightarrow x = 3z$

(2) $\Rightarrow 4z = -2\lambda, \quad \lambda = -2z$

$\left. \begin{array}{l} (1) \Rightarrow y - 2z = -\mu \\ (3) \Rightarrow y = 3\mu \end{array} \right\} \; \mu = \tfrac{1}{2}z, \quad y = \tfrac{3}{2}z$

(4) $\Rightarrow 3z + 2\left(\tfrac{3}{2}z\right) = 6 \Rightarrow z = 1, \quad x = 3, \quad y = \tfrac{3}{2}$

$f\left(3, \tfrac{3}{2}, 1\right) = 6$

20. Maximize $f(x, y, z) = xyz$ subject to $x^2 + z^2 = 5$ and $x - 2y = 0$.

$F(x, y, z, \lambda, \mu) = xyz + \lambda(x^2 + z^2 - 5) + \mu(x - 2y)$

$$F_x = yz + 2x\lambda + \mu = 0$$

$$F_y = xz - 2\mu = 0, \qquad \mu = \frac{xz}{2}$$

$$F_z = xy + 2z\lambda = 0, \qquad \lambda = -\frac{xy}{2z}$$

$$F_\lambda = x^2 + z^2 - 5 = 0, \qquad z = \sqrt{5 - x^2}$$

$$F_\mu = x - 2y = 0, \qquad y = \frac{x}{2}$$

$$F_x = yz + 2x\lambda + \mu = \frac{x\sqrt{5 - x^2}}{2} - \frac{x^3}{2\sqrt{5 - x^2}} + \frac{x\sqrt{5 - x^2}}{2} = 0$$

$$x\sqrt{5 - x^2} = \frac{x^3}{2\sqrt{5 - x^2}}$$

$$2x(5 - x^2) = x^3$$

$$0 = 3x^3 - 10x = x(3x^2 - 10)$$

$$x = 0 \text{ or } \sqrt{10/3}$$

Since x, y, and z are positive, we have $x = \sqrt{10/3}, \quad y = \tfrac{1}{2}\sqrt{10/3}$, and $z = \sqrt{5/3}$.

$$f\left(\sqrt{10/3}, \tfrac{1}{2}\sqrt{10/3}, \sqrt{5/3}\right) = \frac{5\sqrt{15}}{9}$$

21. Maximize $V = xyz$ subject to the constraint $x + 2y + 2z = 108$.

$F(x,\ y,\ z,\ \lambda) = xyz + \lambda(x + 2y + 2z - 108)$

$$\left.\begin{array}{l} F_x = yz + \lambda = 0 \\ F_y = xz + 2\lambda = 0 \\ F_z = xy + 2\lambda = 0 \end{array}\right\} \quad x = 2y, \quad y = z$$

$F_\lambda = x + 2y + 2z - 108 = 0$

$6y = 108, \quad x = 36, \quad y = 18, \quad z = 18$

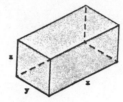

Volume is maximum when dimensions are $36 \times 18 \times 18$ inches.

22. Maximize $V = xyz$ subject to $1.5xy + 2xz + 2yz = C$.

$F(x,\ y,\ z,\ \lambda) = xyz + \lambda(1.5xy + 2xz + 2yz - C)$

$F_x = yz + 1.5y\lambda + 2z\lambda = 0$

$F_y = xz + 1.5x\lambda + 2z\lambda = 0$

$F_z = xy + 2x\lambda + 2y\lambda = 0$

$F_\lambda = 1.5xy + 2xz + 2yz - C = 0$

$xF_x - yF_y = 2xz\lambda - 2yz\lambda = 0, \qquad x = y$

$F_z = x^2 + 4x\lambda = 0, \qquad \lambda = -\dfrac{x}{4}$

$F_y = xz - \dfrac{3x^2}{8} - \dfrac{xz}{2} = 0, \qquad z = \dfrac{3}{4}x$

$F_\lambda = \dfrac{3}{2}x^2 + \dfrac{3}{2}x^2 + \dfrac{3}{2}x^2 = C$

$x^2 = \dfrac{2C}{9} \Rightarrow x = \dfrac{\sqrt{2C}}{3}, \quad y = \dfrac{\sqrt{2C}}{3}, \quad z = \dfrac{\sqrt{2C}}{4}$

23. Minimize $C = 5xy + 3(2xz + 2yz + xy)$ subject to the constraint $xyz = 480$.

$F(x,\ y,\ z,\ \lambda) = 8xy + 6xz + 6yz + \lambda(xyz - 480)$

$$\left.\begin{array}{l} F_x = 8y + 6z + yz\lambda = 0 \\ F_y = 8x + 6z + xz\lambda = 0 \\ F_z = 6x + 6y + xy\lambda = 0 \end{array}\right\} \begin{array}{l} x = y \\ \\ 4y = 3z \end{array}$$

$F_\lambda = xyz - 480 = 0 \Rightarrow \frac{4}{3}y^3 = 480$

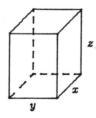

$x = y = \sqrt[3]{360}, \quad z = \frac{4}{3}\sqrt[3]{360}$

$\sqrt[3]{360} \times \sqrt[3]{360} \times \frac{4}{3}\sqrt[3]{360}$ feet

24. Minimize $A = 2\pi rh + 2\pi r^2$ subject to the constraint $\pi r^2 h = V_0$.

$F(r,\ h,\ \lambda) = 2\pi rh + 2\pi r^2 + \lambda(\pi r^2 h - V_0)$

$$\left.\begin{array}{l} F_r = 2\pi h + 4\pi r + 2\pi rh\lambda = 0 \\ F_h = 2\pi r + \pi r^2\lambda = 0 \end{array}\right\} \quad h = 2r$$

$F_\lambda = \pi r^2 h - V_0 = 0 \Rightarrow 2\pi r^3 = V_0$

$r = \sqrt[3]{V_0/2\pi}, \quad h = 2\sqrt[3]{V_0/2\pi}$

25. Maximize $V = (2x)(2y)(2z)$ subject to $\dfrac{x^2}{a^2} + \dfrac{y^2}{b^2} + \dfrac{z^2}{c^2} = 1$.

$$F(x, \ y, \ z, \ \lambda) = 8xyz + \lambda\left(\dfrac{x^2}{a^2} + \dfrac{y^2}{b^2} + \dfrac{z^2}{c^2} - 1\right)$$

$$F_x = 8yz + \dfrac{2\lambda x}{a^2} = 0, \qquad 8xyz + \dfrac{2\lambda x^2}{a^2} = 0$$

$$F_y = 8xz + \dfrac{2\lambda y}{b^2} = 0, \qquad 8xyz + \dfrac{2\lambda y^2}{b^2} = 0$$

$$F_z = 8xy + \dfrac{2\lambda z}{c^2} = 0, \qquad 8xyz + \dfrac{2\lambda z^2}{c^2} = 0$$

$$F_\lambda = \dfrac{x^2}{a^2} + \dfrac{y^2}{b^2} + \dfrac{z^2}{c^2} - 1 = 0$$

From the first three equations, we have $\dfrac{x^2}{a^2} = \dfrac{y^2}{b^2} = \dfrac{z^2}{c^2}$. Substituting into $F_\lambda = 0$ yields

$$\dfrac{3x^2}{a^2} = 1 \quad \text{or} \quad x = \dfrac{a}{\sqrt{3}}$$

$$\dfrac{3y^2}{b^2} = 1 \quad \text{or} \quad y = \dfrac{b}{\sqrt{3}}$$

$$\dfrac{3z^2}{c^2} = 1 \quad \text{or} \quad z = \dfrac{c}{\sqrt{3}}.$$

Therefore, the dimensions of the box are $\dfrac{2\sqrt{3}\,a}{3} \times \dfrac{2\sqrt{3}\,b}{3} \times \dfrac{2\sqrt{3}\,c}{3}$.

26. Maximize $P = xyz$ subject to $x + y + z = S$.

$$F(x, \ y, \ z, \ \lambda) = xyz + \lambda(x + y + z - S)$$

$$\left.\begin{array}{l} F_x = yz + \lambda = 0 \\[4pt] F_y = xz + \lambda = 0 \\[4pt] F_z = xy + \lambda = 0 \end{array}\right\} \begin{array}{l} x = y \\[6pt] y = z \end{array}$$

$$F_\lambda = x + y + z - S = 0 \Rightarrow 3x = S$$

$$x = y = z = \dfrac{S}{3}$$

Therefore, $\quad xyz \le \left(\dfrac{S}{3}\right)\left(\dfrac{S}{3}\right)\left(\dfrac{S}{3}\right), \quad x, \ y, \ z > 0$

$$xyz \le \dfrac{S^3}{27}$$

$$\sqrt[3]{xyz} \le \dfrac{S}{3}$$

$$\sqrt[3]{xyz} \le \dfrac{x + y + z}{3}.$$

27. Using the formula $\text{time} = \dfrac{\text{distance}}{\text{rate}}$, we wish to minimize $T(x,\ y) = \dfrac{\sqrt{d_1{}^2 + x^2}}{v_1} + \dfrac{\sqrt{d_2{}^2 + y^2}}{v_2}$ subject to the constraint $x + y = a$.

$$F(x,\ y,\ \lambda) = \frac{\sqrt{d_1{}^2 + x^2}}{v_1} + \frac{\sqrt{d_2{}^2 + y^2}}{v_2} + \lambda(x + y - a)$$

$$\left. \begin{aligned} F_x &= \frac{x}{v_1\sqrt{d_1{}^2 + x^2}} + \lambda = 0 \\[2mm] F_y &= \frac{y}{v_2\sqrt{d_2{}^2 + y^2}} + \lambda = 0 \end{aligned} \right\} \quad \frac{x}{v_1\sqrt{d_1{}^2 + x^2}} = \frac{y}{v_2\sqrt{d_2{}^2 + y^2}}$$

$$F_\lambda = x + y - a = 0$$

Since $\sin\theta_1 = \dfrac{x}{\sqrt{d_1{}^2 + x^2}}$ and $\sin\theta_2 = \dfrac{y}{\sqrt{d_2{}^2 + y^2}}$, we have $\dfrac{x/\sqrt{d_1{}^2 + x^2}}{v_1} = \dfrac{y/\sqrt{d_2{}^2 + y^2}}{v_2}$ or $\dfrac{\sin\theta_1}{v_1} = \dfrac{\sin\theta_2}{v_2}$.

28. Maximize $T(x,\ y,\ z) = 100 + x^2 + y^2$ subject to $x^2 + y^2 + z^2 = 50$ and $x - z = 0$.

$$F(x,\ y,\ z,\ \lambda,\ \mu) = 100 + x^2 + y^2 + \lambda(x^2 + y^2 + z^2 - 50) + \mu(x - z)$$

$$\begin{aligned} F_x &= 2x + 2x\lambda + \mu = 0, & \mu &= 0 \ (\text{since } \lambda = -1) \\ F_y &= 2y + 2y\lambda = 0, & \lambda &= -1 \\ F_z &= 2z\lambda - \mu = 0, & z &= 0 \ (\text{since } \mu = 0) \\ F_\lambda &= x^2 + y^2 + z^2 - 50 = 0, & y &= \sqrt{50} \ (\text{since } x = z = 0) \\ F_\mu &= x - z = 0, & x &= z \end{aligned}$$

$$x = 0, \quad y = \sqrt{50}, \quad z = 0$$

$$T(0,\ \sqrt{50},\ 0) = 150$$

29. Minimize the square of the distance $f(x,\ y) = x^2 + y^2$ subject to the constraint $g(x,\ y) = 2x + 3y + 1 = 0$.

$$F(x,\ y,\ \lambda) = x^2 + y^2 + \lambda(2x + 3y + 1)$$

$$\left. \begin{aligned} F_x &= 2x + 2\lambda = 0 \\ F_y &= 2y + 3\lambda = 0 \end{aligned} \right\} \quad x = -\lambda, \quad y = -\frac{3\lambda}{2}$$

$$F_\lambda = 2x + 3y + 1 = 0$$

Substituting into $F_\lambda = 0$ yields $2(-\lambda) + 3(-3\lambda/2) + 1 = 0$ or $\lambda = 2/13$. Therefore, the point on the line is $(-2/13,\ -3/13)$ and the desired distance is

$$d = \sqrt{\left(-\tfrac{2}{13}\right)^2 + \left(-\tfrac{3}{13}\right)^2} = \frac{1}{\sqrt{13}} = \frac{\sqrt{13}}{13}.$$

30. Minimize the square of the distance $f(x, y) = (x^2+y-10)^2$ subject to the constraint $g(x, y) = (x-4)^2+y^2-4 = 0$.

$F(x, y, \lambda) = x^2 + (y - 10)^2 + \lambda[(x - 4)^2 + y^2 - 4]$

$$F_x = 2x + 2(x - 4)\lambda = 0, \qquad x = \frac{4\lambda}{1 + \lambda}$$

$$F_y = 2(y - 10) + 2y\lambda = 0, \qquad y = \frac{10}{1 + \lambda}$$

$$F_\lambda = (x - 4)^2 + y^2 - 4 = 0$$

$$F_\lambda = \left(\frac{4\lambda}{1 + \lambda} - 4\right)^2 + \left(\frac{10}{1 + \lambda}\right)^2 - 4 = 0$$

$$\frac{16}{(1 + \lambda)^2} + \frac{100}{(1 + \lambda)^2} = 4$$

$$29 = (1 + \lambda)^2$$

$$\lambda = -1 \pm \sqrt{29}$$

$$\lambda = -1 + \sqrt{29}$$

$$x = \frac{4(-1 + \sqrt{29})}{\sqrt{29}} = 4\left(1 - \frac{1}{\sqrt{29}}\right)$$

$$y = \frac{10}{\sqrt{29}}$$

$$d = \sqrt{16[1 - (1/\sqrt{29})]^2 + [(10/\sqrt{29}) - 10]^2} \approx 8.77$$

31. Minimize the square of the distance $f(x, y, z) = (x - 2)^2 + (y - 1)^2 + (z - 1)^2$ subject to the constraint $g(x, y, z) = x + y + z - 1 = 0$.

$F(x, y, z, \lambda) = (x - 2)^2 + (y - 1)^2 + (z - 1)^2 + \lambda(x + y + z - 1)$

$$\left.\begin{array}{l} F_x = 2(x - 2) + \lambda = 0 \\[4pt] F_y = 2(y - 1) + \lambda = 0 \\[4pt] F_z = 2(z - 1) + \lambda = 0 \end{array}\right\} \quad x = 2 - \frac{\lambda}{2}, \ y = 1 - \frac{\lambda}{2}, \ z = 1 - \frac{\lambda}{2}$$

$$F_\lambda = x + y + z - 1 = 0$$

Substituting into $F_\lambda = 0$ yields

$$\left(2 - \frac{\lambda}{2}\right) + \left(1 - \frac{\lambda}{2}\right) + \left(1 - \frac{\lambda}{2}\right) - 1 = 0 \text{ or } \lambda = 2.$$

Therefore, the point on the plane is $(1, 0, 0)$ and the distance is $\sqrt{3}$.

32. Minimize the square of the distance $f(x, y, z) = (x - 4)^2 + y^2 + z^2$ subject to the constraint $g(x, y, z) = \sqrt{x^2 + y^2} - z = 0$.

$F(x, y, z, \lambda) = (x - 4)^2 + y^2 + z^2 + \lambda(\sqrt{x^2 + y^2} - z)$

$$F_x = 2(x - 4) + \frac{\lambda x}{\sqrt{x^2 + y^2}} = 0$$

$$F_y = 2y + \frac{\lambda y}{\sqrt{x^2 + y^2}} = 0$$

$$F_z = 2z - \lambda = 0, \qquad\qquad \lambda = 2z$$

$$F_\lambda = \sqrt{x^2 + y^2} - z = 0, \qquad \sqrt{x^2 + y^2} = z$$

$$F_x = 2(x - 4) + 2x = 0, \qquad x = 2$$

$$F_y = 2y + 2y = 0, \qquad\qquad y = 0$$

$$d = \sqrt{(2 - 4)^2 + (0)^2 + (2)^2} = 2\sqrt{2}$$

33. Maximize $P(x, y) = 100x^{0.25}y^{0.75}$ subject to $48x + 36y = 100,000$.

$F(x, y, \lambda) = 100x^{0.25}y^{0.75} + \lambda(48x + 36y - 100,000)$

$F_x = 25x^{-0.75}y^{0.75} + 48\lambda = 0, \qquad \left(\dfrac{y}{x}\right)^{0.75} = -\dfrac{48\lambda}{25}$

$F_y = 75x^{0.25}y^{-0.25} + 36\lambda = 0, \qquad \left(\dfrac{x}{y}\right)^{0.25} = -\dfrac{36\lambda}{75}$

$F_\lambda = 48x + 36y - 100,000 = 0, \qquad \left(\dfrac{y}{x}\right)^{0.75}\left(\dfrac{y}{x}\right)^{0.25} = \left(-\dfrac{48\lambda}{25}\right)\left(-\dfrac{75}{36\lambda}\right), \quad \dfrac{y}{x} = 4, \quad y = 4x$

$48x + 36(4x) = 100,000$

$192x = 100,000$

$x = \dfrac{3125}{6}$

$y = \dfrac{6250}{3}$

$P\left(\dfrac{3125}{6}, \dfrac{6250}{3}\right) \approx 147,314$

34. Maximize $P(x, y) = 100x^{0.6}y^{0.4}$ subject to $48x + 36y = 100,000$.

$F(x, y, \lambda) = 100x^{0.6}y^{0.4} + \lambda(48x + 36y - 100,000)$

$F_x = 60x^{-0.4}y^{0.4} + 48\lambda = 0, \qquad \left(\dfrac{y}{x}\right)^{0.4} = -\dfrac{48\lambda}{60}$

$F_y = 40x^{0.6}y^{-0.6} + 36\lambda = 0, \qquad \left(\dfrac{x}{y}\right)^{0.6} = -\dfrac{36\lambda}{40}$

$F_\lambda = 48x + 36y - 100,000 = 0, \qquad \left(\dfrac{y}{x}\right)^{0.4}\left(\dfrac{y}{x}\right)^{0.6} = \left(-\dfrac{48\lambda}{60}\right)\left(-\dfrac{40}{36\lambda}\right), \quad \dfrac{y}{x} = \dfrac{8}{9}, \quad y = \dfrac{8}{9}x$

$48x + 36\left(\dfrac{8}{9}x\right) = 100,000$

$80x = 100,000$

$x = 1250$

$y = \dfrac{10,000}{9}$

$P\left(1250, \dfrac{10,000}{9}\right) \approx 119,247$

35. Minimize $C = 48x + 36y$ subject to $100x^{0.25}y^{0.75} = 20,000$.

$F(x, y, \lambda) = 48x + 36y + \lambda(x^{0.25}y^{0.75} - 200)$

$F_x = 48 + 0.25\lambda x^{-0.75}y^{0.75} = 0, \qquad \left(\dfrac{y}{x}\right)^{0.75} = -\dfrac{48}{0.25\lambda}$

$F_y = 36 + 0.75\lambda x^{0.25}y^{-0.25} = 0, \qquad \left(\dfrac{x}{y}\right)^{0.25} = -\dfrac{36}{0.75\lambda}$

$F_\lambda = x^{0.25}y^{0.75} - 200 = 0, \qquad \dfrac{y}{x} = \left(-\dfrac{48}{0.25\lambda}\right)\left(-\dfrac{0.75\lambda}{36}\right) = 4, \quad y = 4x$

$F_\lambda = x^{0.25}(4x)^{0.75} - 200 = 0$

$x = \dfrac{200}{4^{0.75}} = \dfrac{200}{2\sqrt{2}} = 50\sqrt{2} \approx 71$

$y = 4x = 200\sqrt{2} \approx 283$

$C\left(50\sqrt{2}, 200\sqrt{2}\right) \approx \$13,576.45$

36. Minimize $C = 48x + 36y$ subject to $100x^{0.6}y^{0.4} = 20,000$.

$F(x, y, \lambda) = 48x + 36y + \lambda(100x^{0.6}y^{0.4} - 20,000)$

$F_x = 48 + \lambda(60x^{-0.4}y^{0.4}) = 0,$ $\left(\dfrac{y}{x}\right)^{0.4} = -\dfrac{48}{60\lambda}$

$F_y = 36 + \lambda(40x^{0.6}y^{-0.6}) = 0,$ $\left(\dfrac{x}{y}\right)^{0.6} = -\dfrac{36}{40\lambda}$

$F_\lambda = 100x^{0.6}y^{0.4} - 20,000 = 0,$ $\left(\dfrac{y}{x}\right)^{0.4}\left(\dfrac{y}{x}\right)^{0.6} = \left(-\dfrac{48}{60\lambda}\right)\left(-\dfrac{40\lambda}{36}\right),$ $\dfrac{y}{x} = \dfrac{8}{9},$ $y = \dfrac{8}{9}x$

$F_\lambda = 100x^{0.6}\left(\dfrac{8}{9}x\right)^{0.4} - 20,000 = 0$

$x = \dfrac{200}{(8/9)^{0.4}} \approx 209.65$

$y = \dfrac{8}{9}\left(\dfrac{200}{(8/9)^{0.4}}\right) \approx 186.35$

$C(209.65, 186.35) = \$16,771.94$

37. Maximize $f(x, y, z) = z$ subject to $x^2 + y^2 + z^2 = 36$ and $2x + y - z = 2$.

$F(x, y, z, \lambda, \mu) = z + \lambda(x^2 + y^2 + z^2 - 36) + \mu(2x + y - z - 2)$

$\left.\begin{array}{l} F_x = 2x\lambda + 2\mu = 0 \\[4pt] F_y = 2y\lambda + \mu = 0 \end{array}\right\} \; x = 2y$

$F_z = 1 + 2x\lambda - \mu = 0$

$F_\lambda = x^2 + y^2 + z^2 - 36 = 0$

$F_\mu = 2x + y - z - 2 = 0, \quad z = 2x + y - 2 = 5y - 2$

Substituting into $F_\lambda = 0$ yields

$(2y)^2 + y^2 + (5y - 2)^2 - 36 = 0$

$15y^2 - 10y - 16 = 0$ or $y = \dfrac{5 \pm \sqrt{265}}{15}$.

Therefore, the desired solution is $x = \dfrac{10 + 2\sqrt{265}}{15}$, $y = \dfrac{5 + \sqrt{265}}{15}$ and $z = \dfrac{-1 + \sqrt{265}}{3}$.

38. Maximize $f(x, y, z) = z$ subject to $x^2 + y^2 - z^2 = 0$ and $x + 2z = 4$.

$F(x, y, z, \lambda, \mu) = z + \lambda(x^2 + y^2 - z^2) + \mu(x + 2z - 4)$

$F_x = 2x\lambda + \mu = 0$

$F_y = 2y\lambda = 0, \quad\quad\quad y = 0$

$F_z = 1 - 2z\lambda + 2\mu = 0$

$F_\lambda = x^2 + y^2 - z^2 = 0$

$F_\mu = x + 2z - 4 = 0, \quad\quad x = -2z + 4$

$(-2z + 4)^2 - z^2 = 0$

$3z^2 - 16z + 16 = 0$

$(3z - 4)(z - 4) = 0$

$z = \frac{4}{3}, \quad z = 4, \quad x = \frac{4}{3}, \quad x = -4$

The maximum value of $f(x, y, z) = z$ occurs at the point $(-4, 0, 4)$.

Chapter 14   Review Exercises

1. $\displaystyle\lim_{(x,y)\to(1,1)} \frac{xy}{x^2+y^2} = \frac{1}{2}$

Continuous except at $(0, 0)$.

2. $\displaystyle\lim_{(x,y)\to(1,1)} \frac{xy}{x^2-y^2}$

Does not exist

Continuous except when $y = \pm x$.

3. $\displaystyle\lim_{(x,y)\to(0,0)} \frac{-4x^2 y}{x^4+y^2} = -4\lim_{r\to0} \frac{r^3\cos^2\theta\sin\theta}{r^4\cos^4\theta + r^2\sin^2\theta}$

$\displaystyle = -4\lim_{r\to0} \frac{r\cos^2\theta\sin\theta}{r^2\cos^4\theta + \sin^2\theta} = -4\lim_{r\to0} \frac{\cos^2\theta\sin\theta}{2r\cos^4\theta} = -4\lim_{r\to0} \frac{\sin\theta}{2r\cos^2\theta} = \frac{-4\sin\theta}{0}$

Does not exist

Continuous except at $(0, 0)$.

4. $\displaystyle\lim_{(x,y)\to(0,0)} \frac{y + xe^{-y^2}}{1+x^2} = \frac{0+0}{1+0} = 0$

Continuous everywhere

5. $f(x, y) = e^{x^2+y^2}$

The level curves are of the form

$c = e^{x^2+y^2}$

$\ln c = x^2 + y^2$.

Thus, the level curves are circles centered at the origin.

6. $f(x, y) = \ln xy$

The level curves are of the form

$c = \ln xy$

$e^c = xy$.

The level curves are hyperbolas.

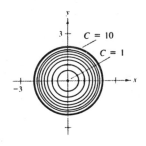

7. $f(x, y) = x^2 - y^2$

The level curves are of the form

$c = x^2 - y^2$

$1 = \dfrac{x^2}{c} - \dfrac{y^2}{c}$.

The level curves are hyperbolas.

8. $f(x, y) = \dfrac{x}{x+y}$

The level curves are of the form

$c = \dfrac{x}{x+y}$

$y = \left(\dfrac{1-c}{c}\right)x$.

The level curves are lines passing through the origin with slope $(1-c)/c$.

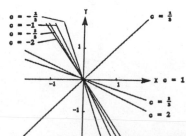

9. $f(x, y) = e^x \cos y$

$$f_x = e^x \cos y$$
$$f_y = -e^x \sin y$$

10. $f(x, y) = \dfrac{xy}{x + y}$

$$f_x = \frac{y(x + y) - xy}{(x + y)^2} = \frac{y^2}{(x + y)^2}$$
$$f_y = \frac{x^2}{(x + y)^2}$$

11. $z = xe^y + ye^x$

$$\frac{\partial z}{\partial x} = e^y + ye^x$$
$$\frac{\partial z}{\partial y} = xe^y + e^x$$

12. $z = \ln(x^2 + y^2 + 1)$

$$\frac{\partial z}{\partial x} = \frac{2x}{x^2 + y^2 + 1}$$
$$\frac{\partial z}{\partial y} = \frac{2y}{x^2 + y^2 + 1}$$

13. $g(x, y) = \dfrac{xy}{x^2 + y^2}$

$$g_x = \frac{y(x^2 + y^2) - xy(2x)}{(x^2 + y^2)^2} = \frac{y(y^2 - x^2)}{(x^2 + y^2)^2}$$
$$g_y = \frac{x(x^2 - y^2)}{(x^2 + y^2)^2}$$

14. $w = \sqrt{x^2 + y^2 + z^2}$

$$\frac{\partial w}{\partial x} = \frac{1}{2}(x^2 + y^2 + z^2)^{-1/2}(2x) = \frac{x}{\sqrt{x^2 + y^2 + z^2}}$$
$$\frac{\partial w}{\partial y} = \frac{y}{\sqrt{x^2 + y^2 + z^2}}$$
$$\frac{\partial w}{\partial z} = \frac{z}{\sqrt{x^2 + y^2 + z^2}}$$

15. $f(x, y, z) = z \arctan \dfrac{y}{x}$

$$f_x = \frac{z}{1 + (y^2/x^2)}\left(-\frac{y}{x^2}\right) = \frac{-yz}{x^2 + y^2}$$
$$f_y = \frac{z}{1 + (y^2/x^2)}\left(\frac{1}{x}\right) = \frac{xz}{x^2 + y^2}$$
$$f_z = \arctan \frac{y}{x}$$

16. $f(x, y, z) = \dfrac{1}{\sqrt{1 - x^2 - y^2 - z^2}}$

$$f_x = -\frac{1}{2}(1 - x^2 - y^2 - z^2)^{-3/2}(-2x)$$
$$= \frac{x}{(1 - x^2 - y^2 - z^2)^{3/2}}$$
$$f_y = \frac{y}{(1 - x^2 - y^2 - z^2)^{3/2}}$$
$$f_z = \frac{z}{(1 - x^2 - y^2 - z^2)^{3/2}}$$

17. $u(x, t) = ce^{-n^2 t} \sin(nx)$

$$\frac{\partial u}{\partial x} = cne^{-n^2 t} \cos(nx)$$
$$\frac{\partial u}{\partial t} = -cn^2 e^{-n^2 t} \sin(nx)$$

18. $u(x, t) = c(\sin akx) \cos kt$

$$\frac{\partial u}{\partial x} = akc(\cos akx) \cos kt$$
$$\frac{\partial u}{\partial t} = -kc(\sin akx) \sin kt$$

19.
$$x^2 y - 2xyz - xz - z^2 = 0$$
$$2xy - 2xy\frac{\partial z}{\partial x} - 2yz - x\frac{\partial z}{\partial x} - z - 2z\frac{\partial z}{\partial x} = 0$$
$$\frac{\partial z}{\partial x} = -\frac{2yz + z - 2xy}{2xy + x + 2z}$$
$$x^2 - 2xy\frac{\partial z}{\partial y} - 2xz - x\frac{\partial z}{\partial y} - 2z\frac{\partial z}{\partial y} = 0$$
$$\frac{\partial z}{\partial y} = \frac{x^2 - 2xz}{2xy + x + 2z}$$

20.
$$xz^2 - y\sin z = 0$$
$$2xz\frac{\partial z}{\partial x} + z^2 - y\cos z\frac{\partial z}{\partial x} = 0$$
$$\frac{\partial z}{\partial x} = \frac{z^2}{y\cos z - 2xz}$$
$$2xz\frac{\partial z}{\partial y} - y\cos z\frac{\partial z}{\partial y} - \sin z = 0$$
$$\frac{\partial z}{\partial y} = \frac{\sin z}{2xz - y\cos z}$$

21. $f(x,\ y) = 3x^2 - xy + 2y^3$
$$f_x = 6x - y$$
$$f_y = -x + 6y^2$$
$$f_{xx} = 6$$
$$f_{yy} = 12y$$
$$f_{xy} = -1$$
$$f_{yx} = -1$$

22. $h(x,\ y) = \dfrac{x}{x+y}$
$$h_x = \frac{y}{(x+y)^2}$$
$$h_y = \frac{-x}{(x+y)^2}$$
$$h_{xx} = \frac{-2y}{(x+y)^3}$$
$$h_{yy} = \frac{2x}{(x+y)^3}$$
$$h_{xy} = \frac{(x+y) - 2y(x+y)}{(x+y)^4} = \frac{x-y}{(x+y)^3}$$
$$h_{yx} = \frac{-(x+y) + 2x(x+y)}{(x+y)^4} = \frac{x-y}{(x+y)^3}$$

23. $h(x,\ y) = x\sin y + y\cos x$
$$h_x = \sin y - y\sin x$$
$$h_y = x\cos y + \cos x$$
$$h_{xx} = -y\cos x$$
$$h_{yy} = -x\sin y$$
$$h_{xy} = \cos y - \sin x$$
$$h_{yx} = \cos y - \sin x$$

24. $g(x,\ y) = \cos(x - 2y)$
$$g_x = -\sin(x - 2y)$$
$$g_y = 2\sin(x - 2y)$$
$$g_{xx} = -\cos(x - 2y)$$
$$g_{yy} = -4\cos(x - 2y)$$
$$g_{xy} = 2\cos(x - 2y)$$
$$g_{yx} = 2\cos(x - 2y)$$

25. $z = x^2 - y^2$
$$\frac{\partial z}{\partial x} = 2x$$
$$\frac{\partial^2 z}{\partial x^2} = 2$$
$$\frac{\partial z}{\partial y} = -2y$$
$$\frac{\partial^2 z}{\partial y^2} = -2$$

Therefore, $\dfrac{\partial^2 z}{\partial x^2} + \dfrac{\partial^2 z}{\partial y^2} = 0$.

26. $z = x^3 - 3xy^2$
$$\frac{\partial z}{\partial x} = 3x^2 - 3y^2$$
$$\frac{\partial^2 z}{\partial x^2} = 6x$$
$$\frac{\partial z}{\partial y} = -6xy$$
$$\frac{\partial^2 z}{\partial y^2} = -6x$$

Therefore, $\dfrac{\partial^2 z}{\partial x^2} + \dfrac{\partial^2 z}{\partial y^2} = 0$.

27. $z = \dfrac{y}{x^2 + y^2}$

$$\frac{\partial z}{\partial x} = \frac{-2xy}{(x^2 + y^2)^2}$$

$$\frac{\partial^2 z}{\partial x^2} = -2y\left[\frac{-4x^2}{(x^2 + y^2)^3} + \frac{1}{(x^2 + y^2)^2}\right]$$

$$= 2y\frac{3x^2 - y^2}{(x^2 + y^2)^3}$$

$$\frac{\partial z}{\partial y} = \frac{(x^2 + y^2) - 2y}{(x^2 + y^2)^2} = \frac{x^2 - y^2}{(x^2 + y^2)^2}$$

$$\frac{\partial^2 z}{\partial y^2} = \frac{(x^2 + y^2)^2(-2y) - 2(x^2 - y^2)(x^2 + y^2)(2y)}{(x^2 + y^2)^4}$$

$$= -2y\frac{3x^2 - y^2}{(x^2 + y^2)^3}$$

Therefore, $\dfrac{\partial^2 z}{\partial x^2} + \dfrac{\partial^2 z}{\partial y^2} = 0.$

28. $z = e^x \sin y$

$$\frac{\partial z}{\partial x} = e^x \sin y$$

$$\frac{\partial^2 z}{\partial x^2} = e^x \sin y$$

$$\frac{\partial z}{\partial y} = e^x \cos y$$

$$\frac{\partial^2 z}{\partial y^2} = -e^x \sin y$$

Therefore, $\dfrac{\partial^2 z}{\partial x^2} + \dfrac{\partial^2 z}{\partial y^2} = 0.$

29. $z = x \sin \dfrac{y}{x}$

$$dz = \frac{\partial z}{\partial x}\,dx + \frac{\partial z}{\partial y}\,dy = \left(\sin\frac{y}{x} - \frac{y}{x}\cos\frac{y}{x}\right)dx + \left(\cos\frac{y}{x}\right)dy$$

30. $z = \dfrac{xy}{\sqrt{x^2 + y^2}}$

$$dz = \frac{\partial z}{\partial x}\,dx + \frac{\partial z}{\partial y}\,dy$$

$$= \left[\frac{\sqrt{x^2 + y^2}\,y - xy(x/\sqrt{x^2 + y^2})}{x^2 + y^2}\right]dx + \left[\frac{\sqrt{x^2 + y^2}\,x - xy(y/\sqrt{x^2 + y^2})}{x^2 + y^2}\right]dy$$

$$= \frac{y^3}{(x^2 + y^2)^{3/2}}\,dx + \frac{x^3}{(x^2 + y^2)^{3/2}}\,dy$$

31. $u = x^2 + y^2 + z^2, \quad x = r\cos t, \quad y = r\sin t, \quad z = t$

$$\frac{\partial u}{\partial r} = \frac{\partial u}{\partial x}\frac{\partial x}{\partial r} + \frac{\partial u}{\partial y}\frac{\partial y}{\partial r} + \frac{\partial u}{\partial z}\frac{\partial z}{\partial r} = 2x\cos t + 2y\sin t + 2z(0) = 2(r\cos^2 t + r\sin^2 t) = 2r$$

$$\frac{\partial u}{\partial t} = \frac{\partial u}{\partial x}\frac{\partial x}{\partial t} + \frac{\partial u}{\partial y}\frac{\partial y}{\partial t} + \frac{\partial u}{\partial z}\frac{\partial z}{\partial t} = 2x(-r\sin t) + 2y(r\cos t) + 2z = 2(-r^2\sin t\cos t + r^2\sin t\cos t) + 2t = 2t$$

Check: $u(r,\ t) = r^2\cos^2 t + r^2\sin^2 t + t^2 = r^2 + t^2$

$$\frac{\partial u}{\partial r} = 2r$$

$$\frac{\partial u}{\partial t} = 2t$$

32. $u = y^2 - x, \quad x = \cos t, \quad y = \sin t$

$$\frac{du}{dt} = \frac{\partial u}{\partial x}\frac{\partial x}{\partial t} + \frac{\partial u}{\partial y}\frac{\partial y}{\partial t} = -1(-\sin t) + 2y(\cos t) = \sin t + 2(\sin t)\cos t = \sin t(1 + 2\cos t)$$

Check: $u = \sin^2 t - \cos t$

$$\frac{du}{dt} = 2\sin t\cos t + \sin t = \sin t(1 + 2\cos t)$$

33. $f(x, y) = x^2 y$

$$\nabla f = 2xy\mathbf{i} + x^2\mathbf{j}$$

$$\nabla f(2, 1) = 4\mathbf{i} + 4\mathbf{j}$$

$$\mathbf{u} = \frac{1}{\sqrt{2}}\mathbf{v} = \frac{\sqrt{2}}{2}\mathbf{i} - \frac{\sqrt{2}}{2}\mathbf{j}$$

$$D_{\mathbf{u}}f(2, 1) = \nabla f(2, 1) \cdot \mathbf{u} = 2\sqrt{2} - 2\sqrt{2} = 0$$

34. $f(x, y) = \frac{1}{4}y^2 - x^2$

$$\nabla f = -2x\mathbf{i} + \frac{1}{2}y\mathbf{j}$$

$$\nabla f(1, 4) = -2\mathbf{i} + 2\mathbf{j}$$

$$\mathbf{u} = \frac{1}{\sqrt{5}}\mathbf{v} = \frac{2\sqrt{5}}{5}\mathbf{i} + \frac{\sqrt{5}}{5}\mathbf{j}$$

$$D_{\mathbf{u}}f(1, 4) = \nabla f(1, 4) \cdot \mathbf{u} = -\frac{4\sqrt{5}}{5} + \frac{2\sqrt{5}}{5} = -\frac{2\sqrt{5}}{5}$$

35. $f(x, y, z) = y^2 + xz$

$$\nabla f = z\mathbf{i} + 2y\mathbf{j} + x\mathbf{k}$$

$$\nabla f(1, 2, 2) = 2\mathbf{i} + 4\mathbf{j} + \mathbf{k}$$

$$\mathbf{u} = \frac{1}{3}\mathbf{v} = \frac{2}{3}\mathbf{i} - \frac{1}{3}\mathbf{j} + \frac{2}{3}\mathbf{k}$$

$$D_{\mathbf{u}}f(1, 2, 2) = \nabla f(1, 2, 2) \cdot \mathbf{u} = \frac{4}{3} - \frac{4}{3} + \frac{2}{3} = \frac{2}{3}$$

36. $f(x, y, z) = 6x^2 + 3xy - 4y^2z$

$$\nabla f = (12x + 3y)\mathbf{i} + (3x - 8yz)\mathbf{j} + (-4y^2)\mathbf{k}$$

$$\nabla f(1, 0, 1) = 12\mathbf{i} + 3\mathbf{j}$$

$$\mathbf{u} = \frac{1}{\sqrt{3}}\mathbf{v}$$

$$= \frac{\sqrt{3}}{3}\mathbf{i} + \frac{\sqrt{3}}{3}\mathbf{j} - \frac{\sqrt{3}}{3}\mathbf{k}$$

$$D_{\mathbf{u}}f(1, 0, 1) = \nabla f(1, 0, 1) \cdot \mathbf{u} = 4\sqrt{3} + \sqrt{3} + 0 = 5\sqrt{3}$$

37. $z = \frac{y}{x^2 + y^2}$

$$\nabla z = -\frac{2xy}{(x^2 + y^2)^2}\mathbf{i} + \frac{x^2 - y^2}{(x^2 + y^2)^2}\mathbf{j}$$

$$\nabla z(1, 1) = -\frac{1}{2}\mathbf{i} = \left\langle -\frac{1}{2}, 0 \right\rangle$$

$$\|\nabla z(1, 1)\| = \frac{1}{2}$$

38. $z = \frac{x^2}{x - y}$

$$\nabla z = \frac{x^2 - 2xy}{(x - y)^2}\mathbf{i} + \frac{x^2}{(x - y)^2}\mathbf{j}$$

$$\nabla z(2, 1) = 4\mathbf{j}$$

$$\|\nabla z(2, 1)\| = 4$$

39. $z = e^{-x}\cos y$

$$\nabla z = -e^{-x}\cos y\mathbf{i} - e^{-x}\sin y\mathbf{j}$$

$$\nabla z\left(0, \frac{\pi}{4}\right) = -\frac{\sqrt{2}}{2}\mathbf{i} - \frac{\sqrt{2}}{2}\mathbf{j} = \left\langle -\frac{\sqrt{2}}{2}, -\frac{\sqrt{2}}{2} \right\rangle$$

$$\left\|\nabla z\left(0, \frac{\pi}{4}\right)\right\| = 1$$

40. $z = x^2 y$

$$\nabla z = 2xy\mathbf{i} + x^2\mathbf{j}$$

$$\nabla z(2, 1) = 4\mathbf{i} + 4\mathbf{j}$$

$$\|\nabla z(2, 1)\| = 4\sqrt{2}$$

41. $F(x, y, z) = x^2 y - z = 0$

$$\nabla F = 2xy\mathbf{i} + x^2\mathbf{j} - \mathbf{k}$$

$$\nabla F(2, 1, 4) = 4\mathbf{i} + 4\mathbf{j} - \mathbf{k}$$

Therefore, the equation of the tangent plane is $4(x - 2) + 4(y - 1) - (z - 4) = 0$ or $4x + 4y - z = 8$, and the equation of the normal line is $\dfrac{x - 2}{4} = \dfrac{y - 1}{4} = \dfrac{z - 4}{-1}$.

42. $F(x, y, z) = y^2 + z^2 - 25 = 0$

$$\nabla F = 2y\mathbf{j} + 2z\mathbf{k}$$

$$\nabla F(2, 3, 4) = 6\mathbf{j} + 8\mathbf{k} = 2(3\mathbf{j} + 4\mathbf{k})$$

Therefore, the equation of the tangent plane is $3(y - 3) + 4(z - 4) = 0$ or $3y + 4z = 25$, and the equation of the normal line is $x = 2$, $\dfrac{y - 3}{3} = \dfrac{z - 4}{4}$.

43. $F(x, y, z) = x^2 + y^2 - 4x + 6y + z + 9 = 0$

$$\nabla F = (2x - 4)\mathbf{i} + (2y + 6)\mathbf{j} + \mathbf{k}$$

$$\nabla F(2, -3, 4) = \mathbf{k}$$

Therefore, the equation of the tangent plane is $z - 4 = 0$ or $z = 4$ and the equation of the normal line is $x = 2$, $y = -3$, $z = 4 + t$.

44. $F(x, y, z) = x^2 + y^2 + z^2 - 9 = 0$

$$\nabla F = 2x\mathbf{i} + 2y\mathbf{j} + 2z\mathbf{k}$$

$$\nabla F(1, 2, 2) = 2\mathbf{i} + 4\mathbf{j} + 4\mathbf{k} = 2(\mathbf{i} + 2\mathbf{j} + 2\mathbf{k})$$

Therefore, the equation of the tangent plane is $(x - 1) + 2(y - 2) + 2(z - 2) = 0$ or $x + 2y + 2z = 9$, and the equation of the normal line is $\dfrac{x - 1}{1} = \dfrac{y - 2}{2} = \dfrac{z - 2}{2}$.

45. $F(x, y, z) = x^2 - y^2 - z = 0$

$G(x, y, z) = 3 - z = 0$

$$\nabla F = 2x\mathbf{i} - 2y\mathbf{j} - \mathbf{k}$$

$$\nabla G = -\mathbf{k}$$

$$\nabla F(2, 1, 3) = 4\mathbf{i} - 2\mathbf{j} - \mathbf{k}$$

$$\nabla F \times \nabla G = \begin{vmatrix} \mathbf{i} & \mathbf{j} & \mathbf{k} \\ 4 & -2 & -1 \\ 0 & 0 & -1 \end{vmatrix} = 2(\mathbf{i} + 2\mathbf{j})$$

Therefore, the equation of the tangent line is $\dfrac{x - 2}{1} = \dfrac{y - 1}{2}$, $z = 3$.

46. $F(x, y, z) = y^2 + z - 25 = 0$

$G(x, y, z) = x - y = 0$

$$\nabla F = 2y\mathbf{i} + \mathbf{k}$$

$$\nabla G = \mathbf{i} - \mathbf{j}$$

$$\nabla F(4, 4, 9) = 8\mathbf{i} + \mathbf{k}$$

$$\nabla F \times \nabla G = \begin{vmatrix} \mathbf{i} & \mathbf{j} & \mathbf{k} \\ 8 & 0 & 1 \\ 1 & -1 & 0 \end{vmatrix} = \mathbf{i} + \mathbf{j} - 8\mathbf{k}$$

Therefore, the equation of the tangent line is $\dfrac{x - 4}{1} = \dfrac{y - 4}{1} = \dfrac{z - 9}{-8}$.

47. $f(x, y) = x^3 - 3xy + y^2$

$$f_x = 3x^2 - 3y = 3(x^2 - y) = 0$$
$$f_y = -3x + 2y = 0$$
$$f_{xx} = 6x$$
$$f_{yy} = 2$$
$$f_{xy} = -3$$

From $f_x = 0$, we have $y = x^2$. Substituting this into $f_y = 0$, we have $-3x + 2x^2 = x(2x - 3) = 0$. Thus, $x = 0$ or $\frac{3}{2}$. At the critical point $(0, 0)$, $f_{xx}f_{yy} - (f_{xy})^2 < 0$. Therefore, $(0, 0, 0)$ is a saddle point. At the critical point $\left(\frac{3}{2}, \frac{9}{4}\right)$, $f_{xx}f_{yy} - (f_{xy})^2 > 0$ and $f_{xx} > 0$. Therefore, $\left(\frac{3}{2}, \frac{9}{4}, -\frac{27}{16}\right)$ is a relative minimum.

48.
$$f(x, y) = 2x^2 + 6xy + 9y^2 + 8x + 14$$
$$f_x = 4x + 6y + 8 = 0$$
$$f_y = 6x + 18y = 0, \quad x = -3y$$
$$4(-3y) + 6y = -8 \Rightarrow y = \frac{4}{3}, \quad x = -4$$
$$f_{xx} = 4$$
$$f_{yy} = 18$$
$$f_{xy} = 6$$
$$f_{xx}f_{yy} - (f_{xy})^2 = 4(18) - (6)^2 = 36 > 0$$

Therefore, $\left(-4, \frac{4}{3}, -2\right)$ is a relative minimum.

49. $f(x, y) = xy + \dfrac{1}{x} + \dfrac{1}{y}$

$$f_x = y - \frac{1}{x^2} = 0, \quad x^2 y = 1$$
$$f_y = x - \frac{1}{y^2} = 0, \quad xy^2 = 1$$

Thus, $x^2 y = xy^2$ or $x = y$ and substitution yields the critical point $(1, 1)$.

$$f_{xx} = \frac{2}{x^3}$$
$$f_{xy} = 1$$
$$f_{yy} = \frac{2}{y^3}$$

At the critical point $(1, 1)$, $f_{xx} = 2 > 0$ and $f_{xx}f_{yy} - (f_{xy})^2 = 3 > 0$. Thus, $(1, 1, 3)$ is a relative minimum.

50. $z = 50(x + y) - (0.1x^3 + 20x + 150) - (0.05y^3 + 20.6y + 125)$

$z_x = 50 - 0.3x^2 - 20 = 0, \quad x = \pm 10$

$z_y = 50 - 0.15y^2 - 20.6 = 0, \quad y = \pm 14$

Critical points: $(10, 14), (10, -14), (-10, 14), (-10, -14)$

$z_{xx} = -0.6x, \quad z_{yy} = -0.3y, \quad z_{xy} = 0$

At $(10, 14), \quad z_{xx}z_{yy} - (z_{xy})^2 = (-6)(-4.2) - 0^2 > 0, \quad z_{xx} < 0.$
$\quad$ $(10, 14, 199.4)$ is a relative maximum.

At $(10, -14), \quad z_{xx}z_{yy} - (z_{xy})^2 = (-6)(4.2) - 0^2 < 0.$
$\quad$ $(10, -14, -349.4)$ is a saddle point.

At $(-10, 14), \quad z_{xx}z_{yy} - (z_{xy})^2 = (6)(-4.2) - 0^2 < 0.$
$\quad$ $(-10, 14, -200.6)$ is a saddle point.

At $(-10, -14), \quad z_{xx}z_{yy} - (x_{xy})^2 = (6)(4.2) - 0^2 > 0, \quad z_{xx} > 0.$
$\quad$ $(-10, -14, -749.4)$ is a relative minimum.

51. $z = x^2y$ subject to $x + 2y - 2 = 0.$

$F(x, y, \lambda) = x^2y + \lambda(x + 2y - 2)$

$\left.\begin{array}{l} F_x = 2xy + \lambda = 0 \\ F_y = x^2 + 2\lambda = 0 \end{array}\right\} \quad \dfrac{x^2}{2} = 2xy \Rightarrow x = 0 \text{ or } x = 4y$

$F_\lambda = x + 2y - 2 = 0$

If $x = 0, \quad y = 1.$ If $x = 4y, \quad 6y = 2, \quad y = \frac{1}{3}, \quad x = \frac{4}{3}.$

$(0, 1, 0)$ is a relative minimum.

$\left(\frac{4}{3}, \frac{1}{3}, \frac{16}{27}\right)$ is a relative maximum.

52. $w = xy + yz + xz$ subject to $x + y + z - 1 = 0.$

$F(x, y, z, \lambda) = xy + yz + xz + \lambda(x + y + z - 1)$

$\left.\begin{array}{l} F_x = y + z + \lambda = 0 \\ F_y = x + z + \lambda = 0 \\ F_z = x + y + \lambda = 0 \end{array}\right\} \left.\begin{array}{l} x = y \\ y = z \end{array}\right\} \ x = y = z$

$F_\lambda = x + y + z - 1 = 0$

$\quad 3x = 1 \Rightarrow x = y = z = \frac{1}{3}$

$\left(\frac{1}{3}, \frac{1}{3}, \frac{1}{3}, \frac{1}{3}\right)$ is a relative maximum.

53. $z^2 = x^2 + y^2$

$2z\, dz = 2x\, dx + 2y\, dy$

$dz = \dfrac{x}{z}\, dx + \dfrac{y}{z}\, dy = \dfrac{5}{13}\left(\dfrac{1}{16}\right) + \dfrac{12}{13}\left(\dfrac{1}{16}\right) = \dfrac{17}{208} \approx 0.082 \text{ in.}$

Percentage error: $\dfrac{dz}{z} = \dfrac{17/208}{13} = 0.0063 = 0.63\%$

54. From the accompanying figure we observe

$$\tan\theta = \frac{h}{x} \text{ or } h = x\tan\theta$$

$$dh = \frac{\partial h}{\partial x}\,dx + \frac{\partial h}{\partial\theta}\,d\theta$$

$$= \tan\theta\,dx + x\sec^2\theta\,d\theta.$$

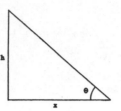

Letting $x = 100$,

$$dx = \pm\frac{1}{2}, \quad \theta = \frac{11\pi}{60}, \text{ and } d\theta = \frac{\pi}{180}.$$

(Note that we express the measurement of the angle in radians.)

The maximum error is approximately

$$dh = \tan\left(\frac{11\pi}{60}\right)\left(\pm\frac{1}{2}\right) + 100\sec^2\left(\frac{11\pi}{60}\right)\left(\frac{\pi}{180}\right) \approx \pm 2.8 \text{ ft.}$$

55. $V = \frac{1}{3}\pi r^2 h$

$$dV = \frac{2}{3}\pi rh\,dr + \frac{1}{3}\pi r^2\,dh = \frac{2}{3}\pi(2)(5)\left(\pm\frac{1}{8}\right) + \frac{1}{3}\pi(2)^2\left(\pm\frac{1}{8}\right) = \pm\frac{5}{6}\pi \pm \frac{1}{6}\pi = \pm\pi \text{ in}^3$$

56. $A = \pi r\sqrt{r^2 + h^2}$

$$dA = \left(\pi\sqrt{r^2 + h^2} + \frac{\pi r^2}{\sqrt{r^2 + h^2}}\right)dr + \frac{\pi rh}{\sqrt{r^2 + h^2}}\,dh$$

$$= \frac{\pi(2r^2 + h^2)}{\sqrt{r^2 + h^2}}\,dr + \frac{\pi rh}{\sqrt{r^2 + h^2}}\,dh = \frac{\pi(8 + 25)}{\sqrt{29}}\left(\pm\frac{1}{8}\right) + \frac{10\pi}{\sqrt{29}}\left(\pm\frac{1}{8}\right) = \pm\frac{43\pi}{8\sqrt{29}}$$

57. $P(x_1,\ x_2) = R - C_1 - C_2$

$$= [225 - 0.4(x_1 + x_2)](x_1 + x_2) - (0.05x_1^2 + 15x_1 + 5400) - (0.03x_2^2 + 15x_2 + 6100)$$

$$= -0.45x_1^2 - 0.43x_2^2 - 0.8x_1 x_2 + 210x_1 + 210x_2 - 11500$$

$$P_{x_1} = -0.9x_1 - 0.8x_2 + 210 = 0$$

$$0.9x_1 + 0.8x_2 = 210$$

$$P_{x_2} = -0.86x_2 - 0.8x_1 + 210 = 0$$

$$0.8x_1 + 0.86x_2 = 210$$

Solving this system yields $x_1 \approx 94$ and $x_2 \approx 157$.

$$P_{x_1 x_1} = -0.9$$

$$P_{x_1 x_2} = -0.8$$

$$P_{x_2 x_2} = -0.86$$

$$P_{x_1 x_1} < 0$$

$$P_{x_1 x_1}P_{x_2 x_2} - (P_{x_1 x_2})^2 > 0$$

Therefore, profit is maximum when $x_1 \approx 94$ and $x_2 \approx 157$.

58. Minimize $C = 0.25x_1^2 + 10x_1 + 0.15x_2^2 + 12x_2$ subject to $x_1 + x_2 = 1000$.

$F(x_1, x_2, \lambda) = 0.25x_1^2 + 10x_1 + 0.15x_2^2 + 12x_2 + \lambda(x_1 + x_2 - 1000)$

$\left. \begin{aligned} F_{x_1} &= 0.5x_1 + 10 + \lambda = 0 \\ F_{x_2} &= 0.3x_2 + 12 + \lambda = 0 \end{aligned} \right\} \quad 0.5x_1 - 0.3x_2 = 2$

$F_\lambda = x_1 + x_2 - 1000 = 0$

$$5x_1 - 3x_2 = 20$$
$$\underline{3x_1 + 3x_2 = 3000}$$
$$8x_1 \qquad = 3020$$

$$x_1 = 377.5$$
$$x_2 = 622.5$$

59. Maximize $f(x, y) = 4x + xy + 2y$ subject to $20x + 4y = 2000$.

$F(x, y, \lambda) = 4x + xy + 2y + \lambda(20x + 4y - 2000)$

$\left. \begin{aligned} F_x &= 4 + y + 20\lambda = 0 \\ F_y &= x + 2 + 4\lambda = 0 \end{aligned} \right\} \quad 5x - y = -6$

$F_\lambda = 20x + 4y - 2000 = 0$

$$5x + y = 500$$
$$\underline{5x - y = -6}$$
$$10x \qquad = 494$$

$$x = 49.4$$
$$y = 253$$

60. Maximize $f(\sigma, \beta, \gamma) = (\sin \sigma)(\sin \beta)(\sin \gamma)$ subject to $\sigma + \beta + \gamma = 180°$.

$F(\sigma, \beta, \gamma, \lambda) = (\sin \sigma)(\sin \beta)(\sin \gamma) + \lambda(\sigma + \beta + \gamma - 180°)$

$\left. \begin{aligned} F_\sigma &= (\cos \sigma)(\sin \beta)(\sin \gamma) + \lambda = 0 \\ F_\beta &= (\sin \sigma)(\cos \beta)(\sin \gamma) + \lambda = 0 \\ F_\gamma &= (\sin \sigma)(\sin \beta)(\cos \gamma) + \lambda = 0 \end{aligned} \right\} \quad \begin{aligned} \cos \sigma \sin \beta &= \sin \sigma \cos \beta \\ \tan \beta &= \tan \sigma \\ \cos \beta \sin \gamma &= \sin \beta \cos \gamma \\ \tan \gamma &= \tan \beta \end{aligned}$

$F_\lambda = \sigma + \beta + \gamma - 180° = 0$

Since $\tan \sigma = \tan \beta = \tan \gamma$ and $\sigma, \beta, \gamma < 180$, then

$$\sigma = \beta = \gamma$$
$$3\sigma = 180°$$
$$\sigma = \beta = \gamma = 60°.$$

Therefore, the triangle is equilateral.

CHAPTER 15
Multiple Integration

Section 15.1 Iterated Integrals and Area in the Plane

1. $\displaystyle\int_0^x (2x - y)\, dy = \left(2xy - \frac{1}{2}y^2\right)\Big]_0^x = \frac{3}{2}x^2$

2. $\displaystyle\int_x^{x^2} \frac{y}{x}\, dy = \frac{1}{2}\frac{y^2}{x}\Big]_x^{x^2} = \frac{1}{2}\left(\frac{x^4}{x} - \frac{x^2}{x}\right) = \frac{x}{2}(x^2 - 1)$

3. $\displaystyle\int_1^{2y} \frac{y}{x}\, dx = y\ln x\Big]_1^{2y} = y\ln 2y - 0 = y\ln 2y$

4. $\displaystyle\int_0^{\cos y} y\, dx = yx\Big]_0^{\cos y} = y\cos y$

5. $\displaystyle\int_0^{\sqrt{4-x^2}} x^2 y\, dy = \frac{1}{2}x^2 y^2\Big]_0^{\sqrt{4-x^2}} = \frac{4x^2 - x^4}{2}$

6. $\displaystyle\int_{x^2}^{\sqrt{x}} (x^2 + y^2)\, dy = \left(x^2 y + \frac{1}{3}y^3\right)\Big]_{x^2}^{\sqrt{x}} = x^2\sqrt{x} + \frac{1}{3}(\sqrt{x})^3 - x^4 - \frac{1}{3}x^6 = x^{5/2} + \frac{1}{3}x^{3/2} - x^4 - \frac{1}{3}x^6$

7. $\displaystyle\int_{e^y}^{y} \frac{y\ln x}{x}\, dx = \frac{1}{2}y\ln^2 x\Big]_{e^y}^{y} = \frac{y}{2}(\ln^2 y - y^2)$

8. $\displaystyle\int_{-\sqrt{1-y^2}}^{\sqrt{1-y^2}} (x^2 + y^2)\, dx = \left(\frac{1}{3}x^3 + y^2 x\right)\Big]_{-\sqrt{1-y^2}}^{\sqrt{1-y^2}} = 2\left[\frac{1}{3}(1-y^2)^{3/2} + y^2(1-y^2)^{1/2}\right] = \frac{2\sqrt{1-y^2}}{3}(1 + 2y^2)$

9. $\displaystyle\int_0^{x^3} ye^{-y/x}\, dy = (-xye^{-y/x})\Big]_0^{x^3} + x\int_0^{x^3} e^{-y/x}\, dy = -x^4 e^{-x^2} - \left[x^2 e^{-y/x}\right]_0^{x^3} = x^2(1 - e^{-x^2} - x^2 e^{-x^2})$

$u = y, \quad du = dy, \quad dv = e^{-y/x}\, dy, \quad v = -xe^{-y/x}$

10. $\displaystyle\int_y^{\pi/2} \sin^3 x \cos y\, dx = \int_y^{\pi/2} (1 - \cos^2 x)\sin x \cos y\, dx = \left(-\cos x + \frac{1}{3}\cos^3 x\right)\cos y\Big]_y^{\pi/2}$

$$= \left(\cos y - \frac{1}{3}\cos^3 y\right)\cos y$$

11. $\displaystyle\int_0^1\int_0^2 (x + y)\, dy\, dx = \int_0^1 \left(xy + \frac{1}{2}y^2\right)\Big]_0^2\, dx = \int_0^1 (2x + 2)\, dx = (x^2 + 2x)\Big]_0^1 = 3$

12. $\displaystyle\int_0^1\int_0^x \sqrt{1-x^2}\, dy\, dx = \int_0^1 y\sqrt{1-x^2}\Big]_0^x\, dx = \int_0^1 x\sqrt{1-x^2}\, dx = -\frac{1}{2}\left(\frac{2}{3}\right)(1-x^2)^{3/2}\Big]_0^1 = \frac{1}{3}$

13. $\displaystyle\int_1^2\int_0^4 (x^2 - 2y^2 + 1)\, dx\, dy = \int_1^2 \left(\frac{1}{3}x^3 - 2xy^2 + x\right)\Big]_0^4\, dy$

$$= \int_1^2 \left(\frac{64}{3} - 8y^2 + 4\right)\, dy = \frac{4}{3}\int_1^2 (19 - 6y^2)\, dy = \frac{4}{3}(19y - 2y^3)\Big]_1^2 = \frac{20}{3}$$

14. $\displaystyle\int_0^1 \int_y^{2y} (1 + 2x^2 + 2y^2)\, dx\, dy = \int_0^1 \left(x + \frac{2}{3}x^3 + 2xy^2 \right) \Big]_y^{2y} dy$

$$= \int_0^1 \left(y + \frac{20}{3}y^3 \right) dy = \left(\frac{1}{2}y^2 + \frac{5}{3}y^4 \right) \Big]_0^1 = \frac{13}{6}$$

15. $\displaystyle\int_0^1 \int_0^{\sqrt{1-y^2}} (x + y)\, dx\, dy = \int_0^1 \left(\frac{1}{2}x^2 + xy \right) \Big]_0^{\sqrt{1-y^2}} dy$

$$= \int_0^1 \left[\frac{1}{2}(1 - y^2) + y\sqrt{1 - y^2} \right] dy = \left[\frac{1}{2}y - \frac{1}{6}y^3 - \frac{1}{2}\left(\frac{2}{3}\right)(1 - y^2)^{3/2} \right]_0^1 = \frac{2}{3}$$

16. $\displaystyle\int_0^2 \int_{3y^2-6y}^{2y-y^2} 3y\, dx\, dy = \int_0^2 3xy \Big]_{3y^2-6y}^{2y-y^2} dy = 3\int_0^2 (8y^2 - 4y^3)\, dy = 3\left(\frac{8}{3}y^3 - y^4 \right) \Big]_0^2 = 16$

17. $\displaystyle\int_0^2 \int_0^{\sqrt{4-y^2}} \frac{2}{\sqrt{4 - y^2}}\, dx\, dy = \int_0^2 \frac{2x}{\sqrt{4 - y^2}} \Big]_0^{\sqrt{4-y^2}} dy = \int_0^2 2\, dy = 2y \Big]_0^2 = 4$

18. $\displaystyle\int_0^{\pi/2} \int_0^{2\cos\theta} r\, dr\, d\theta = \int_0^{\pi/2} \frac{r^2}{2} \Big]_0^{2\cos\theta} d\theta = \int_0^{\pi/2} 2\cos^2\theta\, d\theta = \left(\theta - \frac{1}{2}\sin 2\theta \right) \Big]_0^{\pi/2} = \frac{\pi}{2}$

19. $\displaystyle\int_0^{\pi/2} \int_0^{\sin\theta} \theta r\, dr\, d\theta = \int_0^{\pi/2} \theta \frac{r^2}{2} \Big]_0^{\sin\theta} d\theta = \int_0^{\pi/2} \frac{1}{2}\theta \sin^2\theta\, d\theta = \frac{1}{4}\int_0^{\pi/2} (\theta - \theta\cos 2\theta)\, d\theta$

$$= \frac{1}{4}\left[\frac{\theta^2}{2} - \left(\frac{1}{4}\cos 2\theta + \frac{\theta}{2}\sin 2\theta \right) \right]_0^{\pi/2}$$

$$= \frac{\pi^2}{32} + \frac{1}{8}$$

20. $\displaystyle\int_0^{\pi/4} \int_0^{\cos\theta} 3r^2 \sin\theta\, dr\, d\theta = \int_0^{\pi/4} r^3 \sin\theta \Big]_0^{\cos\theta} d\theta$

$$= \int_0^{\pi/4} \cos^3\theta \sin\theta\, d\theta = -\frac{\cos^4\theta}{4} \Big]_0^{\pi/4} = -\frac{1}{4}\left[\left(\frac{1}{\sqrt{2}}\right)^4 - 1 \right] = \frac{3}{16}$$

21. $\displaystyle\int_1^\infty \int_0^{1/x} y\, dy\, dx = \int_1^\infty \frac{y^2}{2} \Big]_0^{1/x} dx = \frac{1}{2}\int_1^\infty \frac{1}{x^2}\, dx = -\frac{1}{2x} \Big]_1^\infty = 0 + \frac{1}{2} = \frac{1}{2}$

22. $\displaystyle\int_0^3 \int_0^\infty \frac{x^2}{1 + y^2}\, dy\, dx = \int_0^3 x^2 \arctan y \Big]_0^\infty dx = \int_0^3 x^2 \left(\frac{\pi}{2}\right) dx = \frac{\pi}{2} \cdot \frac{x^3}{3} \Big]_0^3 = \frac{9\pi}{2}$

23. $\displaystyle\int_0^\infty \int_0^\infty xye^{-(x^2+y^2)}\, dx\, dy = \int_0^\infty -\frac{1}{2}ye^{-(x^2+y^2)} \Big]_0^\infty dy = \int_0^\infty \frac{1}{2}ye^{-y^2}\, dy = \left(-\frac{1}{4}e^{-y^2} \right) \Big]_0^\infty = \frac{1}{4}$

24. $\displaystyle\int_1^\infty \int_1^\infty \frac{1}{xy}\, dx\, dy = \int_1^\infty \frac{1}{y}\ln x \Big]_1^\infty dy = \int_1^\infty \left[\frac{1}{y}(\infty) - \frac{1}{y}(0) \right] dy$

Diverges

25. $\int_0^4 \int_0^y f(x, y) \, dx \, dy, \quad 0 \le x \le y, \quad 0 \le y \le 4$

$$= \int_0^4 \int_x^4 f(x, y) \, dy \, dx$$

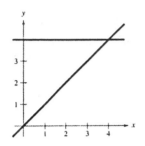

26. $\int_0^4 \int_{\sqrt{y}}^2 f(x, y) \, dx \, dy, \quad \sqrt{y} \le x \le 2, \quad 0 \le y \le 4$

$$= \int_0^2 \int_0^{x^2} f(x, y) \, dy \, dx$$

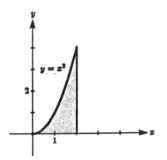

27. $\int_{-\pi/2}^{\pi/2} \int_0^{\cos x} f(x, y) \, dy \, dx, \quad 0 \le y \le \cos x, \quad -\frac{\pi}{2} \le x \le \frac{\pi}{2}$

$$= \int_0^1 \int_{-\arccos y}^{\arccos y} f(x, y) \, dx \, dy$$

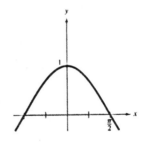

28. $\int_{-1}^1 \int_{x^2}^1 f(x, y) \, dy \, dx, \quad x^2 \le y \le 1, \quad -1 \le x \le 1$

$$= \int_0^1 \int_{-\sqrt{y}}^{\sqrt{y}} f(x, y) \, dx \, dy$$

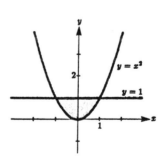

29. $\int_0^1 \int_0^2 dy \, dx = \int_0^2 \int_0^1 dx \, dy = 2$

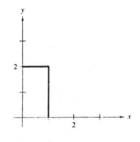

30. $\int_1^2 \int_2^4 dx \, dy = \int_2^4 \int_1^2 dy \, dx = 2$

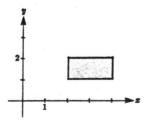

31. $\int_0^1 \int_{-\sqrt{1-y^2}}^{\sqrt{1-y^2}} dx\, dy = \int_{-1}^1 \int_0^{\sqrt{1-x^2}} dy\, dx = \dfrac{\pi}{2}$

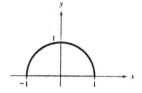

32. $\int_0^2 \int_0^x dy\, dx + \int_2^4 \int_0^{4-x} dy\, dx = \int_0^2 \int_y^{4-y} dx\, dy = 4$

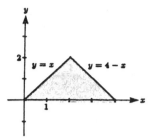

33. $\int_0^2 \int_{x/2}^1 dy\, dx = \int_0^1 \int_0^{2y} dx\, dy = 1$

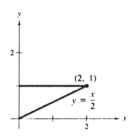

34. $\int_0^4 \int_{\sqrt{x}}^2 dy\, dx = \int_0^2 \int_0^{y^2} dx\, dy = \dfrac{8}{3}$

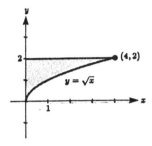

35. $\int_0^1 \int_{y^2}^{\sqrt[3]{y}} dx\, dy = \int_0^1 \int_{x^3}^{\sqrt{x}} dy\, dx = \dfrac{5}{12}$

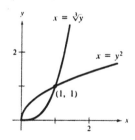

36. $\int_{-2}^2 \int_0^{4-y^2} dx\, dy = \int_0^4 \int_{-\sqrt{4-x}}^{\sqrt{4-x}} dy\, dx = \dfrac{32}{3}$

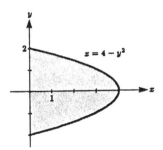

37. $A = \int_0^8 \int_0^3 dy\, dx = \int_0^8 y\Big]_0^3 dx = \int_0^8 3\, dx = 3x\Big]_0^8 = 24$

$A = \int_0^3 \int_0^8 dx\, dy = \int_0^3 x\Big]_0^8 dy = \int_0^3 8\, dy = 8y\Big]_0^3 = 24$

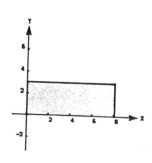

38. $A = \int_1^3 \int_1^3 dy\,dx = \int_1^3 y\Big]_1^3 dx = \int_1^3 2\,dx = 2x\Big]_1^3 = 4$

$A = \int_1^3 \int_1^3 dx\,dy = \int_1^3 x\Big]_1^3 dy = \int_1^3 2\,dy = 2y\Big]_1^3 = 4$

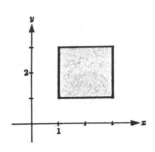

39. $A = \int_0^2 \int_0^{4-x^2} dy\,dx$

$= \int_0^2 y\Big]_0^{4-x^2} dx$

$= \int_0^2 (4 - x^2)\,dx$

$= \left(4x - \dfrac{x^3}{3}\right)\Big]_0^2 = \dfrac{16}{3}$

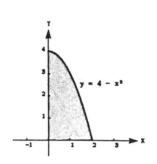

$A = \int_0^4 \int_0^{\sqrt{4-y}} dx\,dy$

$= \int_0^4 x\Big]_0^{\sqrt{4-y}} dy = \int_0^4 \sqrt{4-y}\,dy = -\int_0^4 (4-y)^{1/2}(-1)\,dy = -\dfrac{2}{3}(4-y)^{3/2}\Big]_0^4 = \dfrac{2}{3}(8) = \dfrac{16}{3}$

40. $A = \int_2^5 \int_0^{1/\sqrt{x-1}} dy\,dx = \int_2^5 y\Big]_0^{1/\sqrt{x-1}} dx = \int_2^5 \dfrac{1}{\sqrt{x-1}}\,dx = 2\sqrt{x-1}\Big]_2^5 = 2$

$A = \int_0^{1/2} \int_2^5 dx\,dy + \int_{1/2}^1 \int_2^{1+(1/y^2)} dx\,dy$

$= \int_0^{1/2} x\Big]_2^5 dy + \int_{1/2}^1 x\Big]_2^{1+(1/y^2)} dy$

$= \int_0^{1/2} 3\,dy + \int_{1/2}^1 \left(\dfrac{1}{y^2} - 1\right) dy$

$= 3y\Big]_0^{1/2} + \left(-\dfrac{1}{y} - y\right)\Big]_{1/2}^1 = 2$

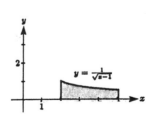

41. $A = \int_{-2}^{1} \int_{x+2}^{4-x^2} dy\, dx$

$= \int_{-2}^{1} y \Big]_{x+2}^{4-x^2} dx = \int_{-2}^{1} (4 - x^2 - x - 2)\, dx = \int_{-2}^{1} (2 - x - x^2)\, dx = \left(2x - \frac{1}{2}x^2 - \frac{1}{3}x^3\right)\Big]_{-2}^{1} = \frac{9}{2}$

$A = \int_{0}^{3} \int_{-\sqrt{4-y}}^{y-2} dx\, dy + 2\int_{3}^{4} \int_{0}^{\sqrt{4-y}} dx\, dy$

$= \int_{0}^{3} x \Big]_{-\sqrt{4-y}}^{y-2} dy + 2\int_{3}^{4} x \Big]_{0}^{\sqrt{4-y}} dy$

$= \int_{0}^{3} (y - 2 + \sqrt{4-y})\, dy + 2\int_{3}^{4} \sqrt{4-y}\, dy$

$= \left(\frac{1}{2}y^2 - 2y - \frac{2}{3}(4-y)^{3/2}\right)\Big]_{0}^{3} - \frac{4}{3}(4-y)^{3/2}\Big]_{3}^{4} = \frac{9}{2}$

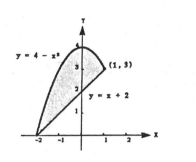

42. $A = \int_{0}^{2} \int_{0}^{\sqrt{4-x^2}} dy\, dx = \int_{0}^{2} \sqrt{4-x^2}\, dx = 4\int_{0}^{\pi/2} \cos^2\theta\, d\theta = 2\int_{0}^{\pi/2} (1 + \cos 2\theta)\, d\theta = 2\left(\theta + \frac{1}{2}\sin 2\theta\right)\Big]_{0}^{\pi/2} = \pi$

$(x = 2\sin\theta,\quad dx = 2\cos\theta\, d\theta,\quad \sqrt{4-x^2} = 2\cos\theta)$

$A = \int_{0}^{2} \int_{0}^{\sqrt{4-y^2}} dx\, dy$

$= \int_{0}^{2} \sqrt{4-y^2}\, dy$

$= 4\int_{0}^{\pi/2} \cos^2\theta\, d\theta$

$= 2\int_{0}^{\pi/2} (1 + \cos 2\theta)\, d\theta$

$= 2\left(\theta + \frac{1}{2}\sin 2\theta\right)\Big]_{0}^{\pi/2} = \pi$

$(y = 2\sin\theta,\quad dy = 2\cos\theta\, d\theta,\quad \sqrt{4-y^2} = 2\cos\theta)$

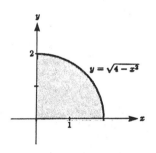

43. $\int_{0}^{4} \int_{0}^{(2-\sqrt{x})^2} dy\, dx = \int_{0}^{4} y \Big]_{0}^{(2-\sqrt{x})^2} dx$

$= \int_{0}^{4} (4 - 4\sqrt{x} + x)\, dx$

$= \left(4x - \frac{8}{3}x\sqrt{x} + \frac{x^2}{2}\right)\Big]_{0}^{4} = \frac{8}{3}$

$\int_{0}^{4} \int_{0}^{(2-\sqrt{y})^2} dx\, dy = \frac{8}{3}$

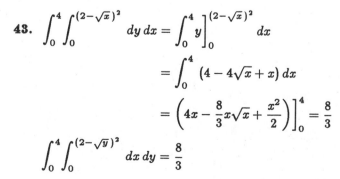

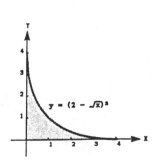

Integration steps are similar to those above.

44. $A = \int_0^1 \int_{x^{3/2}}^x dy\, dx$

$\quad = \int_0^1 y\Big]_{x^{3/2}}^x dx$

$\quad = \int_0^1 (x - x^{3/2})\, dx$

$\quad = \left(\frac{1}{2}x^2 - \frac{2}{5}x^{5/2}\right)\Big]_0^1 = \frac{1}{10}$

$A = \int_0^1 \int_y^{y^{2/3}} dx\, dy = \int_0^1 x\Big]_y^{y^{2/3}} dy = \int_0^1 (y^{2/3} - y)\, dy = \left(\frac{3}{5}y^{5/3} - \frac{1}{2}y^2\right)\Big]_0^1 = \frac{1}{10}$

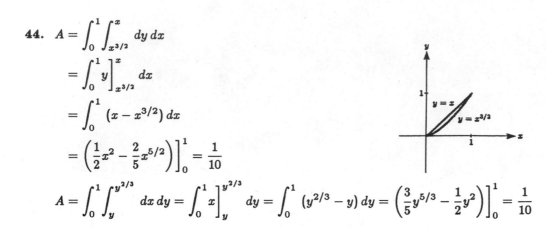

45. $A = \int_0^3 \int_0^{2x/3} dy\, dx + \int_3^5 \int_0^{5-x} dy\, dx$

$\quad = \int_0^3 y\Big]_0^{2x/3} dx + \int_3^5 y\Big]_0^{5-x} dx$

$\quad = \int_0^3 \frac{2x}{3}\, dx + \int_3^5 (5 - x)\, dx$

$\quad = \frac{1}{3}x^2\Big]_0^3 + \left(5x - \frac{1}{2}x^2\right)\Big]_3^5 = 5$

$A = \int_0^2 \int_{3y/2}^{5-y} dx\, dy = \int_0^2 x\Big]_{3y/2}^{5-y} dy = \int_0^2 \left(5 - y - \frac{3y}{2}\right) dy = \int_0^2 \left(5 - \frac{5y}{2}\right) dy = \left(5y - \frac{5}{4}y^2\right)\Big]_0^2 = 5$

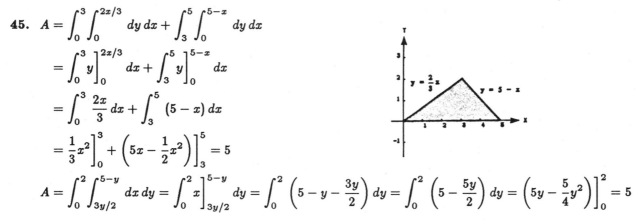

46. $A = \int_0^3 \int_0^x dy\, dx + \int_3^9 \int_0^{9/x} dy\, dx$

$\quad = \int_0^3 y\Big]_0^x dx + \int_3^9 y\Big]_0^{9/x} dx$

$\quad = \int_0^3 x\, dx + \int_3^9 \frac{9}{x}\, dx$

$\quad = \frac{1}{2}x^2\Big]_0^3 + 9\ln x\Big]_3^9$

$\quad = \frac{9}{2} + 9(\ln 9 - \ln 3)$

$\quad = \frac{9}{2}(1 + \ln 9)$

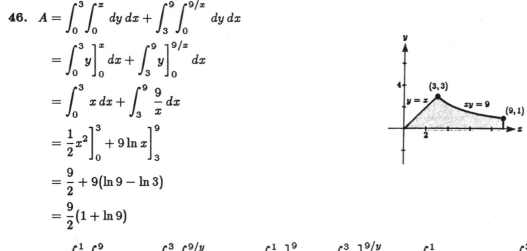

$A = \int_0^1 \int_y^9 dx\, dy + \int_1^3 \int_y^{9/y} dx\, dy = \int_0^1 x\Big]_y^9 dy + \int_1^3 x\Big]_y^{9/y} dy = \int_0^1 (9 - y)\, dy + \int_1^3 \left(\frac{9}{y} - y\right) dy$

$\qquad = \left(9y - \frac{1}{2}y^2\right)\Big]_0^1 + \left(9\ln y - \frac{1}{2}y^2\right)\Big]_1^3$

$\qquad = \frac{9}{2}(1 + \ln 9)$

47. $\dfrac{A}{4} = \displaystyle\int_0^a \int_0^{(b/a)\sqrt{a^2-x^2}} dy\,dx = \int_0^a y\Big]_0^{(b/a)\sqrt{a^2-x^2}} dx = \dfrac{b}{a}\int_0^a \sqrt{a^2-x^2}\,dx$

$$= ab\int_0^{\pi/2} \cos^2\theta\,d\theta$$

$$= \dfrac{ab}{2}\int_0^{\pi/2} (1+\cos 2\theta)\,d\theta$$

$$= \dfrac{ab}{2}\left(\theta + \dfrac{1}{2}\sin 2\theta\right)\Big]_0^{\pi/2}$$

$$= \dfrac{\pi ab}{4}$$

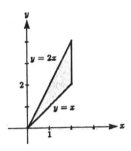

Therefore, $A = \pi ab$.

$(x = a\sin\theta, \quad dx = a\cos\theta\,d\theta)$

$\dfrac{A}{4} = \displaystyle\int_0^b \int_0^{(a/b)\sqrt{b^2-y^2}} dx\,dy = \dfrac{\pi ab}{4}$

Therefore, $A = \pi ab$.

Integration steps are similar to those above.

48. $A = \displaystyle\int_0^2 \int_{y/2}^{y} dx\,dy + \int_2^4 \int_{y/2}^{2} dx\,dy$

$$= \int_0^2 \dfrac{y}{2}\,dy + \int_2^4 \left(2 - \dfrac{y}{2}\right)dy$$

$$= \dfrac{y^2}{4}\Big]_0^2 + \left(2y - \dfrac{y^2}{4}\right)\Big]_2^4$$

$$= 1 + (4 - 3) = 2$$

$A = \displaystyle\int_0^2 \int_x^{2x} dy\,dx = \int_0^2 (2x - x)\,dx = \dfrac{x^2}{2}\Big]_0^2 = 2$

49. $\displaystyle\int_0^2 \int_x^2 x\sqrt{1+y^3}\,dy\,dx = \int_0^2 \int_0^y x\sqrt{1+y^3}\,dx\,dy = \int_0^2 \sqrt{1+y^3}\cdot\dfrac{x^2}{2}\Big]_0^y dy = \dfrac{1}{2}\int_0^2 \sqrt{1+y^3}\,y^2\,dy$

$$= \dfrac{1}{2}\cdot\dfrac{1}{3}\cdot\dfrac{2}{3}(1+y^3)^{3/2}\Big]_0^2$$

$$= \dfrac{1}{9}(27) - \dfrac{1}{9}(1) = \dfrac{26}{9}$$

50. $\displaystyle\int_0^2 \int_x^2 e^{-y^2}\,dy\,dx = \int_0^2 \int_0^y e^{-y^2}\,dx\,dy = \int_0^2 xe^{-y^2}\Big]_0^y dy = \int_0^2 ye^{-y^2}\,dy = -\dfrac{1}{2}e^{-y^2}\Big]_0^2$

$$= -\dfrac{1}{2}(e^{-4}) + \dfrac{1}{2}e^0$$

$$= \dfrac{1}{2}\left(1 - \dfrac{1}{e^4}\right) \approx 0.4908$$

51. $\displaystyle\int_0^1\int_y^1 \sin(x^2)\,dx\,dy = \int_0^1\int_0^x \sin(x^2)\,dy\,dx = \int_0^1 y\sin(x^2)\Big]_0^x dx = \int_0^1 x\sin(x^2)\,dx$

$$= -\frac{1}{2}\cos(x^2)\Big]_0^1$$

$$= -\frac{1}{2}\cos 1 + \frac{1}{2}(1)$$

$$= \frac{1}{2}(1-\cos 1) \approx 0.2298$$

52. $\displaystyle\int_0^2\int_{y^2}^4 \sqrt{x}\sin x\,dx\,dy = \int_0^4\int_0^{\sqrt{x}} \sqrt{x}\sin x\,dy\,dx$

$$= \int_0^4 y\sqrt{x}\sin x\Big]_0^{\sqrt{x}} dx = \int_0^4 x\sin x\,dx = \Big[\sin x - x\cos x\Big]_0^4 = \sin 4 - 4\cos 4 \approx 1.858$$

53. $\displaystyle\int_0^2\int_0^{4-x^2} e^{xy}\,dy\,dx = \int_0^2 \frac{1}{x}e^{xy}\Big]_0^{4-x^2} dx = \int_0^2 \frac{1}{x}[e^{x(4-x^2)} - 1]\,dx \approx 20.565$

54. $\displaystyle\int_0^1\int_y^{2y} \sin(x+y)\,dx\,dy = \int_0^1 -\cos(x+y)\Big]_y^{2y} dy$

$$= \int_0^1 [-\cos(3y) + \cos(2y)]\,dy = \Big[-\frac{1}{3}\sin(3y) + \frac{1}{2}\sin(2y)\Big]_0^1 = -\frac{1}{3}\sin 3 + \frac{1}{2}\sin 2 \approx 0.4076$$

Section 15.2   Double Integrals and Volume

1. $\displaystyle\int_0^2\int_0^1 (1+2x+2y)\,dy\,dx = \int_0^2 (y+2xy+y^2)\Big]_0^1 dx$

$$= \int_0^2 (2+2x)\,dx$$

$$= (2x+x^2)\Big]_0^2$$

$$= 8$$

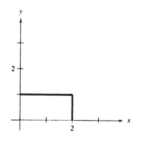

2. $\displaystyle\int_0^\pi\int_0^{\pi/2} \sin^2 x\cos^2 y\,dy\,dx = \int_0^\pi \frac{1}{2}\sin^2 x\left(y+\frac{1}{2}\sin 2y\right)\Big]_0^{\pi/2} dx$

$$= \int_0^\pi \frac{1}{2}\sin^2 x\left(\frac{\pi}{2}\right)\,dx$$

$$= \frac{\pi}{8}\int_0^\pi (1-\cos 2x)\,dx$$

$$= \frac{\pi}{8}\left(x-\frac{1}{2}\sin 2x\right)\Big]_0^\pi$$

$$= \frac{\pi^2}{8}$$

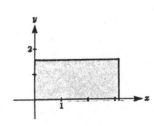

3. $\displaystyle \int_0^6 \int_{y/2}^3 (x+y)\,dx\,dy = \int_0^6 \left(\frac{1}{2}x^2 + xy\right)\Big]_{y/2}^3 dy$

$\displaystyle \qquad\qquad = \int_0^6 \left(\frac{9}{2} + 3y - \frac{5}{8}y^2\right) dy$

$\displaystyle \qquad\qquad = \left(\frac{9}{2}y + \frac{3}{2}y^2 - \frac{5}{24}y^3\right)\Big]_0^6$

$\displaystyle \qquad\qquad = 36$

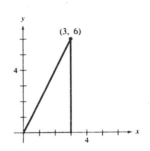

4. $\displaystyle \int_0^1 \int_y^{\sqrt{y}} x^2 y^2\,dx\,dy = \int_0^1 \frac{1}{3}x^3 y^2\Big]_y^{\sqrt{y}} dy$

$\displaystyle \qquad\qquad = \frac{1}{3}\int_0^1 (y^{7/2} - y^5)\,dy$

$\displaystyle \qquad\qquad = \frac{1}{3}\left(\frac{2}{9}y^{9/2} - \frac{1}{6}y^6\right)\Big]_0^1$

$\displaystyle \qquad\qquad = \frac{1}{54}$

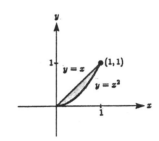

5. $\displaystyle \int_{-a}^a \int_{-\sqrt{a^2-x^2}}^{\sqrt{a^2-x^2}} (x+y)\,dy\,dx = \int_{-a}^a \left(xy + \frac{1}{2}y^2\right)\Big]_{-\sqrt{a^2-x^2}}^{\sqrt{a^2-x^2}} dx$

$\displaystyle \qquad\qquad = \int_{-a}^a 2x\sqrt{a^2 - x^2}\,dx$

$\displaystyle \qquad\qquad = -\frac{2}{3}(a^2 - x^2)^{3/2}\Big]_{-a}^a = 0$

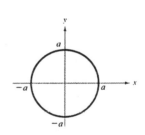

6. $\displaystyle \int_0^1 \int_{y-1}^0 e^{x+y}\,dx\,dy + \int_0^1 \int_0^{1-y} e^{x+y}\,dx\,dy = \int_0^1 e^{x+y}\Big]_{y-1}^0 dy + \int_0^1 e^{x+y}\Big]_0^{1-y} dy$

$\displaystyle \qquad\qquad = \int_0^1 (e - e^{2y-1})\,dy$

$\displaystyle \qquad\qquad = \left(ey - \frac{1}{2}e^{2y-1}\right)\Big]_0^1$

$\displaystyle \qquad\qquad = \frac{1}{2}(e + e^{-1})$

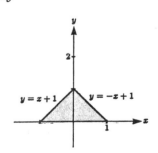

7. $\displaystyle \int_0^5 \int_0^3 xy\,dx\,dy = \int_0^3 \int_0^5 xy\,dy\,dx$

$\displaystyle \qquad\qquad = \int_0^3 \frac{1}{2}xy^2\Big]_0^5 dx$

$\displaystyle \qquad\qquad = \frac{25}{2}\int_0^3 x\,dx$

$\displaystyle \qquad\qquad = \frac{25}{4}x^2\Big]_0^3 = \frac{225}{4}$

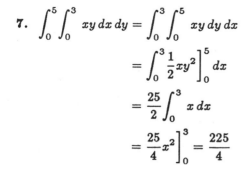

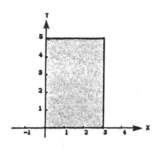

8.
$$\int_0^{\pi/2}\int_{-\pi}^{\pi}\sin x \sin y \, dx \, dy = \int_{-\pi}^{\pi}\int_0^{\pi/2}\sin x \sin y \, dy \, dx$$

$$= \int_{-\pi}^{\pi} -\sin x \cos y \Big]_0^{\pi/2} dx$$

$$= \int_{-\pi}^{\pi}\sin x \, dx$$

$$= 0$$

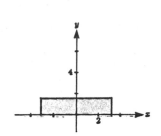

9.
$$\int_0^2\int_{y/2}^{y}\frac{y}{x^2+y^2}\,dx\,dy + \int_2^4\int_{y/2}^{2}\frac{y}{x^2+y^2}\,dx\,dy = \int_0^2\int_x^{2x}\frac{y}{x^2+y^2}\,dy\,dx$$

$$= \frac{1}{2}\int_0^2 \ln(x^2+y^2)\Big]_x^{2x}dx$$

$$= \frac{1}{2}\int_0^2 \left(\ln 5x^2 - \ln 2x^2\right) dx$$

$$= \frac{1}{2}\ln\frac{5}{2}\int_0^2 dx$$

$$= \frac{1}{2}\left(\ln\frac{5}{2}\right)x\Big]_0^2$$

$$= \ln\frac{5}{2}$$

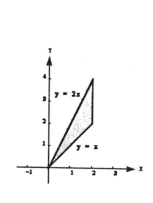

10.
$$\int_0^2\int_{y^2}^{4}\frac{y}{1+x^2}\,dx\,dy = \int_0^4\int_0^{\sqrt{x}}\frac{y}{1+x^2}\,dy\,dx$$

$$= \frac{1}{2}\int_0^4 \frac{y^2}{1+x^2}\Big]_0^{\sqrt{x}}dx$$

$$= \frac{1}{2}\int_0^4 \frac{x}{1+x^2}\,dx$$

$$= \frac{1}{4}\ln(1+x^2)\Big]_0^4$$

$$= \frac{1}{4}\ln(17)$$

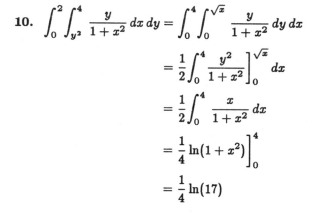

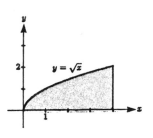

11.
$$\int_0^4\int_0^{3x/4} x \, dy \, dx + \int_4^5\int_0^{\sqrt{25-x^2}} x \, dy \, dx = \int_0^3\int_{4y/3}^{\sqrt{25-y^2}} x \, dx \, dy$$

$$= \int_0^3 \frac{1}{2}x^2\Big]_{4y/3}^{\sqrt{25-y^2}} dy$$

$$= \frac{25}{18}\int_0^3 (9-y^2)\,dy$$

$$= \frac{25}{18}\left(9y - \frac{1}{3}y^3\right)\Big]_0^3 = 25$$

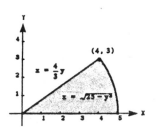

12. $\int_{-2}^{2}\int_{0}^{\sqrt{4-x^2}} (x^2+y^2)\, dy\, dx = \int_{0}^{2}\int_{-\sqrt{4-y^2}}^{\sqrt{4-y^2}} (x^2+y^2)\, dx\, dy$

$$= \int_{-2}^{2} \left(x^2 y + \frac{1}{3}y^3 \right) \Big]_{0}^{\sqrt{4-x^2}} dx$$

$$= \int_{-2}^{2} \left[x^2\sqrt{4-x^2} + \frac{1}{3}(4-x^2)^{3/2} \right] dx$$

$$= \left[-\frac{x}{4}(4-x^2)^{3/2} + \frac{1}{2}\left(x\sqrt{4-x^2} + 4\arcsin\frac{x}{2} \right) \right.$$

$$\left. + \frac{1}{12}\left[x(4-x^2)^{3/2} + 6x\sqrt{4-x^2} + 24\arcsin\frac{x}{2} \right] \right]_{-2}^{2} = 4\pi$$

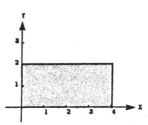

13. $\int_{0}^{4}\int_{0}^{2} \frac{y}{2}\, dy\, dx = \int_{0}^{4} \frac{y^2}{4}\Big]_{0}^{2} dx$

$$= \int_{0}^{4} dx$$

$$= 4$$

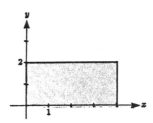

14. $\int_{0}^{4}\int_{0}^{2} (6-2y)\, dy\, dx = \int_{0}^{4} \left[6y - y^2 \right]_{0}^{2} dx$

$$= \int_{0}^{4} 8\, dx$$

$$= 32$$

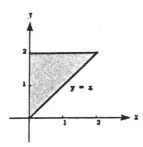

15. $\int_{0}^{2}\int_{0}^{y} (6-x-y)\, dx\, dy = \int_{0}^{2} \left[6x - \frac{x^2}{2} - xy \right]_{0}^{y} dy$

$$= \int_{0}^{2} \left(6y - \frac{3}{2}y^2 \right) dy$$

$$= \left[3y^2 - \frac{1}{2}y^3 \right]_{0}^{2}$$

$$= 8$$

16. $\displaystyle\int_0^2\int_0^x 6\,dy\,dx = \int_0^2 6x\,dx$

$$= 3x^2\Big]_0^2$$

$$= 12$$

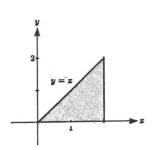

17. $\displaystyle\int_0^6\int_0^{(-2/3)x+4}\left(\frac{12-2x-3y}{4}\right)dy\,dx = \int_0^6\left[\frac{1}{4}\left(12y-2xy-\frac{3}{2}y^2\right)\right]_0^{(-2/3)x+4}dx$

$$= \int_0^6\left(\frac{1}{6}x^2-2x+6\right)dx$$

$$= \left[\frac{1}{18}x^3-x^2+6x\right]_0^6$$

$$= 12$$

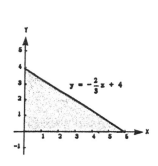

18. $\displaystyle\int_0^1\int_0^{1-x}(1-x-y)\,dy\,dx = \int_0^1\left[y-xy-\frac{y^2}{2}\right]_0^{1-x}dx$

$$= \int_0^1\frac{1}{2}(1-x)^2\,dx$$

$$= -\frac{1}{6}(1-x)^3\Big]_0^1$$

$$= \frac{1}{6}$$

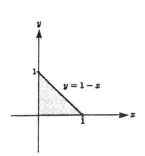

19. $\displaystyle\int_0^1\int_0^y(1-xy)\,dx\,dy = \int_0^1\left[x-\frac{x^2y}{2}\right]_0^y dy$

$$= \int_0^1\left(y-\frac{y^3}{2}\right)dy$$

$$= \left[\frac{y^2}{2}-\frac{y^4}{8}\right]_0^1$$

$$= \frac{3}{8}$$

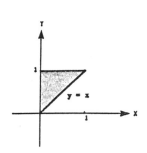

20. $\displaystyle\int_0^2\int_0^y(4-y^2)\,dx\,dy = \int_0^2(4y-y^3)\,dy$

$$= \left[2y^2-\frac{y^4}{4}\right]_0^2$$

$$= 4$$

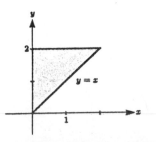

21. $4\displaystyle\int_0^2\int_0^{\sqrt{4-x^2}}(4-x^2-y^2)\,dy\,dx = 4\int_0^2\left[4\sqrt{4-x^2}-x^2\sqrt{4-x^2}-\frac{1}{3}(\sqrt{4-x^2})^3\right]dx$

$$= 4\int_0^2\left[(4-x^2)\sqrt{4-x^2}-\frac{1}{3}(4-x^2)\sqrt{4-x^2}\right]dx$$

$$= \frac{8}{3}\int_0^2\left[4\sqrt{4-x^2}-x^2\sqrt{4-x^2}\right]dx$$

$$= \frac{8}{3}\left[\frac{4}{2}\left(x\sqrt{4-x^2}+4\arcsin\frac{x}{2}\right)\right.$$

$$\left. -\frac{1}{8}\left[x(2x^2-4)\sqrt{4-x^2}+16\arcsin\frac{x}{2}\right]\right]_0^2$$

$$= \frac{8}{3}\left[2(4)\left(\frac{\pi}{2}\right)-\frac{1}{8}(16)\left(\frac{\pi}{2}\right)\right] = 8\pi$$

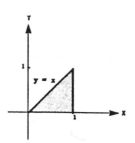

(Use Formulas 37 and 38.)

22. $\displaystyle\int_0^1\int_0^x\sqrt{1-x^2}\,dy\,dx = \int_0^1 x\sqrt{1-x^2}\,dx$

$$= -\frac{1}{3}(1-x^2)^{3/2}\Big]_0^1$$

$$= \frac{1}{3}$$

23. $\displaystyle\int_0^\infty\int_0^\infty\frac{1}{(x+1)^2(y+1)^2}\,dy\,dx = \int_0^\infty\left[-\frac{1}{(x+1)^2(y+1)}\right]_0^\infty dx = \int_0^\infty\frac{1}{(x+1)^2}\,dx = -\frac{1}{(x+1)}\Big]_0^\infty = 1$

24. $\displaystyle\int_0^\infty\int_0^\infty e^{-(x+y)/2}\,dy\,dx = \int_0^\infty\left[-2e^{-(x+y)/2}\right]_0^\infty dx = \int_0^\infty 2e^{-x/2}\,dx = -4e^{-x/2}\Big]_0^\infty = 4$

25. $V = \displaystyle\int_0^1\int_0^x xy\,dy\,dx = \int_0^1\frac{1}{2}xy^2\Big]_0^x dx$

$$= \frac{1}{2}\int_0^1 x^3\,dx$$

$$= \frac{1}{8}x^4\Big]_0^1$$

$$= \frac{1}{8}$$

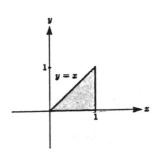

26. $V = \int_0^5 \int_0^x x \, dy \, dx = \int_0^5 xy \Big]_0^x dx$

$$= \int_0^5 x^2 \, dx$$

$$= \frac{1}{3}x^3 \Big]_0^5$$

$$= \frac{125}{3}$$

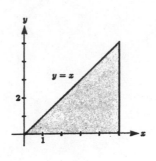

27. $V = \int_0^2 \int_0^4 x^2 \, dy \, dx = \int_0^2 x^2 y \Big]_0^4 dx$

$$= \int_0^2 4x^2 \, dx$$

$$= \frac{4x^3}{3} \Big]_0^2$$

$$= \frac{32}{3}$$

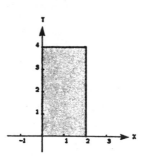

28. $V = 8 \int_0^r \int_0^{\sqrt{r^2-x^2}} \sqrt{r^2 - x^2 - y^2} \, dy \, dx$

$$= 4 \int_0^r \left[y\sqrt{r^2 - x^2 - y^2} + (r^2 - x^2) \arcsin \frac{y}{\sqrt{r^2 - x^2}} \right]_0^{\sqrt{r^2-x^2}} dx$$

$$= 4\left(\frac{\pi}{2}\right) \int_0^r (r^2 - x^2) \, dx$$

$$= 2\pi \left(r^2 x - \frac{1}{3}x^3 \right) \Big]_0^r$$

$$= \frac{4\pi r^3}{3}$$

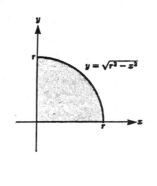

29. Divide the solid into two equal parts.

$$V = 2 \int_0^1 \int_0^x \sqrt{1 - x^2} \, dy \, dx$$

$$= 2 \int_0^1 y\sqrt{1 - x^2} \Big]_0^x dx$$

$$= 2 \int_0^1 x\sqrt{1 - x^2} \, dx$$

$$= -\frac{2}{3}(1 - x^2)^{3/2} \Big]_0^1 = \frac{2}{3}$$

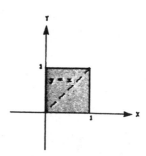

30. $V = \int_0^1 \int_0^{1-x^2} (1-x^2)\, dy\, dx$

$= \int_0^1 y(1-x^2) \Big]_0^{1-x^2} dx$

$= \int_0^1 (1-x^2)^2\, dx$

$= \int_0^1 (1 - 2x^2 + x^4)\, dx$

$= \left(x - \frac{2}{3}x^3 + \frac{1}{5}x^5 \right) \Big]_0^1 = \frac{8}{15}$

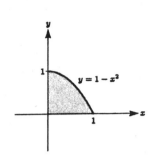

31. $V = \int_0^2 \int_0^{\sqrt{4-x^2}} (x+y)\, dy\, dx$

$= \int_0^2 \left(xy + \frac{1}{2}y^2 \right) \Big]_0^{\sqrt{4-x^2}} dx$

$= \int_0^2 \left(x\sqrt{4-x^2} + 2 - \frac{1}{2}x^2 \right) dx$

$= \left[-\frac{1}{3}(4-x^2)^{3/2} + 2x - \frac{1}{6}x^3 \right]_0^2 = \frac{16}{3}$

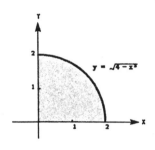

32. $V = \int_0^2 \int_0^{\infty} \frac{1}{1+y^2}\, dy\, dx$

$= \int_0^2 \arctan y \Big]_0^{\infty} dx$

$= \int_0^2 \frac{\pi}{2}\, dx$

$= \frac{\pi x}{2} \Big]_0^2 = \pi$

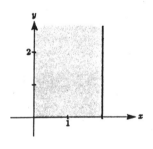

33. $V = 4 \int_0^2 \int_0^{\sqrt{4-x^2}} (4 - x^2 - y^2)\, dy\, dx$

$= 4 \int_0^2 \left(4y - x^2y - \frac{y^3}{3} \right) \Big]_0^{\sqrt{4-x^2}} dx$

$= 4 \int_0^2 \left[4\sqrt{4-x^2} - x^2\sqrt{4-x^2} - \frac{1}{3}(4-x^2)^{3/2} \right] dx, \quad x = 2\sin\theta$

$= 4 \int_0^{\pi/2} \left(8\cos\theta - 8\sin^2\theta\cos\theta - \frac{8}{3}\cos^3\theta \right) 2\cos\theta\, d\theta$

$= 4 \int_0^{\pi/2} \frac{32}{3}\cos^4\theta\, d\theta$

$= 4 \left(\frac{32}{3} \right) \left(\frac{3\pi}{16} \right) = 8\pi$

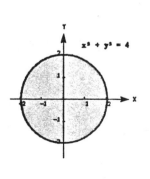

34. $V = \int_0^9 \int_0^{\sqrt{9-y}} \sqrt{9-y}\, dx\, dy$

$\quad = \int_0^9 x\sqrt{9-y} \,\Big]_0^{\sqrt{9-y}} dy$

$\quad = \int_0^9 (9-y)\, dy$

$\quad = \left[9y - \dfrac{y^2}{2} \right]_0^9$

$\quad = \left[81 - \dfrac{81}{2} \right] = \dfrac{81}{2}$

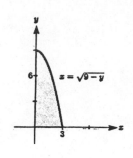

35. $V = 4\int_0^2 \int_0^{\sqrt{4-x^2}} (x^2+y^2)\, dy\, dx$

$\quad = 4\int_0^2 \left[x^2\sqrt{4-x^2} + \dfrac{1}{3}(4-x^2)^{3/2} \right] dx, \quad x = 2\sin\theta$

$\quad = 4\int_0^{\pi/2} \left(16\cos^2\theta - \dfrac{32}{3}\cos^4\theta \right) d\theta$

$\quad = 4\left[16\left(\dfrac{\pi}{4}\right) - \dfrac{32}{3}\left(\dfrac{3\pi}{16}\right) \right]$

$\quad = 8\pi$

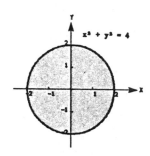

36. $V = \int_0^5 \int_0^{\pi} \sin^2 x\, dx\, dy$

$\quad = \int_0^5 \dfrac{\pi}{2}\, dy$

$\quad = \dfrac{\pi}{2} y \,\Big]_0^5$

$\quad = \dfrac{5\pi}{2}$

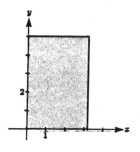

37. Average $= \dfrac{1}{8}\int_0^4 \int_0^2 x\, dy\, dx = \dfrac{1}{8}\int_0^4 2x\, dx = \dfrac{x^2}{8}\,\Big]_0^4 = 2$

38. Average $= \dfrac{1}{8}\int_0^4 \int_0^2 xy\, dy\, dx = \dfrac{1}{8}\int_0^4 2x\, dx = \dfrac{x^2}{8}\,\Big]_0^4 = 2$

39. Average $= \dfrac{1}{4}\int_0^2 \int_0^2 (x^2+y^2)\, dx\, dy = \dfrac{1}{4}\int_0^2 \left[\dfrac{x^3}{3} + xy^2 \right]_0^2 dy = \dfrac{1}{4}\int_0^2 \left(\dfrac{8}{3} + 2y^2 \right) dy$

$\qquad\qquad\qquad\qquad = \dfrac{1}{4}\left(\dfrac{8}{3}y + \dfrac{2}{3}y^3 \right)\Big]_0^2 = \dfrac{8}{3}$

40. Average $= \dfrac{1}{1/2}\int_0^1 \int_0^x e^{x+y}\, dy\, dx = 2\int_0^1 (e^{2x} - e^x)\, dx = 2\left[\dfrac{1}{2}e^{2x} - e^x \right]_0^1 = (e-1)^2$

41. Average $= \dfrac{1}{1250}\displaystyle\int_{300}^{325}\int_{200}^{250}100x^{0.6}y^{0.4}\,dx\,dy = \dfrac{1}{1250}\displaystyle\int_{300}^{325}(100y^{0.4})\dfrac{x^{1.6}}{1.6}\Big]_{200}^{250}dy$

$$= \dfrac{128,844.1}{1250}\int_{300}^{325}y^{0.4}\,dy$$

$$= 103.0753\left[\dfrac{y^{1.4}}{1.4}\right]_{300}^{325}$$

$$\approx 25,645.24$$

42. Average $= \dfrac{1}{150}\displaystyle\int_{45}^{60}\int_{40}^{50}[192x+576y-x^2-5y^2-2xy-5000]\,dx\,dy$

$$= \dfrac{1}{150}\int_{45}^{60}\left[96x^2+576xy-\dfrac{x^3}{3}-5xy^2-x^2y-5000x\right]_{40}^{50}dy$$

$$= \dfrac{1}{150}\int_{45}^{60}\left(\dfrac{48,200}{3}+4860y-50y^2\right)dy$$

$$= \dfrac{1}{150}\left[\dfrac{48,200}{3}y+2430y^2-\dfrac{50}{3}y^3\right]_{45}^{60}$$

$$\approx 13,246.67$$

43. $\displaystyle\int_{0}^{1}\int_{y/2}^{1/2}e^{-x^2}\,dx\,dy = \int_{0}^{1/2}\int_{0}^{2x}e^{-x^2}\,dy\,dx$

$$= \int_{0}^{1/2}2xe^{-x^2}\,dx$$

$$= -e^{-x^2}\Big]_{0}^{1/2}$$

$$= -e^{-1/4}+1$$

$$= 1-e^{-1/4}$$

$$\approx 0.2212$$

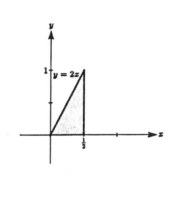

44. $\displaystyle\int_{0}^{1}\int_{0}^{\arccos y}\sin x\sqrt{1+\sin^2 x}\,dx\,dy = \int_{0}^{\pi/2}\int_{0}^{\cos x}\sin x\sqrt{1+\sin^2 x}\,dy\,dx$

$$= \int_{0}^{\pi/2}(1+\sin^2 x)^{1/2}\sin x\cos x\,dx$$

$$= \dfrac{1}{2}\cdot\dfrac{2}{3}(1+\sin^2 x)^{3/2}\Big]_{0}^{\pi/2}$$

$$= \dfrac{1}{3}[2\sqrt{2}-1]$$

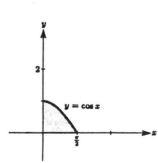

45. $\displaystyle\int_{0}^{\ln 10}\int_{e^x}^{10}\dfrac{1}{\ln y}\,dy\,dx = \int_{1}^{10}\int_{0}^{\ln y}\dfrac{1}{\ln y}\,dx\,dy$

$$= \int_{1}^{10}\dfrac{x}{\ln y}\Big]_{0}^{\ln y}dy$$

$$= \int_{1}^{10}dy$$

$$= y\Big]_{1}^{10} = 9$$

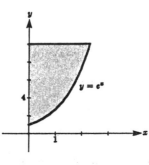

46. $\displaystyle\int_0^2 \int_{x^2}^4 \sqrt{y}\cos y \, dy \, dx = \int_0^4 \int_0^{\sqrt{y}} \sqrt{y}\cos y \, dx \, dy$

$$= \int_0^4 x\sqrt{y}\cos y \Big]_0^{\sqrt{y}} dy$$

$$= \int_0^4 y\cos y \, dy$$

$$= \Big[y\sin y + \cos y \Big]_0^4$$

$$= 4\sin 4 + \cos 4 - 1$$

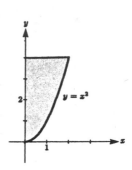

47.

$(x_i,\ y_i)$	$\left(\frac{1}{2}, \frac{1}{2}\right)$	$\left(\frac{3}{2}, \frac{1}{2}\right)$	$\left(\frac{5}{2}, \frac{1}{2}\right)$	$\left(\frac{7}{2}, \frac{1}{2}\right)$	$\left(\frac{1}{2}, \frac{3}{2}\right)$	$\left(\frac{3}{2}, \frac{3}{2}\right)$	$\left(\frac{5}{2}, \frac{3}{2}\right)$	$\left(\frac{7}{2}, \frac{3}{2}\right)$
$(x_i + y_i)\Delta x_i \Delta y_i$	1	2	3	4	2	3	4	5

$$\sum_{i=1}^8 f(x_i,\ y_i)\Delta x_i \Delta y_i = 24$$

$$\int_0^4 \int_0^2 (x+y)\, dy\, dx = \int_0^4 \left(xy + \frac{1}{2}y^2 \right)\Big]_0^2 dx = \int_0^4 (2x+2)\, dx = (x^2 + 2x)\Big]_0^4 = 24$$

48.

$(x_i,\ y_i)$	$\left(\frac{1}{2}, \frac{1}{2}\right)$	$\left(\frac{3}{2}, \frac{1}{2}\right)$	$\left(\frac{5}{2}, \frac{1}{2}\right)$	$\left(\frac{7}{2}, \frac{1}{2}\right)$	$\left(\frac{1}{2}, \frac{3}{2}\right)$	$\left(\frac{3}{2}, \frac{3}{2}\right)$	$\left(\frac{5}{2}, \frac{3}{2}\right)$	$\left(\frac{7}{2}, \frac{3}{2}\right)$
$(x_i y_i)\Delta x_i \Delta y_i$	$\frac{1}{4}$	$\frac{3}{4}$	$\frac{5}{4}$	$\frac{7}{4}$	$\frac{3}{4}$	$\frac{9}{4}$	$\frac{15}{4}$	$\frac{21}{4}$

$$\sum_{i=1}^8 f(x_i,\ y_i)\Delta x_i \Delta y_i = 16$$

$$\int_0^4 \int_0^2 xy \, dy\, dx = \int_0^4 \frac{1}{2}xy^2 \Big]_0^2 dx = \int_0^4 2x \, dx = x^2 \Big]_0^4 = 16$$

49.

$(x_i,\ y_i)$	$\left(\frac{1}{2}, \frac{1}{2}\right)$	$\left(\frac{3}{2}, \frac{1}{2}\right)$	$\left(\frac{5}{2}, \frac{1}{2}\right)$	$\left(\frac{7}{2}, \frac{1}{2}\right)$	$\left(\frac{1}{2}, \frac{3}{2}\right)$	$\left(\frac{3}{2}, \frac{3}{2}\right)$	$\left(\frac{5}{2}, \frac{3}{2}\right)$	$\left(\frac{7}{2}, \frac{3}{2}\right)$
$(x_i^2 + y_i^2)\Delta x_i \Delta y_i$	$\frac{1}{2}$	$\frac{5}{2}$	$\frac{13}{2}$	$\frac{25}{2}$	$\frac{5}{2}$	$\frac{9}{2}$	$\frac{17}{2}$	$\frac{29}{2}$

$$\sum_{i=1}^8 f(x_i,\ y_i)\Delta x_i \Delta y_i = \frac{104}{2} = 52$$

$$\int_0^4 \int_0^2 (x^2 + y^2)\, dy\, dx = \int_0^4 \left(x^2 y + \frac{1}{3}y^3 \right)\Big]_0^2 dx = \int_0^4 \left(2x^2 + \frac{8}{3} \right) dx = \left(\frac{2}{3}x^3 + \frac{8}{3}x \right)\Big]_0^4$$

$$= \frac{160}{3} = 53\tfrac{1}{3}$$

50.

$(x_i, \ y_i)$	$\left(\dfrac{1}{2}, \dfrac{1}{2}\right)$	$\left(\dfrac{3}{2}, \dfrac{1}{2}\right)$	$\left(\dfrac{5}{2}, \dfrac{1}{2}\right)$	$\left(\dfrac{7}{2}, \dfrac{1}{2}\right)$	$\left(\dfrac{1}{2}, \dfrac{3}{2}\right)$	$\left(\dfrac{3}{2}, \dfrac{3}{2}\right)$	$\left(\dfrac{5}{2}, \dfrac{3}{2}\right)$	$\left(\dfrac{7}{2}, \dfrac{3}{2}\right)$
$\dfrac{\Delta x_i \Delta y_i}{(x_i + 1)(y_i + 1)}$	$\dfrac{4}{9}$	$\dfrac{4}{15}$	$\dfrac{4}{21}$	$\dfrac{4}{27}$	$\dfrac{4}{15}$	$\dfrac{4}{25}$	$\dfrac{4}{35}$	$\dfrac{4}{45}$

$$\sum_{i=1}^{8} f(x_i, \ y_i)\Delta x_i \Delta y_i \approx 1.67958$$

$$\int_0^4 \int_0^2 \frac{1}{(x+1)(y+1)} \, dy \, dx = \int_0^4 \frac{1}{x+1} \ln|y+1| \Big]_0^2 \, dx = (\ln 3) \ln|x+1| \Big]_0^4 = (\ln 3)(\ln 5) \approx 1.76815$$

51. Line through $(0, 0)$ and $(2, 1)$: $y = \dfrac{x}{2}$

Line through $(0, 5)$ and $(2, 1)$: $y = -2x + 5$

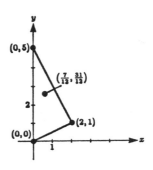

$$\int_R\!\!\int y \, dA = \int_0^2 \int_{x/2}^{-2x+5} y \, dy \, dx$$

$$= \int_0^2 \frac{y^2}{2} \Big]_{x/2}^{-2x+5} \, dx$$

$$= \frac{1}{2}\int_0^2 \left[(-2x+5)^2 - \frac{x^2}{4} \right] dx$$

$$= \frac{1}{2}\left[-\frac{1}{6}(-2x+5)^3 - \frac{x^3}{12} \right]_0^2$$

$$= \frac{1}{2}\left[-\frac{1}{6} - \frac{4}{6} \right] - \frac{1}{2}\left[-\frac{125}{6} \right] = 10$$

$$\int_R\!\!\int xy \, dA = \int_0^2 \int_{x/2}^{-2x+5} xy \, dy \, dx = \frac{1}{2}\int_0^2 x\left[(-2x+5)^2 - \frac{x^2}{4} \right] dx$$

$$= \frac{1}{2}\int_0^2 \left[4x^3 - 20x^2 + 25x - \frac{x^3}{4} \right] dx$$

$$= \frac{1}{2}\left[x^4 - \frac{20x^3}{3} + \frac{25x^2}{2} - \frac{x^4}{16} \right]_0^2$$

$$= \frac{1}{2}\left[16 - \frac{160}{3} + 50 - 1 \right] = \frac{35}{6}$$

$$\int_R\!\!\int y^2 \, dA = \int_0^2 \int_{x/2}^{-2x+5} y^2 \, dy \, dx = \frac{1}{3}\int_0^2 \left[(-2x+5)^3 - \frac{x^3}{8} \right] dx$$

$$= \frac{1}{3}\left[-\frac{1}{8}(-2x+5)^4 - \frac{x^4}{32} \right]_0^2$$

$$= \frac{1}{3}\left[\left(-\frac{1}{8} - \frac{1}{2} \right) - \left(-\frac{625}{8} \right) \right] = \frac{155}{6}$$

Therefore, $x_p = \dfrac{\displaystyle\int_R\!\!\int xy \, dA}{\displaystyle\int_R\!\!\int y \, dA} = \dfrac{35/6}{10} = \dfrac{7}{12}$ and $y_p = \dfrac{\displaystyle\int_R\!\!\int y^2 \, dA}{\displaystyle\int_R\!\!\int y \, dA} = \dfrac{155/6}{10} = \dfrac{31}{12}.$

52. Line through $(0, 0)$ and $(3, 1)$: $y = \dfrac{x}{3}$

Line through $(0, 7)$ and $(3, 1)$: $y = -2x + 7$

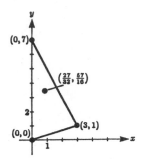

$$\iint_R y\, dA = \int_0^3 \int_{x/3}^{-2x+7} y\, dy\, dx$$

$$= \frac{1}{2} \int_0^3 \left[(-2x - 7)^2 - \frac{x^2}{9} \right] dx$$

$$= \frac{1}{2} \left[-\frac{1}{6}(-2x + 7)^3 - \frac{x^3}{27} \right]_0^3$$

$$= \frac{1}{2} \left[-\frac{1}{6} - 1 \right] - \frac{1}{2}\left[-\frac{343}{6} \right] = 28$$

$$\iint_R xy\, dA = \int_0^3 \int_{x/3}^{-2x+7} xy\, dy\, dx = \frac{1}{2}\int_0^3 x\left[(-2x - 7)^2 - \frac{x^2}{9} \right] dx$$

$$= \frac{1}{2}\int_0^3 \left[4x^3 - 28x^2 + 49x - \frac{x^3}{9} \right] dx$$

$$= \frac{1}{2}\left[x^4 - \frac{28x^3}{3} + \frac{49x^2}{2} - \frac{x^4}{36} \right]_0^3$$

$$= \frac{1}{2}\left[81 - 252 + \frac{441}{2} - \frac{9}{4} \right] = \frac{189}{8}$$

$$\iint_R y^2\, dA = \int_0^3 \int_{x/3}^{-2x+7} y^2\, dy\, dx = \frac{1}{3}\int_0^2 \left[(-2x + 7)^3 - \frac{x^3}{27} \right] dy\, dx$$

$$= \frac{1}{3}\left[-\frac{1}{8}(-2x + 7)^4 - \frac{x^4}{108} \right]_0^3$$

$$= \frac{1}{3}\left[\left(-\frac{1}{8} - \frac{3}{4} \right) - \left(-\frac{2401}{8} \right) \right] = \frac{399}{4}$$

Therefore, $x_p = \dfrac{\displaystyle\iint_R xy\, dA}{\displaystyle\iint_R y\, dA} = \dfrac{189/8}{28} = \dfrac{27}{32}$ and $y_p = \dfrac{\displaystyle\iint_R y^2\, dA}{\displaystyle\iint_R y\, dA} = \dfrac{399/4}{28} = \dfrac{57}{16}$.

53. f is a continuous function such that $0 \le f(x, y) \le 1$ over a region R of area 1. Let $f(m, n) =$ the minimum value of f over R and $f(M, N) =$ the maximum value of f over R. Then

$$f(m, n)\iint_R dA \le \iint_R f(x, y)\, dA \le f(M, N)\iint_R dA.$$

Since $\displaystyle\iint_R dA = 1$ and $0 \le f(m, n) \le f(M, N) \le 1$, we have

$$0 \le f(m, n)(1) \le \iint_R f(x, y)\, dA \le f(M, N)(1) \le 1.$$

Therefore, $0 \le \displaystyle\iint_R f(x, y)\, dA \le 1$.

54. $\dfrac{x}{a} + \dfrac{y}{b} + \dfrac{z}{c} = 1$

$z = c\left(1 - \dfrac{x}{a} - \dfrac{y}{b}\right)$

$V = \displaystyle\iint\limits_R f(x,\,y)\,dA = \int_0^a \int_0^{b[1-(x/a)]} c\left(1 - \dfrac{x}{a} - \dfrac{y}{b}\right) dy\,dx$

$\qquad = \displaystyle\int_0^a c\left(y - \dfrac{xy}{a} - \dfrac{y^2}{2b}\right)\bigg]_0^{b[1-(x/a)]} dx$

$\qquad = \displaystyle\int_0^a c\left[b\left(1 - \dfrac{x}{a}\right) - \dfrac{xb}{a}\left(1 - \dfrac{x}{a}\right) - \dfrac{b^2}{2b}\left(1 - \dfrac{x}{a}\right)^2\right] dx$

$\qquad = c\left[-\dfrac{ab}{2}\left(1 - \dfrac{x}{a}\right)^2 - \dfrac{x^2 b}{2a} + \dfrac{x^3 b}{3a^2} + \dfrac{ab}{6}\left(1 - \dfrac{x}{a}\right)^3\right]_0^a$

$\qquad = c\left[\left(-\dfrac{ab}{2} + \dfrac{ab}{3}\right) - \left(-\dfrac{ab}{2} + \dfrac{ab}{6}\right)\right] = \dfrac{abc}{6}$

55. $V = \displaystyle\int_0^2 \int_0^{-0.5x+1} \dfrac{2}{(1+x^2)+y^2}\,dy\,dx = \int_0^2 \dfrac{2}{\sqrt{1+x^2}} \arctan\dfrac{y}{\sqrt{1+x^2}}\bigg]_0^{-0.5x+1} dx$

$\qquad = \displaystyle\int_0^2 \dfrac{2}{\sqrt{1+x^2}} \arctan\left(\dfrac{-0.5x+1}{\sqrt{1+x^2}}\right) dx$

$\qquad \approx 1.2315$

56. $\displaystyle\int_0^{16} \int_0^{4-\sqrt{y}} \ln(1+x+y)\,dx\,dy = \int_0^{16} (1+x+y)\left[-1 + \ln(1+x+y)\right]_0^{4-\sqrt{y}} dy$

$\qquad = \displaystyle\int_0^{16} [1+(4-\sqrt{y})+y][-1 + \ln[1+(4-\sqrt{y})+y]]\,dy$

$\qquad = \displaystyle\int_0^{16} [5 - \sqrt{y} + y][-1 + \ln(5 - \sqrt{y} + y)]\,dy$

$\qquad \approx 231.65$

Section 15.3   Change of Variables: Polar Coordinates

1. $\displaystyle\int_0^{2\pi} \int_0^6 3r^2 \sin\theta\,dr\,d\theta = \int_0^{2\pi} r^3 \sin\theta\bigg]_0^6 d\theta$

$\qquad = \displaystyle\int_0^{2\pi} 216 \sin\theta\,d\theta$

$\qquad = -216\cos\theta\bigg]_0^{2\pi} = 0$

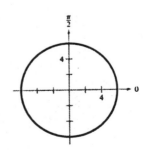

2. $\displaystyle\int_0^{\pi/4}\int_0^4 r^2\sin\theta\cos\theta\,dr\,d\theta = \int_0^{\pi/4}\frac{r^3}{3}\sin\theta\cos\theta\Big]_0^4\,d\theta$

$$= \left(\frac{64}{3}\right)\frac{\sin^2\theta}{2}\Big]_0^{\pi/4}$$

$$= \frac{16}{3}$$

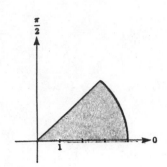

3. $\displaystyle\int_0^{\pi/2}\int_2^3\sqrt{9-r^2}\,r\,dr\,d\theta = \int_0^{\pi/2}-\frac{1}{3}(9-r^2)^{3/2}\Big]_2^3\,d\theta$

$$= \frac{5\sqrt5}{3}\theta\Big]_0^{\pi/2}$$

$$= \frac{5\sqrt5\,\pi}{6}$$

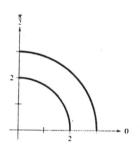

4. $\displaystyle\int_0^{\pi/2}\int_0^3 re^{-r^2}\,dr\,d\theta = \int_0^{\pi/2}-\frac{1}{2}e^{-r^2}\Big]_0^3\,d\theta$

$$= -\frac{1}{2}(e^{-9}-1)\theta\Big]_0^{\pi/2}$$

$$= \frac{\pi}{4}\left(1-\frac{1}{e^9}\right)$$

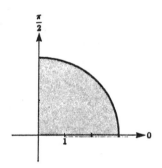

5. $\displaystyle\int_0^{\pi/2}\int_0^{1+\sin\theta}\theta\,dr\,d\theta = \int_0^{\pi/2}(\theta+\theta\sin\theta)\,d\theta$

$$= \left[\frac{\theta^2}{2}+\sin\theta-\theta\cos\theta\right]_0^{\pi/2}$$

$$= \frac{\pi^2}{8}+1$$

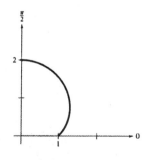

6. $\displaystyle\int_0^{\pi/2}\int_0^{1-\cos\theta}\sin\theta\,dr\,d\theta = \int_0^{\pi/2}(\sin\theta-\sin\theta\cos\theta)\,d\theta$

$$= \left[-\cos\theta+\frac{\cos^2\theta}{2}\right]_0^{\pi/2}$$

$$= \frac{1}{2}$$

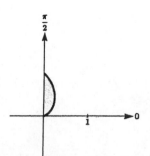

7. $A = \displaystyle\int_0^\pi \int_0^{6\cos\theta} r\, dr\, d\theta = \int_0^\pi 18\cos^2\theta\, d\theta = 9\int_0^\pi (1+\cos 2\theta)\, d\theta = 9\left(\theta + \dfrac{1}{2}\sin 2\theta\right)\Big]_0^\pi = 9\pi$

8. $A = \displaystyle\int_0^{2\pi} \int_2^4 r\, dr\, d\theta = \int_0^{2\pi} 6\, d\theta = 12\pi$

9. $\displaystyle\int_0^{2\pi} \int_0^{1+\cos\theta} r\, dr\, d\theta = \frac{1}{2}\int_0^{2\pi} (1 + 2\cos\theta + \cos^2\theta)\, d\theta = \frac{1}{2}\int_0^{2\pi}\left(1 + 2\cos\theta + \frac{1+\cos 2\theta}{2}\right) d\theta$

$$= \frac{1}{2}\left[\theta + 2\sin\theta + \frac{1}{2}\left(\theta + \frac{1}{2}\sin 2\theta\right)\right]_0^{2\pi} = \frac{3\pi}{2}$$

10. $\displaystyle\int_0^{2\pi} \int_0^{3-2\sin\theta} r\, dr\, d\theta = \frac{1}{2}\int_0^{2\pi}(9 - 12\sin\theta + 4\sin^2\theta)\, d\theta = \frac{1}{2}\int_0^{2\pi}[9 - 12\sin\theta + 2(1-\cos 2\theta)]\, d\theta$

$$= \frac{1}{2}\left[9\theta + 12\cos\theta + 2\theta - \sin 2\theta\right]_0^{2\pi}$$

$$= \frac{1}{2}[22\pi + 12 - 12] = 11\pi$$

11. $3\displaystyle\int_0^{\pi/3} \int_0^{2\sin 3\theta} r\, dr\, d\theta = \frac{3}{2}\int_0^{\pi/3} 4\sin^2 3\theta\, d\theta = 3\int_0^{\pi/3}(1-\cos 6\theta)\, d\theta = 3\left[\theta - \frac{1}{6}\sin 6\theta\right]_0^{\pi/3} = \pi$

12. $8\displaystyle\int_0^{\pi/4} \int_0^{3\cos 2\theta} r\, dr\, d\theta = 4\int_0^{\pi/4} 9\cos^2 2\theta\, d\theta = 18\int_0^{\pi/4}(1+\cos 4\theta)\, d\theta = 18\left[\theta + \frac{1}{4}\sin 4\theta\right]_0^{\pi/4} = \frac{9\pi}{2}$

13. $\displaystyle\int_0^a \int_0^{\sqrt{a^2-y^2}} y\, dx\, dy = \int_0^{\pi/2} \int_0^a r^2\sin\theta\, dr\, d\theta = \frac{a^3}{3}\int_0^{\pi/2}\sin\theta\, d\theta = \frac{a^3}{3}(-\cos\theta)\Big]_0^{\pi/2} = \frac{a^3}{3}$

14. $\displaystyle\int_0^a \int_0^{\sqrt{a^2-x^2}} x\, dy\, dx = \int_0^{\pi/2} \int_0^a r^2\cos\theta\, dr\, d\theta = \frac{a^3}{3}\int_0^{\pi/2}\cos\theta\, d\theta = \frac{a^3}{3}\sin\theta\Big]_0^{\pi/2} = \frac{a^3}{3}$

15. $\displaystyle\int_0^3 \int_0^{\sqrt{9-x^2}} \arctan\frac{y}{x}\, dy\, dx = \int_0^{\pi/2} \int_0^3 r\theta\, dr\, d\theta = \int_0^{\pi/2} \frac{9}{2}\theta\, d\theta = \frac{9}{4}\left(\frac{\pi}{2}\right)^2 = \frac{9\pi^2}{16}$

16. $\displaystyle\int_0^2 \int_y^{\sqrt{8-y^2}} \sqrt{x^2+y^2}\, dx\, dy = \int_0^{\pi/4} \int_0^{2\sqrt{2}} r^2\, dr\, d\theta$

$$= \int_0^{\pi/4} \frac{(2\sqrt{2})^3}{3}\, d\theta = \frac{(2\sqrt{2})^3}{3}\theta\Big]_0^{\pi/4} = \frac{(2\sqrt{2})^3}{3}\cdot\frac{\pi}{4} = \frac{4\sqrt{2}\,\pi}{3}$$

17. $\displaystyle\int_0^2 \int_0^{\sqrt{2x-x^2}} xy\, dy\, dx = \int_0^{\pi/2} \int_0^{2\cos\theta} r^3\cos\theta\sin\theta\, dr\, d\theta = 4\int_0^{\pi/2}\cos^5\theta\sin\theta\, d\theta = -\frac{4\cos^6\theta}{6}\Big]_0^{\pi/2} = \frac{2}{3}$

18. $\displaystyle\int_0^4 \int_0^{\sqrt{4y-y^2}} x^2\, dx\, dy = \int_0^{\pi/2} \int_0^{4\sin\theta} r^3\cos^2\theta\, dr\, d\theta$

$$= \int_0^{\pi/2} 64\sin^4\theta\cos^2\theta\, d\theta = 64\int_0^{\pi/2}(\sin^4\theta - \sin^6\theta)\, d\theta$$

$$= \frac{64}{6}\left[\sin^5\theta\cos\theta - \frac{\sin^3\theta\cos\theta}{4} + \frac{3}{8}(\theta - \sin\theta\cos\theta)\right]_0^{\pi/2} = 2\pi$$

19. $\displaystyle\int_0^2 \int_0^x \sqrt{x^2+y^2}\, dy\, dx + \int_2^{2\sqrt{2}} \int_0^{\sqrt{8-x^2}} \sqrt{x^2+y^2}\, dy\, dx = \int_0^{\pi/4} \int_0^{2\sqrt{2}} r^2\, dr\, d\theta$

$\displaystyle\qquad\qquad = \int_0^{\pi/4} \frac{16\sqrt{2}}{3}\, d\theta$

$\displaystyle\qquad\qquad = \frac{4\sqrt{2}\,\pi}{3}$

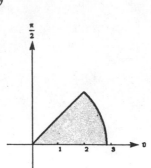

20. $\displaystyle\int_0^{(5\sqrt{2})/2} \int_0^x xy\, dy\, dx + \int_{(5\sqrt{2})/2}^5 \int_0^{\sqrt{25-x^2}} xy\, dy\, dx = \int_0^{\pi/4} \int_0^5 r^3 \sin\theta\cos\theta\, dr\, d\theta$

$\displaystyle\qquad\qquad = \int_0^{\pi/4} \frac{625}{4} \sin\theta\cos\theta\, d\theta$

$\displaystyle\qquad\qquad = \frac{625}{8} \sin^2\theta \Big]_0^{\pi/4}$

$\displaystyle\qquad\qquad = \frac{625}{16}$

21. $\displaystyle\int_0^2 \int_0^{\sqrt{4-x^2}} (x+y)\, dy\, dx = \int_0^{\pi/2} \int_0^2 (r\cos\theta + r\sin\theta)r\, dr\, d\theta = \int_0^{\pi/2} \int_0^2 (\cos\theta + \sin\theta)r^2\, dr\, d\theta$

$\displaystyle\qquad\qquad = \frac{8}{3} \int_0^{\pi/2} (\cos\theta + \sin\theta)\, d\theta$

$\displaystyle\qquad\qquad = \frac{8}{3}(\sin\theta - \cos\theta)\Big]_0^{\pi/2}$

$\displaystyle\qquad\qquad = \frac{16}{3}$

22. $\displaystyle\int_0^2 \int_0^{\sqrt{4-x^2}} e^{-(x^2+y^2)}\, dy\, dx = \int_0^{\pi/2} \int_0^2 e^{-r^2} r\, dr\, d\theta = \frac{1}{2} \int_0^{\pi/2} (1-e^{-4})\, d\theta = \frac{\pi}{4}(1-e^{-4})$

23. $\displaystyle\int_0^1 \int_0^{\sqrt{1-x^2}} \arctan\frac{y}{x}\, dy\, dx = \int_0^{\pi/2} \int_0^1 \theta r\, dr\, d\theta = \frac{1}{2} \int_0^{\pi/2} \theta\, d\theta = \frac{1}{4}\theta^2 \Big]_0^{\pi/2} = \frac{\pi^2}{16}$

24. $\displaystyle\int_0^3 \int_0^{\sqrt{9-x^2}} (9-x^2-y^2)\, dy\, dx = \int_0^{\pi/2} \int_0^3 (9-r^2)r\, dr\, d\theta = \int_0^{\pi/2} \int_0^3 (9r-r^3)\, dr\, d\theta$

$\displaystyle\qquad\qquad = \int_0^{\pi/2} \left(\frac{9}{2}r^2 - \frac{1}{4}r^4\right)\Big]_0^3\, d\theta$

$\displaystyle\qquad\qquad = \frac{81}{4} \int_0^{\pi/2} d\theta = \frac{81\pi}{8}$

25. $\displaystyle V = \int_0^{\pi/2} \int_0^1 (r\cos\theta)(r\sin\theta)r\, dr\, d\theta = \frac{1}{2} \int_0^{\pi/2} \int_0^1 r^3 \sin 2\theta\, dr\, d\theta = \frac{1}{8} \int_0^{\pi/2} \sin 2\theta\, d\theta$

$\displaystyle\qquad\qquad = -\frac{1}{16} \cos 2\theta \Big]_0^{\pi/2} = \frac{1}{8}$

26. $V = 4 \int_0^{\pi/2} \int_0^2 (r^2 + 1) r \, dr \, d\theta = 24 \int_0^{\pi/2} d\theta = 12\pi$

27. $V = \int_0^{2\pi} \int_0^5 r^2 \, dr \, d\theta = \dfrac{250\pi}{3}$

28. $V = \int_0^{2\pi} \int_2^4 r^2 \, dr \, d\theta = \dfrac{112\pi}{3}$

29. $V = \int_0^{\pi} \int_0^{4\cos\theta} \sqrt{16 - r^2}\, r \, dr \, d\theta = \int_0^{\pi} -\dfrac{1}{3}(\sqrt{16 - r^2})^3 \Big]_0^{4\cos\theta} d\theta = -\dfrac{1}{3} \int_0^{\pi} (64\sin^3\theta - 64)\, d\theta$

$$= \dfrac{64}{3} \int_0^{\pi} [1 - \sin\theta(1 - \cos^2\theta)]\, d\theta$$

$$= \dfrac{64}{3} \left[\theta + \cos\theta - \dfrac{\cos^3\theta}{3} \right]_0^{\pi}$$

$$= \dfrac{64}{9}(3\pi - 4)$$

30. $V = \int_0^{2\pi} \int_1^4 \sqrt{16 - r^2}\, r \, dr \, d\theta = \int_0^{2\pi} -\dfrac{1}{3}(\sqrt{16 - r^2})^3 \Big]_1^4 d\theta = \int_0^{2\pi} 5\sqrt{15}\, d\theta = 10\sqrt{15}\,\pi$

31. $V = \int_0^{2\pi} \int_a^4 \sqrt{16 - r^2}\, r \, dr \, d\theta = \int_0^{2\pi} -\dfrac{1}{3}(\sqrt{16 - r^2})^3 \Big]_a^4 d\theta = \dfrac{1}{3}(\sqrt{16 - a^2})^3 (2\pi)$

One-half the volume of the hemisphere is $\dfrac{64\pi}{3}$.

$$\dfrac{2\pi}{3}(16 - a^2)^{3/2} = \dfrac{64\pi}{3}$$

$$(16 - a^2)^{3/2} = 32$$

$$16 - a^2 = 32^{2/3}$$

$$a^2 = 16 - 32^{2/3} = 16 - 8\sqrt[3]{2}$$

$$a = \sqrt{4(4 - 2\sqrt[3]{2})} = 2\sqrt{4 - 2\sqrt[3]{2}} \approx 2.4332$$

32. $x^2 + y^2 + z^2 = a^2 \Rightarrow z = \sqrt{a^2 - (x^2 + y^2)} = \sqrt{a^2 - r^2}$

$$V = 8 \int_0^{\pi/2} \int_0^a \sqrt{a^2 - r^2}\, r \, dr \, d\theta \quad \text{(8 times the volume in the first octant)}$$

$$= 8 \int_0^{\pi/2} -\dfrac{1}{2} \cdot \dfrac{2}{3}(a^2 - r^2)^{3/2} \Big]_0^a d\theta$$

$$= 8 \int_0^{\pi/2} \dfrac{a^3}{3}\, d\theta = \dfrac{8a^3}{3}\theta \Big]_0^{\pi/2}$$

$$= \dfrac{4\pi a^3}{3}$$

33. $I^2 = \int_{-\infty}^{\infty} \int_{-\infty}^{\infty} e^{-(x^2+y^2)/2}\, dA = 4 \int_0^{\pi/2} \int_0^{\infty} e^{-r^2/2}\, r \, dr \, d\theta = 4 \int_0^{\pi/2} -e^{-r^2/2} \Big]_0^{\infty} d\theta = 4 \int_0^{\pi/2} d\theta = 2\pi$

Therefore, $I = \sqrt{2\pi}$.

34. $I^2 = \left(\int_{-\infty}^{\infty} e^{-x^2} \, dx \right) \left(\int_{-\infty}^{\infty} e^{-y^2} \, dy \right)$

$= \int_{-\infty}^{\infty} \int_{-\infty}^{\infty} e^{-(x^2+y^2)} \, dA = 4 \int_{0}^{\pi/2} \int_{0}^{\infty} e^{-r^2} r \, dr \, d\theta = 4 \int_{0}^{\pi/2} \left[-\frac{1}{2} e^{-r^2} \right]_{0}^{\infty} d\theta = 2 \int_{0}^{\pi/2} d\theta = \pi$

Therefore, $I = \sqrt{\pi}$.

35. $A = \int_{0}^{\Delta\theta} \int_{0}^{r_2} r \, dr \, d\theta - \int_{0}^{\Delta\theta} \int_{0}^{r_1} r \, dr \, d\theta$

$= \int_{0}^{\Delta\theta} \frac{r_2{}^2}{2} \, d\theta - \int_{0}^{\Delta\theta} \frac{r_1{}^2}{2} \, d\theta = \frac{r_2{}^2}{2} \Delta\theta - \frac{r_1{}^2}{2} \Delta\theta = \frac{r_2{}^2 - r_1{}^2}{2} \Delta\theta = \frac{(r_2 + r_1)}{2} (r_2 - r_1) \Delta\theta = r \, \Delta r \, \Delta\theta$

Section 15.4   Center of Mass and Moments of Inertia

1. (a) $m = \int_{0}^{a} \int_{0}^{b} k \, dy \, dx = kab$

$M_x = \int_{0}^{a} \int_{0}^{b} ky \, dy \, dx = \frac{kab^2}{2}$

$M_y = \int_{0}^{a} \int_{0}^{b} kx \, dy \, dx = \frac{ka^2 b}{2}$

$\overline{x} = \frac{M_y}{m} = \frac{ka^2 b/2}{kab} = \frac{a}{2}$

$\overline{y} = \frac{M_x}{m} = \frac{kab^2/2}{kab} = \frac{b}{2}$

(b) $m = \int_{0}^{a} \int_{0}^{b} ky \, dy \, dx = \frac{kab^2}{2}$

$M_x = \int_{0}^{a} \int_{0}^{b} ky^2 \, dy \, dx = \frac{kab^3}{3}$

$M_y = \int_{0}^{a} \int_{0}^{b} kxy \, dy \, dx = \frac{ka^2 b^2}{4}$

$\overline{x} = \frac{M_y}{m} = \frac{ka^2 b^2/4}{kab^2/2} = \frac{a}{2}$

$\overline{y} = \frac{M_x}{m} = \frac{kab^3/3}{kab^2/2} = \frac{2}{3} b$

2. (a) $m = \int_{0}^{a} \int_{0}^{b} kxy \, dy \, dx = \frac{ka^2 b^2}{4}$

$M_x = \int_{0}^{a} \int_{0}^{b} kxy^2 \, dy \, dx = \frac{ka^2 b^3}{6}$

$M_y = \int_{0}^{a} \int_{0}^{b} kx^2 y \, dy \, dx = \frac{ka^3 b^2}{6}$

$\overline{x} = \frac{M_y}{m} = \frac{ka^3 b^2/6}{ka^2 b^2/4} = \frac{2}{3} a$

$\overline{y} = \frac{M_x}{m} = \frac{ka^2 b^3/6}{ka^2 b^2/4} = \frac{2}{3} b$

(b) $m = \int_{0}^{a} \int_{0}^{b} k(x^2 + y^2) \, dy \, dx = \frac{kab}{3}(a^2 + b^2)$

$M_x = \int_{0}^{a} \int_{0}^{b} k(x^2 y + y^3) \, dy \, dx$

$= \frac{kab^2}{12}(2a^2 + 3b^2)$

$M_y = \int_{0}^{a} \int_{0}^{b} k(x^3 + xy^2) \, dy \, dx$

$= \frac{ka^2 b}{12}(3a^2 + 2b^2)$

$\overline{x} = \frac{M_y}{m} = \frac{(ka^2 b/12)(3a^2 + 2b^2)}{(kab/3)(a^2 + b^2)}$

$= \frac{a}{4} \left(\frac{3a^2 + 2b^2}{a^2 + b^2} \right)$

$\overline{y} = \frac{M_x}{m} = \frac{(kab^2/12)(2a^2 + 3b^2)}{(kab/3)(a^2 + b^2)} = \frac{b}{4} \left(\frac{2a^2 + 3b^2}{a^2 + b^2} \right)$

3. **(a)** $m = \dfrac{k}{2}bh$

$\bar{x} = \dfrac{b}{2}$ by symmetry

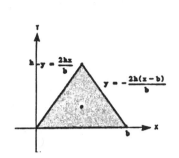

$$M_x = \int_0^{b/2}\int_0^{2hx/b} ky\, dy\, dx + \int_{b/2}^{b}\int_0^{-2h(x-b)/b} ky\, dy\, dx$$

$$= \frac{kbh^2}{12} + \frac{kbh^2}{12} = \frac{kbh^2}{6}$$

$$\bar{y} = \frac{M_x}{m} = \frac{kbh^2/6}{kbh/2} = \frac{h}{3}$$

(b) $m = \displaystyle\int_0^{b/2}\int_0^{2hx/b} ky\, dy\, dx + \int_{b/2}^{b}\int_0^{-2h(x-b)/b} ky\, dy\, dx = \dfrac{kbh^2}{6}$

$$M_x = \int_0^{b/2}\int_0^{2hx/b} ky^2\, dy\, dx + \int_{b/2}^{b}\int_0^{-2h(x-b)/b} ky^2\, dy\, dx = \frac{kbh^3}{12}$$

$$M_y = \int_0^{b/2}\int_0^{2hx/b} kxy\, dy\, dx + \int_{b/2}^{b}\int_0^{-2h(x-b)/b} kxy\, dy\, dx = \frac{kb^2h^2}{12}$$

$$\bar{x} = \frac{M_y}{m} = \frac{kb^2h^2/12}{kbh^2/6} = \frac{b}{2}$$

$$\bar{y} = \frac{M_x}{m} = \frac{kbh^3/12}{kbh^2/6} = \frac{h}{2}$$

4. **(a)** $m = \dfrac{a^2 k}{2}$

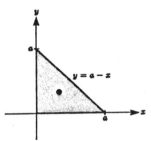

$$M_x = \int_0^{a}\int_0^{a-x} ky\, dy\, dx = \frac{ka^3}{6}$$

$M_y = M_x$ by symmetry

$$\bar{x} = \bar{y} = \frac{M_x}{m} = \frac{ka^3/6}{ka^2/2} = \frac{a}{3}$$

(b) $m = \displaystyle\int_0^{a}\int_0^{a-x} (x^2 + y^2)\, dy\, dx$

$$= \int_0^{a}\left(x^2 y + \frac{1}{3}y^3 \right)\Big]_0^{a-x} dx = \int_0^{a}\left[ax^2 - x^3 + \frac{1}{3}(a-x)^3 \right] dx = \frac{a^4}{6}$$

$$M_y = \int_0^{a}\int_0^{a-x} (x^3 + xy^2)\, dy\, dx$$

$$= \int_0^{a}\left(ax^3 - x^4 + \frac{1}{3}a^3 x - a^2 x^2 + ax^3 - \frac{1}{3}x^4 \right) dx = \frac{1}{3}\int_0^{a}(a^3 x - 3a^2 x^2 + 6ax^3 - 4x^4)\, dx = \frac{a^5}{15}$$

$$\bar{x} = \frac{M_y}{m} = \frac{a^5/15}{a^4/6} = \frac{2a}{5}$$

$$\bar{y} = \frac{2a}{5} \text{ by symmetry}$$

5. (a) $\bar{x} = 0$ by symmetry

$$m = \frac{\pi a^2 k}{2}$$

$$M_x = \int_{-a}^{a} \int_{0}^{\sqrt{a^2-x^2}} ky \, dy \, dx = \frac{2a^3 k}{3}$$

$$\bar{y} = \frac{M_x}{m} = \frac{2a^3 k}{3} \cdot \frac{2}{\pi a^2 k} = \frac{4a}{3\pi}$$

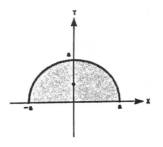

(b) $$m = \int_{-a}^{a} \int_{0}^{\sqrt{a^2-x^2}} k(a-y)y \, dy \, dx = \frac{a^4 k}{24}(16 - 3\pi)$$

$$M_x = \int_{-a}^{a} \int_{0}^{\sqrt{a^2-x^2}} k(a-y)y^2 \, dy \, dx = \frac{a^5 k}{120}(15\pi - 32)$$

$$M_y = \int_{-a}^{a} \int_{0}^{\sqrt{a^2-x^2}} kx(a-y)y \, dy \, dx = 0$$

$$\bar{x} = \frac{M_y}{m} = 0$$

$$\bar{y} = \frac{M_x}{m} = \frac{a}{5}\left[\frac{15\pi - 32}{16 - 3\pi}\right]$$

6. (a) $$m = \int_{0}^{a} \int_{0}^{\sqrt{a^2-x^2}} k \, dy \, dx = \frac{k\pi a^2}{4}$$

$$M_y = \int_{0}^{a} \int_{0}^{\sqrt{a^2-x^2}} kx \, dy \, dx = k\int_{0}^{a} x\sqrt{a^2 - x^2} \, dx = -\frac{k}{3}(a^2 - x^2)^{3/2}\Big]_{0}^{a} = \frac{ka^3}{3}$$

$$\bar{x} = \frac{M_y}{m} = \frac{ka^3/3}{k\pi a^2/4} = \frac{4a}{3\pi}$$

$$\bar{y} = \frac{4a}{3\pi} \text{ by symmetry}$$

(b) $$m = \int_{0}^{a} \int_{0}^{\sqrt{a^2-x^2}} k(x^2 + y^2) \, dy \, dx = \int_{0}^{\pi/2} \int_{0}^{a} kr^3 \, dr \, d\theta = \frac{ka^4 \pi}{8}$$

$$M_x = \int_{0}^{a} \int_{0}^{\sqrt{a^2-x^2}} k(x^2 + y^2)y \, dy \, dx = \int_{0}^{\pi/2} \int_{0}^{a} kr^4 \sin\theta \, dr \, d\theta = \frac{ka^5}{5}$$

$M_y = M_x$ by symmetry

$$\bar{x} = \bar{y} = \frac{M_x}{m} = \frac{ka^5}{5} \cdot \frac{8}{ka^4 \pi} = \frac{8a}{5\pi}$$

7. $$m = \int_{0}^{4} \int_{0}^{\sqrt{x}} kxy \, dy \, dx = \frac{32k}{3}$$

$$M_x = \int_{0}^{4} \int_{0}^{\sqrt{x}} kxy^2 \, dy \, dx = \frac{256k}{21}$$

$$M_y = \int_{0}^{4} \int_{0}^{\sqrt{x}} kx^2 y \, dy \, dx = 32k$$

$$\bar{x} = \frac{M_y}{m} = \frac{32k}{1} \cdot \frac{3}{32k} = 3$$

$$\bar{y} = \frac{M_x}{m} = \frac{256k}{21} \cdot \frac{3}{32k} = \frac{8}{7}$$

8. $m = \int_0^4 \int_0^{x^2} kx \, dy \, dx = 64k$

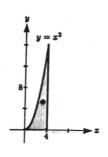

$M_x = \int_0^4 \int_0^{x^2} kxy \, dy \, dx = \dfrac{1024k}{3}$

$M_y = \int_0^4 \int_0^{x^2} kx^2 \, dy \, dx = \dfrac{1024k}{5}$

$\bar{x} = \dfrac{M_y}{m} = \dfrac{1024k}{5} \cdot \dfrac{1}{64k} = \dfrac{16}{5}$

$\bar{y} = \dfrac{M_x}{m} = \dfrac{1024k}{3} \cdot \dfrac{1}{64k} = \dfrac{16}{3}$

9. $m = \int_0^2 \int_0^{e^{-x}} ky \, dy \, dx = \dfrac{k}{4}\left(1 - e^{-4}\right)$

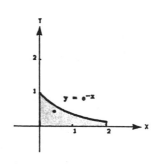

$M_x = \int_0^2 \int_0^{e^{-x}} ky^2 \, dy \, dx = \dfrac{k}{9}\left(1 - e^{-6}\right)$

$M_y = \int_0^2 \int_0^{e^{-x}} kxy \, dy \, dx = \dfrac{k\left(1 - 5e^{-4}\right)}{8}$

$\bar{x} = \dfrac{M_y}{m} = \dfrac{k\left(e^4 - 5\right)}{8e^4} \cdot \dfrac{4e^4}{k\left(e^4 - 1\right)} = \dfrac{e^4 - 5}{2\left(e^4 - 1\right)}$

$\bar{y} = \dfrac{M_x}{m} = \dfrac{k\left(e^6 - 1\right)}{9e^6} \cdot \dfrac{4e^4}{k\left(e^4 - 1\right)} = \dfrac{4}{9}\left[\dfrac{e^6 - 1}{e^6 - e^2}\right]$

10. $m = \int_1^e \int_0^{\ln x} \dfrac{k}{x} \, dy \, dx = \dfrac{k}{2}$

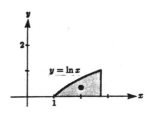

$M_x = \int_1^e \int_0^{\ln x} \dfrac{k}{x} y \, dy \, dx = \dfrac{k}{6}$

$M_y = \int_1^e \int_0^{\ln x} k \, dy \, dx = k$

$\bar{x} = \dfrac{M_y}{m} = \dfrac{k}{1} \cdot \dfrac{2}{k} = 2$

$\bar{y} = \dfrac{M_x}{m} = \dfrac{k}{6} \cdot \dfrac{2}{k} = \dfrac{1}{3}$

11. $\bar{x} = 0$ by symmetry

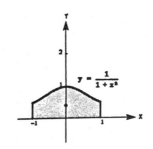

$m = \int_{-1}^1 \int_0^{1/(1+x^2)} k \, dy \, dx = \dfrac{k\pi}{2}$

$M_x = \int_{-1}^1 \int_0^{1/(1+x^2)} ky \, dy \, dx = \dfrac{k}{8}\left(2 + \pi\right)$

$\bar{y} = \dfrac{M_x}{m} = \dfrac{k}{8}\left(2 + \pi\right) \cdot \dfrac{2}{k\pi} = \dfrac{2 + \pi}{4\pi}$

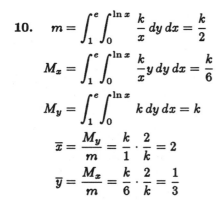

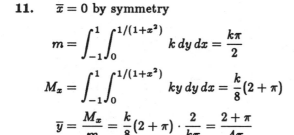

12. $m = \int_1^4 \int_0^{4/x} kx^2 \, dy \, dx = 30k$

$M_x = \int_1^4 \int_0^{4/x} kx^2 y \, dy \, dx = 24k$

$M_y = \int_1^4 \int_0^{4/x} kx^3 \, dy \, dx = 84k$

$\bar{x} = \dfrac{M_y}{m} = \dfrac{84k}{30k} = \dfrac{14}{5}$

$\bar{y} = \dfrac{M_x}{m} = \dfrac{24k}{30k} = \dfrac{4}{5}$

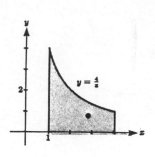

13. $\bar{y} = 0$ by symmetry

$m = \int_{-4}^4 \int_0^{16-y^2} kx \, dx \, dy = \dfrac{8192k}{15}$

$M_y = \int_{-4}^4 \int_0^{16-y^2} kx^2 \, dx \, dy = \dfrac{524288k}{105}$

$\bar{x} = \dfrac{M_y}{m} = \dfrac{524288k}{105} \cdot \dfrac{15}{8192k} = \dfrac{64}{7}$

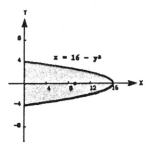

14. $\bar{x} = 0$ by symmetry

$m = \int_{-3}^3 \int_0^{9-x^2} ky^2 \, dy \, dx = \dfrac{23328k}{35}$

$M_x = \int_{-3}^3 \int_0^{9-x^2} ky^3 \, dy \, dx = \dfrac{139968k}{35}$

$\bar{y} = \dfrac{M_x}{m} = \dfrac{139968k}{35} \cdot \dfrac{35}{23328k} = 6$

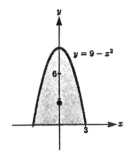

15. $\bar{x} = \dfrac{L}{2}$ by symmetry

$m = \int_0^L \int_0^{\sin \pi x/L} ky \, dy \, dx = \dfrac{kL}{4}$

$M_x = \int_0^L \int_0^{\sin \pi x/L} ky^2 \, dy \, dx = \dfrac{4kL}{9\pi}$

$\bar{y} = \dfrac{M_x}{m} = \dfrac{4kL}{9\pi} \cdot \dfrac{4}{kL} = \dfrac{16}{9\pi}$

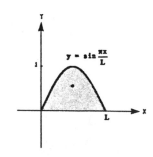

16. $m = \int_0^{L/2} \int_0^{\cos \pi x/L} k \, dy \, dx = \dfrac{kL}{\pi}$

$M_x = \int_0^{L/2} \int_0^{\cos \pi x/L} ky \, dy \, dx = \dfrac{kL}{8}$

$M_y = \int_0^{L/2} \int_0^{\cos \pi x/L} kx \, dy \, dx = \dfrac{L^2(\pi - 2)k}{2\pi^2}$

$\bar{x} = \dfrac{M_y}{m} = \dfrac{L^2(\pi - 2)k}{2\pi^2} \cdot \dfrac{\pi}{kL} = \dfrac{L(\pi - 2)}{2\pi}$

$\bar{y} = \dfrac{M_x}{m} = \dfrac{kL}{8} \cdot \dfrac{\pi}{kL} = \dfrac{\pi}{8}$

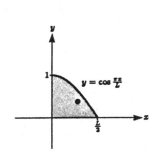

17. $m = \dfrac{\pi a^2 k}{8}$

$M_x = \int\!\!\int_R ky \, dA = \int_0^{\pi/4} \int_0^a kr^2 \sin \theta \, dr \, d\theta = \dfrac{ka^3(2 - \sqrt{2})}{6}$

$M_y = \int\!\!\int_R kx \, dA = \int_0^{\pi/4} \int_0^a kr^2 \cos \theta \, dr \, d\theta = \dfrac{ka^3 \sqrt{2}}{6}$

$\bar{x} = \dfrac{M_y}{m} = \dfrac{ka^3 \sqrt{2}}{6} \cdot \dfrac{8}{\pi a^2 k} = \dfrac{4a\sqrt{2}}{3\pi}$

$\bar{y} = \dfrac{M_x}{m} = \dfrac{ka^3(2 - \sqrt{2})}{6} \cdot \dfrac{8}{\pi a^2 k} = \dfrac{4a(2 - \sqrt{2})}{3\pi}$

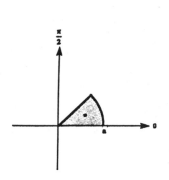

18. $m = \int\!\!\int_R k\sqrt{x^2 + y^2} \, dA = \int_0^{\pi/4} \int_0^a kr^2 \, dr \, d\theta = \dfrac{ka^3 \pi}{12}$

$M_x = \int\!\!\int_R k\sqrt{x^2 + y^2} \, y \, dA = \int_0^{\pi/4} \int_0^a kr^3 \sin \theta \, d\theta = \dfrac{ka^4(2 - \sqrt{2})}{8}$

$M_y = \int\!\!\int_R k\sqrt{x^2 + y^2} \, x \, dA = \int_0^{\pi/4} \int_0^a kr^3 \cos \theta \, d\theta = \dfrac{ka^4 \sqrt{2}}{8}$

$\bar{x} = \dfrac{M_y}{m} = \dfrac{ka^4 \sqrt{2}}{8} \cdot \dfrac{12}{ka^3 \pi} = \dfrac{3\sqrt{2}\, a}{2\pi}$

$\bar{y} = \dfrac{M_x}{m} = \dfrac{ka^4(2 - \sqrt{2})}{8} \cdot \dfrac{12}{ka^3 \pi} = \dfrac{3(2 - \sqrt{2})a}{2\pi}$

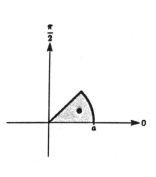

19. $\bar{y} = 0$ by symmetry

$m = \int\!\!\int_R k \, dA = \int_{-\pi/6}^{\pi/6} \int_0^{2\cos 3\theta} kr \, dr \, d\theta = \dfrac{k\pi}{3}$

$M_y = \int\!\!\int_R kx \, dA$

$= \int_{-\pi/6}^{\pi/6} \int_0^{2\cos 3\theta} kr^2 \cos \theta \, dr \, d\theta = \int_{-\pi/6}^{\pi/6} \dfrac{8k}{3} \cos^3 3\theta \cos \theta \, d\theta \approx 1.17k$

$\bar{x} = \dfrac{M_y}{m} \approx 1.17k \left(\dfrac{3}{\pi k} \right) \approx 1.12$

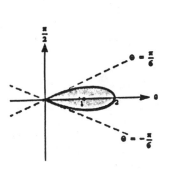

20. $\bar{y} = 0$ by symmetry

$$m = \iint_R k\,dA = \int_0^{2\pi}\int_0^{1+\cos\theta} kr\,dr\,d\theta = \frac{3\pi k}{2}$$

$$M_y = \iint_R kx\,dA = \int_0^{2\pi}\int_0^{1+\cos\theta} kr^2\cos\theta\,dr\,d\theta$$

$$= \frac{k}{3}\int_0^{2\pi}\cos\theta(1 + 3\cos\theta + 3\cos^2\theta + \cos^3\theta)\,d\theta$$

$$= \frac{k}{3}\int_0^{2\pi}\left[\cos\theta + \frac{3}{2}(1+\cos\theta) + 3\cos\theta(1-\sin^2\theta) + \frac{1}{4}(1+\cos 2\theta)^2\right]d\theta$$

$$= \frac{5k\pi}{4}$$

$$\bar{x} = \frac{M_y}{m} = \frac{5k\pi}{4}\cdot\frac{2}{3k\pi} = \frac{5}{6}$$

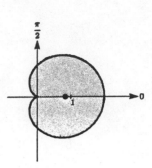

21. $m = bh$

$$I_x = \int_0^b\int_0^h y^2\,dy\,dx = \frac{bh^3}{3}$$

$$I_y = \int_0^b\int_0^h x^2\,dy\,dx = \frac{b^3h}{3}$$

$$\bar{\bar{x}} = \sqrt{\frac{I_y}{m}} = \sqrt{\frac{b^3h}{3}\cdot\frac{1}{bh}} = \sqrt{\frac{b^2}{3}} = \frac{b}{\sqrt{3}}$$

$$\bar{\bar{y}} = \sqrt{\frac{I_x}{m}} = \sqrt{\frac{bh^3}{3}\cdot\frac{1}{bh}} = \sqrt{\frac{h^2}{3}} = \frac{h}{\sqrt{3}}$$

22. $m = \int_0^b\int_0^{h-(hx/b)} dy\,dx = \frac{bh}{2}$

$$I_x = \int_0^b\int_0^{h-(hx/b)} y^2\,dy\,dx = \frac{bh^3}{12}$$

$$I_y = \int_0^b\int_0^{h-(hx/b)} x^2\,dy\,dx = \frac{b^3h}{12}$$

$$\bar{\bar{x}} = \sqrt{\frac{I_y}{m}} = \sqrt{\frac{b^3h/12}{bh/2}} = \frac{b}{\sqrt{6}}$$

$$\bar{\bar{y}} = \sqrt{\frac{I_x}{m}} = \sqrt{\frac{bh^3/12}{bh/2}} = \frac{h}{\sqrt{6}}$$

23. $m = \pi r^2$

$$I_x = \iint_R y^2\,dA = \int_0^{2\pi}\int_0^r r^3\sin^2\theta\,dr\,d\theta = \frac{r^4\pi}{4}$$

$$I_y = \iint_R x^2\,dA = \int_0^{2\pi}\int_0^r r^3\cos^2\theta\,dr\,d\theta = \frac{r^4\pi}{4}$$

$$I_0 = I_x + I_y = \frac{r^4\pi}{4} + \frac{r^4\pi}{4} = \frac{r^4\pi}{2}$$

$$\bar{\bar{x}} = \bar{\bar{y}} = \sqrt{\frac{I_x}{m}} = \sqrt{\frac{r^4\pi}{4}\cdot\frac{1}{\pi r^2}} = \frac{r}{2}$$

24. $m = \frac{\pi r^2}{2}$

$$I_x = \iint_R y^2\,dA = \int_0^\pi\int_0^r r^3\sin^2\theta\,dr\,d\theta = \frac{r^4\pi}{8}$$

$$I_y = \iint_R x^2\,dA = \int_0^\pi\int_0^r r^3\cos^2\theta\,dr\,d\theta = \frac{r^4\pi}{8}$$

$$I_0 = I_x + I_y = \frac{r^4\pi}{8} + \frac{r^4\pi}{8} = \frac{r^4\pi}{4}$$

$$\bar{\bar{x}} = \bar{\bar{y}} = \sqrt{\frac{I_x}{m}} = \sqrt{\frac{r^4\pi}{8}\cdot\frac{2}{\pi r^2}} = \frac{r}{2}$$

25. $m = \dfrac{\pi r^2}{4}$

$$I_x = \int\!\!\!\int_R y^2\, dA = \int_0^{\pi/2}\!\!\int_0^r r^3 \sin^2\theta\, dr\, d\theta = \frac{\pi r^4}{16}$$

$$I_y = \int\!\!\!\int_R x^2\, dA = \int_0^{\pi/2}\!\!\int_0^r r^3 \cos^2\theta\, dr\, d\theta = \frac{\pi r^4}{16}$$

$$I_0 = I_x + I_y = \frac{\pi r^4}{16} + \frac{\pi r^4}{16} = \frac{\pi r^4}{8}$$

$$\bar{\bar{x}} = \bar{\bar{y}} = \sqrt{\frac{I_x}{m}} = \sqrt{\frac{\pi r^4}{16}\cdot\frac{4}{\pi r^2}} = \frac{r}{2}$$

26. $m = \pi ab$

$$I_x = 4\int_0^a\!\!\int_0^{b/a\sqrt{a^2-x^2}} y^2\, dy\, dx$$

$$= 4\int_0^a \frac{b}{3a}(a^2 - x^2)^{3/2}\, dx$$

$$= \frac{4b}{3a}\int_0^a \left[a^2\sqrt{a^2-x^2} - x^2\sqrt{a^2-x^2}\,\right] dx$$

$$= \frac{4b}{3a}\left[\frac{a^2}{2}\left(x\sqrt{a^2-x^2} + a^2\arcsin\frac{x}{a}\right) - \frac{1}{8}\left[x(2x^2 - a^2)\sqrt{a^2-x^2} + a^4\arcsin\frac{x}{a}\right]\right]_0^a = \frac{a^3 b\pi}{4}$$

$$I_y = 4\int_0^b\!\!\int_0^{a/b\sqrt{b^2-y^2}} x^2\, dx\, dy = \frac{ab^3\pi}{4}$$

$$I_0 = I_x + I_y = \frac{a^3 b\pi}{4} + \frac{ab^3\pi}{4} = \frac{ab\pi}{4}(a^2 + b^2)$$

$$\bar{\bar{x}} = \sqrt{\frac{I_y}{m}} = \sqrt{\frac{ab^3\pi}{4}\cdot\frac{1}{\pi ab}} = \frac{b}{2}$$

$$\bar{\bar{y}} = \sqrt{\frac{I_x}{m}} = \sqrt{\frac{a^3 b\pi}{4}\cdot\frac{1}{\pi ab}} = \frac{a}{2}$$

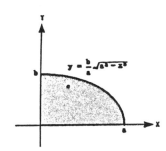

$y = \dfrac{b}{a}\sqrt{a^2-x^2}$

27. $\rho = ky$

$$m = k\int_0^a\!\!\int_0^b y\, dy\, dx = \frac{kab^2}{2}$$

$$I_x = k\int_0^a\!\!\int_0^b y^3\, dy\, dx = \frac{kab^4}{4}$$

$$I_y = k\int_0^a\!\!\int_0^b x^2 y\, dy\, dx = \frac{ka^3 b^2}{6}$$

$$I_0 = I_x + I_y = \frac{3kab^4 + 2kb^2 a^3}{12}$$

$$\bar{\bar{x}} = \sqrt{\frac{I_y}{m}} = \sqrt{\frac{ka^3 b^2/6}{kab^2/2}} = \sqrt{\frac{a^2}{3}} = \frac{a}{\sqrt{3}}$$

$$\bar{\bar{y}} = \sqrt{\frac{I_x}{m}} = \sqrt{\frac{kab^4/4}{kab^2/2}} = \sqrt{\frac{b^2}{2}} = \frac{b}{\sqrt{2}}$$

28. $\rho = ky$

$$m = 2k\int_0^a\!\!\int_0^{\sqrt{a^2-x^2}} y\, dy\, dx$$

$$= k\int_0^a (a^2 - x^2)\, dx = \frac{2ka^3}{3}$$

$$I_x = k\int_{-a}^a\!\!\int_0^{\sqrt{a^2-x^2}} y^3\, dy\, dx = \frac{4ka^5}{15}$$

$$I_y = k\int_{-a}^a\!\!\int_0^{\sqrt{a^2-x^2}} x^2 y\, dy\, dx = \frac{2ka^5}{15}$$

$$I_0 = I_x + I_y = \frac{2ka^5}{5}$$

$$\bar{\bar{x}} = \sqrt{\frac{I_y}{m}} = \sqrt{\frac{2ka^5/15}{2ka^3/3}} = \sqrt{\frac{a^2}{5}} = \frac{a}{\sqrt{5}}$$

$$\bar{\bar{y}} = \sqrt{\frac{I_x}{m}} = \sqrt{\frac{4ka^5/15}{2ka^3/3}} = \sqrt{\frac{2a^2}{5}} = \frac{2a}{\sqrt{10}}$$

29. $\rho = kx$

$$m = k\int_0^2 \int_0^{4-x^2} x\,dy\,dx = 4k$$

$$I_x = k\int_0^2 \int_0^{4-x^2} xy^2\,dy\,dx = \frac{32k}{3}$$

$$I_y = k\int_0^2 \int_0^{4-x^2} x^3\,dy\,dx = \frac{16k}{3}$$

$$I_0 = I_x + I_y = 16k$$

$$\overline{\overline{x}} = \sqrt{\frac{I_y}{m}} = \sqrt{\frac{16k/3}{4k}} = \sqrt{\frac{4}{3}} = \frac{2}{\sqrt{3}}$$

$$\overline{\overline{y}} = \sqrt{\frac{I_x}{m}} = \sqrt{\frac{32k/3}{4k}} = \sqrt{\frac{8}{3}} = \frac{4}{\sqrt{6}}$$

30. $\rho = kxy$

$$m = k\int_0^1 \int_{x^2}^x xy\,dy\,dx = \frac{k}{2}\int_0^1 (x^3 - x^5)\,dx = \frac{k}{24}$$

$$I_x = k\int_0^1 \int_{x^2}^x xy^3\,dy\,dx = \frac{k}{4}\int_0^1 (x^5 - x^9)\,dx = \frac{k}{60}$$

$$I_y = k\int_0^1 \int_{x^2}^x x^3 y\,dy\,dx = \frac{k}{2}\int_0^1 (x^5 - x^7)\,dx = \frac{k}{48}$$

$$I_0 = I_x + I_y = \frac{9k}{240} = \frac{3k}{80}$$

$$\overline{\overline{x}} = \sqrt{\frac{I_y}{m}} = \sqrt{\frac{k/48}{k/24}} = \frac{1}{\sqrt{2}}$$

$$\overline{\overline{y}} = \sqrt{\frac{I_x}{m}} = \sqrt{\frac{k/60}{k/24}} = \sqrt{\frac{2}{5}}$$

31. $\rho = kxy$

$$m = \int_0^4 \int_0^{\sqrt{x}} kxy\,dy\,dx = \frac{32k}{3}$$

$$I_x = \int_0^4 \int_0^{\sqrt{x}} kxy^3\,dy\,dx = 16k$$

$$I_y = \int_0^4 \int_0^{\sqrt{x}} kx^3 y\,dy\,dx = \frac{512k}{5}$$

$$I_0 = I_x + I_y = \frac{592k}{5}$$

$$\overline{\overline{x}} = \sqrt{\frac{I_y}{m}} = \sqrt{\frac{512k}{5}\cdot\frac{3}{32k}} = \sqrt{\frac{48}{5}} = \frac{4\sqrt{15}}{5}$$

$$\overline{\overline{y}} = \sqrt{\frac{I_x}{m}} = \sqrt{\frac{16k}{1}\cdot\frac{3}{32k}} = \sqrt{\frac{3}{2}} = \frac{\sqrt{6}}{2}$$

32. $\rho = x^2 + y^2$

$$m = \int_0^1 \int_{x^2}^{\sqrt{x}} (x^2 + y^2)\,dy\,dx = \frac{6}{35}$$

$$I_x = \int_0^1 \int_{x^2}^{\sqrt{x}} (x^2 + y^2)y^2\,dy\,dx = \frac{158}{2079}$$

$$I_y = \int_0^1 \int_{x^2}^{\sqrt{x}} (x^2 + y^2)x^2\,dy\,dx = \frac{158}{2079}$$

$$I_0 = I_x + I_y = \frac{316}{2079}$$

$$\overline{\overline{x}} = \sqrt{\frac{I_y}{m}} = \sqrt{\frac{158}{2079}\cdot\frac{35}{6}} = \sqrt{\frac{395}{891}}$$

$$\overline{\overline{y}} = \sqrt{\frac{I_x}{m}} = \overline{\overline{x}} = \sqrt{\frac{395}{891}}$$

33. $\rho = kx$

$$m = \int_0^1 \int_{x^2}^{\sqrt{x}} kx\,dy\,dx = \frac{3k}{20}$$

$$I_x = \int_0^1 \int_{x^2}^{\sqrt{x}} kxy^2\,dy\,dx = \frac{3k}{56}$$

$$I_y = \int_0^1 \int_{x^2}^{\sqrt{x}} kx^3\,dy\,dx = \frac{k}{18}$$

$$I_0 = I_x + I_y = \frac{55k}{504}$$

$$\overline{\overline{x}} = \sqrt{\frac{I_y}{m}} = \sqrt{\frac{k}{18}\cdot\frac{20}{3k}} = \frac{\sqrt{30}}{9}$$

$$\overline{\overline{y}} = \sqrt{\frac{I_x}{m}} = \sqrt{\frac{3k}{56}\cdot\frac{20}{3k}} = \frac{\sqrt{70}}{14}$$

34. $\rho = ky$

$$m = 2\int_0^2 \int_{x^3}^{4x} ky\,dy\,dx = \frac{512k}{21}$$

$$I_x = 2\int_0^2 \int_{x^3}^{4x} ky^3\,dy\,dx = \frac{32768k}{65}$$

$$I_y = 2\int_0^2 \int_{x^3}^{4x} kx^2 y\,dy\,dx = \frac{2048k}{45}$$

$$I_0 = I_x + I_y = \frac{321536k}{585}$$

$$\overline{\overline{x}} = \sqrt{\frac{I_y}{m}} = \sqrt{\frac{2048k}{45}\cdot\frac{21}{512k}} = \sqrt{\frac{28}{15}} = \frac{2\sqrt{105}}{15}$$

$$\overline{\overline{y}} = \sqrt{\frac{I_x}{m}} = \sqrt{\frac{32768k}{65}\cdot\frac{21}{512k}} = \frac{8\sqrt{1365}}{65}$$

35. $I = 2k \displaystyle\int_{-b}^{b}\int_{0}^{\sqrt{b^2-x^2}} (x-a)^2 \, dy \, dx = 2k \int_{-b}^{b} (x-a)^2 \sqrt{b^2-x^2} \, dx$

$= 2k \left[\displaystyle\int_{-b}^{b} x^2 \sqrt{b^2-x^2} \, dx - 2a \int_{-b}^{b} x\sqrt{b^2-x^2} \, dx + a^2 \int_{-b}^{b} \sqrt{b^2-x^2} \, dx \right]$

$= 2k \left[\dfrac{\pi b^4}{8} + 0 + \dfrac{\pi a^2 b^2}{2} \right] = \dfrac{k\pi b^2}{4}(b^2 + 4a^2)$

36. $I = \displaystyle\int_{0}^{4}\int_{0}^{2} k(x-6)^2 \, dy \, dx = \int_{0}^{4} 2k(x-6)^2 \, dx = \dfrac{2k}{3}(x-6)^3 \Big]_{0}^{4} = \dfrac{416k}{3}$

37. $I = \displaystyle\int_{0}^{4}\int_{0}^{\sqrt{x}} kx(x-6)^2 \, dy \, dx = \int_{0}^{4} kx\sqrt{x}(x^2 - 12x + 36) \, dx = k\left[\dfrac{2}{9}x^{9/2} - \dfrac{24}{7}x^{7/2} + \dfrac{72}{5}x^{5/2} \right]_{0}^{4} = \dfrac{42752k}{315}$

38. $I = \displaystyle\int_{-a}^{a}\int_{0}^{\sqrt{a^2-x^2}} ky(y-a)^2 \, dy \, dx$

$= \displaystyle\int_{-a}^{a} k\left[\dfrac{y^4}{4} - \dfrac{2ay^3}{3} + \dfrac{a^2 y^2}{2} \right]_{0}^{\sqrt{a^2-x^2}} dx$

$= \displaystyle\int_{-a}^{a} k\left[\dfrac{1}{4}(a^4 - 2a^2 x^2 + x^4) - \dfrac{2a}{3}(a^2\sqrt{a^2-x^2} - x^2\sqrt{a^2-x^2}) + \dfrac{a^2}{2}(a^2 - x^2) \right] dx$

$= k\left[\dfrac{1}{4}\left(a^4 x - \dfrac{2a^2 x^3}{3} + \dfrac{a^5}{5} \right) - \dfrac{2a}{3}\left[\dfrac{a^2}{2}\left(x\sqrt{a^2-x^2} + a^2 \arcsin\dfrac{x}{a} \right) \right. \right.$

$\left. \left. - \dfrac{1}{8}\left(x(2x^2 - a^2)\sqrt{a^2-x^2} + a^4 \arcsin\dfrac{x}{a} \right) \right] + \dfrac{a^2}{2}\left(a^2 x - \dfrac{x^3}{3} \right) \right]_{-a}^{a}$

$= 2k\left[\dfrac{1}{4}\left(a^5 - \dfrac{2}{3}a^5 + \dfrac{1}{5}a^5 \right) - \dfrac{2a}{3}\left(\dfrac{a^4 \pi}{4} - \dfrac{a^4 \pi}{16} \right) + \dfrac{a^2}{2}\left(a^3 - \dfrac{a^3}{3} \right) \right]$

$= 2k\left(\dfrac{7a^5}{15} - \dfrac{a^5 \pi}{8} \right) = ka^5\left(\dfrac{56 - 15\pi}{60} \right)$

39. $I = \displaystyle\int_{0}^{a}\int_{0}^{\sqrt{a^2-x^2}} k(a-y)(y-a)^2 \, dy \, dx$

$= \displaystyle\int_{0}^{a}\int_{0}^{\sqrt{a^2-x^2}} k(a-y)^3 \, dy \, dx$

$= \displaystyle\int_{0}^{a} -\dfrac{k}{4}(a-y)^4 \Big]_{0}^{\sqrt{a^2-x^2}} dx$

$= -\dfrac{k}{4} \displaystyle\int_{0}^{a} (a^4 - 4a^3 y + 6a^2 y^2 - 4ay^3 + y^4) \Big]_{0}^{\sqrt{a^2-x^2}} dx$

$= -\dfrac{k}{4} \displaystyle\int_{0}^{a} [a^4 - 4a^3\sqrt{a^2-x^2} + 6a^2(a^2 - x^2) - 4a(a^2 - x^2)\sqrt{a^2-x^2} + (a^4 - 2a^2 x^2 + x^4) - a^4] \, dx$

$= -\dfrac{k}{4} \displaystyle\int_{0}^{a} [7a^4 - 8a^2 x^2 + x^4 - 8a^3\sqrt{a^2-x^2} + 4ax^2\sqrt{a^2-x^2}] \, dx$

$= -\dfrac{k}{4}\left[7a^4 x - \dfrac{8a^2}{3}x^3 + \dfrac{x^5}{5} - 4a^3\left(x\sqrt{a^2-x^2} + a^2 \arcsin\dfrac{x}{a} \right) + \dfrac{a}{2}\left(x(2x^2 - a^2)\sqrt{a^2-x^2} + a^4 \arcsin\dfrac{x}{a} \right) \right]_{0}^{a}$

$= -\dfrac{k}{4}\left(7a^5 - \dfrac{8}{3}a^5 + \dfrac{1}{5}a^5 - 2a^5\pi + \dfrac{1}{4}a^5\pi \right) = a^5 k\left(\dfrac{7\pi}{16} - \dfrac{17}{15} \right)$

40. $I = \int_{-2}^{2}\int_{0}^{4-x^2} k(y-2)^2\, dy\, dx = \int_{-2}^{2} \frac{k}{3}(y-1)^3 \Big]_{0}^{4-x^2} dx = \int_{-2}^{2} \frac{k}{3}[(2-x^2)^3 + 8]\, dx$

$$= \frac{k}{3}\int_{-2}^{2} (16 - 12x^2 + 6x^4 - x^6)\, dx$$

$$= \frac{k}{3}\left(16x - 4x^3 + \frac{6}{5}x^5 - \frac{1}{7}x^7\right)\Big]_{-2}^{2}$$

$$= \frac{2k}{3}\left(32 - 32 + \frac{192}{5} - \frac{128}{7}\right)$$

$$= \frac{1408k}{105}$$

41. $m = \int_{1}^{10}\int_{0}^{\ln x} \sqrt{x}\, dy\, dx = \int_{1}^{10} \sqrt{x}\ln x\, dx = \left[\frac{2}{3}x^{3/2}\ln x - \frac{4}{9}x^{3/2}\right]_{1}^{10} \approx 34.9326$

$M_x = \int_{1}^{10}\int_{0}^{\ln x} y\sqrt{x}\, dy\, dx = \int_{1}^{10} \frac{1}{2}(\ln x)^2\sqrt{x}\, dx \approx 32.5985$

$M_y = \int_{1}^{10}\int_{0}^{\ln x} x\sqrt{x}\, dy\, dx = \int_{1}^{10} x\sqrt{x}\ln x\, dx \approx 240.8201$

$\bar{x} = \dfrac{M_y}{m} \approx 6.8938$

$\bar{y} = \dfrac{M_x}{m} \approx 0.9332$

42. $I = \int_{-\sqrt{2}}^{\sqrt{2}}\int_{-\sqrt{2-x^2}}^{\sqrt{2-x^2}} 1.6(x-3)^2\, dy\, dx = 3.2\int_{-\sqrt{2}}^{\sqrt{2}}\int_{0}^{\sqrt{2-x^2}} (x-3)^2\, dy\, dx = 3.2\int_{-\sqrt{2}}^{\sqrt{2}} (x-3)^2\sqrt{2-x^2}\, dx \approx 95.5018$

43. Orient the xy-coordinate system so that L is along the y-axis and R is in the first quadrant. Then the volume of the solid is

$$V = \int\!\!\int_{R} 2\pi x\, dA$$

$$= 2\pi \int\!\!\int_{R} x\, dA$$

$$= 2\pi \left(\frac{\displaystyle\int\!\!\int_{R} x\, dA}{\displaystyle\int\!\!\int_{R} dA}\right)\int\!\!\int_{R} dA$$

$$= 2\pi\bar{x}A.$$

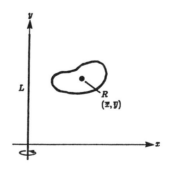

By our positioning, $\bar{x} = r$. Therefore, $V = 2\pi r A$.

Section 15.5   Surface Area

1. $f(x, y) = 2x + 2y$

R = triangle with vertices $(0, 0, 0)$, $(2, 0, 0)$, $(0, 2, 0)$

$f_x = 2, \quad f_y = 2$

$\sqrt{1 + (f_x)^2 + (f_y)^2} = 3$

$S = \int_0^2 \int_0^{2-x} 3 \, dy \, dx = 3 \int_0^2 (2 - x) \, dx = 3 \left(2x - \frac{x^2}{2} \right) \Big]_0^2 = 6$

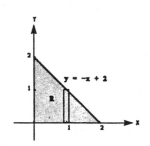

2. $f(x, y) = 10 + 2x - 3y$

R = square with vertices $(0, 0, 0)$, $(2, 0, 0)$, $(0, 2, 0)$, $(2, 2, 0)$

$f_x = 2, \quad f_y = -3$

$\sqrt{1 + (f_x)^2 + (f_y)^2} = \sqrt{14}$

$S = \int_0^2 \int_0^2 \sqrt{14} \, dy \, dx = \int_0^2 2\sqrt{14} \, dx = 4\sqrt{14}$

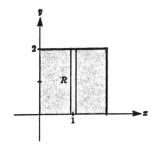

3. $f(x, y) = 8 + 2x + 2y$

$R = \{(x, y) : x^2 + y^2 \le 4\}$

$f_x = 2, \quad f_y = 2$

$\sqrt{1 + (f_x)^2 + (f_y)^2} = 3$

$S = \int_{-2}^2 \int_{-\sqrt{4-x^2}}^{\sqrt{4-x^2}} 3 \, dy \, dx$

$= \int_{-2}^2 6\sqrt{4 - x^2} \, dx = 3 \left(x\sqrt{4 - x^2} + 4 \arcsin \frac{x}{2} \right) \Big]_{-2}^2 = 12\pi$

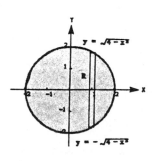

4. $f(x, y) = 10 + 2x - 3y$

$R = \{(x, y) : x^2 + y^2 \le 9\}$

$f_x = 2, \quad f_y = -3$

$\sqrt{1 + (f_x)^2 + (f_y)^2} = \sqrt{14}$

$S = \int_{-3}^3 \int_{-\sqrt{9-x^2}}^{\sqrt{9-x^2}} \sqrt{14} \, dy \, dx = \int_{-3}^3 2\sqrt{14}\sqrt{9 - x^2} \, dx$

$= \sqrt{14} \left(x\sqrt{9 - x^2} + 9 \arcsin \frac{x}{3} \right) \Big]_{-3}^3 = 9\sqrt{14}\,\pi$

5. $f(x, y) = 9 - x^2$

R = square with vertices $(0, 0, 0)$, $(3, 0, 0)$, $(0, 3, 0)$, $(3, 3, 0)$

$f_x = -2x$, $f_y = 0$

$\sqrt{1 + (f_x)^2 + (f_y)^2} = \sqrt{1 + 4x^2}$

$$S = \int_0^3 \int_0^3 \sqrt{1 + 4x^2}\, dy\, dx = \int_0^3 3\sqrt{1 + 4x^2}\, dx$$

$$= \frac{3}{4}\left(2x\sqrt{1 + 4x^2} + \ln|2x + \sqrt{1 + 4x^2}|\right)\Big]_0^3$$

$$= \frac{3}{4}(6\sqrt{37} + \ln|6 + \sqrt{37}|)$$

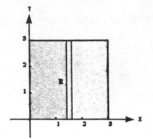

6. $f(x, y) = y^2$

R = square with vertices $(0, 0, 0)$, $(3, 0, 0)$, $(0, 3, 0)$, $(3, 3, 0)$

$f_x = 0$, $f_y = 2y$

$\sqrt{1 + (f_x)^2 + (f_y)^2} = \sqrt{1 + 4y^2}$

$$S = \int_0^3 \int_0^3 \sqrt{1 + 4y^2}\, dx\, dy = \int_0^3 3\sqrt{1 + 4y^2}\, dy$$

$$= \frac{3}{4}\left(2y\sqrt{1 + 4y^2} + \ln|2y + \sqrt{1 + 4y^2}|\right)\Big]_0^3$$

$$= \frac{3}{4}(6\sqrt{37} + \ln|6 + \sqrt{37}|)$$

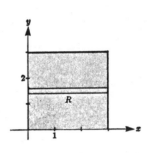

7. $f(x, y) = 2y + x^2$

R = triangle with vertices $(0, 0, 0)$, $(1, 0, 0)$, $(1, 1, 0)$

$f_x = 2x$, $f_y = 2$

$\sqrt{1 + (f_x)^2 + (f_y)^2} = \sqrt{5 + 4x^2}$

$$S = \int_0^1 \int_0^x \sqrt{5 + 4x^2}\, dy\, dx$$

$$= \int_0^1 x\sqrt{5 + 4x^2}\, dx = \frac{1}{12}(5 + 4x^2)^{3/2}\Big]_0^1 = \frac{1}{12}(27 - 5\sqrt{5})$$

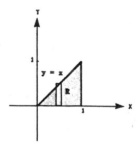

8. $f(x, y) = 2x + y^2$

R = triangle with vertices $(0, 0, 0)$, $(2, 0, 0)$, $(0, 2, 0)$

$f_x = 2$, $f_y = 2y$

$\sqrt{1 + (f_x)^2 + (f_y)^2} = \sqrt{5 + 4y^2}$

$$S = \int_0^2 \int_0^{2-y} \sqrt{5 + 4y^2}\, dx\, dx = \int_0^2 \sqrt{5 + 4y^2}(2 - y)\, dy$$

$$= \int_0^2 2\sqrt{5 + 4y^2}\, dy - \int_0^2 \sqrt{5 + 4y^2}\, y\, dy$$

$$= \left[\frac{1}{2}(2y\sqrt{5 + 4y^2} + 5\ln|2y + \sqrt{5 + 4y^2}|) - \frac{1}{2}(5 + 4y^2)^{3/2}\right]_0^2$$

$$= \left[\frac{1}{2}(4\sqrt{21} + 5\ln|4 + \sqrt{21}|) - \frac{1}{12}(21\sqrt{21})\right] - \left[\frac{5}{2}\ln\sqrt{5} - \frac{1}{12}(5\sqrt{5})\right]$$

$$= \frac{5}{2}\ln\left(\frac{4 + \sqrt{21}}{\sqrt{5}}\right) + \left(\frac{3\sqrt{21} + 5\sqrt{5}}{12}\right)$$

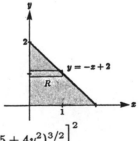

9. $f(x, y) = 2 + x^{3/2}$

R = quadrangle with vertices $(0, 0, 0)$, $(0, 4, 0)$, $(3, 4, 0)$, $(3, 0, 0)$

$f_x = \dfrac{3}{2}x^{1/2}$, $\quad f_y = 0$

$\sqrt{1 + (f_x)^2 + (f_y)^2} = \sqrt{1 + (9/4)x} = \dfrac{\sqrt{4 + 9x}}{2}$

$S = \displaystyle\int_0^3 \int_0^4 \dfrac{\sqrt{4 + 9x}}{2} \, dy \, dx = \int_0^3 4\left(\dfrac{\sqrt{4 + 9x}}{2}\right) dx$

$\qquad = \dfrac{4}{27}(4 + 9x)^{3/2}\Big]_0^3$

$\qquad = \dfrac{4}{27}(31\sqrt{31} - 8)$

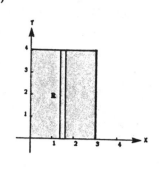

10. $f(x) = 2 + \dfrac{2}{3}x^{3/2}$

$R = \{(x, y) : 0 \le x \le 1, \ 0 \le y \le 1 - x\}$

$f_x = x^{1/2}$, $\quad f_y = 0$

$\sqrt{1 + (f_x)^2 + (f_y)^2} = \sqrt{1 + x}$

$S = \displaystyle\int_0^1 \int_0^{1-y} \sqrt{1 + x} \, dx \, dy = \int_0^1 \dfrac{2}{3}(1 + x)^{3/2}\Big]_0^{1-y} dy$

$\qquad = \dfrac{2}{3}\displaystyle\int_0^1 [(2 - y)^{3/2} - 1] \, dy$

$\qquad = \dfrac{2}{3}\left[-\dfrac{2}{5}(2 - y)^{5/2} - y\right]_0^1$

$\qquad = \dfrac{2}{3}\left[\left(-\dfrac{2}{5} - 1\right) - \left(-\dfrac{2}{5}[4\sqrt{2}]\right)\right]$

$\qquad = \dfrac{2}{3}\left(\dfrac{8\sqrt{2} - 7}{5}\right)$

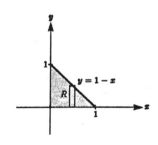

11. $f(x, y) = \ln|\sec x|$

$R = \left\{(x, y) : 0 \le x \le \dfrac{\pi}{4}, \ 0 \le y \le \tan x\right\}$

$f_x = \tan x$, $\quad f_y = 0$

$\sqrt{1 + (f_x)^2 + (f_y)^2} = \sqrt{1 + \tan^2 x} = \sec x$

$S = \displaystyle\int_0^{\pi/4} \int_0^{\tan x} \sec x \, dy \, dx$

$\quad = \displaystyle\int_0^{\pi/4} \sec x \tan x \, dx = \sec x\Big]_0^{\pi/4} = \sqrt{2} - 1$

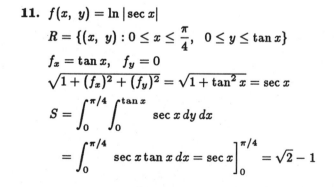

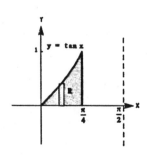

12. $f(x, y) = 4 + x^2 - y^2$

$R = \{(x, y) : x^2 + y^2 \le 1\}$

$f_x = 2x$, $\quad f_y = -2y$

$\sqrt{1 + (f_x)^2 + (f_y)^2} = \sqrt{1 + 4x^2 + 4y^2}$

$S = \displaystyle\int_{-1}^1 \int_{-\sqrt{1-x^2}}^{\sqrt{1-x^2}} \sqrt{1 + 4x^2 + 4y^2} \, dy \, dx$

$\quad = \displaystyle\int_0^{2\pi} \int_0^1 \sqrt{1 + 4r^2}\, r \, dr \, d\theta = \dfrac{(5\sqrt{5} - 1)\pi}{6}$

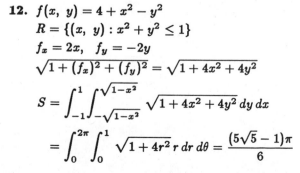

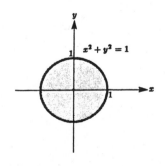

13. $f(x, y) = 4 - x^2 - y^2$

$R = \{(x, y) : 0 \le f(x, y)\}$

$0 \le 4 - x^2 - y^2, \quad x^2 + y^2 \le 4$

$f_x = -2x, \quad f_y = -2y$

$\sqrt{1 + (f_x)^2 + (f_y)^2} = \sqrt{1 + 4x^2 + 4y^2}$

$S = \displaystyle\int_{-2}^{2} \int_{-\sqrt{4-x^2}}^{\sqrt{4-x^2}} \sqrt{1 + 4x^2 + 4y^2}\, dy\, dx$

$= \displaystyle\int_{0}^{2\pi} \int_{0}^{2} \sqrt{1 + 4r^2}\, r\, dr\, d\theta = \dfrac{(17\sqrt{17} - 1)\pi}{6}$

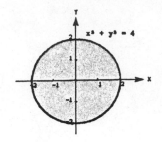

14. $f(x, y) = x^2 + y^2$

$R = \{(x, y) : 0 \le f(x, y) \le 16\}$

$0 \le x^2 + y^2 \le 16$

$f_x = 2x, \quad f_y = 2y$

$\sqrt{1 + (f_x)^2 + (f_y)^2} = \sqrt{1 + 4x^2 + 4y^2}$

$S = \displaystyle\int_{-4}^{4} \int_{-\sqrt{16-x^2}}^{\sqrt{16-x^2}} \sqrt{1 + 4x^2 + 4y^2}\, dy\, dx$

$= \displaystyle\int_{0}^{2\pi} \int_{0}^{4} \sqrt{1 + 4r^2}\, r\, dr\, d\theta = \dfrac{(65\sqrt{65} - 1)\pi}{6}$

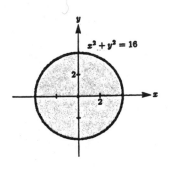

15. $f(x, y) = \sqrt{x^2 + y^2}$

$R = \{(x, y) : 0 \le f(x, y) \le 1\}$

$0 \le \sqrt{x^2 + y^2} \le 1, \quad x^2 + y^2 \le 1$

$f_x = \dfrac{x}{\sqrt{x^2 + y^2}}, \quad f_y = \dfrac{y}{\sqrt{x^2 + y^2}}$

$\sqrt{1 + (f_x)^2 + (f_y)^2} = \sqrt{1 + \dfrac{x^2}{x^2 + y^2} + \dfrac{y^2}{x^2 + y^2}} = \sqrt{2}$

$S = \displaystyle\int_{-1}^{1} \int_{-\sqrt{1-x^2}}^{\sqrt{1-x^2}} \sqrt{2}\, dy\, dx = \int_{0}^{2\pi} \int_{0}^{1} \sqrt{2}\, r\, dr\, d\theta = \sqrt{2}\,\pi$

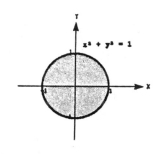

16. $f(x, y) = xy$

$R = \{(x, y) : x^2 + y^2 \le 16\}$

$f_x = y, \quad f_y = x$

$\sqrt{1 + (f_x)^2 + (f_y)^2} = \sqrt{1 + y^2 + x^2}$

$S = \displaystyle\int_{-4}^{4} \int_{-\sqrt{16-x^2}}^{\sqrt{16-x^2}} \sqrt{1 + y^2 + x^2}\, dy\, dx$

$= \displaystyle\int_{0}^{2\pi} \int_{0}^{4} \sqrt{1 + r^2}\, r\, dr\, d\theta = \dfrac{2\pi}{3}(17\sqrt{17} - 1)$

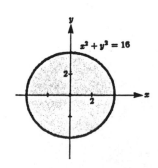

17. $f(x, y) = \sqrt{a^2 - x^2 - y^2}$

$R = \{(x, y) : x^2 + y^2 \le b^2, \quad b < a\}$

$f_x = \dfrac{-x}{\sqrt{a^2 - x^2 - y^2}}, \quad f_y = \dfrac{-y}{\sqrt{a^2 - x^2 - y^2}}$

$\sqrt{1 + (f_x)^2 + (f_y)^2} = \sqrt{1 + \dfrac{x^2}{a^2 - x^2 - y^2} + \dfrac{y^2}{a^2 - x^2 - y^2}} = \dfrac{a}{\sqrt{a^2 - x^2 - y^2}}$

$S = \displaystyle\int_{-b}^{b} \int_{-\sqrt{b^2 - x^2}}^{\sqrt{b^2 - x^2}} \dfrac{a}{\sqrt{a^2 - x^2 - y^2}} \, dy \, dx$

$\quad = \displaystyle\int_{0}^{2\pi} \int_{0}^{b} \dfrac{a}{\sqrt{a^2 - r^2}} \, r \, dr \, d\theta = 2\pi a(a - \sqrt{a^2 - b^2})$

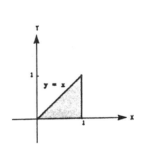

18. See Exercise 17.

$S = \displaystyle\int_{-a}^{a} \int_{-\sqrt{a^2 - x^2}}^{\sqrt{a^2 - x^2}} \dfrac{a}{\sqrt{a^2 - x^2 - y^2}} \, dy \, dx = \int_{0}^{2\pi} \int_{0}^{a} \dfrac{a}{\sqrt{a^2 - r^2}} \, r \, dr \, d\theta = 2\pi a^2$

19. $f(x, y) = \sqrt{1 - x^2}$

$S = \displaystyle\iint_R \sqrt{1 + (f_x)^2 + (f_y)^2} \, dA = 16 \int_0^1 \int_0^x \sqrt{1 + \dfrac{x^2}{1 - x^2}} \, dy \, dx$

$\quad = 16 \displaystyle\int_0^1 \int_0^x \dfrac{1}{\sqrt{1 - x^2}} \, dy \, dx$

$\quad = 16 \displaystyle\int_0^1 \dfrac{x}{\sqrt{1 - x^2}} \, dx$

$\quad = -16\sqrt{1 - x^2} \Big]_0^1 = 16$

20. $f(x, y) = k\sqrt{x^2 + y^2}$

$\sqrt{1 + (f_x)^2 + (f_y)^2} = \sqrt{1 + \dfrac{k^2 x^2}{x^2 + y^2} + \dfrac{k^2 y^2}{x^2 + y^2}} = \sqrt{k^2 + 1}$

$S = \displaystyle\iint_R \sqrt{1 + (f_x)^2 + (f_y)^2} \, dA = \iint_R \sqrt{k^2 + 1} \, dA = \sqrt{k^2 + 1} \iint_R dA = A\sqrt{k^2 + 1} = \pi r^2 \sqrt{k^2 + 1}$

21. $f(x, y) = x^3 - 3xy + y^3$

$R = $ square with vertices $(1, 1, 0), (-1, 1, 0), (-1, -1, 0), (1, -1, 0)$

$f_x = 3x^2 - 3y = 3(x^2 - y), \quad f_y = -3x + 3y^2 = 3(y^2 - x)$

$S = \displaystyle\int_{-1}^{1} \int_{-1}^{1} \sqrt{1 + 9(x^2 - y)^2 + 9(y^2 - x)^2} \, dy \, dx$

22. $f(x, y) = e^{-x} \sin y$

$R = \{(x, y) : 0 \le x \le 4, \quad 0 \le y \le x\}$

$f_x = -e^{-x} \sin y, \quad f_y = e^{-x} \cos y$

$\sqrt{1 + (f_x)^2 + (f_y)^2} = \sqrt{1 + e^{-2x} \sin^2 y + e^{-2x} \cos^2 y} = \sqrt{1 + e^{-2x}}$

$S = \displaystyle\int_0^4 \int_0^x \sqrt{1 + e^{-2x}} \, dy \, dx$

23. $f(x, y) = e^{-x} \sin y$

$R = \{(x, y) : x^2 + y^2 \le 4\}$

See Exercise 22.

$$S = \int_{-2}^{2} \int_{-\sqrt{4-x^2}}^{\sqrt{4-x^2}} \sqrt{1 + e^{-2x}} \, dy \, dx$$

24. $f(x, y) = x^2 - 3xy - y^2$

$R = \{(x, y) : 0 \le x \le 4, \ 0 \le y \le x\}$

$f_x = 2x - 3y, \quad f_y = -3x - 2y = -(3x + 2y)$

$$\sqrt{1 + (f_x)^2 + (f_y)^2} = \sqrt{1 + (2x - 3y)^2 + (3x + 2y)^2}$$

$$= \sqrt{1 + 13(x^2 + y^2)}$$

$$S = \int_{0}^{4} \int_{0}^{x} \sqrt{1 + 13(x^2 + y^2)} \, dy \, dx$$

25. $f(x, y) = e^{xy}$

$R = \{(x, y) : 0 \le x \le 4, \ 0 \le y \le 10\}$

$f_x = y e^{xy}, \quad f_y = x e^{xy}$

$$\sqrt{1 + (f_x)^2 + (f_y)^2} = \sqrt{1 + y^2 e^{2xy} + x^2 e^{2xy}} = \sqrt{1 + e^{2xy}(x^2 + y^2)}$$

$$S = \int_{0}^{4} \int_{0}^{10} \sqrt{1 + e^{2xy}(x^2 + y^2)} \, dy \, dx$$

26. $f(x, y) = \cos(x^2 + y^2)$

$R = \left\{(x, y) : x^2 + y^2 \le \dfrac{\pi}{2}\right\}$

$f_x = -2x \sin(x^2 + y^2), \quad f_y = -2y \sin(x^2 + y^2)$

$$\sqrt{1 + (f_x)^2 + (f_y)^2} = \sqrt{1 + 4x^2 \sin^2(x^2 + y^2) + 4y^2 \sin^2(x^2 + y^2)} = \sqrt{1 + 4[\sin^2(x^2 + y^2)](x^2 + y^2)}$$

$$S = \int_{-\sqrt{\pi/2}}^{\sqrt{\pi/2}} \int_{-\sqrt{(\pi/2)-x^2}}^{\sqrt{(\pi/2)-x^2}} \sqrt{1 + 4(x^2 + y^2) \sin^2(x^2 + y^2)} \, dy \, dx$$

27. $f(x, y) = e^x$

$R = \{(x, y) : 0 \le x \le 1, \ 0 \le y \le 1\}$

$f_x = e^x, \quad f_y = 0$

$$\sqrt{1 + (f_x)^2 + (f_y)^2} = \sqrt{1 + e^{2x}}$$

$$S = \int_{0}^{1} \int_{0}^{1} \sqrt{1 + e^{2x}} \, dy \, dx$$

$$= \int_{0}^{1} \sqrt{1 + e^{2x}} \, dx \approx 2.0035$$

28. $f(x, y) = \dfrac{2}{5} y^{5/2}$

$R = \{(x, y) : 0 \le x \le 1, \ 0 \le y \le 1\}$

$f_x = 0, \quad f_y = y^{3/2}$

$$\sqrt{1 + (f_x)^2 + (f_y)^2} = \sqrt{1 + y^3}$$

$$S = \int_{0}^{1} \int_{0}^{1} \sqrt{1 + y^3} \, dx \, dy$$

$$= \int_{0}^{1} \sqrt{1 + y^3} \, dy \approx 1.1114$$

29. $f(x, y) = 4 - x^2 - y^2$

$R = \{(x, y) : 0 \le x \le 1, \ 0 \le y \le 1\}$

$f_x = -2x, \quad f_y = -2y$

$$\sqrt{1 + (f_x)^2 + (f_y)^2} = \sqrt{1 + 4x^2 + 4y^2}$$

$$S = \int_{0}^{1} \int_{0}^{1} \sqrt{(1 + 4x^2) + 4y^2} \, dy \, dx$$

$$= \int_{0}^{1} \frac{1}{4} \left[2y\sqrt{1 + 4x^2 + 4y^2} + (1 + 4x^2) \ln \left| 2y + \sqrt{1 + 4x^2 + 4y^2} \right| \right]_{0}^{1} dx$$

$$= \frac{1}{4} \int_{0}^{1} \left[2\sqrt{5 + 4x^2} + (1 + 4x^2) \ln(2 + \sqrt{5 + 4x^2}) - (1 + 4x^2) \ln \sqrt{1 + 4x^2} \right] dx$$

$$= \frac{1}{4} \int_{0}^{1} \left[2\sqrt{5 + 4x^2} + (1 + 4x^2) \ln \left[\frac{2 + \sqrt{5 + 4x^2}}{\sqrt{1 + 4x^2}} \right] \right] dx$$

$$\approx 1.8616$$

30. $f(x, y) = \dfrac{2}{3}x^{3/2} + \cos x$

$R = \{(x, y) : 0 \le x \le 1, \ 0 \le y \le 1\}$

$f_x = x^{1/2} - \sin x, \quad f_y = 0$

$\sqrt{1 + (f_x)^2 + (f_y)^2} = \sqrt{1 + (\sqrt{x} - \sin x)^2}$

$S = \displaystyle\int_0^1 \int_0^1 \sqrt{1 + (\sqrt{x} - \sin x)^2}\, dy\, dx = \int_0^1 \sqrt{1 + (\sqrt{x} - \sin x)^2}\, dx \approx 1.2136$

31. (a) $V = \displaystyle\int\int_R f(x, y)\, dA$

$\qquad = 8 \displaystyle\int\int_R \sqrt{625 - x^2 - y^2}\, dA \quad$ where R is the region in the first quadrant

$\qquad = 8 \displaystyle\int_0^{\pi/2} \int_4^{25} \sqrt{625 - r^2}\, r\, dr\, d\theta$

$\qquad = -4 \displaystyle\int_0^{\pi/2} \dfrac{2}{3}(625 - r^2)^{3/2}\Big]_4^{25} d\theta$

$\qquad = -\dfrac{8}{3}[0 - 609\sqrt{609}] \cdot \dfrac{\pi}{2}$

$\qquad = 812\pi\sqrt{609}\ \text{cm}^3$

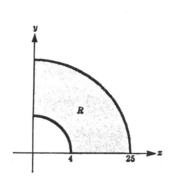

(b) $A = \displaystyle\int\int_R \sqrt{1 + (f_x)^2 + (f_y)^2}\, dA = 8\int\int_R \sqrt{1 + \dfrac{x^2}{625 - x^2 - y^2} + \dfrac{y^2}{625 - x^2 - y^2}}\, dA$

$\qquad = 8 \displaystyle\int\int_R \dfrac{25}{\sqrt{625 - x^2 - y^2}}\, dA$

$\qquad = 8 \displaystyle\int_0^{\pi/2} \int_4^{25} \dfrac{25}{\sqrt{625 - r^2}} r\, dr\, d\theta$

$\qquad = \left[-200\sqrt{625 - r^2}\right]_4^{25} \cdot \dfrac{\pi}{2}$

$\qquad = 100\pi\sqrt{609}\ \text{cm}^2$

32. Position the rectangle so that it lies in the first quadrant along the x- and y-axes. Then for any point (x, y) in the region, we have the relationship $\tan\theta = z/x$ or $z = x\tan\theta$; $\partial z/\partial x = \tan\theta$, $\partial z/\partial y = 0$.

$A = \displaystyle\int\int_R \sqrt{1 + \left(\dfrac{\partial z}{\partial x}\right)^2 + \left(\dfrac{\partial z}{\partial y}\right)^2}\, dA$

$\quad = \displaystyle\int_0^{\Delta x} \int_0^{\Delta y} \sqrt{1 + \tan^2\theta}\, dy\, dx$

$\quad = \displaystyle\int_0^{\Delta x} \int_0^{\Delta y} \sec\theta\, dy\, dx$

$\quad = (\sec\theta)\Delta y\, \Delta x$

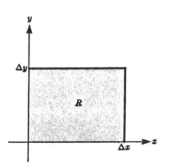

33. $f(x, y) = 4 - x^2 - y^2$

$f_x = -2x, \quad f_y = -2y$

$$S = \int_{-2}^{2} \int_{-2}^{2} \sqrt{1 + 4x^2 + 4y^2} \, dy \, dx$$

$$= \frac{1}{4} \int_{-2}^{2} \left[2y\sqrt{1 + 4x^2 + 4y^2} + (1 + 4x^2) \ln \left| 2y + \sqrt{1 + 4x^2 + 4y^2} \right| \right]_{-2}^{2} \, dx$$

$$= \frac{1}{4} \int_{-2}^{2} \left\{ \left[4\sqrt{17 + 4x^2} + (1 + 4x^2) \ln \left| 4 + \sqrt{17 + 4x^2} \right| \right] - \left[-4\sqrt{17 + 4x^2} + (1 + 4x^2) \ln \left| -4 + \sqrt{17 + 4x^2} \right| \right] \right\} \, dx$$

$$= \frac{1}{4} \int_{-2}^{2} \left[8\sqrt{17 + 4x^2} + (1 + 4x^2) \ln \left| \frac{4 + \sqrt{17 + 4x^2}}{-4 + \sqrt{17 + 4x^2}} \right| \right] \, dx$$

$$\approx 52.0183$$

Section 15.6   Triple Integrals and Applications

1. $\displaystyle\int_{0}^{3} \int_{0}^{2} \int_{0}^{1} (x + y + z) \, dx \, dy \, dz = \int_{0}^{3} \int_{0}^{2} \left(\frac{1}{2}x^2 + xy + xz \right) \Big]_{0}^{1} \, dy \, dx = \int_{0}^{3} \int_{0}^{2} \left(\frac{1}{2} + y + z \right) \, dy \, dz$

$$= \int_{0}^{3} \left(\frac{1}{2}y + \frac{1}{2}y^2 + yz \right) \Big]_{0}^{2} \, dz$$

$$= (3z + z^2) \Big]_{0}^{3} = 18$$

2. $\displaystyle\int_{-1}^{1} \int_{-1}^{1} \int_{-1}^{1} x^2 y^2 z^2 \, dx \, dy \, dz = \frac{1}{3} \int_{-1}^{1} \int_{-1}^{1} x^3 y^2 z^2 \Big]_{-1}^{1} \, dy \, dz = \frac{2}{3} \int_{-1}^{1} \int_{-1}^{1} y^2 z^2 \, dy \, dz = \frac{2}{9} \int_{-1}^{1} y^3 z^2 \Big]_{-1}^{1} \, dz$

$$= \frac{4}{9} \int_{-1}^{1} z^2 \, dz$$

$$= \frac{4}{27} z^3 \Big]_{-1}^{1} = \frac{8}{27}$$

3. $\displaystyle\int_{0}^{1} \int_{0}^{x} \int_{0}^{xy} x \, dz \, dy \, dx = \int_{0}^{1} \int_{0}^{x} xz \Big]_{0}^{xy} \, dy \, dx = \int_{0}^{1} \int_{0}^{x} x^2 y \, dy \, dx = \int_{0}^{1} \frac{x^2 y^2}{2} \Big]_{0}^{x} \, dx = \int_{0}^{1} \frac{x^4}{2} \, dx$

$$= \frac{x^5}{10} \Big]_{0}^{1} = \frac{1}{10}$$

4. $\displaystyle\int_{0}^{4} \int_{0}^{\pi} \int_{0}^{1-x} x \sin y \, dz \, dy \, dx = \int_{0}^{4} \int_{0}^{\pi} (x \sin y) z \Big]_{0}^{1-x} \, dy \, dx$

$$= \int_{0}^{4} \int_{0}^{\pi} x(1 - x) \sin y \, dy \, dx$$

$$= \int_{0}^{4} -x(1 - x) \cos y \Big]_{0}^{\pi} \, dx = \int_{0}^{4} 2x(1 - x) \, dx = \left[x^2 - \frac{2x^3}{3} \right]_{0}^{4} = -\frac{80}{3}$$

5. $\displaystyle\int_1^4\int_0^1\int_0^x 2ze^{-x^2}\,dy\,dx\,dz = \int_1^4\int_0^1 (2ze^{-x^2})y\Big]_0^x dx\,dz$

$$= \int_1^4\int_0^1 2xze^{-x^2}\,dx\,dz$$

$$= \int_1^4 -ze^{-x^2}\Big]_0^1 dz = \int_1^4 z(1-e^{-1})\,dz = (1-e^{-1})\frac{z^2}{2}\Big]_1^4 = \frac{15}{2}\left(1-\frac{1}{e}\right)$$

6. $\displaystyle\int_1^4\int_1^{e^2}\int_0^{1/xz} \ln z\,dy\,dz\,dx = \int_1^4\int_1^{e^2} (\ln z)y\Big]_0^{1/xz} dz\,dx = \int_1^4\int_1^{e^2} \frac{\ln z}{xz}\,dz\,dx = \int_1^4 \frac{1}{x}\frac{(\ln z)^2}{2}\Big]_1^{e^2} dx$

$$= \int_1^4 \frac{2}{x}\,dx$$

$$= 2\ln|x|\Big]_1^4 = 2\ln 4$$

7. $\displaystyle\int_0^9\int_0^{y/3}\int_0^{\sqrt{y^2-9x^2}} z\,dz\,dx\,dy = \frac{1}{2}\int_0^9\int_0^{y/3} (y^2-9x^2)\,dx\,dy$

$$= \frac{1}{2}\int_0^9 (xy^2-3x^3)\Big]_0^{y/3} dy = \frac{2}{18}\int_0^9 y^3\,dy = \frac{1}{36}y^4\Big]_0^9 = \frac{729}{4}$$

8. $\displaystyle\int_0^{\sqrt2}\int_0^{\sqrt{2-x^2}}\int_{2x^2+y^2}^{4-y^2} y\,dz\,dy\,dx = \int_0^{\sqrt2}\int_0^{\sqrt{2-x^2}} (4y-2x^2y-2y^3)\,dy\,dx$

$$= \int_0^{\sqrt2}\left(2y^2-x^2y^2-\frac{1}{2}y^4\right)\Big]_0^{\sqrt{2-x^2}} dx$$

$$= \int_0^{\sqrt2}\left(2-2x^2+\frac{1}{2}x^4\right)dx$$

$$= \left(2x-\frac{2}{3}x^3+\frac{1}{10}x^5\right)\Big]_0^{\sqrt2}$$

$$= \frac{16\sqrt2}{15}$$

9. $\displaystyle\int_0^2\int_{-\sqrt{4-x^2}}^{\sqrt{4-x^2}}\int_0^{x^2} x\,dz\,dy\,dx = \int_0^2\int_{-\sqrt{4-x^2}}^{\sqrt{4-x^2}} x^3\,dy\,dx = \int_0^2 x^3y\Big]_{-\sqrt{4-x^2}}^{\sqrt{4-x^2}} dx = 2\int_0^2 x^3\sqrt{4-x^2}\,dx$

$$= 2\left(-\frac{1}{5}x^2-\frac{8}{15}\right)(4-x^2)^{3/2}\Big]_0^2$$

$$= \frac{128}{15}$$

10. $\displaystyle\int_0^{\pi/2}\int_0^{y/2}\int_0^{1/y} \sin y\,dz\,dx\,dy = \int_0^{\pi/2}\int_0^{y/2} \frac{\sin y}{y}\,dx\,dy = \frac{1}{2}\int_0^{\pi/2} \sin y\,dy = -\frac{1}{2}\cos y\Big]_0^{\pi/2} = \frac{1}{2}$

11. Plane: $3x + 6y + 4z = 12$

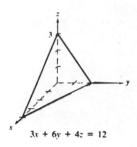

$$3x + 6y + 4z = 12$$

$$\int_0^3 \int_0^{(12-4z)/3} \int_0^{(12-4z-3x)/6} dy\, dx\, dz$$

12. Top plane: $x + y + z = 10$
Side cylinder: $x^2 + y^2 = 16$

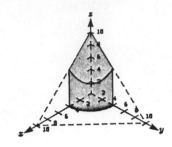

$$\int_0^4 \int_0^{\sqrt{16-y^2}} \int_0^{10-x-y} dz\, dx\, dy$$

13. Top cylinder: $y^2 + z^2 = 1$
Side plane: $x = y$

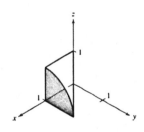

$$\int_0^1 \int_0^x \int_0^{\sqrt{1-y^2}} dz\, dy\, dx$$

14. Elliptic cone: $4x^2 + z^2 = y^2$

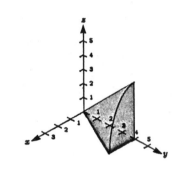

$$\int_0^4 \int_z^4 \int_0^{\sqrt{y^2-z^2}/2} dx\, dy\, dz$$

15. $Q = \{(x, y, z) : 0 \le x \le 1, \ 0 \le y \le x, \ 0 \le z \le 3\}$

$$\iiint\limits_Q xyz\, dV = \int_0^3 \int_0^1 \int_y^1 xyz\, dx\, dy\, dz = \int_0^3 \int_0^1 \int_0^x xyz\, dy\, dx\, dz$$

$$= \int_0^1 \int_0^3 \int_y^1 xyz\, dx\, dz\, dy$$

$$= \int_0^1 \int_0^3 \int_0^x xyz\, dy\, dz\, dx$$

$$= \int_0^1 \int_y^1 \int_0^3 xyz\, dz\, dx\, dy$$

$$= \int_0^1 \int_0^x \int_0^3 xyz\, dz\, dy\, dx$$

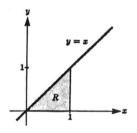

16. $Q = \{(x,\ y,\ z) : 0 \le x \le 2,\quad x^2 \le y \le 4,\quad 0 \le z \le 2 - x\}$

$$\iiint\limits_{Q} xyz\, dV = \int_0^2 \int_{x^2}^4 \int_0^{2-x} xyz\, dz\, dy\, dx$$

$$= \int_0^4 \int_0^{\sqrt{y}} \int_0^{2-x} xyz\, dz\, dx\, dy$$

$$= \int_0^2 \int_0^{2-x} \int_{x^2}^4 xyz\, dy\, dz\, dx$$

$$= \int_0^2 \int_0^{2-z} \int_{x^2}^4 xyz\, dy\, dx\, dz$$

$$= \int_0^2 \int_0^{(2-z)^2} \int_0^{\sqrt{y}} xyz\, dx\, dy\, dz + \int_0^2 \int_{(2-z)^2}^4 \int_0^{2-z} xyz\, dx\, dy\, dz$$

$$= \int_0^4 \int_0^{2-\sqrt{y}} \int_0^{\sqrt{y}} xyz\, dx\, dz\, dy + \int_0^4 \int_{2-\sqrt{y}}^2 \int_0^{2-z} dx\, dz\, dy$$

17. $\displaystyle \int_{-2}^2 \int_0^{4-y^2} \int_0^x dz\, dx\, dy = \int_{-2}^2 \int_0^{4-y^2} x\, dx\, dy = \frac{1}{2}\int_{-2}^2 (4-y^2)^2\, dy = \int_0^2 (16 - 8y^2 + y^4)\, dy$

$$= \left(16y - \frac{8}{3}y^3 + \frac{1}{5}y^5\right)\Bigg]_0^2 = \frac{256}{15}$$

18. $\displaystyle \int_0^1 \int_0^1 \int_0^{xy} dz\, dy\, dx = \int_0^1 \int_0^1 xy\, dy\, dx = \int_0^1 \frac{x}{2}\, dx = \frac{x^2}{4}\Bigg]_0^1 = \frac{1}{4}$

19. $\displaystyle 8\int_0^r \int_0^{\sqrt{r^2-x^2}} \int_0^{\sqrt{r^2-x^2-y^2}} dx\, dy\, dx = 8\int_0^r \int_0^{\sqrt{r^2-x^2}} \sqrt{r^2-x^2-y^2}\, dy\, dx$

$$= 4\int_0^r \left[y\sqrt{r^2-x^2-y^2} + (r^2-x^2)\arcsin\left(\frac{y}{\sqrt{r^2-x^2}}\right)\right]_0^{\sqrt{r^2-x^2}} dx$$

$$= 4\left(\frac{\pi}{2}\right)\int_0^r (r^2-x^2)\, dx$$

$$= 2\pi\left(r^2 x - \frac{1}{3}x^3\right)\Bigg]_0^r = \frac{4}{3}\pi r^3$$

20. $\displaystyle 4\int_0^3 \int_0^{\sqrt{9-x^2}} \int_0^{9-x^2-y^2} dz\, dy\, dx = 4\int_0^3 \int_0^{\sqrt{9-x^2}} (9 - x^2 - y^2)\, dy\, dx$

$$= 4\int_0^3 \left(9y - x^2 y - \frac{1}{3}y^3\right)\Bigg]_0^{\sqrt{9-x^2}} dx$$

$$= 4\int_0^3 \left[9\sqrt{9-x^2} - x^2\sqrt{9-x^2} - \frac{1}{3}(9-x^2)^{3/2}\right] dx$$

$$= \frac{8}{3}\int_0^3 (9-x^2)^{3/2}\, dx$$

$$= \frac{2}{3}\left[x(9-x^2)^{3/2} + \frac{27x}{2}\sqrt{9-x^2} + \frac{243}{2}\arcsin\frac{x}{3}\right]_0^3 = \frac{81\pi}{2}$$

21. $\displaystyle\int_0^2 \int_0^{4-x^2} \int_0^{4-x^2} dz\,dy\,dx = \int_0^2 (4-x^2)^2\,dx = \int_0^2 (16 - 8x^2 + x^4)\,dx = \left(16x - \frac{8}{3}x^3 + \frac{1}{5}x^5\right)\Big]_0^2 = \frac{256}{15}$

22. $\displaystyle\int_0^2 \int_0^{2-x} \int_0^{9-x^2} dz\,dy\,dx = \int_0^2 \int_0^{2-x} (9-x^2)\,dy\,dx = \int_0^2 (9-x^2)(2-x)\,dx = \int_0^2 (18 - 9x - 2x^2 + x^3)\,dx$

$$= \left(18x - \frac{9}{2}x^2 - \frac{2}{3}x^3 + \frac{1}{4}x^4\right)\Big]_0^2 = \frac{50}{3}$$

23. $\displaystyle m = k\int_0^6 \int_0^{4-(2x/3)} \int_0^{2-(y/2)-(x/3)} dz\,dy\,dx = 8k$

$\displaystyle M_{yz} = k\int_0^6 \int_0^{4-(2x/3)} \int_0^{2-(y/2)-(x/3)} x\,dz\,dy\,dx = 12k$

$\displaystyle \overline{x} = \frac{M_{yz}}{m} = \frac{12k}{8k} = \frac{3}{2}$

24. $\displaystyle m = k\int_0^6 \int_0^{(12-2x)/3} \int_0^{(12-2x-3y)/6} y\,dz\,dy\,dx$

$\displaystyle = 8k$

$\displaystyle M_{xz} = k\int_0^6 \int_0^{(12-2x)/3} \int_0^{(12-2x-3y)/6} y^2\,dz\,dy\,dx$

$\displaystyle = \frac{64k}{5}$

$\displaystyle \overline{y} = \frac{M_{xz}}{m} = \frac{8k}{5}$

25. $\displaystyle m = k\int_0^4 \int_0^4 \int_0^{4-x} x\,dz\,dy\,dx = k\int_0^4 \int_0^4 x(4-x)\,dy\,dx = 4k\int_0^4 (4x - x^2)\,dx = \frac{128k}{3}$

$\displaystyle M_{xy} = k\int_0^4 \int_0^4 \int_0^{4-x} xz\,dz\,dy\,dx = k\int_0^4 \int_0^4 x\frac{(4-x)^2}{2}\,dy\,dx = 2k\int_0^4 (16x - 8x^2 + x^3)\,dx = \frac{128k}{3}$

$\displaystyle \overline{z} = \frac{M_{xy}}{m} = 1$

26. $\displaystyle m = k\int_0^b \int_0^{a[1-(y/b)]} \int_0^{c[1-(y/b)-(x/a)]} dz\,dx\,dy = \frac{kabc}{6}$

$\displaystyle M_{xz} = k\int_0^b \int_0^{a[1-(y/b)]} \int_0^{c[1-(y/b)-(x/a)]} y\,dz\,dx\,dy = \frac{kab^2c}{24}$

$\displaystyle \overline{y} = \frac{M_{xz}}{m} = \frac{kab^2c/24}{kabc/6} = \frac{b}{4}$

27. $\displaystyle m = k\int_0^b \int_0^b \int_0^b xy\,dz\,dy\,dx = \frac{kb^5}{4}$

$\displaystyle M_{yz} = k\int_0^b \int_0^b \int_0^b x^2 y\,dz\,dy\,dx = \frac{kb^6}{6}$

$\displaystyle M_{xz} = k\int_0^b \int_0^b \int_0^b xy^2\,dz\,dy\,dx = \frac{kb^6}{6}$

$\displaystyle M_{xy} = k\int_0^b \int_0^b \int_0^b xyz\,dz\,dy\,dx = \frac{kb^6}{8}$

$\displaystyle \overline{x} = \frac{M_{yz}}{m} = \frac{kb^6/6}{kb^5/4} = \frac{2b}{3}$

$\displaystyle \overline{y} = \frac{M_{xz}}{m} = \frac{kb^6/6}{kb^5/4} = \frac{2b}{3}$

$\displaystyle \overline{z} = \frac{M_{xy}}{m} = \frac{kb^6/8}{kb^5/4} = \frac{b}{2}$

28. $\displaystyle m = k\int_0^a \int_0^b \int_0^c z\,dz\,dy\,dx = \frac{kabc^2}{2}$

$\displaystyle M_{xy} = k\int_0^a \int_0^b \int_0^c z^2\,dz\,dy\,dx = \frac{kabc^3}{3}$

$\displaystyle M_{yz} = k\int_0^a \int_0^b \int_0^c xz\,dz\,dy\,dx = \frac{ka^2bc^2}{4}$

$\displaystyle M_{xz} = k\int_0^a \int_0^b \int_0^c yz\,dz\,dy\,dx = \frac{kab^2c^2}{4}$

$\displaystyle \overline{x} = \frac{M_{yz}}{m} = \frac{ka^2bc^2/4}{kabc^2/2} = \frac{a}{2}$

$\displaystyle \overline{y} = \frac{M_{xz}}{m} = \frac{kab^2c^2/4}{kabc^2/2} = \frac{b}{2}$

$\displaystyle \overline{z} = \frac{M_{xy}}{m} = \frac{kabc^3/3}{kabc^2/2} = \frac{2c}{3}$

29. $m = \dfrac{1}{3}k\pi r^2 h$

$\bar{x} = \bar{y} = 0$ by symmetry

$M_{xy} = 4k \displaystyle\int_0^r \int_0^{\sqrt{r^2-x^2}} \int_{h\sqrt{x^2+y^2}/r}^{h} z\,dz\,dy\,dx$

$= \dfrac{3kh^2}{r^2} \displaystyle\int_0^r \int_0^{\sqrt{r^2-x^2}} (r^2 - x^2 - y^2)\,dy\,dx$

$= \dfrac{4kh^2}{3r^2} \displaystyle\int_0^r (r^2 - x^2)^{3/2}\,dx$

$= \dfrac{k\pi r^2 h^2}{4}$

$\bar{z} = \dfrac{M_{xy}}{m} = \dfrac{k\pi r^2 h^2/4}{k\pi r^2 h/3} = \dfrac{3h}{4}$

30. $m = 2k \displaystyle\int_0^2 \int_0^{\sqrt{4-x^2}} \int_0^y dz\,dy\,dx$

$= k \displaystyle\int_0^2 (4 - x^2)\,dx = \dfrac{16k}{3}$

$M_{yz} = 2k \displaystyle\int_0^2 \int_0^{\sqrt{4-x^2}} \int_0^y x\,dz\,dy\,dx = 4k$

$M_{xz} = 2k \displaystyle\int_0^2 \int_0^{\sqrt{4-x^2}} \int_0^y y\,dz\,dy\,dx = 2k\pi$

$M_{xy} = 2k \displaystyle\int_0^2 \int_0^{\sqrt{4-x^2}} \int_0^y z\,dz\,dy\,dx = k\pi$

$\bar{x} = \dfrac{M_{yz}}{m} = \dfrac{4k}{16k/3} = \dfrac{3}{4}$

$\bar{y} = \dfrac{M_{xz}}{m} = \dfrac{2k\pi}{16k/3} = \dfrac{3\pi}{8}$

$\bar{z} = \dfrac{M_{xy}}{m} = \dfrac{k\pi}{16k/3} = \dfrac{3\pi}{16}$

31. $m = \dfrac{128k\pi}{3}$

$\bar{x} = \bar{y} = 0$ by symmetry

$z = \sqrt{4^2 - x^2 - y^2}$

$M_{xy} = 4k \displaystyle\int_0^4 \int_0^{\sqrt{4^2-x^2}} \int_0^{\sqrt{4^2-x^2-y^2}} z\,dz\,dy\,dx$

$= 2k \displaystyle\int_0^4 \int_0^{\sqrt{4^2-x^2}} (4^2 - x^2 - y^2)\,dy\,dx$

$= 2k \displaystyle\int_0^4 \left(16y - x^2 y - \dfrac{1}{3}y^3\right)\Big]_0^{\sqrt{4^2-x^2}} dx$

$= \dfrac{4k}{3} \displaystyle\int_0^4 (4^2 - x^2)^{3/2}\,dx$

$= \dfrac{1024k}{3} \displaystyle\int_0^{\pi/2} \cos^4\theta\,d\theta \quad (\text{let } x = 4\sin\theta)$

$= 64\pi k \quad$ by Wallis's Formula

$\bar{z} = \dfrac{M_{xy}}{m} = \dfrac{64k\pi}{1} \cdot \dfrac{3}{128k\pi} = \dfrac{3}{2}$

32. $\bar{x} = 0$

$m = 2k \displaystyle\int_0^2 \int_0^1 \int_0^{1/(y^2+1)} dz\,dy\,dx$

$= 2k \displaystyle\int_0^2 \int_0^1 \dfrac{1}{y^2+1}\,dy\,dx$

$= 2k \left(\dfrac{\pi}{4}\right) \displaystyle\int_0^2 dx = k\pi$

$M_{xz} = 2k \displaystyle\int_0^2 \int_0^1 \int_0^{1/(y^2+1)} y\,dz\,dy\,dx$

$= 2k \displaystyle\int_0^2 \int_0^1 \dfrac{y}{y^2+1}\,dy\,dx$

$= k \displaystyle\int_0^2 (\ln 2)\,dx = k\ln 4$

$M_{xy} = 2k \displaystyle\int_0^2 \int_0^1 \int_0^{1/(y^2+1)} z\,dz\,dy\,dx$

$= k \displaystyle\int_0^2 \int_0^1 \dfrac{1}{(y^2+1)^2}\,dy\,dx$

$= k \displaystyle\int_0^2 \left[\dfrac{y}{2(y^2+1)} + \dfrac{1}{2}\arctan y\right]_0^1 dx$

$= k \left(\dfrac{1}{4} + \dfrac{\pi}{8}\right) \displaystyle\int_0^2 dx$

$= k \left(\dfrac{1}{2} + \dfrac{\pi}{4}\right)$

$\bar{y} = \dfrac{M_{xz}}{m} = \dfrac{k\ln 4}{k\pi} = \dfrac{\ln 4}{\pi}$

$\bar{z} = \dfrac{M_{xy}}{m} = k\left(\dfrac{1}{2} + \dfrac{\pi}{4}\right)/k\pi = \dfrac{2+\pi}{4\pi}$

33. (a) $I_x = k \int_0^a \int_0^a \int_0^a (y^2 + z^2) \, dx \, dy \, dz = ka \int_0^a \int_0^a (y^2 + z^2) \, dy \, dz = ka \int_0^a \left(\frac{1}{3}y^3 + z^2 y \right) \Big]_0^a dz$

$$= ka \int_0^a \left(\frac{1}{3}a^3 + az^2 \right) dz$$

$$= ka \left(\frac{1}{3}a^3 z + \frac{1}{3}az^3 \right) \Big]_0^a = \frac{2ka^5}{3}$$

$I_x = I_y = I_z = \dfrac{2ka^5}{3}$ by symmetry

(b) $I_x = k \int_0^a \int_0^a \int_0^a (y^2 + z^2)xyz \, dx \, dy \, dz = \frac{ka^2}{2} \int_0^a \int_0^a (y^3 z + yz^3) \, dy \, dz = \frac{ka^2}{2} \int_0^a \left(\frac{y^4 z}{4} + \frac{y^2 z^3}{2} \right) \Big]_0^a dz$

$$= \frac{ka^4}{8} \int_0^a (a^2 z + 2z^3) \, dz$$

$$= \frac{ka^4}{8} \left(\frac{a^2 z^2}{2} + \frac{2z^4}{4} \right) \Big]_0^a = \frac{ka^8}{8}$$

$I_x = I_y = I_z = \dfrac{ka^8}{8}$ by symmetry

34. (a) $I_{xy} = k \int_{-a/2}^{a/2} \int_{-a/2}^{a/2} \int_{-a/2}^{a/2} z^2 \, dz \, dy \, dx = \dfrac{ka^5}{12}$

$I_{xz} = I_{yz} = \dfrac{ka^5}{12}$ by symmetry

$I_x = I_y = I_z = \dfrac{ka^5}{12} + \dfrac{ka^5}{12} = \dfrac{ka^5}{6}$

(b) $I_{xy} = k \int_{-a/2}^{a/2} \int_{-a/2}^{a/2} \int_{-a/2}^{a/2} z^2(x^2 + y^2) \, dz \, dy \, dx = \frac{a^3 k}{12} \int_{-a/2}^{a/2} \int_{-a/2}^{a/2} (x^2 + y^2) \, dy \, dx = \frac{a^7 k}{72}$

$I_{xz} = k \int_{-a/2}^{a/2} \int_{-a/2}^{a/2} \int_{-a/2}^{a/2} y^2(x^2 + y^2) \, dz \, dy \, dx = ka \int_{-a/2}^{a/2} \int_{-a/2}^{a/2} (x^2 y^2 + y^4) \, dy \, dx = \frac{7ka^7}{360}$

$I_{yz} = I_{xz}$ by symmetry

$I_x = I_{xy} + I_{xz} = \dfrac{a^7 k}{30}$

$I_y = I_{xy} + I_{yz} = \dfrac{a^7 k}{30}$

$I_z = I_{yz} + I_{xz} = \dfrac{7ka^7}{180}$

35. (a) $I_x = k \int_0^4 \int_0^4 \int_0^{4-x} (y^2 + z^2)\, dz\, dy\, dx = k \int_0^4 \int_0^4 \left[y^2(4-x) + \frac{1}{3}(4-x)^3 \right] dy\, dx$

$$= k \int_0^4 \left[\frac{y^3}{3}(4-x) + \frac{y}{3}(4-x)^3 \right]_0^4 dx$$

$$= k \int_0^4 \left[\frac{64}{3}(4-x) + \frac{4}{3}(4-x)^3 \right] dx$$

$$= k \left[-\frac{32}{3}(4-x)^2 - \frac{1}{3}(4-x)^4 \right]_0^4$$

$$= 256k$$

$I_y = k \int_0^4 \int_0^4 \int_0^{4-x} (x^2 + z^2)\, dz\, dy\, dx = k \int_0^4 \int_0^4 \left[x^2(4-x) + \frac{1}{3}(4-x)^3 \right] dy\, dx$

$$= 4k \int_0^4 \left[4x^2 - x^3 + \frac{1}{3}(4-x)^3 \right] dx$$

$$= 4k \left[\frac{4}{3}x^3 - \frac{1}{4}x^4 - \frac{1}{12}(4-x)^4 \right]_0^4$$

$$= \frac{512k}{3}$$

$I_z = k \int_0^4 \int_0^4 \int_0^{4-x} (x^2 + y^2)\, dz\, dy\, dx = k \int_0^4 \int_0^4 (x^2 + y^2)(4-x)\, dy\, dx$

$$= k \int_0^4 \left(x^2 y + \frac{y^3}{3} \right)(4-x) \Bigg]_0^4 dx$$

$$= k \int_0^4 \left(4x^2 + \frac{64}{3} \right)(4-x)\, dx$$

$$= k \int_0^4 \left(16x^2 - 4x^3 - \frac{64}{3}x + \frac{256}{3} \right) dx$$

$$= k \left(\frac{16}{3}x^3 - x^4 - \frac{32}{3}x^2 + \frac{256}{3}x \right) \Bigg]_0^4$$

$$= 256k$$

35. (b) $I_x = k \displaystyle\int_0^4 \int_0^4 \int_0^{4-x} y(y^2 + z^2)\, dz\, dy\, dx = k \int_0^4 \int_0^4 \left[y^3(4-x) + \frac{1}{3}y(4-x)^3 \right] dy\, dx$

$$= k \int_0^4 \left[\frac{y^4}{4}(4-x) + \frac{y^2}{6}(4-x)^3 \right]_0^4 dx$$

$$= k \int_0^4 \left[64(4-x) + \frac{8}{3}(4-x)^3 \right] dx$$

$$= k \left[-32(4-x)^2 - \frac{2}{3}(4-x)^4 \right]_0^4$$

$$= \frac{2048k}{3}$$

$I_y = k \displaystyle\int_0^4 \int_0^4 \int_0^{4-x} y(x^2 + z^2)\, dz\, dy\, dx = k \int_0^4 \int_0^4 \left[x^2 y(4-x) + \frac{1}{3}y(4-x)^3 \right] dy\, dx$

$$= 8k \int_0^4 \left[4x^2 - x^3 + \frac{1}{3}(4-x)^3 \right] dx$$

$$= 8k \left[\frac{4}{3}x^3 - \frac{1}{4}x^4 - \frac{1}{12}(4-x)^4 \right]_0^4$$

$$= \frac{1024k}{3}$$

$I_z = k \displaystyle\int_0^4 \int_0^4 \int_0^{4-x} y(x^2 + y^2)\, dz\, dy\, dx = k \int_0^4 \int_0^4 (x^2 y + y^3)(4-x)\, dx$

$$= k \int_0^4 \left(\frac{x^2 y^2}{2} + \frac{y^4}{4} \right)(4-x) \Big]_0^4 dx$$

$$= k \int_0^4 (8x^2 + 64)(4-x)\, dx$$

$$= 8k \int_0^4 (32 - 8x + 4x^2 - x^3)\, dx$$

$$= 8k \left(32x - 4x^2 + \frac{4}{3}x^3 - \frac{1}{4}x^4 \right) \Big]_0^4$$

$$= \frac{2048k}{3}$$

36. (a) $I_{xy} = k\int_0^4 \int_0^2 \int_0^{4-y^2} z^3\, dz\, dy\, dx = k\int_0^4 \int_0^2 \frac{1}{4}(4-y^2)^4\, dy\, dx$

$$= \frac{k}{4}\int_0^4 \int_0^2 (256 - 256y^2 + 96y^4 - 16y^6 + y^8)\, dy\, dx$$

$$= \frac{k}{4}\int_0^4 \left(256y - \frac{256y^3}{3} + \frac{96y^5}{5} - \frac{16y^7}{7} + \frac{y^9}{9}\right)\Bigg]_0^2 dx$$

$$= k\int_0^4 \frac{49152}{945}\, dx = \frac{65536k}{315}$$

$I_{xz} = k\int_0^4 \int_0^2 \int_0^{4-y^2} y^2 z\, dz\, dy\, dx = k\int_0^4 \int_0^2 \frac{1}{2}y^2(4-y^2)^2\, dy\, dx = k\int_0^4 \int_0^2 \frac{1}{2}(16y^2 - 8y^4 + y^6)\, dy\, dx$

$$= \frac{k}{2}\int_0^4 \left(\frac{16y^3}{3} - \frac{8y^5}{5} + \frac{y^7}{7}\right)\Bigg]_0^2 dx$$

$$= \frac{k}{2}\int_0^4 \frac{1024}{105}\, dx = \frac{2048k}{105}$$

$I_{yz} = k\int_0^4 \int_0^2 \int_0^{4-y^2} x^2 z\, dz\, dy\, dx = k\int_0^4 \int_0^2 \frac{1}{2}x^2(4-y^2)^2\, dy\, dx = k\int_0^4 \int_0^2 \frac{1}{2}x^2(16 - 8y^2 + y^4)\, dy\, dx$

$$= \frac{k}{2}\int_0^4 x^2\left(16y - \frac{8y^3}{3} + \frac{y^5}{5}\right)\Bigg]_0^2 dx$$

$$= \frac{k}{2}\int_0^4 \frac{256}{15}x^2\, dx = \frac{8192k}{45}$$

$$I_x = I_{xz} + I_{xy} = \frac{2048k}{9}, \quad I_y = I_{yz} + I_{xy} = \frac{8192k}{21}, \quad I_z = I_{yz} + I_{xz} = \frac{63488k}{315}$$

(b) $I_{xy} = \int_0^4 \int_0^2 \int_0^{4-y^2} z^2(4-z)\, dz\, dy\, dx = k\int_0^4 \int_0^2 \int_0^{4-y^2} 4z^2\, dz\, dy\, dx - k\int_0^4 \int_0^2 \int_0^{4-y^2} z^3\, dz\, dy\, dx$

$$= \frac{32768k}{105} - \frac{65536k}{315} = \frac{32768k}{315}$$

$I_{xz} = k\int_0^4 \int_0^2 \int_0^{4-y^2} y^2(4-z)\, dz\, dy\, dx = k\int_0^4 \int_0^2 \int_0^{4-y^2} 4y^2\, dz\, dy\, dx - k\int_0^4 \int_0^2 \int_0^{4-y^2} y^2 z\, dz\, dy\, dx$

$$= \frac{1024k}{15} - \frac{2048k}{105} = \frac{1024k}{21}$$

$I_{yz} = k\int_0^4 \int_0^2 \int_0^{4-y^2} x^2(4-z)\, dz\, dy\, dx = k\int_0^4 \int_0^2 \int_0^{4-y^2} 4x^2\, dz\, dy\, dx - k\int_0^4 \int_0^2 \int_0^{4-y^2} x^2 z\, dz\, dy\, dx$

$$= \frac{4096k}{9} - \frac{8192k}{45} = \frac{4096k}{15}$$

$$I_x = I_{xz} + I_{xy} = \frac{48128k}{315}, \quad I_y = I_{yz} + I_{xy} = \frac{118784k}{315}, \quad I_z = I_{xz} + I_{yz} = \frac{11264k}{35}$$

37. $I_{xy} = k \displaystyle\int_{-L/2}^{L/2}\int_{-a}^{a}\int_{-\sqrt{a^2-x^2}}^{\sqrt{a^2-x^2}} z^2\, dz\, dx\, dy$

$\qquad = k \displaystyle\int_{-L/2}^{L/2}\int_{-a}^{a} \frac{2}{3}(a^2 - x^2)\sqrt{a^2 - x^2}\, dx\, dy$

$\qquad = \dfrac{2}{3} \displaystyle\int_{-L/2}^{L/2} \left[\frac{a^2}{2}\left(x\sqrt{a^2 - x^2} + a^2 \arcsin\frac{x}{a} \right) - \frac{1}{8}\left[x(2x^2 - a^2)\sqrt{x^2 - a^2} + a^4 \arcsin\frac{x}{a} \right] \right]_{-a}^{a} dy$

$\qquad = \dfrac{2k}{3} \displaystyle\int_{-L/2}^{L/2} 2\left(\frac{a^4\pi}{4} - \frac{a^4\pi}{16} \right) dy = \frac{a^4\pi L k}{4}$

Since $m = \pi a^2 L k$, $\quad I_{xy} = \dfrac{ma^2}{4}$.

$I_{xz} = k \displaystyle\int_{-L/2}^{L/2}\int_{-a}^{a}\int_{-\sqrt{a^2-x^2}}^{\sqrt{a^2-x^2}} y^2\, dz\, dx\, dy = 2k \displaystyle\int_{-L/2}^{L/2}\int_{-a}^{a} y^2 \sqrt{a^2 - x^2}\, dx\, dy$

$\qquad\qquad\qquad = 2k \displaystyle\int_{-L/2}^{L/2} \frac{y^2}{2}\left(x\sqrt{a^2 - x^2} + a^2 \arcsin\frac{x}{a} \right)\Big]_{-a}^{a} dy$

$\qquad\qquad\qquad = k\pi a^2 \displaystyle\int_{-L/2}^{L/2} y^2\, dy$

$\qquad\qquad\qquad = \dfrac{2k\pi a^2}{3}\left(\frac{L^3}{8} \right) = \frac{1}{12}mL^2$

$I_{yz} = k \displaystyle\int_{-L/2}^{L/2}\int_{-a}^{a}\int_{-\sqrt{a^2-x^2}}^{\sqrt{a^2-x^2}} x^2\, dz\, dx\, dy = 2k \displaystyle\int_{-L/2}^{L/2}\int_{-a}^{a} x^2 \sqrt{a^2 - x^2}\, dx\, dy$

$\qquad\qquad\qquad = 2k \displaystyle\int_{-L/2}^{L/2} \frac{1}{8}\left[x(2x^2 - a^2)\sqrt{a^2 - x^2} + a^4 \arcsin\frac{x}{a} \right]_{-a}^{a} dy$

$\qquad\qquad\qquad = \dfrac{ka^4\pi}{4} \displaystyle\int_{-L/2}^{L/2} dy$

$\qquad\qquad\qquad = \dfrac{ka^4\pi L}{4} = \frac{ma^2}{4}$

$I_x = I_{xy} + I_{xz} = \dfrac{ma^2}{4} + \dfrac{mL^2}{12} = \dfrac{m}{12}(3a^2 + L^2)$

$I_y = I_{xy} + I_{yz} = \dfrac{ma^2}{4} + \dfrac{ma^2}{4} = \dfrac{ma^2}{2}$

$I_z = I_{xz} + I_{yz} = \dfrac{mL^2}{12} + \dfrac{ma^2}{4} = \dfrac{m}{12}(3a^2 + L^2)$

38. $I_{xy} = \int_{-c/2}^{c/2}\int_{-a/2}^{a/2}\int_{-b/2}^{b/2} z^2\, dz\, dy\, dx = \frac{b^3}{12}\int_{-c/2}^{c/2}\int_{-a/2}^{a/2} dy\, dx = \frac{1}{12}b^2(abc) = \frac{1}{12}mb^2$

$I_{xz} = \int_{-c/2}^{c/2}\int_{-a/2}^{a/2}\int_{-b/2}^{b/2} y^2\, dz\, dy\, dx = b\int_{-c/2}^{c/2}\int_{-a/2}^{a/2} y^2\, dy\, dx = \frac{ba^3}{12}\int_{-c/2}^{c/2} dx = \frac{ba^3 c}{12} = \frac{1}{12}a^2(abc) = \frac{1}{12}ma^2$

$I_{yz} = \int_{-c/2}^{c/2}\int_{-a/2}^{a/2}\int_{-b/2}^{b/2} x^2\, dz\, dy\, dx = ab\int_{-c/2}^{c/2} x^2\, dx = \frac{abc^3}{12} = \frac{1}{12}c^2(abc) = \frac{1}{12}mc^2$

$I_x = I_{xy} + I_{xz} = \frac{1}{12}m(a^2 + b^2)$

$I_y = I_{xy} + I_{yz} = \frac{1}{12}m(b^2 + c^2)$

$I_z = I_{xz} + I_{yz} = \frac{1}{12}m(a^2 + c^2)$

39. $\int_{-1}^{1}\int_{-1}^{1}\int_{0}^{1-x} (x^2 + y^2)\sqrt{x^2 + y^2 + z^2}\, dz\, dy\, dx$ **40.** $\int_{-1}^{1}\int_{-\sqrt{1-x^2}}^{\sqrt{1-x^2}}\int_{0}^{4-x^2-y^2} kx^2(x^2 + y^2)\, dz\, dy\, dx$

41. $\int_{0}^{2}\int_{0}^{\sqrt{4-x^2}}\int_{1}^{4} \frac{x^2 \sin y}{z}\, dz\, dy\, dx = \int_{0}^{2}\int_{0}^{\sqrt{4-x^2}} \left[x^2 \sin y \ln|z|\right]_{1}^{4} dy\, dx$

$$= \int_{0}^{2}\left[x^2 \ln 4(-\cos y)\right]_{0}^{\sqrt{4-x^2}} dx = \int_{0}^{2} x^2 \ln 4[1 - \cos\sqrt{4-x^2}\,]\, dx \approx 2.4416$$

42. $\int_{0}^{3}\int_{0}^{2-(2y/3)}\int_{0}^{6-2y-3z} ze^{-x^2y^2}\, dx\, dz\, dy = \int_{0}^{6}\int_{0}^{(6-x)/2}\int_{0}^{(6-x-2y)/3} ze^{-x^2y^2}\, dz\, dy\, dx$

$$= \int_{0}^{6}\int_{0}^{3-(x/2)} \frac{1}{2}\left(\frac{6-x-2y}{3}\right)^2 e^{-x^2y^2}\, dy\, dx$$

Computer approximation is 2.118. Function is not integrable.

Section 15.7   Triple Integrals in Cylindrical and Spherical Coordinates

1. $\int_{0}^{4}\int_{0}^{\pi/2}\int_{0}^{2} r\cos\theta\, dr\, d\theta\, dz = \int_{0}^{4}\int_{0}^{\pi/2}\left[\frac{r^2}{2}\cos\theta\right]_{0}^{2} d\theta\, dz = \int_{0}^{4}\int_{0}^{\pi/2} 2\cos\theta\, d\theta\, dz$

$$= \int_{0}^{4}\left[2\sin\theta\right]_{0}^{\pi/2} dz$$

$$= \int_{0}^{4} 2\, dz = 8$$

2. $\int_{0}^{\pi/4}\int_{0}^{2}\int_{0}^{2-r} rz\, dz\, dr\, d\theta = \int_{0}^{\pi/4}\int_{0}^{2}\left[\frac{rz^2}{2}\right]_{0}^{2-r} dr\, d\theta = \frac{1}{2}\int_{0}^{\pi/4}\int_{0}^{2}(4r - 4r^2 + r^3)\, dr\, d\theta$

$$= \frac{1}{2}\int_{0}^{\pi/4}\left[2r^2 - \frac{4r^3}{3} + \frac{r^4}{4}\right]_{0}^{2} d\theta$$

$$= \frac{2}{3}\int_{0}^{\pi/4} d\theta = \frac{\pi}{6}$$

3. $\displaystyle\int_0^{\pi/2}\int_0^{2\cos^2\theta}\int_0^{4-r^2} r\sin\theta\,dz\,dr\,d\theta = \int_0^{\pi/2}\int_0^{2\cos^2\theta} r(4-r^2)\sin\theta\,dr\,d\theta = \int_0^{\pi/2}\left(2r^2-\frac{r^4}{4}\right)\sin\theta\Big]_0^{2\cos^2\theta}d\theta$

$$= \int_0^{\pi/2}[8\cos^4\theta - 4\cos^8\theta]\sin\theta\,d\theta$$

$$= \left[-\frac{8\cos^5\theta}{5}+\frac{4\cos^9\theta}{9}\right]_0^{\pi/2} = \frac{52}{45}$$

4. $\displaystyle\int_0^4\int_0^z\int_0^{\pi/2} re^r\,d\theta\,dr\,dz = \frac{\pi}{2}\int_0^4\int_0^z re^r\,dr\,dz = \frac{\pi}{2}\int_0^4\left[re^r-e^r\right]_0^z dz = \frac{\pi}{2}\int_0^4 (ze^z - e^z+1)\,dz$

$$= \frac{\pi}{2}\left[ze^z - e^z - e^z + z\right]_0^4$$

$$= \frac{\pi}{2}(2e^4+6) = \pi(e^4+3)$$

5. $\displaystyle\int_0^\pi\int_0^{\pi/2}\int_0^2 e^{-\rho^3}\rho^2\,d\rho\,d\theta\,d\phi = \int_0^\pi\int_0^{\pi/2} -\frac{1}{3}e^{-\rho^3}\Big]_0^2 d\theta\,d\phi = \int_0^\pi\int_0^{\pi/2}\frac{1}{3}(1-e^{-8})\,d\theta\,d\phi = \frac{\pi^2}{6}(1-e^{-8})$

6. $\displaystyle\int_0^{\pi/2}\int_0^\pi\int_0^{\sin\theta} (2\cos\phi)\rho^2\,d\rho\,d\theta\,d\phi = \frac{2}{3}\int_0^{\pi/2}\int_0^\pi \cos\phi(\sin^3\theta)\,d\theta\,d\phi = \frac{2}{3}\int_0^{\pi/2}\int_0^\pi \cos\phi[\sin\theta(1-\cos^2\theta)]\,d\theta\,d\phi$

$$= \frac{2}{3}\int_0^{\pi/2}\left[\cos\phi\left(-\cos\theta+\frac{\cos^3\theta}{3}\right)\right]_0^{\pi/2} d\phi$$

$$= \frac{2}{3}\int_0^{\pi/2}\frac{2}{3}\cos\phi\,d\phi$$

$$= \frac{4}{9}\sin\phi\Big]_0^{\pi/2} = \frac{4}{9}$$

7. $\displaystyle\int_0^{2\pi}\int_0^{\pi/4}\int_0^{\cos\phi} \rho^2\sin\phi\,d\rho\,d\theta\,d\phi = \frac{1}{3}\int_0^{2\pi}\int_0^{\pi/4} \cos^3\phi\sin\phi\,d\theta\,d\phi = \frac{\pi}{12}\int_0^{2\pi}\cos^3\phi\sin\phi\,d\phi$

$$= -\frac{\pi}{48}\cos^4\phi\Big]_0^{2\pi} = 0$$

8. $\displaystyle\int_0^{\pi/4}\int_0^{\pi/4}\int_0^{\cos\theta} \rho^2\sin\phi\cos\phi\,d\rho\,d\theta\,d\phi = \frac{1}{3}\int_0^{\pi/4}\int_0^{\pi/4}\cos^3\theta\sin\phi\cos\phi\,d\theta\,d\phi$

$$= \frac{1}{3}\int_0^{\pi/4}\int_0^{\pi/4}\sin\phi\cos\phi[\cos\theta(1-\sin^2\theta)]\,d\theta\,d\phi$$

$$= \frac{1}{3}\int_0^{\pi/4}\sin\phi\cos\phi\left[\sin\theta - \frac{\sin^3\theta}{3}\right]_0^{\pi/4}d\phi$$

$$= \frac{5\sqrt{2}}{36}\int_0^{\pi/4}\sin\phi\cos\phi\,d\phi$$

$$= \frac{5\sqrt{2}}{36}\frac{\sin^2\phi}{2}\Big]_0^{\pi/4} = \frac{5\sqrt{2}}{144}$$

9. $\displaystyle\int_0^{\pi/2}\int_0^3\int_0^{e^{-r^2}} r\,dz\,dr\,d\theta = \int_0^{\pi/2}\int_0^3 re^{-r^2}\,dr\,d\theta$

$$= \int_0^{\pi/2} -\frac{1}{2}e^{-r^2}\Big]_0^3\,d\theta$$

$$= \int_0^{\pi/2} \frac{1}{2}(1-e^{-9})\,d\theta$$

$$= \frac{\pi}{4}(1-e^{-9})$$

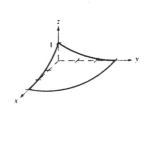

10. $\displaystyle\int_0^{2\pi}\int_0^{\sqrt{3}}\int_0^{3-r^2} r\,dz\,dr\,d\theta = \int_0^{2\pi}\int_0^{\sqrt{3}} r(3-r^2)\,dr\,d\theta$

$$= \int_0^{2\pi} \left(\frac{3r^2}{2}-\frac{r^4}{4}\right)\Big]_0^{\sqrt{3}}\,d\theta$$

$$= \int_0^{2\pi} \frac{9}{4}\,d\theta = \frac{9\pi}{2}$$

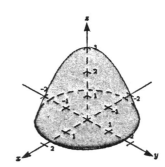

11. $\displaystyle 4\int_0^{\pi/2}\int_{\pi/6}^{\pi/2}\int_0^4 \rho^2\sin\phi\,d\rho\,d\phi\,d\theta = \frac{256}{3}\int_0^{\pi/2}\int_{\pi/6}^{\pi/2}\sin\phi\,d\phi\,d\theta$

$$= \frac{256}{3}\int_0^{\pi/2} -\cos\phi\Big]_{\pi/6}^{\pi/2}\,d\theta$$

$$= \frac{128\sqrt{3}}{3}\int_0^{\pi/2}\,d\theta$$

$$= \frac{64\sqrt{3}\,\pi}{3}$$

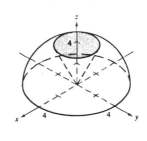

12. $\displaystyle\int_0^{2\pi}\int_0^\pi\int_2^5 \rho^2\sin\phi\,d\rho\,d\phi\,d\theta = \frac{117}{3}\int_0^{2\pi}\int_0^\pi \sin\phi\,d\phi\,d\theta$

$$= \frac{117}{3}\int_0^{2\pi} -\cos\phi\Big]_0^\pi\,d\theta$$

$$= \frac{468\pi}{3}$$

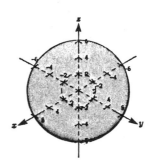

13. (a) $\displaystyle\int_0^{2\pi}\int_0^2\int_{r^2}^4 r^2\cos\theta\,dz\,dr\,d\theta = 0$

(b) $\displaystyle\int_0^{2\pi}\int_0^{\arctan(1/2)}\int_4^{4\sec\phi} \rho^3\sin^2\phi\cos\theta\,d\rho\,d\phi\,d\theta + \int_0^{2\pi}\int_{\arctan(1/2)}^{\pi/2}\int_{2\sqrt{5}}^{\cot\phi\csc\phi} \rho^3\sin^2\phi\cos\theta\,d\rho\,d\phi\,d\theta$

14. (a) $\displaystyle\int_0^{\pi/2}\int_0^2\int_0^{\sqrt{16-r^2}} r^2\,dz\,dr\,d\theta = \frac{8\pi^2}{3} - 2\pi\sqrt{3}$

(b) $\displaystyle\int_0^{\pi/2}\int_0^{\pi/6}\int_0^4 \rho^3\sin^2\phi\,d\rho\,d\phi\,d\theta + \int_0^{\pi/2}\int_{\pi/6}^{\pi/2}\int_4^{2\csc\phi} \rho^3\sin^2\phi\,d\rho\,d\phi\,d\theta = \frac{8\pi^2}{3} - 2\pi\sqrt{3}$

15. (a) $\displaystyle\int_0^{2\pi}\int_0^a\int_a^{a+\sqrt{a^2-r^2}} r^2\cos\theta\,dz\,dr\,d\theta = 0$

　　(b) $\displaystyle\int_0^{\pi/4}\int_0^{2\pi}\int_{a\sec\phi}^{2a\cos\phi} \rho^3\sin^2\phi\cos\theta\,d\rho\,d\theta\,d\phi = 0$

16. (a) $\displaystyle\int_0^{\pi/2}\int_0^1\int_0^{\sqrt{1-r^2}} r\sqrt{r^2+z^2}\,dz\,dr\,d\theta = \frac{\pi}{8}$

　　(b) $\displaystyle\int_0^{\pi/2}\int_0^{\pi/2}\int_0^1 \rho^3\sin\phi\,d\rho\,d\phi\,d\theta = \frac{\pi}{8}$

17.　$z = h - \dfrac{h}{r_0}\sqrt{x^2+y^2} = \dfrac{h}{r_0}(r_0-r)$

$$V = 4\int_0^{\pi/2}\int_0^{r_0}\int_0^{h(r_0-r)/r_0} r\,dz\,dr\,d\theta$$

$$= \frac{4h}{r_0}\int_0^{\pi/2}\int_0^{r_0}(r_0 r - r^2)\,dr\,d\theta$$

$$= \frac{4h}{r_0}\int_0^{\pi/2}\frac{r_0^3}{6}\,d\theta$$

$$= \frac{4h}{r_0}\left(\frac{r_0^3}{6}\right)\left(\frac{\pi}{2}\right) = \frac{1}{3}\pi r_0^2 h$$

18.　　$\bar{x} = \bar{y} = 0$ by symmetry

$$m = \frac{1}{3}\pi r_0^2 hk \text{ from Exercise 17}$$

$$M_{xy} = 4k\int_0^{\pi/2}\int_0^{r_0}\int_0^{h(r_0-r)/r_0} zr\,dz\,dr\,d\theta$$

$$= \frac{2kh^2}{r_0^2}\int_0^{\pi/2}\int_0^{r_0}(r_0^2 r - 2r_0 r^2 + r^3)\,dr\,d\theta$$

$$= \frac{2kh^2}{r_0^2}\left(\frac{r_0^4}{12}\right)\left(\frac{\pi}{2}\right) = \frac{kr_0^2 h^2 \pi}{12}$$

$$\bar{z} = \frac{M_{xy}}{m} = \frac{kr_0^2 h^2 \pi}{12}\left(\frac{3}{\pi r_0^2 hk}\right) = \frac{h}{4}$$

19.　　$\rho = k\sqrt{x^2+y^2} = kr$

　　　$\bar{x} = \bar{y} = 0$ by symmetry

$$m = 4k\int_0^{\pi/2}\int_0^{r_0}\int_0^{h(r_0-r)/r_0} r^2\,dz\,dr\,d\theta = \frac{4kh}{r_0}\int_0^{\pi/2}\int_0^{r_0}(r_0 r^2 - r^3)\,dr\,d\theta = \frac{4kh}{r_0}\left(\frac{r_0^4}{12}\right)\left(\frac{\pi}{2}\right) = \frac{1}{6}k\pi r_0^3 h$$

$$M_{xy} = 4k\int_0^{\pi/2}\int_0^{r_0}\int_0^{h(r_0-r)/r_0} r^2 z\,dz\,dr\,d\theta = \frac{2kh^2}{r_0^2}\int_0^{\pi/2}\int_0^{r_0} r^2(r_0-r)^2\,dr\,d\theta$$

$$= \frac{2kh^2}{r_0^2}\int_0^{\pi/2}\int_0^{r_0}(r_0^2 r^2 - 2r_0 r^3 + r^4)\,dr\,d\theta$$

$$= \frac{2kh^2}{r_0^2}\left(\frac{r_0^5}{30}\right)\left(\frac{\pi}{2}\right) = \frac{1}{30}k\pi r_0^3 h^2$$

$$\bar{z} = \frac{M_{xy}}{m} = \frac{k\pi r_0^3 h^2/30}{k\pi r_0^3 h/6} = \frac{h}{5}$$

20.　　$\rho = kz$

　　　$\bar{x} = \bar{y} = 0$ by symmetry

$$m = 4k\int_0^{\pi/2}\int_0^{r_0}\int_0^{h(r_0-r)/r_0} zr\,dz\,dr\,d\theta$$

$$= \frac{2kh^2}{r_0^2}\int_0^{\pi/2}\int_0^{r_0} r(r_0-r)^2\,dr\,d\theta = \frac{2kh^2}{r_0^2}\int_0^{\pi/2}\int_0^{r_0}(r_0^2 r - 2r_0 r^2 + r^3)\,dr\,d\theta$$

$$= \frac{2kh^2}{r_0^2}\int_0^{\pi/2}\left(\frac{1}{2}r_0^2 r^2 - \frac{2}{3}r_0 r^3 + \frac{1}{4}r^4\right)\Bigg]_0^{r_0} d\theta = \frac{2kh^2}{r_0^2}\left(\frac{r_0^4}{12}\right)\left(\frac{\pi}{2}\right) = \frac{1}{12}k\pi r_0^2 h^2$$

$$M_{xy} = 4k\int_0^{\pi/2}\int_0^{r_0}\int_0^{h(r_0-r)/r_0} z^2 r\,dz\,dr\,d\theta$$

$$= \frac{4kh^3}{3r_0^3}\int_0^{\pi/2}\int_0^{r_0}(r_0^3 r - 3r_0^2 r^2 + 3r_0 r^3 - r^4)\,dr\,d\theta = \frac{4kh^3}{3r_0^3}\left(\frac{r_0^5}{20}\right)\left(\frac{\pi}{2}\right) = \frac{1}{30}k\pi r_0^2 h^3$$

$$\bar{z} = \frac{M_{xy}}{m} = \frac{k\pi r_0^2 h^3/30}{k\pi r_0^2 h^2/12} = \frac{2h}{5}$$

21. $I_z = 4k \int_0^{\pi/2} \int_0^{r_0} \int_0^{h(r_0-r)/r_0} r^3 \, dz \, dr \, d\theta$

$= \dfrac{4kh}{r_0} \int_0^{\pi/2} \int_0^{r_0} (r_0 r^3 - r^4) \, dr \, d\theta$

$= \dfrac{4kh}{r_0} \left(\dfrac{r_0^5}{20} \right) \left(\dfrac{\pi}{2} \right)$

$= \dfrac{1}{10} k\pi r_0^4 h$

$= \left(\dfrac{1}{3} k\pi r_0^2 h \right) \left(\dfrac{3}{10} r_0^2 \right)$

$= \dfrac{3}{10} m r_0^2$

22. $I_z = 4k \int_0^{\pi/2} \int_0^{r_0} \int_0^{(r_0-r)/r_0} r^5 \, dz \, dr \, d\theta$

$= \dfrac{4kh}{r_0} \int_0^{\pi/2} \int_0^{r_0} (r_0 r^5 - r^6) \, dr \, d\theta$

$= \dfrac{4kh}{r_0} \left(\dfrac{r_0^7}{42} \right) \left(\dfrac{\pi}{2} \right)$

$= \dfrac{k r_0^6 h \pi}{21}$

$= \left(\dfrac{1}{3} k\pi r_0^2 h \right) \left(\dfrac{1}{7} r_0^4 \right)$

$= \dfrac{1}{7} m r_0^4$

23. $m = k(\pi b^2 h - \pi a^2 h) = k\pi h (b^2 - a^2)$

$I_z = 4k \int_0^{\pi/2} \int_a^b \int_0^h r^3 \, dz \, dr \, d\theta$

$= 4kh \int_0^{\pi/2} \int_a^b r^3 \, dr \, d\theta$

$= kh \int_0^{\pi/2} (b^4 - a^4) \, d\theta$

$= \dfrac{k\pi (b^4 - a^4) h}{2}$

$= \dfrac{k\pi (b^2 - a^2)(b^2 + a^2) h}{2}$

$= \dfrac{1}{2} m (a^2 + b^2)$

24. $m = k\pi a^2 h$

$I_z = 2k \int_0^{\pi/2} \int_0^{r_0 \sin \theta} \int_0^h r^3 \, dz \, dr \, d\theta$

$= \dfrac{2kh}{4} \int_0^{\pi/2} (2a \sin \theta)^4 \, d\theta$

$= 8k a^4 h \int_0^{\pi/2} \sin^4 \theta \, d\theta$

$= 8k a^4 h \dfrac{3\pi}{16}$ by Wallis's Formula

$= \dfrac{3}{2} k\pi a^4 h$

$= \dfrac{3}{2} m a^2$

25. $V = 4 \int_0^{\pi/2} \int_0^{a \cos \theta} \int_0^{\sqrt{a^2 - r^2}} r \, dz \, dr \, d\theta = 4 \int_0^{\pi/2} \int_0^{a \cos \theta} r\sqrt{a^2 - r^2} \, dr \, d\theta = \dfrac{4}{3} a^3 \int_0^{\pi/2} (1 - \sin^3 \theta) \, d\theta$

$= \dfrac{4}{3} a^3 \left[\theta + \dfrac{1}{3} \cos \theta (\sin^2 \theta + 2) \right]_0^{\pi/2}$

$= \dfrac{4}{3} a^3 \left(\dfrac{\pi}{2} - \dfrac{2}{3} \right)$

26. $V = \dfrac{2}{3} \pi (4)^3 + 4 \left[\int_0^{\pi/2} \int_0^{2\sqrt{2}} \int_0^r r \, dz \, dr \, d\theta + \int_0^{\pi/2} \int_{2\sqrt{2}}^4 \int_0^{\sqrt{16 - r^2}} r \, dz \, dr \, d\theta \right]$

(volume of lower hemisphere) + 4(volume in the first octant)

$V = \dfrac{128\pi}{3} + 4 \left[\int_0^{\pi/2} \int_0^{2\sqrt{2}} r^2 \, dr \, d\theta + \int_0^{\pi/2} \int_{2\sqrt{2}}^4 r\sqrt{16 - r^2} \, dr \, d\theta \right]$

$= \dfrac{128\pi}{3} + 4 \left[\dfrac{8\sqrt{2}\,\pi}{3} + \int_0^{\pi/2} -\dfrac{1}{3}(16 - r^2)^{3/2} \Big]_{2\sqrt{2}}^4 \, d\theta \right]$

$= \dfrac{128\pi}{3} + 4 \left[\dfrac{8\sqrt{2}\,\pi}{3} + \dfrac{8\sqrt{2}\,\pi}{3} \right]$

$= \dfrac{128\pi}{3} + \dfrac{64\sqrt{2}\,\pi}{3} = \dfrac{64\pi}{3} (2 + \sqrt{2})$

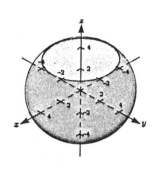

27. $\displaystyle\int_0^{2\pi}\int_{-\pi/2}^{\pi/2}\int_0^{4\sin\phi}\rho^2\sin\phi\,d\rho\,d\phi\,d\theta = \int_0^{2\pi}\int_{-\pi/2}^{\pi/2}\frac{64}{3}\sin^4\phi\,d\phi\,d\theta$

$$= \frac{64}{3}\int_0^{2\pi}\int_{-\pi/2}^{\pi/2}\left(\frac{1-\cos 2\phi}{2}\right)^2 d\phi\,d\theta$$

$$= \frac{16}{3}\int_0^{2\pi}\int_{-\pi/2}^{\pi/2}\left(1-2\cos 2\phi+\frac{1+\cos 4\phi}{2}\right)d\phi\,d\theta$$

$$= \frac{16}{3}\int_0^{2\pi}\left[\phi-\sin 2\phi+\frac{1}{2}\phi+\frac{1}{8}\sin 4\phi\right]_{-\pi/2}^{\pi/2}d\theta$$

$$= \frac{16}{3}\cdot\frac{3\pi}{2}\int_0^{2\pi}d\theta$$

$$= (8\pi)(2\pi) = 16\pi^2$$

28. $\displaystyle V = 8\int_{\pi/4}^{\pi/2}\int_0^{\pi/2}\int_a^b\rho^2\sin\phi\,d\rho\,d\theta\,d\phi = \frac{8}{3}(b^3-a^3)\int_{\pi/4}^{\pi/2}\int_0^{\pi/2}\sin\phi\,d\theta\,d\phi$

$$= \frac{4\pi}{3}(b^3-a^3)\int_{\pi/4}^{\pi/2}\sin\phi\,d\phi$$

$$= \frac{4\pi}{3}(b^3-a^3)(-\cos\phi)\Big]_{\pi/4}^{\pi/2}$$

$$= \frac{2\sqrt{2}\,\pi}{3}(b^3-a^3)$$

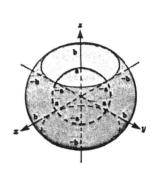

29. $\displaystyle m = 8k\int_0^{\pi/2}\int_0^{\pi/2}\int_0^a\rho^3\sin\phi\,d\rho\,d\theta\,d\phi$

$$= 2ka^4\int_0^{\pi/2}\int_0^{\pi/2}\sin\phi\,d\theta\,d\phi$$

$$= k\pi a^4\int_0^{\pi/2}\sin\phi\,d\phi$$

$$= k\pi a^4(-\cos\phi)\Big]_0^{\pi/2}$$

$$= k\pi a^4$$

30. $\displaystyle m = 8k\int_0^{\pi/2}\int_0^{\pi/2}\int_0^a\rho^3\sin^2\phi\,d\rho\,d\theta\,d\phi$

$$= 2ka^4\int_0^{\pi/2}\int_0^{\pi/2}\sin^2\phi\,d\theta\,d\phi$$

$$= k\pi a^4\int_0^{\pi/2}\sin^2\phi\,d\phi$$

$$= k\pi a^4\left(\frac{1}{2}\phi-\frac{1}{4}\sin 2\phi\right)\Big]_0^{\pi/2}$$

$$= k\pi a^4\frac{\pi}{4} = \frac{1}{4}k\pi^2 a^4$$

31. $m = \dfrac{2}{3}k\pi r^3$

$\bar{x} = \bar{y} = 0$ by symmetry

$M_{xy} = 4k \displaystyle\int_0^{\pi/2} \int_0^{\pi/2} \int_0^r \rho^3 \cos\phi \sin\phi \, d\rho \, d\theta \, d\phi$

$= \dfrac{1}{2}kr^4 \displaystyle\int_0^{\pi/2} \int_0^{\pi/2} \sin 2\phi \, d\theta \, d\phi$

$= \dfrac{kr^4\pi}{4} \displaystyle\int_0^{\pi/2} \sin 2\phi \, d\phi$

$= -\dfrac{1}{8}k\pi r^4 \cos 2\phi \Big]_0^{\pi/2}$

$= \dfrac{1}{4}k\pi r^4$

$\bar{z} = \dfrac{M_{xy}}{m} = \dfrac{k\pi r^4/4}{2k\pi r^3/3} = \dfrac{3r}{8}$

32. $\bar{x} = \bar{y} = 0$ by symmetry

$m = k\left(\dfrac{2}{3}\pi R^3 - \dfrac{2}{3}\pi r^3\right) = \dfrac{2}{3}k\pi(R^3 - r^3)$

$M_{xy} = 4k \displaystyle\int_0^{\pi/2} \int_0^{\pi/2} \int_r^R \rho^3 \cos\phi \sin\phi \, d\rho \, d\theta \, d\phi$

$= \dfrac{1}{2}k(R^4 - r^4) \displaystyle\int_0^{\pi/2} \int_0^{\pi/2} \sin 2\phi \, d\theta \, d\phi$

$= \dfrac{1}{4}k\pi(R^4 - r^4) \displaystyle\int_0^{\pi/2} \sin 2\phi \, d\phi$

$= -\dfrac{1}{8}k\pi(R^4 - r^4) \cos 2\phi \Big]_0^{\pi/2}$

$= \dfrac{1}{4}k\pi(R^4 - r^4)$

$\bar{z} = \dfrac{M_{xy}}{m} = \dfrac{k\pi(R^4 - r^4)/4}{2k\pi(R^3 - r^3)/3} = \dfrac{3(R^4 - r^4)}{8(R^3 - r^3)}$

33. $I_z = 4k \displaystyle\int_{\pi/4}^{\pi/2} \int_0^{\pi/2} \int_0^{\cos\phi} \rho^4 \sin^3\phi \, d\rho \, d\theta \, d\phi$

$= \dfrac{4}{5}k \displaystyle\int_{\pi/4}^{\pi/2} \int_0^{\pi/2} \cos^5\phi \sin^3\phi \, d\theta \, d\phi$

$= \dfrac{2}{5}k\pi \displaystyle\int_{\pi/4}^{\pi/2} \cos^5\phi(1 - \cos^2\phi)\sin\phi \, d\phi$

$= \dfrac{2}{5}k\pi\left(-\dfrac{1}{6}\cos^6\phi + \dfrac{1}{8}\cos^8\phi\right)\Big]_{\pi/4}^{\pi/2}$

$= \dfrac{k\pi}{192}$

34. $I_z = 4k \displaystyle\int_0^{\pi/2} \int_0^{\pi/2} \int_r^R \rho^4 \sin^3\phi \, d\rho \, d\theta \, d\phi$

$= \dfrac{4k}{5}(R^5 - r^5) \displaystyle\int_0^{\pi/2} \int_0^{\pi/2} \sin^3\phi \, d\theta \, d\phi$

$= \dfrac{2k\pi}{5}(R^5 - r^5) \displaystyle\int_0^{\pi/2} \sin\phi(1 - \cos^2\phi) \, d\phi$

$= \dfrac{2k\pi}{5}(R^5 - r^5)\left(-\cos\phi + \dfrac{\cos^3\phi}{3}\right)\Big]_0^{\pi/2}$

$= \dfrac{4k\pi}{15}(R^5 - r^5)$

Section 15.8   Change of Variables: Jacobians

1. $x = -\dfrac{1}{2}(u - v)$

$y = \dfrac{1}{2}(u + v)$

$\dfrac{\partial x}{\partial u}\dfrac{\partial y}{\partial v} - \dfrac{\partial y}{\partial u}\dfrac{\partial x}{\partial v} = \left(-\dfrac{1}{2}\right)\left(\dfrac{1}{2}\right) - \left(\dfrac{1}{2}\right)\left(\dfrac{1}{2}\right)$

$= -\dfrac{1}{2}$

2. $x = au + bv$

$y = cu + dv$

$\dfrac{\partial x}{\partial u}\dfrac{\partial y}{\partial v} - \dfrac{\partial y}{\partial u}\dfrac{\partial x}{\partial v} = ad - cb$

3. $x = u - v^2$

$y = u + v$

$\dfrac{\partial x}{\partial u}\dfrac{\partial y}{\partial v} - \dfrac{\partial y}{\partial u}\dfrac{\partial x}{\partial v} = (1)(1) - (1)(-2v) = 1 + 2v$

4. $x = u - uv$

$y = uv$

$\dfrac{\partial x}{\partial u}\dfrac{\partial y}{\partial v} - \dfrac{\partial y}{\partial u}\dfrac{\partial x}{\partial v} = (1 - v)u - v(-u) = u$

5.
$$x = u \cos \theta - v \sin \theta$$
$$y = u \sin \theta + v \cos \theta$$
$$\frac{\partial x}{\partial u} \frac{\partial y}{\partial v} - \frac{\partial y}{\partial u} \frac{\partial x}{\partial v} = \cos^2 \theta + \sin^2 \theta = 1$$

6.
$$x = u + a$$
$$y = v + a$$
$$\frac{\partial x}{\partial u} \frac{\partial y}{\partial v} - \frac{\partial y}{\partial u} \frac{\partial x}{\partial v} = (1)(1) - (0)(0) = 1$$

7.
$$x = e^u \sin v$$
$$y = e^u \cos v$$
$$\frac{\partial x}{\partial u} \frac{\partial y}{\partial v} - \frac{\partial y}{\partial u} \frac{\partial x}{\partial v} = (e^u \sin v)(-e^u \sin v) - (e^u \cos v)(e^u \cos v) = -e^{2u}$$

8.
$$x = \frac{u}{v}$$
$$y = u + v$$
$$\frac{\partial x}{\partial u} \frac{\partial y}{\partial v} - \frac{\partial y}{\partial u} \frac{\partial x}{\partial v} = \left(\frac{1}{v}\right)(1) - (1)\left(-\frac{u}{v^2}\right) = \frac{1}{v} + \frac{u}{v^2} = \frac{u+v}{v^2}$$

9. $x = 3u + 2v$

$y = 3v$

$v = \dfrac{y}{3}$

$u = \dfrac{x - 2v}{3} = \dfrac{x - 2(y/3)}{3} = \dfrac{x}{3} - \dfrac{2y}{9}$

(x, y)	(u, v)
$(0, 0)$	$(0, 0)$
$(3, 0)$	$(1, 0)$
$(2, 3)$	$(0, 1)$

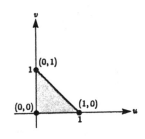

10. $x = 4u + v$

$y = u + 2v$

$u = \dfrac{2x - y}{7}$

$v = \dfrac{-x + 4y}{7}$

(x, y)	(u, v)
$(0, 0)$	$(0, 0)$
$(1, 2)$	$(0, 1)$
$(5, 3)$	$(1, 1)$
$(4, 1)$	$(1, 0)$

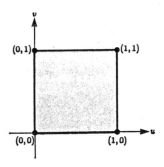

11. $x = \dfrac{1}{2}(u + v), \quad y = \dfrac{1}{2}(u - v)$

$$\frac{\partial x}{\partial u} \frac{\partial y}{\partial v} - \frac{\partial y}{\partial u} \frac{\partial x}{\partial v} = \left(\frac{1}{2}\right)\left(-\frac{1}{2}\right) - \left(\frac{1}{2}\right)\left(\frac{1}{2}\right) = -\frac{1}{2}$$

$$\int_R\!\!\int 48xy \, dx \, dy = \int_0^2 \int_0^2 48\left[\frac{1}{2}(u+v)\right]\left[\frac{1}{2}(u-v)\right]\left(\frac{1}{2}\right) dv \, du$$

$$= 6\int_0^2 \int_0^2 (u^2 - v^2) \, dv \, du = 6\int_0^2 \left(2u^2 - \frac{8}{3}\right) du = 6\left(\frac{2}{3}u^3 - \frac{8}{3}u\right)\Bigg]_0^2 = 0$$

12. $x = \dfrac{1}{2}(u + v)$

$y = \dfrac{1}{2}(u - v)$

$\dfrac{\partial x}{\partial u} \dfrac{\partial y}{\partial v} - \dfrac{\partial y}{\partial u} \dfrac{\partial x}{\partial v} = -\dfrac{1}{2}$ (Same as in Exercise 11)

$\displaystyle \iint_R 4(x^2 + y^2)\, dx\, dy = \int_{-1}^{1}\int_{-1}^{1} 4\left[\frac{1}{4}(u+v)^2 + \frac{1}{4}(u-v)^2\right]\left(\frac{1}{2}\right) dv\, du$

$\displaystyle = \int_{-1}^{1}\int_{-1}^{1} (u^2 + v^2)\, dv\, du = \int_{-1}^{1} 2\left(u^2 + \frac{1}{3}\right) du = 2\left(\frac{u^3}{3} + \frac{u}{3}\right)\Bigg]_{-1}^{1} = \frac{8}{3}$

13. $x = \dfrac{1}{2}(u + v)$

$y = \dfrac{1}{2}(u - v)$

$\dfrac{\partial x}{\partial u} \dfrac{\partial y}{\partial v} - \dfrac{\partial y}{\partial u} \dfrac{\partial x}{\partial v} = -\dfrac{1}{2}$ (Same as in Exercise 11)

$\displaystyle \iint_R 4(x + y)e^{x-y}\, dy\, dx = \int_{0}^{2}\int_{u-2}^{0} 4ue^v \left(\frac{1}{2}\right) dv\, du$

$\displaystyle = \int_{0}^{2} 2u(1 - e^{u-2})\, du$

$\displaystyle = 2\left[\frac{u^2}{2} - ue^{u-2} + e^{u-2}\right]_{0}^{2}$

$\displaystyle = 2(1 - e^{-2})$

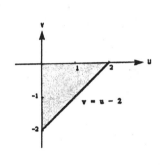

14. $x = u + v$

$y = u$

$\dfrac{\partial x}{\partial u} \dfrac{\partial y}{\partial v} - \dfrac{\partial y}{\partial u} \dfrac{\partial x}{\partial v} = (1)(0) - (1)(1) = -1$

$\displaystyle \iint_R y(x - y)\, dx\, dy = \int_{0}^{3}\int_{0}^{4} uv(1)\, dv\, du$

$\displaystyle = \int_{0}^{3} 8u\, du$

$\displaystyle = 36$

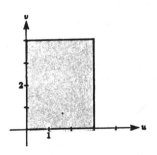

15. $x = u, \quad y = uv, \quad u = x, \quad v = \dfrac{y}{x}$

$$\frac{\partial x}{\partial u}\frac{\partial y}{\partial v} - \frac{\partial y}{\partial u}\frac{\partial x}{\partial v} = u$$

$$\iint_R \frac{\sqrt{x+y}}{x}\,dy\,dx = \int_0^1 \int_0^4 \frac{\sqrt{u+uv}}{u} u\,du\,dv$$

$$= \int_0^1 \int_0^4 \sqrt{u(1+v)}\,du\,dv$$

$$= \int_0^1 \sqrt{1+v}\left(\frac{2}{3}u^{3/2}\right)\Big]_0^4 dv$$

$$= \frac{16}{3}\left[\frac{2}{3}(1+v)^{3/2}\right]_0^1$$

$$= \frac{32}{9}[2\sqrt{2} - 1]$$

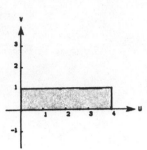

16. $x = \dfrac{u}{v}$

$y = v$

$$\frac{\partial x}{\partial u}\frac{\partial y}{\partial v} - \frac{\partial y}{\partial u}\frac{\partial x}{\partial v} = \frac{1}{v}$$

$$\iint_R y\sin xy\,dy\,dx = \int_1^4 \int_1^4 v(\sin u)\frac{1}{v}\,dv\,du = \int_1^4 3\sin u\,du = -3\cos u\Big]_1^4 = 3(\cos 1 - \cos 4) \approx 3.5818$$

17. $u = x + y = 4, \qquad v = x - y = 0$

$\quad\;\, u = x + y = 8, \qquad v = x - y = 4$

$\quad\;\, x = \dfrac{1}{2}(u + v), \qquad y = \dfrac{1}{2}(u - v)$

$$\frac{\partial(x,\,y)}{\partial(u,\,v)} = -\frac{1}{2}$$

$$\iint_R (x+y)e^{x-y}\,dA = \int_4^8 \int_0^4 ue^v\left(\frac{1}{2}\right)\,dv\,du$$

$$= \frac{1}{2}\int_4^8 u(e^4 - 1)\,du$$

$$= \frac{1}{4}u^2(e^4 - 1)\Big]_4^8$$

$$= 12(e^4 - 1)$$

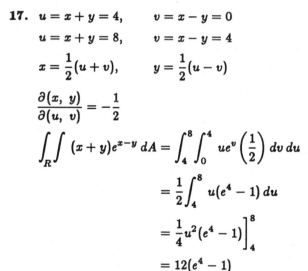

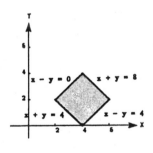

18. $u = x + y = \pi, \qquad v = x - y = 0$

$\quad u = x + y = 2\pi, \qquad v = x - y = \pi$

$\quad x = \dfrac{1}{2}(u + v), \qquad y = \dfrac{1}{2}(u - v)$

$\quad \dfrac{\partial(x,\,y)}{\partial(u,\,v)} = -\dfrac{1}{2}$

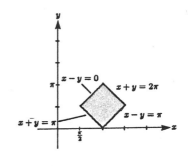

$$\iint_R (x+y)^2 \sin^2(x-y)\,dA = \int_0^\pi \int_\pi^{2\pi} u^2 \sin^2 v \left(\frac{1}{2}\right) du\,dv$$

$$= \int_0^\pi \frac{1}{2}\left(\frac{u^3}{3}\right) \frac{1 - \cos 2v}{2} \Big]_\pi^{2\pi} dv$$

$$= \frac{7\pi^3}{12}\left(v + \frac{1}{2}\sin 2v\right)\Big]_0^\pi$$

$$= \frac{7\pi^4}{12}$$

19. $u = x + 4y = 0, \qquad v = x - y = 0$

$\quad u = x + 4y = 5, \qquad v = x - y = 5$

$\quad x = \dfrac{1}{5}(u + 4v), \qquad y = \dfrac{1}{5}(u - v)$

$\quad \dfrac{\partial x}{\partial u}\dfrac{\partial y}{\partial v} - \dfrac{\partial y}{\partial u}\dfrac{\partial x}{\partial v} = \left(\dfrac{1}{5}\right)\left(-\dfrac{1}{5}\right) - \left(\dfrac{1}{5}\right)\left(\dfrac{4}{5}\right) = -\dfrac{1}{5}$

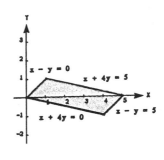

$$\iint_R \sqrt{(x-y)(x+4y)}\,dA = \int_0^5 \int_0^5 \sqrt{uv}\left(\frac{1}{5}\right) du\,dv$$

$$= \int_0^5 \frac{1}{5}\left(\frac{2}{3}\right)u^{3/2}\sqrt{v}\,\Big]_0^5 dv$$

$$= \frac{2\sqrt{5}}{3}\left(\frac{2}{3}\right)v^{3/2}\Big]_0^5$$

$$= \frac{100}{9}$$

20. $u = 3x + 2y = 0, \qquad v = 2y - x = 0$

$\quad u = 3x + 2y = 16, \qquad v = 2y - x = 8$

$\quad x = \dfrac{1}{4}(u - v), \qquad y = \dfrac{1}{8}(u + 3v)$

$\quad \dfrac{\partial x}{\partial u}\dfrac{\partial y}{\partial v} - \dfrac{\partial y}{\partial u}\dfrac{\partial x}{\partial v} = \left(\dfrac{1}{4}\right)\left(\dfrac{3}{8}\right) - \left(\dfrac{1}{8}\right)\left(-\dfrac{1}{4}\right) = \dfrac{1}{8}$

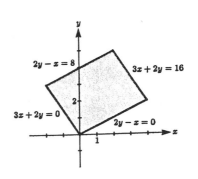

$$\iint_R (3x+2y)^2 \sqrt{2y-x}\,dA = \int_0^8 \int_0^{16} u^2 \sqrt{v}\left(\frac{1}{8}\right) du\,dv$$

$$= \int_0^8 \frac{1}{8}\left(\frac{u^3}{3}\right)\sqrt{v}\,\Big]_0^{16} dv$$

$$= \frac{512}{3}\left(\frac{2}{3}\right)v^{3/2}\Big]_0^8$$

$$= \frac{16384\sqrt{2}}{9}$$

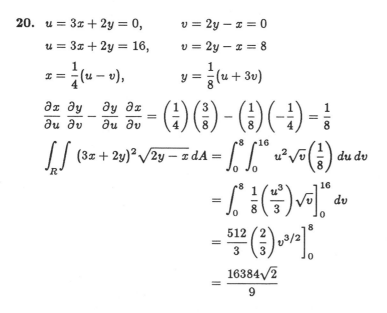

21. $u = x + y, \quad v = x - y, \quad x = \dfrac{1}{2}(u + v), \quad y = \dfrac{1}{2}(u - v)$

$$\frac{\partial x}{\partial u}\frac{\partial y}{\partial v} - \frac{\partial y}{\partial u}\frac{\partial x}{\partial v} = -\frac{1}{2}$$

$$\int\int_R \sqrt{x + y}\, dA = \int_0^a \int_{-u}^u \sqrt{u}\left(\frac{1}{2}\right) dv\, du = \int_0^a u\sqrt{u}\, du = \frac{2}{5}u^{5/2}\bigg]_0^a = \frac{2}{5}a^{5/2}$$

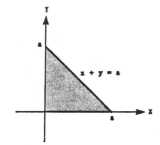
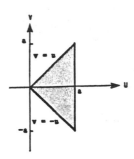

22. $u = x = 1, \qquad v = xy = 1$

$u = x = 4, \qquad v = xy = 4$

$x = u, \qquad\qquad y = \dfrac{v}{u}$

$$\frac{\partial x}{\partial u}\frac{\partial y}{\partial v} - \frac{\partial y}{\partial u}\frac{\partial x}{\partial v} = \frac{1}{u}$$

$$\int\int_R \frac{xy}{1 + x^2 y^2}\, dA = \int_1^4 \int_1^4 \frac{v}{1 + v^2}\left(\frac{1}{u}\right) dv\, du$$

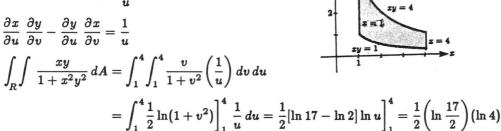

$$= \int_1^4 \frac{1}{2}\ln(1 + v^2)\bigg]_1^4 \frac{1}{u}\, du = \frac{1}{2}[\ln 17 - \ln 2]\ln u\bigg]_1^4 = \frac{1}{2}\left(\ln \frac{17}{2}\right)(\ln 4)$$

23. $\qquad \dfrac{x^2}{a^2} + \dfrac{y^2}{b^2} = 1, \quad x = au, \quad y = bv$

$$\frac{(au)^2}{a^2} + \frac{(bv)^2}{b^2} = 1$$

$$u^2 + v^2 = 1$$

(a) $\dfrac{x^2}{a^2} + \dfrac{y^2}{b^2} = 1$ $\qquad\qquad\qquad\qquad u^2 + v^2 = 1$

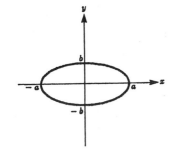
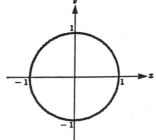

(b) $\dfrac{\partial(x,\ y)}{\partial(u,\ v)} = \dfrac{\partial x}{\partial u}\dfrac{\partial y}{\partial v} - \dfrac{\partial y}{\partial u}\dfrac{\partial x}{\partial v} = (a)(b) - (0)(0) = ab$

24. (a) $f(x, y) = 16 - x^2 - y^2$

$R : \dfrac{x^2}{16} + \dfrac{y^2}{9} \le 1$

$V = \displaystyle\iint_R f(x, y)\, dA$

Let $x = 4u$ and $y = 3v$.

$$\iint_R (16 - x^2 - y^2)\, dA = \int_{-1}^{1} \int_{-\sqrt{1-u^2}}^{\sqrt{1-u^2}} (16 - 16u^2 - 9v^2) 12\, dv\, du$$

Let $u = r\cos\theta, \quad v = r\sin\theta$.

$$= \int_0^{2\pi} \int_0^1 (16 - 16r^2\cos^2\theta - 9r^2\sin^2\theta) 12r\, dr\, d\theta$$

$$= 12 \int_0^{2\pi} \left[8r^2 - 4r^4\cos^2\theta - \frac{9}{4}r^4\sin^2\theta \right]_0^1 d\theta$$

$$= 12 \int_0^{2\pi} \left[8 - 4\cos^2\theta - \frac{9}{4}\sin^2\theta \right] d\theta$$

$$= 12 \int_0^{2\pi} \left[8 - 4\left(\frac{1 + \cos 2\theta}{2} \right) - \frac{9}{4}\left(\frac{1 - \cos 2\theta}{2} \right) \right] d\theta$$

$$= 12 \int_0^{2\pi} \left[\frac{39}{8} - \frac{7}{8}\cos 2\theta \right] d\theta$$

$$= 12 \left[\frac{39}{8}\theta - \frac{7}{16}\sin 2\theta \right]_0^{2\pi}$$

$$= 12 \left[\frac{39\pi}{4} \right]$$

$$= 117\pi$$

(b) $f(x, y) = A\cos\left[\dfrac{\pi}{2}\sqrt{\dfrac{x^2}{a^2} + \dfrac{y^2}{b^2}} \right]$

$R : \dfrac{x^2}{a^2} + \dfrac{y^2}{b^2} \le 1$

Let $x = au$ and $y = bv$.

$$\iint_R f(x, y)\, dA = \int_{-1}^{1} \int_{-\sqrt{1-u^2}}^{\sqrt{1-u^2}} A\cos\left[\frac{\pi}{2}\sqrt{u^2 + v^2} \right] ab\, dv\, du$$

Let $u = r\cos\theta, \quad v = r\sin\theta$.

$$Aab \int_0^{2\pi} \int_0^1 \cos\left[\frac{\pi}{2}r \right] r\, dr\, d\theta = Aab \left[\frac{2r}{\pi}\sin\left(\frac{\pi r}{2} \right) + \frac{4}{\pi^2}\cos\left(\frac{\pi r}{2} \right) \right]_0^1 (2\pi)$$

$$= 2\pi Aab \left[\left(\frac{2}{\pi} + 0 \right) - \left(0 + \frac{4}{\pi^2} \right) \right]$$

$$= \frac{4(\pi - 2)Aab}{\pi}$$

25. $x = u(1-v), \quad y = uv(1-w), \quad z = uvw$

$$\frac{\partial(x,\, y,\, z)}{\partial(u,\, v,\, w)} = \begin{vmatrix} 1-v & -u & 0 \\ v(1-w) & u(1-w) & -uv \\ vw & uw & uv \end{vmatrix} = (1-v)[u^2 v(1-w) + u^2 vw] + u[uv^2(1-w) + uv^2 w]$$

$$= (1-v)(u^2 v) + u(uv^2)$$

$$= u^2 v$$

26. $x = 4u - v, \quad y = 4v - w, \quad z = u + w$

$$\frac{\partial(x,\, y,\, z)}{\partial(u,\, v,\, w)} = \begin{vmatrix} 4 & -1 & 0 \\ 0 & 4 & -1 \\ 1 & 0 & 1 \end{vmatrix} = 17$$

27. $x = \rho \sin\phi \cos\theta, \quad y = \rho \sin\phi \sin\theta, \quad z = \rho \cos\phi$

$$\frac{\partial(x,\, y,\, z)}{\partial(\rho,\, \theta,\, \phi)} = \begin{vmatrix} \sin\phi \cos\theta & -\rho \sin\phi \sin\theta & \rho \cos\phi \cos\theta \\ \sin\phi \sin\theta & \rho \sin\phi \cos\theta & \rho \cos\phi \sin\theta \\ \cos\phi & 0 & -\rho \sin\phi \end{vmatrix}$$

$$= \cos\phi[-\rho^2 \sin\phi \cos\phi \sin^2\theta - \rho^2 \sin\phi \cos\phi \cos^2\theta] - \rho\sin\phi[\rho \sin^2\phi \cos^2\theta + \rho \sin^2\phi \sin^2\theta]$$

$$= \cos\phi[-\rho^2 \sin\phi \cos\phi(\sin^2\theta + \cos^2\theta)] - \rho\sin\phi[\rho \sin^2\phi(\cos^2\theta + \sin^2\theta)]$$

$$= -\rho^2 \sin\phi \cos^2\phi - \rho^2 \sin^3\phi$$

$$= -\rho^2 \sin\phi(\cos^2\phi + \sin^2\phi)$$

$$= -\rho^2 \sin\phi$$

28. $x = r\cos\theta, \quad y = r\sin\theta, \quad z = z$

$$\frac{\partial(x,\, y,\, z)}{\partial(r,\, \theta,\, z)} = \begin{vmatrix} \cos\theta & -r\sin\theta & 0 \\ \sin\theta & r\cos\theta & 0 \\ 0 & 0 & 1 \end{vmatrix} = 1[r\cos^2\theta + r\sin^2\theta] = r$$

Chapter 15   Review Exercises

1. $\displaystyle\int_0^1 \int_0^{1+x} (3x + 2y)\, dy\, dx = \int_0^1 (3xy + y^2)\Big]_0^{1+x} dx = \int_0^1 (4x^2 + 5x + 1)\, dx = \left(\frac{4}{3}x^3 + \frac{5}{2}x^2 + x\right)\Big]_0^1 = \frac{29}{6}$

2. $\displaystyle\int_0^2 \int_{x^2}^{2x} (x^2 + 2y)\, dy\, dx = \int_0^2 (x^2 y + y^2)\Big]_{x^2}^{2x} dx = \int_0^2 (4x^2 + 2x^3 - 2x^4)\, dx = \left(\frac{4}{3}x^3 + \frac{1}{2}x^4 - \frac{2}{5}x^5\right)\Big]_0^2 = \frac{88}{15}$

3. $\displaystyle\int_0^3 \int_0^{\sqrt{9-x^2}} 4x\, dy\, dx = \int_0^3 4x\sqrt{9 - x^2}\, dx = -\frac{4}{3}(9 - x^2)^{3/2}\Big]_0^3 = 36$

4. $\displaystyle\int_0^{\sqrt{3}} \int_{2-\sqrt{4-y^2}}^{2+\sqrt{4-y^2}} dx\, dy = 2\int_0^{\sqrt{3}} \sqrt{4 - y^2}\, dy = \left(y\sqrt{4 - y^2} + 4\arcsin\frac{y}{2}\right)\Big]_0^{\sqrt{3}} = \sqrt{3} + \frac{4\pi}{3}$

5. $\displaystyle\int_{-2}^4 \int_{y^2/4}^{(y/2)+2} (x - y)\, dx\, dy = \int_{-2}^4 \left(\frac{1}{2}x^2 - xy\right)\Big]_{y^2/4}^{(y/2)+2} dy = \frac{1}{32}\int_{-2}^4 (64 - 32y - 12y^2 + 8y^3 - y^4)\, dy$

$$= \frac{1}{32}\left(64y - 16y^2 - 4y^3 + 2y^4 - \frac{1}{5}y^5\right)\Big]_{-2}^4 = \frac{27}{5}$

6. $\displaystyle\int_{-2}^{2}\int_{0}^{4-y^2} (8x - 2y^2)\, dx\, dy = \int_{-2}^{2} (4x^2 - 2y^2 x)\Big]_{0}^{4-y^2} dy = \int_{-2}^{2} (6y^4 - 40y^2 + 64)\, dy$

$$= 4\int_{0}^{2} (3y^4 - 20y^2 + 32)\, dy$$

$$= 4\left(\frac{3}{5}y^5 - \frac{20}{3}y^3 + 32y\right)\Big]_{0}^{2} = \frac{1792}{15}$$

7. $\displaystyle\int_{0}^{\pi/4}\int_{0}^{h\sec\theta} r^2\, dr\, d\theta = \frac{h^3}{3}\int_{0}^{\pi/4} \sec^3\theta\, d\theta = \frac{h^3}{6}\Big[\sec\theta\tan\theta + \ln|\sec\theta + \tan\theta|\Big]_{0}^{\pi/4} = \frac{h^3}{6}[\sqrt{2} + \ln(\sqrt{2} + 1)]$

8. $\displaystyle\int_{0}^{\pi/2}\int_{0}^{4} r^3\, dr\, d\theta = \int_{0}^{\pi/2} \frac{r^4}{4}\Big]_{0}^{4} d\theta = \int_{0}^{\pi/2} 64\, d\theta = 32\pi$

9. $\displaystyle\int_{0}^{2\pi}\int_{0}^{3}\int_{r^2}^{9} r^2\, dz\, dr\, d\theta = \int_{0}^{2\pi}\int_{0}^{3} (9r^2 - r^4)\, dr\, d\theta = \int_{0}^{2\pi} \left(3r^3 - \frac{r^5}{5}\right)\Big]_{0}^{3} d\theta = \frac{162}{5}\int_{0}^{2\pi} d\theta = \frac{324\pi}{5}$

10. $\displaystyle\int_{0}^{2\pi}\int_{0}^{2}\int_{0}^{r^2/2} r^3\, dz\, dr\, d\theta = \frac{1}{2}\int_{0}^{2\pi}\int_{0}^{2} r^5\, dr\, d\theta = \frac{16}{3}\int_{0}^{2\pi} d\theta = \frac{32\pi}{3}$

11. $\displaystyle\int_{-1}^{1}\int_{-\sqrt{1-x^2}}^{\sqrt{1-x^2}}\int_{-\sqrt{1-x^2-y^2}}^{\sqrt{1-x^2-y^2}} (x^2 + y^2)\, dz\, dy\, dx = \int_{0}^{2\pi}\int_{0}^{1}\int_{-\sqrt{1-r^2}}^{\sqrt{1-r^2}} r^3\, dz\, dr\, d\theta$

$$= 2\int_{0}^{2\pi}\int_{0}^{1} r^3\sqrt{1-r^2}\, dr\, d\theta$$

$$= 2\int_{0}^{2\pi}\left[-\frac{1}{3}r^2(1-r^2)^{3/2} - \frac{2}{15}(1-r^2)^{5/2}\right]_{0}^{1} d\theta$$

$$= 2\int_{0}^{2\pi} \frac{2}{15}\, d\theta = \frac{8\pi}{15}$$

12. $\displaystyle\int_{0}^{\pi/2}\int_{0}^{\pi/2}\int_{0}^{5} \rho\sin\phi\, d\rho\, d\theta\, d\phi = \frac{25}{2}\int_{0}^{\pi/2}\int_{0}^{\pi/2} \sin\phi\, d\theta\, d\phi = \frac{25\pi}{4}\int_{0}^{\pi/2} \sin\phi\, d\phi = \frac{25\pi}{4}(-\cos\phi)\Big]_{0}^{\pi/2} = \frac{25\pi}{4}$

13. $\displaystyle\int_{0}^{a}\int_{0}^{b}\int_{0}^{c} (x^2 + y^2 + z^2)\, dx\, dy\, dz = \int_{0}^{a}\int_{0}^{b} \left(\frac{1}{3}c^3 + cy^2 + cz^2\right) dy\, dz = \int_{0}^{a} \left(\frac{1}{3}bc^3 + \frac{1}{3}b^3 c + bcz^2\right) dz$

$$= \frac{1}{3}abc^3 + \frac{1}{3}ab^3 c + \frac{1}{3}a^3 bc$$

$$= \frac{1}{3}abc(a^2 + b^2 + c^2)$$

14. $\displaystyle\int_0^2\int_0^{\sqrt{4-x^2}}\int_0^{\sqrt{4-x^2-y^2}} xyz\,dz\,dy\,dx = \frac{1}{2}\int_0^2\int_0^{\sqrt{4-x^2}} xy(4-x^2-y^2)\,dy\,dx$

$$= \frac{1}{2}\int_0^2\int_0^{\sqrt{4-x^2}} [x(4-x^2)y - xy^3]\,dy\,dx$$

$$= \frac{1}{2}\int_0^2\left[\frac{1}{2}x(4-x^2)y^2 - \frac{1}{4}xy^4\right]_0^{\sqrt{4-x^2}}\,dx$$

$$= \frac{1}{2}\int_0^2\left[\frac{1}{2}x(4-x^2)^2 - \frac{1}{4}x(4-x^2)^2\right]\,dx$$

$$= \frac{1}{8}\int_0^2 x(4-x^2)^2\,dx$$

$$= -\frac{1}{48}(4-x^2)^3\Big]_0^2 = \frac{4}{3}$$

15. $\displaystyle\int_0^3\int_0^{(3-x)/3} f(x,\ y)\,dy\,dx = \int_0^1\int_0^{3-3y} f(x,\ y)\,dx\,dy$

$$A = \int_0^1\int_0^{3-3y} dx\,dy = \int_0^1 (3-3y)\,dy = \left(3y - \frac{3}{2}y^2\right)\Big]_0^1 = \frac{3}{2}$$

16. $\displaystyle\int_0^2\int_0^x f(x,\ y)\,dy\,dx + \int_2^3\int_0^{6-2x} f(x,\ y)\,dy\,dx = \int_0^2\int_y^{(6-y)/2} f(x,\ y)\,dx\,dy$

$$A = \int_0^2\int_y^{(6-y)/2} dx\,dy = \frac{1}{2}\int_0^2 (6-3y)\,dy = \frac{1}{2}\left(6y - \frac{3}{2}y^2\right)\Big]_0^2 = 3$$

17. $\displaystyle\int_{-5}^3\int_{-\sqrt{25-x^2}}^{\sqrt{25-x^2}} f(x,\ y)\,dy\,dx = \int_{-5}^{-4}\int_{-\sqrt{25-y^2}}^{\sqrt{25-y^2}} f(x,\ y)\,dx\,dy$

$$+ \int_{-4}^4\int_{-\sqrt{25-y^2}}^3 f(x,\ y)\,dx\,dy + \int_4^5\int_{-\sqrt{25-y^2}}^{\sqrt{25-y^2}} f(x,\ y)\,dx\,dy$$

$$A = 2\int_{-5}^3\int_0^{\sqrt{25-x^2}} dy\,dx$$

$$= 2\int_{-5}^3 \sqrt{25-x^2}\,dx = \left(x\sqrt{25-x^2} + 25\arcsin\frac{x}{5}\right)\Big]_{-5}^3 = \frac{25\pi}{2} + 12 + 25\arcsin\frac{3}{5} \approx 67.36$$

18. $\displaystyle\int_0^4\int_{x^2-2x}^{6x-x^2} f(x,\ y)\,dy\,dx = \int_{-1}^0\int_{1-\sqrt{1+y}}^{1+\sqrt{1+y}} f(x,\ y)\,dx\,dy + \int_0^8\int_{3-\sqrt{9-y}}^{1+\sqrt{1+y}} f(x,\ y)\,dx\,dy + \int_8^9\int_{3-\sqrt{9-y}}^{3+\sqrt{9-y}} f(x,\ y)\,dx\,dy$

$$A = \int_0^4\int_{x^2-2x}^{6x-x^2} dy\,dx = \int_0^4 (8x-2x^2)\,dx = \left(4x^2 - \frac{2}{3}x^3\right)\Big]_0^4 = \frac{64}{3}$$

19. $\displaystyle A = 4\int_0^1\int_0^{x\sqrt{1-x^2}} dy\,dx = 4\int_0^1 x\sqrt{1-x^2}\,dx = -\frac{4}{3}(1-x^2)^{3/2}\Big]_0^1 = \frac{4}{3}$

$$A = 4\int_0^{1/2}\int_{\sqrt{(1-\sqrt{1-4y^2})/2}}^{\sqrt{(1+\sqrt{1-4y^2})/2}} dx\,dy$$

20. $\displaystyle A = \int_0^2\int_0^{y^2+1} dx\,dy = \int_0^1\int_0^2 dy\,dx + \int_1^5\int_{\sqrt{x-1}}^2 dy\,dx = \frac{14}{3}$

21. $A = \int_2^5 \int_{x-3}^{\sqrt{x-1}} dy\, dx + 2\int_1^2 \int_0^{\sqrt{x-1}} dy\, dx = \int_{-1}^2 \int_{y^2+1}^{y+3} dx\, dy = \dfrac{9}{2}$

22. $A = \int_0^3 \int_{-y}^{2y-y^2} dx\, dy = \int_{-3}^0 \int_{-x}^{1+\sqrt{1-x}} dy\, dx + \int_0^1 \int_{1-\sqrt{1-x}}^{1+\sqrt{1-x}} dy\, dx = \dfrac{9}{2}$

23. $V = \int_0^4 \int_0^{x^2+4} (x^2 - y + 4)\, dy\, dx = \int_0^4 \left(x^2 y - \dfrac{1}{2}y^2 + 4y\right)\Big]_0^{x^2+4} dx = \int_0^4 \left(\dfrac{1}{2}x^4 + 4x^2 + 8\right) dx$

$$= \left(\dfrac{1}{10}x^5 + \dfrac{4}{3}x^3 + 8x\right)\Big]_0^4 = \dfrac{3296}{15}$$

24. $V = \int_0^3 \int_0^x (x + y)\, dy\, dx = \int_0^3 \left(xy + \dfrac{1}{2}y^2\right)\Big]_0^x dx = \dfrac{3}{2}\int_0^3 x^2\, dx = \dfrac{1}{2}x^3\Big]_0^3 = \dfrac{27}{2}$

25. $V = 4\int_0^h \int_0^{\pi/2} \int_1^{\sqrt{1+z^2}} r\, dr\, d\theta\, dz = 2\int_0^h \int_0^{\pi/2} (1 + z^2 - 1)\, d\theta\, dz = \pi \int_0^h z^2\, dz = \pi\left(\dfrac{1}{3}z^3\right)\Big]_0^h = \dfrac{\pi h^3}{3}$

26. $V = 8\int_0^{\pi/2} \int_b^R \sqrt{R^2 - r^2}\, r\, dr\, d\theta = -\dfrac{8}{3}\int_0^{\pi/2} (R^2 - r^2)^{3/2}\Big]_b^R d\theta = \dfrac{8}{3}(R^2 - b^2)^{3/2} \int_0^{\pi/2} d\theta = \dfrac{4}{3}\pi(R^2 - b^2)^{3/2}$

27. $V = 2\int_0^{\pi/2} \int_0^{2\cos\theta} \int_0^{\sqrt{4-r^2}} r\, dz\, dr\, d\theta = 2\int_0^{\pi/2} \int_0^{2\cos\theta} r\sqrt{4-r^2}\, dr\, d\theta = -\int_0^{\pi/2} \dfrac{2}{3}(4-r^2)^{3/2}\Big]_0^{2\cos\theta} d\theta$

$$= \dfrac{16}{3}\int_0^{\pi/2} (1 - \sin^3 \theta)\, d\theta$$

$$= \dfrac{16}{3}\left[\theta + \cos\theta - \dfrac{1}{3}\cos^3\theta\right]_0^{\pi/2}$$

$$= \dfrac{16}{3}\left(\dfrac{\pi}{2} - \dfrac{2}{3}\right)$$

28. $V = 2\int_0^{\pi/2} \int_0^{2\sin\theta} \int_0^{16-r^2} r\, dz\, dr\, d\theta = 2\int_0^{\pi/2} \int_0^{2\sin\theta} r(16 - r^2)\, dr\, d\theta$

$$= 2\int_0^{\pi/2} (32\sin^2\theta - 4\sin^4\theta)\, d\theta$$

$$= 8\int_0^{\pi/2} (8\sin^2\theta - \sin^4\theta)\, d\theta$$

$$= 8\left[4\theta - 2\sin 2\theta + \dfrac{1}{4}\sin^3\theta \cos\theta - \dfrac{3}{4}\left(\dfrac{1}{2}\theta - \dfrac{1}{4}\sin 2\theta\right)\right]_0^{\pi/2}$$

$$= \dfrac{29\pi}{2}$$

29. (a) $m = k \int_0^1 \int_{2x^3}^{2x} xy \, dy \, dx = \dfrac{k}{4}$

$M_x = k \int_0^1 \int_{2x^3}^{2x} xy^2 \, dy \, dx = \dfrac{16k}{55}$

$M_y = k \int_0^1 \int_{2x^3}^{2x} x^2 y \, dy \, dx = \dfrac{8k}{45}$

$\bar{x} = \dfrac{M_y}{m} = \dfrac{32}{45}$

$\bar{y} = \dfrac{M_x}{m} = \dfrac{64}{55}$

(b) $m = k \int_0^1 \int_{2x^3}^{2x} (x^2 + y^2) \, dy \, dx = \dfrac{17k}{30}$

$M_x = k \int_0^1 \int_{2x^3}^{2x} y(x^2 + y^2) \, dy \, dx = \dfrac{392k}{585}$

$M_y = k \int_0^1 \int_{2x^3}^{2x} x(x^2 + y^2) \, dy \, dx = \dfrac{156k}{385}$

$\bar{x} = \dfrac{M_y}{m} = \dfrac{936}{1309}$

$\bar{y} = \dfrac{M_x}{m} = \dfrac{784}{663}$

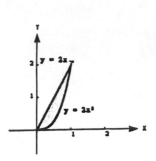

30. $m = k \int_0^L \int_0^{(h/2)[2-(x/L)-(x^2/L^2)]} dy \, dx = \dfrac{kh}{2} \int_0^L \left(2 - \dfrac{x}{L} - \dfrac{x^2}{L^2} \right) dx = \dfrac{7khL}{12}$

$M_x = k \int_0^L \int_0^{(h/2)[2-(x/L)-(x^2/L^2)]} y \, dy \, dx$

$\quad = \dfrac{kh^2}{8} \int_0^L \left(2 - \dfrac{x}{L} - \dfrac{x^2}{L^2} \right)^2 dx$

$\quad = \dfrac{kh^2}{8} \int_0^L \left[4 - \dfrac{4x}{L} - \dfrac{3x^2}{L^2} + \dfrac{2x^3}{L^3} + \dfrac{x^4}{L^4} \right] dx$

$\quad = \dfrac{kh^2}{8} \left[4x - \dfrac{2x^2}{L} - \dfrac{x^3}{L^2} + \dfrac{x^4}{2L^3} + \dfrac{x^5}{5L^4} \right]_0^L$

$\quad = \dfrac{kh^2}{8} \cdot \dfrac{17L}{10} = \dfrac{17kh^2 L}{80}$

$M_y = k \int_0^L \int_0^{(h/2)[2-(x/L)-(x^2/L^2)]} x \, dy \, dx$

$\quad = \dfrac{kh}{2} \int_0^L \left(2x - \dfrac{x^2}{L} - \dfrac{x^3}{L^2} \right) dx = \dfrac{kh}{2} \left[x^2 - \dfrac{x^3}{3L} - \dfrac{x^4}{4L^2} \right]_0^L = \dfrac{kh}{2} \cdot \dfrac{5L^2}{12} = \dfrac{5khL^2}{24}$

$\bar{x} = \dfrac{M_y}{m} = \dfrac{5khL^2}{24} \cdot \dfrac{12}{7khL} = \dfrac{5L}{14}$

$\bar{y} = \dfrac{M_x}{m} = \dfrac{17kh^2 L}{80} \cdot \dfrac{12}{7khL} = \dfrac{51h}{140}$

31. $m = 4k \int_{\pi/4}^{\pi/2} \int_0^{\pi/2} \int_0^{\cos\phi} \rho^2 \sin\phi \, d\rho \, d\theta \, d\phi$

$= \frac{4}{3} k \int_{\pi/4}^{\pi/2} \int_0^{\pi/2} \cos^3\phi \sin\phi \, d\theta \, d\phi = \frac{2}{3} k\pi \int_{\pi/4}^{\pi/2} \cos^3\phi \sin\phi \, d\phi = -\frac{2}{3} k\pi \left(\frac{1}{4} \cos^4\phi \right) \Big]_{\pi/4}^{\pi/2} = \frac{k\pi}{24}$

$M_{xy} = 4k \int_{\pi/4}^{\pi/2} \int_0^{\pi/2} \int_0^{\cos\phi} \rho^3 \cos\phi \sin\phi \, d\rho \, d\theta \, d\phi$

$= k \int_{\pi/4}^{\pi/2} \int_0^{\pi/2} \cos^5\phi \sin\phi \, d\theta \, d\phi = \frac{1}{2} k\pi \int_{\pi/4}^{\pi/2} \cos^5\phi \sin\phi \, d\phi = -\frac{1}{12} k\pi \cos^6\phi \Big]_{\pi/4}^{\pi/2} = \frac{k\pi}{96}$

$\bar{z} = \frac{M_{xy}}{m} = \frac{k\pi/96}{k\pi/24} = \frac{1}{4}$

$\bar{x} = \bar{y} = 0$ by symmetry

32. $m = 2k \int_0^{\pi/2} \int_0^a \int_0^{cr\sin\theta} r \, dz \, dr \, d\theta = 2kc \int_0^{\pi/2} \int_0^a r^2 \sin\theta \, dr \, d\theta = \frac{2}{3} kca^3 \int_0^{\pi/2} \sin\theta \, d\theta = \frac{2}{3} kca^3$

$M_{xz} = 2k \int_0^{\pi/2} \int_0^a \int_0^{cr\sin\theta} r^2 \sin\theta \, dz \, dr \, d\theta = 2kc \int_0^{\pi/2} \int_0^a r^3 \sin^2\theta \, dr \, d\theta = \frac{1}{2} kca^4 \int_0^{\pi/2} \sin^2\theta \, d\theta = \frac{1}{8} \pi kca^4$

$M_{xy} = 2k \int_0^{\pi/2} \int_0^a \int_0^{cr\sin\theta} rz \, dz \, dr \, d\theta = kc^2 \int_0^{\pi/2} \int_0^a r^3 \sin^2\theta \, dr \, d\theta = \frac{1}{4} kc^2 a^4 \int_0^{\pi/2} \sin^2\theta \, d\theta = \frac{1}{16} \pi kc^2 a^4$

$\bar{x} = 0$

$\bar{y} = \frac{M_{xz}}{m} = \frac{\pi kca^4/8}{2kca^3/3} = \frac{3\pi a}{16}$

$\bar{z} = \frac{M_{xy}}{m} = \frac{\pi kc^2 a^4/16}{2kca^3/3} = \frac{3\pi ca}{32}$

33. $m = k \int_0^{\pi/2} \int_0^{\pi/2} \int_0^a \rho^2 \sin\phi \, d\rho \, d\theta \, d\phi = \frac{k\pi a^3}{6}$

$M_{xy} = k \int_0^{\pi/2} \int_0^{\pi/2} \int_0^a (\rho\cos\phi)\rho^2 \sin\phi \, d\rho \, d\theta \, d\phi = \frac{k\pi a^4}{16}$

$\bar{x} = \bar{y} = \bar{z} = \frac{M_{xy}}{m} = \frac{k\pi a^4}{16} \left(\frac{6}{k\pi a^3} \right) = \frac{3a}{8}$

34.
$$m = \frac{500\pi}{3} - \int_0^3 \int_0^{2\pi} \int_4^{\sqrt{25-r^2}} r \, dz \, d\theta \, dr = \frac{500\pi}{3} - \int_0^3 \int_0^{2\pi} \left(r\sqrt{25-r^2} - 4r\right) d\theta \, dr$$

$$= \frac{500\pi}{3} - 2\pi\left[-\frac{1}{3}(25-r^2)^{3/2} - 2r^2\right]_0^3$$

$$= \frac{500\pi}{3} - 2\pi\left[-\frac{64}{3} - 18 + \frac{125}{3}\right] = \frac{500\pi}{3} - \frac{14\pi}{3} = 162\pi$$

$\bar{x} = \bar{y} = 0$ by symmetry

$$M_{xy} = \int_0^{2\pi} \int_0^3 \int_{-\sqrt{25-r^2}}^4 zr \, dz \, dr \, d\theta + \int_0^{2\pi} \int_3^5 \int_{-\sqrt{25-r^2}}^{\sqrt{25-r^2}} zr \, dz \, dr \, d\theta$$

$$= \int_0^{2\pi} \int_0^3 \left[8 - \frac{1}{2}(25-r^2)\right] r \, dr \, d\theta + 0 = \int_0^{2\pi} \int_0^3 \left[\frac{1}{2}r^3 - \frac{9}{2}r\right] dr \, d\theta = \int_0^{2\pi} \left[\frac{1}{8}r^4 - \frac{9}{4}r^2\right]_0^3 d\theta$$

$$= -\frac{81}{8}\theta\Big]_0^{2\pi} = -\frac{81}{4}\pi$$

$$\bar{z} = \frac{M_{xy}}{m} = -\frac{81\pi}{4} \frac{1}{162\pi} = -\frac{1}{8}$$

35.
$$S = \iint_R \sqrt{1 + (f_x)^2 + (f_y)^2} \, dA$$

$$= 4\int_0^4 \int_0^{\sqrt{16-x^2}} \sqrt{1 + 4x^2 + 4y^2} \, dy \, dx = 4\int_0^{\pi/2} \int_0^4 \sqrt{1 + 4r^2} \, r \, dr \, d\theta = \frac{1}{3}(65^{3/2} - 1)\theta\Big]_0^{\pi/2} = \frac{\pi}{6}(65^{3/2} - 1)$$

36. $f(x, y) = 16 - x - y^2$

$R = \{(x, y) : 0 \le x \le 2, \ 0 \le y \le x\}$

$f_x = -1, \quad f_y = -2y$

$$\sqrt{1 + (f_x)^2 + (f_y)^2} = \sqrt{2 + 4y^2}$$

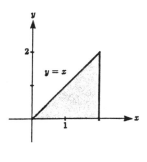

$$S = \int_0^2 \int_y^2 \sqrt{2 + 4y^2} \, dx \, dy = \int_0^2 \left[2\sqrt{2 + 4y^2} - y\sqrt{2 + 4y^2}\right] dy$$

$$= \left[\frac{1}{2}\left(2y\sqrt{2 + 4y^2} + 2\ln|2y + \sqrt{2 + 4y^2}|\right) - \frac{1}{12}(2 + 4y^2)^{3/2}\right]_0^2$$

$$= \frac{1}{2}\left[4\sqrt{18} + 2\ln|4 + \sqrt{18}| - \frac{1}{12}(18\sqrt{18})\right] - \left[\ln\sqrt{2} - \frac{2\sqrt{2}}{12}\right]$$

$$= 6\sqrt{2} + \ln|4 + 3\sqrt{2}| - \frac{9\sqrt{2}}{2} - \ln\sqrt{2} + \frac{\sqrt{2}}{6} = \frac{5\sqrt{2}}{3} + \ln|2\sqrt{2} + 3|$$

37. $I_z = 4k\int_0^{\pi/2} \int_3^4 \int_0^{16-r^2} r^3 \, dz \, dr \, d\theta = 4k\int_0^{\pi/2} \int_3^4 (16r^3 - r^5) \, dr \, d\theta = \frac{833\pi k}{3}$

38. $I_z = k\int_0^\pi \int_0^{2\pi} \int_0^a \rho^2 \sin^2 \phi (\rho)\rho^2 \sin\phi \, d\rho \, d\theta \, d\phi = \frac{4k\pi a^6}{9}$

39. Since $\rho = 6\sin\phi$ represents (in the yz-plane) a circle of radius 3 centered at $(0, 3, 0)$, the integral represents the volume of the torus formed by revolving $(0 < \theta < 2\pi)$ this circle about the z-axis.

CHAPTER 16
Vector Analysis

Section 16.1 Vector Fields

1. $\mathbf{F}(x, y) = \mathbf{i} + \mathbf{j}$

$\|\mathbf{F}\| = \sqrt{2}$

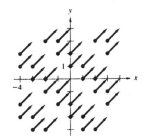

2. $\mathbf{F}(x, y) = 2\mathbf{i}$

$\|\mathbf{F}\| = 2$

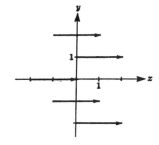

3. $\mathbf{F}(x, y) = x\mathbf{j}$

$\|\mathbf{F}\| = x = c$

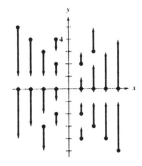

4. $\mathbf{F}(x, y) = y\mathbf{i}$

$\|\mathbf{F}\| = y = c$

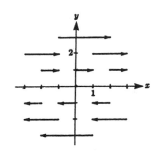

5. $\mathbf{F}(x, y) = x\mathbf{i} + y\mathbf{j}$

$\|\mathbf{F}\| = \sqrt{x^2 + y^2} = c$

$x^2 + y^2 = c^2$

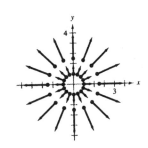

6. $\mathbf{F}(x, y) = -x\mathbf{i} + y\mathbf{j}$

$\|\mathbf{F}\| = \sqrt{x^2 + y^2} = c$

$x^2 + y^2 = c^2$

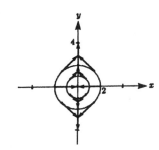

7. $\mathbf{F}(x,\ y) = x\mathbf{i} + 3y\mathbf{j}$

$\|\mathbf{F}\| = \sqrt{x^2 + 9y^2} = c$

$\dfrac{x^2}{c^2} + \dfrac{y^2}{c^2/9} = 1$

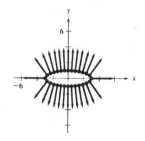

8. $\mathbf{F}(x,\ y) = y\mathbf{i} - x\mathbf{j}$

$\|\mathbf{F}\| = \sqrt{y^2 + x^2} = c$

$x^2 + y^2 = c^2$

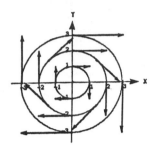

9. $\mathbf{F}(x,\ y) = \dfrac{x}{\sqrt{x^2 + y^2}}\mathbf{i} + \dfrac{y}{\sqrt{x^2 + y^2}}\mathbf{j}$

$\|\mathbf{F}\| = \sqrt{\dfrac{x^2}{x^2 + y^2} + \dfrac{y^2}{x^2 + y^2}} = 1$

10. $\mathbf{F}(x,\ y) = 4x\mathbf{i} + y\mathbf{j}$

$\|\mathbf{F}\| = \sqrt{16x^2 + y^2} = c$

$\dfrac{x^2}{c^2/16} + \dfrac{y^2}{c^2} = 1$

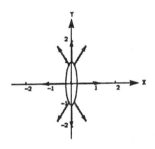

11. $\mathbf{F}(x,\ y,\ z) = 3y\mathbf{j}$

$\|\mathbf{F}\| = 3y$

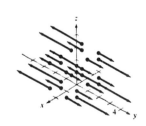

12. $\mathbf{F}(x,\ y) = \mathbf{i} + (x^2 + y^2)\mathbf{j}$

$\|\mathbf{F}\| = \sqrt{1 + (x^2 + y^2)^2} = c$

$(x^2 + y^2)^2 = c^2 - 1$

$x^2 + y^2 = \sqrt{c^2 - 1}$

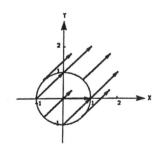

13. $\mathbf{F}(x, y, z) = \mathbf{i} + \mathbf{j} + \mathbf{k}$

$\|\mathbf{F}\| = \sqrt{3}$

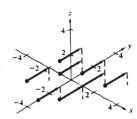

14. $\mathbf{F}(x, y, z) = x\mathbf{i} + y\mathbf{j} + z\mathbf{k}$

$\|\mathbf{F}\| = \sqrt{x^2 + y^2 + z^2} = c$

$x^2 + y^2 + z^2 = c^2$

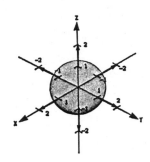

15. $f(x, y) = 5x^2 + 3xy + 10y^2$

$f_x(x, y) = 10x + 3y$

$f_y(x, y) = 3x + 20y$

$\mathbf{F}(x, y) = (10x + 3y)\mathbf{i} + (3x + 20y)\mathbf{j}$

16. $f(x, y) = \sin 3x \cos 4y$

$f_x(x, y) = 3\cos 3x \cos 4y$

$f_y(x, y) = -4\sin 3x \sin 4y$

$\mathbf{F}(x, y) = 3\cos 3x \cos 4y\,\mathbf{i} - 4\sin 3x \sin 4y\,\mathbf{j}$

17. $f(x, y, z) = z - ye^{x^2}$

$f_x(x, y, z) = -2xye^{x^2}$

$f_y(x, y, z) = -e^{x^2}$

$f_z = 1$

$\mathbf{F}(x, y, z) = -2xye^{x^2}\mathbf{i} - e^{x^2}\mathbf{j} + \mathbf{k}$

18. $f(x, y, z) = \dfrac{y}{z} + \dfrac{z}{x} - \dfrac{xz}{y}$

$f_x(x, y, z) = -\dfrac{z}{x^2} - \dfrac{z}{y}$

$f_y(x, y, z) = \dfrac{1}{z} + \dfrac{xz}{y^2}$

$f_z(x, y, z) = -\dfrac{y}{z^2} + \dfrac{1}{x} - \dfrac{x}{y}$

$\mathbf{F}(x, y, z) = \left(-\dfrac{z}{x^2} - \dfrac{z}{y}\right)\mathbf{i} + \left(\dfrac{1}{z} + \dfrac{xz}{y^2}\right)\mathbf{j} + \left(-\dfrac{y}{z^2} + \dfrac{1}{x} - \dfrac{x}{y}\right)\mathbf{k}$

19. $g(x, y, z) = xy\ln(x + y)$

$g_x(x, y, z) = y\ln(x + y) + \dfrac{xy}{x + y}$

$g_y(x, y, z) = x\ln(x + y) + \dfrac{xy}{x + y}$

$g_z(x, y, z) = 0$

$\mathbf{G}(x, y, z) = \left[\dfrac{xy}{x + y} + y\ln(x + y)\right]\mathbf{i} + \left[\dfrac{xy}{x + y} + x\ln(x + y)\right]\mathbf{j}$

20. $g(x, y, z) = x \arcsin yz$

$g_x(x, y, z) = \arcsin yz$

$g_y(x, y, z) = \dfrac{xz}{\sqrt{1 - y^2 z^2}}$

$g_z(x, y, z) = \dfrac{xy}{\sqrt{1 - y^2 z^2}}$

$\mathbf{G}(x, y, z) = (\arcsin yz)\mathbf{i} + \dfrac{xz}{\sqrt{1 - y^2 z^2}}\mathbf{j} + \dfrac{xy}{\sqrt{1 - y^2 z^2}}\mathbf{k}$

21. $\mathbf{F}(x, y) = 2xy\mathbf{i} + x^2\mathbf{j}$

$\dfrac{\partial}{\partial y}[2xy] = 2x$

$\dfrac{\partial}{\partial x}[x^2] = x^2$

Conservative

$f_x(x, y) = 2xy$

$f_y(x, y) = 2x$

$f(x, y) = x^2 y + k$

22. $\mathbf{F}(x, y) = \dfrac{1}{y^2}(y\mathbf{i} - 2x\mathbf{j}) = \dfrac{1}{y}\mathbf{i} - \dfrac{2x}{y^2}\mathbf{j}$

$\dfrac{\partial}{\partial y}\left[\dfrac{1}{y}\right] = -\dfrac{1}{y^2}$

$\dfrac{\partial}{\partial x}\left[-\dfrac{2x}{y^2}\right] = -\dfrac{2}{y^2}$

Not conservative

23. $\mathbf{F}(x, y) = xe^{x^2 y}(2y\mathbf{i} + x\mathbf{j})$

$\dfrac{\partial}{\partial y}[2xye^{x^2 y}] = 2xe^{x^2 y} + 2x^3 ye^{x^2 y}$

$\dfrac{\partial}{\partial x}[x^2 e^{x^2 y}] = 2xe^{x^2 y} + 2x^3 ye^{x^2 y}$

Conservative

$f_x(x, y) = 2xye^{x^2 y}$

$f_y(x, y) = x^2 e^{x^2 y}$

$f(x, y) = e^{x^2 y} + k$

24. $\mathbf{F}(x, y) = 2xy^3\mathbf{i} + 3y^2 x^2\mathbf{j}$

$\dfrac{\partial}{\partial y}[2xy^3] = 6xy^2$

$\dfrac{\partial}{\partial x}[3y^2 x^2] = 6y^2 x$

Conservative

$f_x(x, y) = 2xy^3$

$f_y(x, y) = 3y^2 x^2$

$f(x, y) = x^2 y^3 + k$

25. $\mathbf{F}(x, y) = \dfrac{x}{x^2 + y^2}\mathbf{i} + \dfrac{y}{x^2 + y^2}\mathbf{j}$

$\dfrac{\partial}{\partial y}\left[\dfrac{x}{x^2 + y^2}\right] = -\dfrac{2xy}{(x^2 + y^2)^2}$

$\dfrac{\partial}{\partial x}\left[\dfrac{y}{x^2 + y^2}\right] = -\dfrac{2xy}{(x^2 + y^2)^2}$

Conservative

$f_x(x, y) = \dfrac{x}{x^2 + y^2}$

$f_y = \dfrac{y}{x^2 + y^2}$

$f(x, y) = \dfrac{1}{2}\ln(x^2 + y^2) + k$

26. $\mathbf{F}(x, y) = \dfrac{2y}{x}\mathbf{i} - \dfrac{x^2}{y^2}\mathbf{j}$

$\dfrac{\partial}{\partial y}\left[\dfrac{2y}{x}\right] = \dfrac{2}{x}$

$\dfrac{\partial}{\partial x}\left[-\dfrac{x^2}{y^2}\right] = -\dfrac{2x}{y^2}$

Not conservative

27. $\mathbf{F}(x,\ y) = e^x(\cos y\,\mathbf{i} + \sin y\,\mathbf{j})$

$\dfrac{\partial}{\partial y}[e^x \cos y] = -e^x \sin y$

$\dfrac{\partial}{\partial x}[e^x \sin y] = e^x \sin y$

Not conservative

28. $\mathbf{F}(x,\ y) = \dfrac{2x}{(x^2 + y^2)^2}\mathbf{i} + \dfrac{2y}{(x^2 + y^2)^2}\mathbf{j}$

$\dfrac{\partial}{\partial y}\left[\dfrac{2x}{(x^2 + y^2)^2}\right] = -\dfrac{8xy}{(x^2 + y^2)^3}$

$\dfrac{\partial}{\partial x}\left[\dfrac{2y}{(x^2 + y^2)^2}\right] = -\dfrac{8xy}{(x^2 + y^2)^3}$

Conservative

$f_x(x,\ y) = \dfrac{2x}{(x^2 + y^2)^2}$

$f_y(x,\ y) = \dfrac{2y}{(x^2 + y^2)^2}$

$f(x,\ y) = -\dfrac{1}{x^2 + y^2} + k$

29. $\mathbf{F}(x,\ y,\ z) = xyz\,\mathbf{i} + y\,\mathbf{j} + z\,\mathbf{k},\quad (1,\ 2,\ 1)$

$\text{curl } \mathbf{F} = \begin{vmatrix} \mathbf{i} & \mathbf{j} & \mathbf{k} \\ \dfrac{\partial}{\partial x} & \dfrac{\partial}{\partial y} & \dfrac{\partial}{\partial z} \\ xyz & y & z \end{vmatrix} = xy\,\mathbf{j} - xz\,\mathbf{k}$

$\text{curl } \mathbf{F}\,(1,\ 2,\ 1) = 2\mathbf{j} - \mathbf{k}$

30. $\mathbf{F}(x,\ y,\ z) = x^2 z\,\mathbf{i} - 2xz\,\mathbf{j} + yz\,\mathbf{k},\quad (2,\ -1,\ 3)$

$\text{curl } \mathbf{F} = \begin{vmatrix} \mathbf{i} & \mathbf{j} & \mathbf{k} \\ \dfrac{\partial}{\partial x} & \dfrac{\partial}{\partial y} & \dfrac{\partial}{\partial z} \\ x^2 z & -2xz & yz \end{vmatrix}$

$= (z + 2x)\mathbf{i} - (0 - x^2)\mathbf{j} + (-2z - 0)\mathbf{k}$

$= (z + 2x)\mathbf{i} + x^2\,\mathbf{j} - 2z\,\mathbf{k}$

$\text{curl } \mathbf{F}\,(2,\ -1,\ 3) = 7\mathbf{i} + 4\mathbf{j} - 6\mathbf{k}$

31. $\mathbf{F}(x,\ y,\ z) = e^x \sin y\,\mathbf{i} - e^x \cos y\,\mathbf{j},\quad (0,\ 0,\ 3)$

$\text{curl } \mathbf{F} = \begin{vmatrix} \mathbf{i} & \mathbf{j} & \mathbf{k} \\ \dfrac{\partial}{\partial x} & \dfrac{\partial}{\partial y} & \dfrac{\partial}{\partial z} \\ e^x \sin y & -e^x \cos y & 0 \end{vmatrix} = -2e^x \cos y\,\mathbf{k}$

$\text{curl } \mathbf{F}\,(0,\ 0,\ 3) = -2\mathbf{k}$

32. $\mathbf{F}(x,\ y,\ z) = e^{-xyz}(\mathbf{i} + \mathbf{j} + \mathbf{k}),\quad (3,\ 2,\ 0)$

$\text{curl } \mathbf{F} = \begin{vmatrix} \mathbf{i} & \mathbf{j} & \mathbf{k} \\ \dfrac{\partial}{\partial x} & \dfrac{\partial}{\partial y} & \dfrac{\partial}{\partial z} \\ e^{-xyz} & e^{-xyz} & e^{-xyz} \end{vmatrix}$

$= (-xz + xy)e^{-xyz}\mathbf{i} - (-yz + xy)e^{-xyz}\mathbf{j} + (-yz + xz)e^{-xyz}\mathbf{k}$

$\text{curl } \mathbf{F}\,(3,\ 2,\ 0) = 6\mathbf{i} - 6\mathbf{j}$

33. $\mathbf{F}(x,\ y,\ z) = \arctan\left(\dfrac{x}{y}\right)\mathbf{i} + \ln\sqrt{x^2 + y^2}\,\mathbf{j} + \mathbf{k}$

$\text{curl } \mathbf{F} = \begin{vmatrix} \mathbf{i} & \mathbf{j} & \mathbf{k} \\ \dfrac{\partial}{\partial x} & \dfrac{\partial}{\partial y} & \dfrac{\partial}{\partial z} \\ \arctan\left(\dfrac{x}{y}\right) & \dfrac{1}{2}\ln(x^2 + y^2) & 1 \end{vmatrix} = \left[\dfrac{x}{x^2 + y^2} - \dfrac{(-x/y^2)}{1 + (x/y)^2}\right]\mathbf{k} = \dfrac{2x}{x^2 + y^2}\mathbf{k}$

34. $F(x, y, z) = \dfrac{yz}{y-z}\mathbf{i} + \dfrac{xz}{x-z}\mathbf{j} + \dfrac{xy}{x-y}\mathbf{k}$

$$\text{curl } F = \begin{vmatrix} \mathbf{i} & \mathbf{j} & \mathbf{k} \\ \dfrac{\partial}{\partial x} & \dfrac{\partial}{\partial y} & \dfrac{\partial}{\partial z} \\ \dfrac{yz}{y-z} & \dfrac{xz}{x-z} & \dfrac{xy}{x-y} \end{vmatrix}$$

$$= \left[\dfrac{x^2}{(x-y)^2} - \dfrac{x^2}{(x-z)^2} \right]\mathbf{i} - \left[\dfrac{-y^2}{(x-y)^2} - \dfrac{y^2}{(y-z)^2} \right]\mathbf{j} + \left[\dfrac{-z^2}{(x-z)^2} - \dfrac{-z^2}{(y-z)^2} \right]\mathbf{k}$$

$$= x^2\left[\dfrac{1}{(x-y)^2} - \dfrac{1}{(x-z)^2} \right]\mathbf{i} + y^2\left[\dfrac{1}{(x-y)^2} + \dfrac{1}{(y-z)^2} \right]\mathbf{j} + z^2\left[\dfrac{1}{(y-z)^2} - \dfrac{1}{(x-z)^2} \right]\mathbf{k}$$

35. $F(x, y, z) = \sin(x-y)\mathbf{i} + \sin(y-z)\mathbf{j} + \sin(z-x)\mathbf{k}$

$$\text{curl } F = \begin{vmatrix} \mathbf{i} & \mathbf{j} & \mathbf{k} \\ \dfrac{\partial}{\partial x} & \dfrac{\partial}{\partial y} & \dfrac{\partial}{\partial z} \\ \sin(x-y) & \sin(y-z) & \sin(z-x) \end{vmatrix} = \cos(y-z)\mathbf{i} + \cos(z-x)\mathbf{j} + \cos(x-y)\mathbf{k}$$

36. $F(x, y, z) = \sqrt{x^2 + y^2 + z^2}\,(\mathbf{i} + \mathbf{j} + \mathbf{k})$

$$\text{curl } F = \begin{vmatrix} \mathbf{i} & \mathbf{j} & \mathbf{k} \\ \dfrac{\partial}{\partial x} & \dfrac{\partial}{\partial y} & \dfrac{\partial}{\partial z} \\ \sqrt{x^2+y^2+z^2} & \sqrt{x^2+y^2+z^2} & \sqrt{x^2+y^2+z^2} \end{vmatrix} = \dfrac{(y-z)\mathbf{i} + (z-x)\mathbf{j} + (x-y)\mathbf{k}}{\sqrt{x^2+y^2+z^2}}$$

37. $F(x, y, z) = \sin y\,\mathbf{i} - x\cos y\,\mathbf{j} + \mathbf{k}$

$$\text{curl } F = \begin{vmatrix} \mathbf{i} & \mathbf{j} & \mathbf{k} \\ \dfrac{\partial}{\partial x} & \dfrac{\partial}{\partial y} & \dfrac{\partial}{\partial z} \\ \sin y & -x\cos y & 1 \end{vmatrix} = -2\cos y\,\mathbf{k} \neq 0$$

Not conservative

38. $F(x, y, z) = e^z(y\mathbf{i} + x\mathbf{j} + \mathbf{k})$

$$\text{curl } F = \begin{vmatrix} \mathbf{i} & \mathbf{j} & \mathbf{k} \\ \dfrac{\partial}{\partial x} & \dfrac{\partial}{\partial y} & \dfrac{\partial}{\partial z} \\ ye^z & xe^z & e^z \end{vmatrix} = -xe^z\mathbf{i} + ye^z\,\mathbf{j} \neq 0$$

Not conservative

39. $F(x, y, z) = e^z(y\mathbf{i} + x\mathbf{j} + xy\mathbf{k})$

$$\text{curl } F = \begin{vmatrix} \mathbf{i} & \mathbf{j} & \mathbf{k} \\ \dfrac{\partial}{\partial x} & \dfrac{\partial}{\partial y} & \dfrac{\partial}{\partial z} \\ ye^z & xe^z & xye^z \end{vmatrix} = 0$$

Conservative

$f_x(x, y, z) = ye^z$

$f_y(x, y, z) = xe^z$

$f_z(x, y, z) = xye^z$

$f(x, y, z) = xye^z + k$

40. $F(x, y, z) = 3x^2y^2z\mathbf{i} + 2x^3yz\mathbf{j} + x^3y^2\mathbf{k}$

$$\text{curl } F = \begin{vmatrix} \mathbf{i} & \mathbf{j} & \mathbf{k} \\ \dfrac{\partial}{\partial x} & \dfrac{\partial}{\partial y} & \dfrac{\partial}{\partial z} \\ 3x^2y^2z & 2x^3yz & x^3y^2 \end{vmatrix} = 0$$

Conservative

$f_x(x, y, z) = 3x^2y^2z$

$f_y(x, y, z) = 2x^3yz$

$f_z(x, y, z) = x^3y^2$

$f(x, y, z) = x^3y^2z + k$

41. $F(x, y, z) = \dfrac{1}{y}i - \dfrac{x}{y^2}j + (2z - 1)k$

$$\text{curl } F = \begin{vmatrix} i & j & k \\ \dfrac{\partial}{\partial x} & \dfrac{\partial}{\partial y} & \dfrac{\partial}{\partial z} \\ \dfrac{1}{y} & -\dfrac{x}{y^2} & 2z - 1 \end{vmatrix} = 0$$

Conservative

$f_x(x, y, z) = \dfrac{1}{y}$

$f_y(x, y, z) = -\dfrac{x}{y^2}$

$f_z(x, y, z) = 2z - 1$

$f(x, y, z) = \displaystyle\int \dfrac{1}{y}\, dx = \dfrac{x}{y} + g(y, z) + k$

$f(x, y, z) = \displaystyle\int -\dfrac{x}{y^2}\, dy = \dfrac{x}{y} + h(x, z) + k$

$f(x, y, z) = \displaystyle\int (2z - 1)\, dz = z^2 - z + p(x, y) + k$

$f(x, y, z) = \dfrac{x}{y} + z^2 - z + k$

42. $F(x, y, z) = \dfrac{x}{x^2 + y^2}i + \dfrac{y}{x^2 + y^2}j + k$

$$\text{curl } F = \begin{vmatrix} i & j & k \\ \dfrac{\partial}{\partial x} & \dfrac{\partial}{\partial y} & \dfrac{\partial}{\partial z} \\ \dfrac{x}{x^2 + y^2} & \dfrac{y}{x^2 + y^2} & 1 \end{vmatrix} = O$$

Conservative

$f_x(x, y, z) = \dfrac{x}{x^2 + y^2}$

$f_y(x, y, z) = \dfrac{y}{x^2 + y^2}$

$f_z(x, y, z) = 1$

$f(x, y, z) = \displaystyle\int \dfrac{x}{x^2 + y^2}\, dx$

$\qquad = \dfrac{1}{2}\ln(x^2 + y^2) + g(y, z) + k$

$f(x, y, z) = \displaystyle\int \dfrac{y}{x^2 + y^2}\, dy$

$\qquad = \dfrac{1}{2}\ln(x^2 + y^2) + h(x, z) + k$

$f(x, y, z) = \displaystyle\int dz = z + p(x, y) + k$

$f(x, y, z) = \dfrac{1}{2}\ln(x^2 + y^2) + z + k$

43. $F(x, y, z) = i + 2xj + 3yk$

$G(x, y, z) = xi - yj + zk$

$$F \times G = \begin{vmatrix} i & j & k \\ 1 & 2x & 3y \\ x & -y & z \end{vmatrix} = (2xz + 3y^2)i - (z - 3xy)j + (-y - 2x^2)k$$

$$\text{curl } (F \times G) = \begin{vmatrix} i & j & k \\ \dfrac{\partial}{\partial x} & \dfrac{\partial}{\partial y} & \dfrac{\partial}{\partial z} \\ 2xz + 3y^2 & 3xy - z & -y - 2x^2 \end{vmatrix} = (-1 + 1)i - (-4x - 2x)j + (3y - 6y)k = 6xj - 3yk$$

44. $F(x, y, z) = xi - zk$

$G(x, y, z) = x^2i + yj + z^2k$

$$F \times G = \begin{vmatrix} i & j & k \\ x & 0 & -z \\ x^2 & y & z^2 \end{vmatrix} = yzi - (xz^2 + x^2z)j + xyk$$

$$\text{curl } (F \times G) = \begin{vmatrix} i & j & k \\ \dfrac{\partial}{\partial x} & \dfrac{\partial}{\partial y} & \dfrac{\partial}{\partial z} \\ yz & -xz^2 - x^2z & xy \end{vmatrix}$$

$\qquad = (x + 2xz + x^2)i - (y - y)j + (-z^2 - 2xz - z)k$

$\qquad = x(x + 2z + 1)i - z(z + 2x + 1)k$

45. $\mathbf{F}(x,\ y,\ z) = xyz\mathbf{i} + y\mathbf{j} + z\mathbf{k}$

$$\text{curl } \mathbf{F} = \begin{vmatrix} \mathbf{i} & \mathbf{j} & \mathbf{k} \\ \dfrac{\partial}{\partial x} & \dfrac{\partial}{\partial y} & \dfrac{\partial}{\partial z} \\ xyz & y & z \end{vmatrix} = xy\mathbf{j} - xz\mathbf{k}$$

$$\text{curl(curl } \mathbf{F}) = \begin{vmatrix} \mathbf{i} & \mathbf{j} & \mathbf{k} \\ \dfrac{\partial}{\partial x} & \dfrac{\partial}{\partial y} & \dfrac{\partial}{\partial z} \\ 0 & xy & -xz \end{vmatrix} = z\mathbf{j} + y\mathbf{k}$$

46. $\mathbf{F}(x,\ y,\ z) = x^2 z\mathbf{i} - 2xz\mathbf{j} + yz\mathbf{k}$

$$\text{curl } \mathbf{F} = \begin{vmatrix} \mathbf{i} & \mathbf{j} & \mathbf{k} \\ \dfrac{\partial}{\partial x} & \dfrac{\partial}{\partial y} & \dfrac{\partial}{\partial z} \\ x^2 z & -2xz & yz \end{vmatrix} = (z+2x)\mathbf{i} + x^2\mathbf{j} - 2z\mathbf{k}$$

$$\text{curl(curl } \mathbf{F}) = \begin{vmatrix} \mathbf{i} & \mathbf{j} & \mathbf{k} \\ \dfrac{\partial}{\partial x} & \dfrac{\partial}{\partial y} & \dfrac{\partial}{\partial z} \\ z+2x & x^2 & -2z \end{vmatrix} = \mathbf{j} + 2x\mathbf{k}$$

47. $\mathbf{F}(x,\ y,\ z) = 6x^2\mathbf{i} - xy^2\mathbf{j}$

$\text{div } \mathbf{F}(x,\ y,\ z) = \dfrac{\partial}{\partial x}[6x^2] + \dfrac{\partial}{\partial y}[-xy^2] = 12x - 2xy$

48. $\mathbf{F}(x,\ y,\ z) = xe^x\mathbf{i} + ye^y\mathbf{j}$

$\text{div } \mathbf{F}(x,\ y,\ z) = \dfrac{\partial}{\partial x}[xe^x] + \dfrac{\partial}{\partial y}[ye^y] = xe^x + e^x + ye^y + e^y = e^x(x+1) + e^y(y+1)$

49. $\mathbf{F}(x,\ y,\ z) = \sin x\mathbf{i} + \cos y\mathbf{j} + z^2\mathbf{k}$

$\text{div } \mathbf{F}(x,\ y,\ z) = \dfrac{\partial}{\partial x}[\sin x] + \dfrac{\partial}{\partial y}[\cos y] + \dfrac{\partial}{\partial z}[z^2] = \cos x - \sin y + 2z$

50. $\mathbf{F}(x,\ y,\ z) = \ln(x^2 + y^2)\mathbf{i} + xy\mathbf{j} + \ln(y^2 + z^2)\mathbf{k}$

$\text{div } \mathbf{F}(x,\ y,\ z) = \dfrac{\partial}{\partial x}[\ln(x^2 + y^2)] + \dfrac{\partial}{\partial y}[xy] + \dfrac{\partial}{\partial z}[\ln(y^2 + z^2)] = \dfrac{2x}{x^2 + y^2} + x + \dfrac{2z}{y^2 + z^2}$

51. $\mathbf{F}(x,\ y,\ z) = xyz\mathbf{i} + y\mathbf{j} + z\mathbf{k}$
$\text{div } \mathbf{F}(x,\ y,\ z) = yz + 1 + 1 = yz + 2$
$\text{div } \mathbf{F}(1,\ 2,\ 1) = 4$

52. $\mathbf{F}(x,\ y,\ z) = x^2 z\mathbf{i} - 2xz\mathbf{j} + yz\mathbf{k}$
$\text{div } \mathbf{F}(x,\ y,\ z) = 2xz + y$
$\text{div } \mathbf{F}(2,\ -1,\ 3) = 11$

53. $\mathbf{F}(x,\ y,\ z) = e^x \sin y\mathbf{i} - e^x \cos y\mathbf{j}$
$\text{div } \mathbf{F}(x,\ y,\ z) = e^x \sin y + e^x \sin y$
$\text{div } \mathbf{F}(0,\ 0,\ 3) = 0$

54. $\mathbf{F}(x,\ y,\ z) = e^{-xyz}(\mathbf{i} + \mathbf{j} + \mathbf{k})$
$\text{div } \mathbf{F}(x,\ y,\ z) = e^{-xyz}(-yz - xz - xy)$
$\text{div } \mathbf{F}(3,\ 2,\ 0) = -6$

55. $\mathbf{F}(x,\ y,\ z) = \mathbf{i} + 2x\mathbf{j} + 3y\mathbf{k}$
$\mathbf{G}(x,\ y,\ z) = x\mathbf{i} - y\mathbf{j} + z\mathbf{k}$

$$\mathbf{F} \times \mathbf{G} = \begin{vmatrix} \mathbf{i} & \mathbf{j} & \mathbf{k} \\ 1 & 2x & 3y \\ x & -y & z \end{vmatrix}$$

$$= (2xz + 3y^2)\mathbf{i} - (z - 3xy)\mathbf{j} + (-y - 2x^2)\mathbf{k}$$

$\text{div } (\mathbf{F} \times \mathbf{G}) = 2z + 3x$

56. $\mathbf{F}(x,\ y,\ z) = x\mathbf{i} - z\mathbf{k}$
$\mathbf{G}(x,\ y,\ z) = x^2\mathbf{i} + y\mathbf{j} + z^2\mathbf{k}$

$$\mathbf{F} \times \mathbf{G} = \begin{vmatrix} \mathbf{i} & \mathbf{j} & \mathbf{k} \\ x & 0 & -z \\ x^2 & y & z^2 \end{vmatrix} = -yz\mathbf{i} - (xz^2 + x^2 z)\mathbf{j} + xy\mathbf{k}$$

$\text{div } (\mathbf{F} \times \mathbf{G}) = 0$

57. $F(x, y, z) = xyz\mathbf{i} + y\mathbf{j} + z\mathbf{k}$

$$\text{curl } \mathbf{F} = \begin{vmatrix} \mathbf{i} & \mathbf{j} & \mathbf{k} \\ \dfrac{\partial}{\partial x} & \dfrac{\partial}{\partial y} & \dfrac{\partial}{\partial z} \\ xyz & y & z \end{vmatrix} = xy\mathbf{j} - xz\mathbf{k}$$

div $(\text{curl } \mathbf{F}) = x - x = 0$

58. $F(x, y, z) = x^2 z\mathbf{i} - 2xz\mathbf{j} + yz\mathbf{k}$

$$\text{curl } \mathbf{F} = \begin{vmatrix} \mathbf{i} & \mathbf{j} & \mathbf{k} \\ \dfrac{\partial}{\partial x} & \dfrac{\partial}{\partial y} & \dfrac{\partial}{\partial z} \\ x^2 z & -2xz & yz \end{vmatrix} = (z+2x)\mathbf{i} + x^2\mathbf{j} - 2z\mathbf{k}$$

div $(\text{curl } \mathbf{F}) = 2 - 2 = 0$

59. Let $\mathbf{F} = M\mathbf{i} + N\mathbf{j} + P\mathbf{k}$ and $\mathbf{G} = Q\mathbf{i} + R\mathbf{j} + S\mathbf{k}$ where M, N, P, Q, R, and S have continuous partial derivatives.

$\mathbf{F} + \mathbf{G} = (M + Q)\mathbf{i} + (N + R)\mathbf{j} + (P + S)\mathbf{k}$

$$\text{curl } (\mathbf{F} + \mathbf{G}) = \begin{vmatrix} \mathbf{i} & \mathbf{j} & \mathbf{k} \\ \dfrac{\partial}{\partial x} & \dfrac{\partial}{\partial y} & \dfrac{\partial}{\partial z} \\ M + Q & N + R & P + S \end{vmatrix}$$

$$= \left[\frac{\partial}{\partial y}(P + S) - \frac{\partial}{\partial z}(N + R)\right]\mathbf{i} - \left[\frac{\partial}{\partial x}(P + S) - \frac{\partial}{\partial z}(M + Q)\right]\mathbf{j} + \left[\frac{\partial}{\partial x}(N + R) + \frac{\partial}{\partial y}(M + Q)\right]\mathbf{k}$$

$$= \left(\frac{\partial P}{\partial y} - \frac{\partial N}{\partial z}\right)\mathbf{i} - \left(\frac{\partial P}{\partial x} - \frac{\partial M}{\partial z}\right)\mathbf{j} + \left(\frac{\partial N}{\partial x} - \frac{\partial M}{\partial y}\right)\mathbf{k}$$

$$+ \left(\frac{\partial S}{\partial y} - \frac{\partial R}{\partial z}\right)\mathbf{i} - \left(\frac{\partial S}{\partial x} - \frac{\partial Q}{\partial z}\right)\mathbf{j} + \left(\frac{\partial R}{\partial x} - \frac{\partial Q}{\partial y}\right)\mathbf{k}$$

$$= \text{curl } \mathbf{F} + \text{curl } \mathbf{G}$$

60. Let $f(x, y, z)$ be a scalar function whose second partial derivatives are continuous.

$$\nabla f = \frac{\partial f}{\partial x}\mathbf{i} + \frac{\partial f}{\partial y}\mathbf{j} + \frac{\partial f}{\partial z}\mathbf{k}$$

$$\text{curl}(\nabla f) = \begin{vmatrix} \mathbf{i} & \mathbf{j} & \mathbf{k} \\ \dfrac{\partial}{\partial x} & \dfrac{\partial}{\partial y} & \dfrac{\partial}{\partial z} \\ \dfrac{\partial f}{\partial x} & \dfrac{\partial f}{\partial y} & \dfrac{\partial f}{\partial z} \end{vmatrix} = \left(\frac{\partial^2 f}{\partial y \partial z} - \frac{\partial^2 f}{\partial z \partial y}\right)\mathbf{i} - \left(\frac{\partial^2 f}{\partial x \partial z} - \frac{\partial^2 f}{\partial z \partial x}\right)\mathbf{j} + \left(\frac{\partial^2 f}{\partial x \partial y} - \frac{\partial^2 f}{\partial y \partial x}\right)\mathbf{k} = \mathbf{O}$$

61. Let $\mathbf{F} = M\mathbf{i} + N\mathbf{j} + P\mathbf{k}$ and $\mathbf{G} = R\mathbf{i} + S\mathbf{j} + T\mathbf{k}$.

$$\text{div } (\mathbf{F} + \mathbf{G}) = \frac{\partial}{\partial x}(M + R) + \frac{\partial}{\partial y}(N + S) + \frac{\partial}{\partial z}(P + T) = \frac{\partial M}{\partial x} + \frac{\partial R}{\partial x} + \frac{\partial N}{\partial y} + \frac{\partial S}{\partial y} + \frac{\partial P}{\partial z} + \frac{\partial T}{\partial z}$$

$$= \left[\frac{\partial M}{\partial x} + \frac{\partial N}{\partial y} + \frac{\partial P}{\partial z}\right] + \left[\frac{\partial R}{\partial x} + \frac{\partial S}{\partial y} + \frac{\partial T}{\partial z}\right]$$

$$= \text{div } \mathbf{F} + \text{div } \mathbf{G}$$

62. Let $\mathbf{F} = M\mathbf{i} + N\mathbf{j} + P\mathbf{k}$ and $\mathbf{G} = R\mathbf{i} + S\mathbf{j} + T\mathbf{k}$.

$$\mathbf{F} \times \mathbf{G} = \begin{vmatrix} \mathbf{i} & \mathbf{j} & \mathbf{k} \\ M & N & P \\ R & S & T \end{vmatrix} = (NT - PS)\mathbf{i} - (MT - PR)\mathbf{j} + (MS - NR)\mathbf{k}$$

$$\text{div } (\mathbf{F} \times \mathbf{G}) = \frac{\partial}{\partial x}(NT - PS) + \frac{\partial}{\partial y}(PR - MT) + \frac{\partial}{\partial z}(MS - NR)$$

$$= N\frac{\partial T}{\partial x} + T\frac{\partial N}{\partial x} - P\frac{\partial S}{\partial x} - S\frac{\partial P}{\partial x} + P\frac{\partial R}{\partial y} + R\frac{\partial P}{\partial y} - M\frac{\partial T}{\partial y} - T\frac{\partial M}{\partial y} + M\frac{\partial S}{\partial z} + S\frac{\partial M}{\partial z} - N\frac{\partial R}{\partial z} - R\frac{\partial N}{\partial z}$$

$$= \left[\left(\frac{\partial P}{\partial y} - \frac{\partial N}{\partial z}\right)R + \left(\frac{\partial M}{\partial z} - \frac{\partial P}{\partial x}\right)S + \left(\frac{\partial N}{\partial x} - \frac{\partial M}{\partial y}\right)T\right]$$

$$- \left[M\left(\frac{\partial T}{\partial y} - \frac{\partial S}{\partial z}\right) + N\left(\frac{\partial R}{\partial z} - \frac{\partial T}{\partial x}\right) + P\left(\frac{\partial S}{\partial x} - \frac{\partial R}{\partial y}\right)\right]$$

$$= (\mathbf{curl \ F}) \cdot \mathbf{G} - \mathbf{F} \cdot (\mathbf{curl \ G})$$

63. Let $\mathbf{F} = M\mathbf{i} + N\mathbf{j} + P\mathbf{k} = \nabla f$.

$$\nabla f + (\nabla \times \mathbf{F}) = \nabla f + \mathbf{curl \ F} = \left(M + \frac{\partial P}{\partial y} - \frac{\partial N}{\partial z}\right)\mathbf{i} - \left(-N + \frac{\partial P}{\partial x} - \frac{\partial M}{\partial z}\right)\mathbf{j} + \left(P + \frac{\partial N}{\partial x} - \frac{\partial M}{\partial y}\right)\mathbf{k}$$

$$\nabla \times [\nabla f + (\nabla \times \mathbf{F})] = \begin{vmatrix} \mathbf{i} & \mathbf{j} & \mathbf{k} \\ \dfrac{\partial}{\partial x} & \dfrac{\partial}{\partial y} & \dfrac{\partial}{\partial z} \\ M + \dfrac{\partial P}{\partial y} - \dfrac{\partial N}{\partial z} & N - \dfrac{\partial P}{\partial x} + \dfrac{\partial M}{\partial z} & P + \dfrac{\partial N}{\partial x} - \dfrac{\partial M}{\partial y} \end{vmatrix}$$

$$= \left(\frac{\partial P}{\partial y} + \frac{\partial^2 N}{\partial y \partial x} - \frac{\partial^2 M}{\partial y^2} - \frac{\partial N}{\partial z} + \frac{\partial^2 P}{\partial z \partial x} - \frac{\partial^2 M}{\partial z^2}\right)\mathbf{i}$$

$$- \left(\frac{\partial P}{\partial x} + \frac{\partial^2 N}{\partial x^2} - \frac{\partial^2 M}{\partial x \partial y} - \frac{\partial M}{\partial z} - \frac{\partial^2 P}{\partial z \partial y} + \frac{\partial^2 N}{\partial z^2}\right)\mathbf{j}$$

$$+ \left(\frac{\partial N}{\partial x} - \frac{\partial^2 P}{\partial x^2} + \frac{\partial^2 M}{\partial x \partial z} - \frac{\partial M}{\partial y} - \frac{\partial^2 P}{\partial y^2} + \frac{\partial^2 N}{\partial z \partial y}\right)\mathbf{k}$$

$$= \left(\frac{\partial^2 N}{\partial y \partial x} - \frac{\partial^2 M}{\partial y^2} + \frac{\partial^2 P}{\partial z \partial x} - \frac{\partial^2 M}{\partial z^2}\right)\mathbf{i}$$

$$- \left(\frac{\partial^2 N}{\partial x^2} - \frac{\partial^2 M}{\partial x \partial y} - \frac{\partial^2 P}{\partial z \partial y} + \frac{\partial^2 N}{\partial z^2}\right)\mathbf{j}$$

$$+ \left(-\frac{\partial^2 P}{\partial x^2} + \frac{\partial^2 M}{\partial x \partial z} - \frac{\partial^2 P}{\partial y^2} + \frac{\partial^2 N}{\partial z \partial y}\right)\mathbf{k}$$

$$\left[\text{Since } \frac{\partial P}{\partial y} = \frac{\partial N}{\partial z}, \ \frac{\partial P}{\partial x} = \frac{\partial M}{\partial z}, \text{ and } \frac{\partial N}{\partial x} = \frac{\partial M}{\partial y}.\right]$$

$$= \begin{vmatrix} \mathbf{i} & \mathbf{j} & \mathbf{k} \\ \dfrac{\partial}{\partial x} & \dfrac{\partial}{\partial y} & \dfrac{\partial}{\partial z} \\ \dfrac{\partial P}{\partial y} - \dfrac{\partial N}{\partial z} & \dfrac{\partial M}{\partial z} - \dfrac{\partial P}{\partial x} & \dfrac{\partial N}{\partial x} - \dfrac{\partial M}{\partial y} \end{vmatrix} = \nabla \times (\nabla \times \mathbf{F})$$

64. Let $\mathbf{F} = M\mathbf{i} + N\mathbf{j} + P\mathbf{k} = \nabla f$.

$$\nabla \times (f\mathbf{F}) = \begin{vmatrix} \mathbf{i} & \mathbf{j} & \mathbf{k} \\ \dfrac{\partial}{\partial x} & \dfrac{\partial}{\partial y} & \dfrac{\partial}{\partial z} \\ fM & fN & fP \end{vmatrix} = \left(\frac{\partial f}{\partial y} P + f\frac{\partial P}{\partial y} - \frac{\partial f}{\partial z} N - f\frac{\partial N}{\partial z} \right)\mathbf{i}$$

$$- \left(\frac{\partial f}{\partial x} P + f\frac{\partial P}{\partial x} - \frac{\partial f}{\partial z} M - f\frac{\partial M}{\partial z} \right)\mathbf{j}$$

$$+ \left(\frac{\partial f}{\partial x} N + f\frac{\partial N}{\partial x} - \frac{\partial f}{\partial y} M - f\frac{\partial M}{\partial y} \right)\mathbf{k}$$

$$= f\left[\left(\frac{\partial P}{\partial y} - \frac{\partial N}{\partial z} \right)\mathbf{i} - \left(\frac{\partial P}{\partial x} - \frac{\partial M}{\partial z} \right)\mathbf{j} + \left(\frac{\partial N}{\partial x} - \frac{\partial M}{\partial y} \right)\mathbf{k} \right] + \begin{vmatrix} \mathbf{i} & \mathbf{j} & \mathbf{k} \\ \dfrac{\partial f}{\partial x} & \dfrac{\partial f}{\partial y} & \dfrac{\partial f}{\partial z} \\ M & N & P \end{vmatrix}$$

$$= f[\nabla \times \mathbf{F}] + (\nabla f) \times \mathbf{F}$$

65. Let $\mathbf{F} = M\mathbf{i} + N\mathbf{j} + P\mathbf{k}$, then $f\mathbf{F} = fM\mathbf{i} + fN\mathbf{j} + fP\mathbf{k}$.

$$\text{div } (f\mathbf{F}) = \frac{\partial}{\partial x}(fM) + \frac{\partial}{\partial y}(fN) + \frac{\partial}{\partial z}(fP) = f\frac{\partial M}{\partial x} + M\frac{\partial f}{\partial x} + f\frac{\partial N}{\partial y} + N\frac{\partial f}{\partial y} + f\frac{\partial P}{\partial z} + P\frac{\partial f}{\partial z}$$

$$= f\left(\frac{\partial M}{\partial x} + \frac{\partial N}{\partial y} + \frac{\partial N}{\partial z} \right) + \left(\frac{\partial f}{\partial x}M + \frac{\partial f}{\partial y}N + \frac{\partial f}{\partial z}P \right)$$

$$= f \text{ div } \mathbf{F} + \nabla f \cdot \mathbf{F}$$

66. Let $\mathbf{F} = M\mathbf{i} + N\mathbf{j} + P\mathbf{k}$.

$$\textbf{curl } \mathbf{F} = \left(\frac{\partial P}{\partial y} - \frac{\partial N}{\partial z} \right)\mathbf{i} - \left(\frac{\partial P}{\partial x} - \frac{\partial M}{\partial z} \right)\mathbf{j} + \left(\frac{\partial N}{\partial x} - \frac{\partial M}{\partial y} \right)\mathbf{k}$$

$$\text{div } (\textbf{curl } \mathbf{F}) = \frac{\partial}{\partial x}\left[\frac{\partial P}{\partial y} - \frac{\partial N}{\partial z} \right] - \frac{\partial}{\partial y}\left[\frac{\partial P}{\partial x} - \frac{\partial M}{\partial z} \right] + \frac{\partial}{\partial z}\left[\frac{\partial N}{\partial x} - \frac{\partial M}{\partial y} \right]$$

$$= \frac{\partial^2 P}{\partial x \partial y} - \frac{\partial^2 N}{\partial x \partial z} - \frac{\partial^2 P}{\partial y \partial x} + \frac{\partial^2 M}{\partial y \partial z} + \frac{\partial^2 N}{\partial z \partial x} - \frac{\partial^2 M}{\partial z \partial y} = 0$$

In Exercises 67–70, $\mathbf{F}(x, y, z) = x\mathbf{i} + y\mathbf{j} + z\mathbf{k}$ and $f(x, y, z) = \|\mathbf{F}(x, y, z)\| = \sqrt{x^2 + y^2 + z^2}$.

67. $\ln f = \dfrac{1}{2}\ln(x^2 + y^2 + z^2)$

$$\nabla(\ln f) = \frac{x}{x^2 + y^2 + z^2}\mathbf{i} + \frac{y}{x^2 + y^2 + z^2}\mathbf{j} + \frac{z}{x^2 + y^2 + z^2}\mathbf{k} = \frac{x\mathbf{i} + y\mathbf{j} + z\mathbf{k}}{x^2 + y^2 + z^2} = \frac{\mathbf{F}}{f^2}$$

68. $\dfrac{1}{f} = \dfrac{1}{\sqrt{x^2 + y^2 + z^2}}$

$$\nabla\left(\frac{1}{f} \right) = \frac{-x}{(x^2 + y^2 + z^2)^{3/2}}\mathbf{i} + \frac{-y}{(x^2 + y^2 + z^2)^{3/2}}\mathbf{j} + \frac{-z}{(x^2 + y^2 + z^2)^{3/2}}\mathbf{k} = \frac{-(x\mathbf{i} + y\mathbf{j} + z\mathbf{k})}{(\sqrt{x^2 + y^2 + z^2})^3} = -\frac{\mathbf{F}}{f^3}$$

69. $f^n = (\sqrt{x^2 + y^2 + z^2})^n$

$$\nabla f^n = n(\sqrt{x^2 + y^2 + z^2})^{n-1} \frac{x}{\sqrt{x^2 + y^2 + z^2}} \mathbf{i} + n(\sqrt{x^2 + y^2 + z^2})^{n-1} \frac{y}{\sqrt{x^2 + y^2 + z^2}} \mathbf{j}$$

$$+ n(\sqrt{x^2 + y^2 + z^2})^{n-1} \frac{z}{\sqrt{x^2 + y^2 + z^2}} \mathbf{k}$$

$$= n(\sqrt{x^2 + y^2 + z^2})^{n-2}(x\mathbf{i} + y\mathbf{j} + z\mathbf{k})$$

$$= n f^{n-2} \mathbf{F}$$

70. $w = \dfrac{1}{f} = \dfrac{1}{\sqrt{x^2 + y^2 + z^2}}$

$$\frac{\partial w}{\partial x} = -\frac{x}{(x^2 + y^2 + z^2)^{3/2}}$$

$$\frac{\partial w}{\partial y} = -\frac{y}{(x^2 + y^2 + z^2)^{3/2}}$$

$$\frac{\partial w}{\partial z} = -\frac{z}{(x^2 + y^2 + z^2)^{3/2}}$$

$$\frac{\partial^2 w}{\partial x^2} = \frac{2x^2 - y^2 - z^2}{(x^2 + y^2 + z^2)^{5/2}}$$

$$\frac{\partial^2 w}{\partial y^2} = \frac{2y^2 - x^2 - z^2}{(x^2 + y^2 + z^2)^{5/2}}$$

$$\frac{\partial^2 w}{\partial z^2} = \frac{2z^2 - x^2 - y^2}{(x^2 + y^2 + z^2)^{5/2}}$$

$$\nabla^2 w = \frac{\partial^2 w}{\partial x^2} + \frac{\partial^2 w}{\partial y^2} + \frac{\partial^2 w}{\partial z^2} = 0$$

Therefore $w = \dfrac{1}{f}$ is harmonic.

71. $\mathbf{F}(x,\ y) = M(x,\ y)\mathbf{i} + N(x,\ y)\mathbf{j} = \dfrac{m}{(x^2 + y^2)^{5/2}}[3xy\mathbf{i} + (2y^2 - x^2)\mathbf{j}]$

$$M = \frac{3mxy}{(x^2 + y^2)^{5/2}} = 3mxy(x^2 + y^2)^{-5/2}$$

$$\frac{\partial M}{\partial y} = 3mxy\left[-\frac{5}{2}(x^2 + y^2)^{-7/2}(2y)\right] + (x^2 + y^2)^{-5/2}(3mx)$$

$$= 3mx(x^2 + y^2)^{-7/2}[-5y^2 + (x^2 + y^2)] = \frac{3mx(x^2 - 4y^2)}{(x^2 + y^2)^{7/2}}$$

$$N = \frac{m(2y^2 - x^2)}{(x^2 + y^2)^{5/2}} = m(2y^2 - x^2)(x^2 + y^2)^{-5/2}$$

$$\frac{\partial N}{\partial x} = m(2y^2 - x^2)\left[-\frac{5}{2}(x^2 + y^2)^{-7/2}(2x)\right] + (x^2 + y^2)^{-5/2}(-2mx)$$

$$= mx(x^2 + y^2)^{-7/2}[(2y^2 - x^2)(-5) + (x^2 + y^2)(-2)]$$

$$= mx(x^2 + y^2)^{-7/2}(3x^2 - 12y^2) = \frac{3mx(x^2 - 4y^2)}{(x^2 + y^2)^{7/2}}$$

Therefore, $\dfrac{\partial N}{\partial x} = \dfrac{\partial M}{\partial y}$ and $\mathbf{F}$ is conservative.

Section 16.2   Line Integrals

1.
$$x^2 + y^2 = 9$$
$$\frac{x^2}{9} + \frac{y^2}{9} = 1$$
$$\cos^2 t + \sin^2 t = 1$$
$$\cos^2 t = \frac{x^2}{9}$$
$$\sin^2 t = \frac{y^2}{9}$$
$$x = 3\cos t$$
$$y = 3\sin t$$
$$\mathbf{r}(t) = 3\cos t\mathbf{i} + 3\sin t\mathbf{j}$$
$$0 \le t \le 2\pi$$

2.
$$\frac{x^2}{16} + \frac{y^2}{9} = 1$$
$$\cos^2 t + \sin^2 t = 1$$
$$\cos^2 t = \frac{x^2}{16}$$
$$\sin^2 t = \frac{y^2}{9}$$
$$x = 4\cos t$$
$$y = 3\sin t$$
$$\mathbf{r}(t) = 4\cos t\mathbf{i} + 3\sin t\mathbf{j}$$
$$0 \le t \le 2\pi$$

3. $\mathbf{r}(t) = \begin{cases} t\mathbf{i}, & 0 \le t \le 3 \\ 3\mathbf{i} + (t-3)\mathbf{j}, & 3 \le t \le 6 \\ (9-t)\mathbf{i} + 3\mathbf{j}, & 6 \le t \le 9 \\ (12-t)\mathbf{j}, & 9 \le t \le 12 \end{cases}$

4. $\mathbf{r}(t) = \begin{cases} t\mathbf{i} + \frac{3}{4}t\mathbf{j}, & 0 \le t \le 4 \\ 4\mathbf{i} + (7-t)\mathbf{j}, & 4 \le t \le 7 \\ (11-t)\mathbf{i}, & 7 \le t \le 11 \end{cases}$

5. $\mathbf{r}(t) = \begin{cases} t\mathbf{i} + \sqrt{t}\,\mathbf{j}, & 0 \le t \le 1 \\ (2-t)\mathbf{i} + (2-t)\mathbf{j}, & 1 \le t \le 2 \end{cases}$

6. $\mathbf{r}(t) = \begin{cases} t\mathbf{i} + t^2\mathbf{j}, & 0 \le t \le 2 \\ (4-t)\mathbf{i} + 4\mathbf{j}, & 2 \le t \le 4 \\ (8-t)\mathbf{j}, & 4 \le t \le 8 \end{cases}$

7. $\mathbf{r}(t) = 4t\mathbf{i} + 3t\mathbf{j}, \quad 0 \le t \le 2$

$$\int_C (x-y)\,ds = \int_0^2 (4t - 3t)\sqrt{(4)^2 + (3)^2}\,dt = \int_0^2 5t\,dt = \frac{5t^2}{2}\Big]_0^2 = 10$$

8. $\mathbf{r}(t) = t\mathbf{i} + (1-t)\mathbf{j}, \quad 0 \le t \le 1$

$$\int_C 4xy\,ds = \int_0^1 4t(1-t)\sqrt{1+1}\,dt = 4\sqrt{2}\int_0^1 (t - t^2)\,dt = 4\sqrt{2}\left(\frac{t^2}{2} - \frac{t^3}{3}\right)\Big]_0^1 = \frac{4\sqrt{2}}{6} = \frac{2\sqrt{2}}{3}$$

9. $\mathbf{r}(t) = \sin t\mathbf{i} + \cos t\mathbf{j} + 8t\mathbf{k}, \quad 0 \le t \le \frac{\pi}{2}$

$$\int_C (x^2 + y^2 + z^2)\,ds = \int_0^{\pi/2} (\sin^2 t + \cos^2 t + 64t^2)\sqrt{(\cos t)^2 + (-\sin t)^2 + 64}\,dt$$

$$= \int_0^{\pi/2} \sqrt{65}(1 + 64t^2)\,dt$$

$$= \sqrt{65}\left(t + \frac{64t^3}{3}\right)\Big]_0^{\pi/2}$$

$$= \sqrt{65}\left(\frac{\pi}{2} + \frac{8\pi^3}{3}\right)$$

$$= \frac{\sqrt{65}\pi}{6}(3 + 16\pi^2)$$

10. $\mathbf{r}(t) = 3\mathbf{i} + 12t\mathbf{j} + 5t\mathbf{k}, \quad 0 \le t \le 2$

$$\int_C 8xyz \, ds = \int_0^2 8(3)(12t)(5t)\sqrt{0 + 144 + 25} \, dt = \int_0^2 18720t^2 \, dt = 6240t^3 \Big]_0^2 = 49920$$

11. $\mathbf{r}(t) = t\mathbf{i}, \quad 0 \le t \le 3$

$$\int_C (x^2 + y^2) \, ds = \int_0^3 [t^2 + 0^2]\sqrt{1 + 0} \, dt$$

$$= \int_0^3 t^2 \, dt$$

$$= \frac{1}{3}t^3 \Big]_0^3$$

$$= 9$$

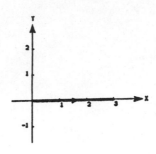

12. $\mathbf{r}(t) = t\mathbf{j}, \quad 1 \le t \le 10$

$$\int_C (x^2 + y^2) \, ds = \int_1^{10} [0 + t^2]\sqrt{0 + 1} \, dt$$

$$= \int_1^{10} t^2 \, dt$$

$$= \frac{1}{3}t^3 \Big]_1^{10}$$

$$= 333$$

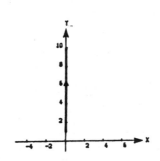

13. $\mathbf{r}(t) = \cos t\mathbf{i} + \sin t\mathbf{j}, \quad 0 \le t \le \dfrac{\pi}{2}$

$$\int_C (x^2 + y^2) \, ds = \int_0^{\pi/2} [\cos^2 t + \sin^2 t]\sqrt{(-\sin t)^2 + (\cos t)^2} \, dt$$

$$= \int_0^{\pi/2} dt$$

$$= \frac{\pi}{2}$$

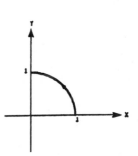

14. $\mathbf{r}(t) = 2\cos t\mathbf{i} + 2\sin t\mathbf{j}, \quad 0 \le t \le \dfrac{\pi}{2}$

$$\int_C (x^2 + y^2) \, ds = \int_0^{\pi/2} [4\cos^2 t + 4\sin^2 t]\sqrt{(-2\sin t)^2 + (2\cos t)^2} \, dt$$

$$= \int_0^{\pi/2} 8 \, dt$$

$$= 4\pi$$

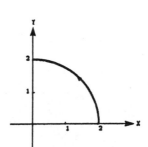

15. $r(t) = ti + tj, \quad 0 \leq t \leq 1$

$$\int_C (x + 4\sqrt{y})\, ds = \int_0^1 (t + 4\sqrt{t})\sqrt{1+1}\, dt$$

$$= \sqrt{2}\left(\frac{t^2}{2} + \frac{8}{3}t^{3/2}\right)\Big]_0^1$$

$$= \frac{19\sqrt{2}}{6}$$

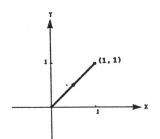

16. $r(t) = ti + 3tj, \quad 0 \leq t \leq 3$

$$\int_C (x + 4\sqrt{y})\, ds = \int_0^3 (t + 4\sqrt{3t})\sqrt{1+9}\, dt$$

$$= \sqrt{10}\left(\frac{t^2}{2} + \frac{8\sqrt{3}}{3}t^{3/2}\right)\Big]_0^3$$

$$= \frac{\sqrt{10}}{6}(27 + 144)$$

$$= \frac{57\sqrt{10}}{2}$$

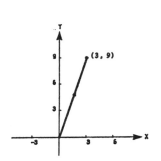

17. $r(t) = \begin{cases} ti, & 0 \leq t \leq 1 \\ (2-t)i + (t-1)j, & 1 \leq t \leq 2 \\ (3-t)j, & 2 \leq t \leq 3 \end{cases}$

$$\int_{C_1} (x + 4\sqrt{y})\, ds = \int_0^1 t\, dt = \frac{1}{2}$$

$$\int_{C_2} (x + 4\sqrt{y})\, ds = \int_1^2 [(2-t) + 4\sqrt{t-1}]\sqrt{1+1}\, dt$$

$$= \sqrt{2}\left[2t - \frac{t^2}{2} + \frac{8}{3}(t-1)^{3/2}\right]_1^2 = \frac{19\sqrt{2}}{6}$$

$$\int_{C_3} (x + 4\sqrt{y})\, ds = \int_2^3 4\sqrt{3-t}\, dt = -\frac{8}{3}(3-t)^{3/2}\Big]_2^3 = \frac{8}{3}$$

$$\int_C (x + 4\sqrt{y})\, ds = \frac{1}{2} + \frac{19\sqrt{2}}{6} + \frac{8}{3} = \frac{19 + 19\sqrt{2}}{6} = \frac{19(1+\sqrt{2})}{6}$$

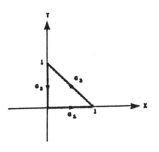

18. $r(t) = \begin{cases} t\mathbf{i}, & 0 \le t \le 1 \\ \mathbf{i} + (t-1)\mathbf{j}, & 1 \le t \le 2 \\ (3-t)\mathbf{i} + \mathbf{j}, & 2 \le t \le 3 \\ (4-t)\mathbf{j}, & 3 \le t \le 4 \end{cases}$

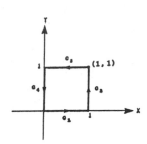

$\displaystyle\int_{C_1} \left(x + 4\sqrt{y}\,\right) ds = \int_0^1 t\,dt = \frac{1}{2}$

$\displaystyle\int_{C_2} \left(x + 4\sqrt{y}\,\right) ds = \int_1^2 \left(1 + 4\sqrt{t-1}\,\right) dt = \left[t + \frac{8}{3}(t-1)^{3/2} \right]_1^2 = \frac{11}{3}$

$\displaystyle\int_{C_3} \left(x + 4\sqrt{y}\,\right) ds = \int_2^3 [(3-t) + 4]\,dt = \left[7t - \frac{t^2}{2} \right]_2^3 = \frac{9}{2}$

$\displaystyle\int_{C_4} \left(x + 4\sqrt{y}\,\right) ds = \int_3^4 4\sqrt{4-t}\,dt = -\frac{8}{3}(4-t)^{3/2}\bigg]_3^4 = \frac{8}{3}$

$\displaystyle\int_C \left(x + 4\sqrt{y}\,\right) ds = \frac{1}{2} + \frac{11}{3} + \frac{9}{2} + \frac{8}{3} = \frac{34}{3}$

19. $\mathbf{F}(x, y) = xy\mathbf{i} + y\mathbf{j}$

$\quad C: \ \mathbf{r}(t) = 4t\mathbf{i} + t\mathbf{j}, \ \ 0 \le t \le 1$

$\quad \mathbf{F}(t) = 4t^2\mathbf{i} + t\mathbf{j}$

$\quad \mathbf{r}'(t) = 4\mathbf{i} + \mathbf{j}$

$\quad \displaystyle\int_C \mathbf{F} \cdot d\mathbf{r} = \int_0^1 (16t^2 + t)\,dt = \left[\frac{16}{3}t^3 + \frac{1}{2}t^2 \right]_0^1 = \frac{35}{6}$

20. $\mathbf{F}(x, y) = xy\mathbf{i} + y\mathbf{j}$

$\quad C: \ \mathbf{r}(t) = 4\cos t\,\mathbf{i} + 4\sin t\,\mathbf{j}, \ \ 0 \le t \le \frac{\pi}{2}$

$\quad \mathbf{F}(t) = 16\sin t\cos t\,\mathbf{i} + 4\sin t\,\mathbf{j}$

$\quad \mathbf{r}'(t) = -4\sin t\,\mathbf{i} + 4\cos t\,\mathbf{j}$

$\quad \displaystyle\int_C \mathbf{F} \cdot d\mathbf{r} = \int_0^{\pi/2} \left(-64\sin^2 t\cos t + 16\sin t\cos t\right) dt$

$\quad\quad = \left[-\frac{64}{3}\sin^3 t + 8\sin^2 t \right]_0^{\pi/2} = -\frac{40}{3}$

21. $\mathbf{F}(x, y) = 3x\mathbf{i} + 4y\mathbf{j}$

$\quad C: \ \mathbf{r}(t) = 2\cos t\,\mathbf{i} + 2\sin t\,\mathbf{j}, \ \ 0 \le t \le \frac{\pi}{2}$

$\quad \mathbf{F}(t) = 6\cos t\,\mathbf{i} + 8\sin t\,\mathbf{j}$

$\quad \mathbf{r}'(t) = -2\sin t\,\mathbf{i} + 2\cos t\,\mathbf{j}$

$\quad \displaystyle\int_C \mathbf{F} \cdot d\mathbf{r} = \int_0^{\pi/2} \left(-12\sin t\cos t + 16\sin t\cos t\right) dt$

$\quad\quad = 2\sin^2 t\bigg]_0^{\pi/2} = 2$

22. $\mathbf{F}(x, y) = 3x\mathbf{i} + 4y\mathbf{j}$

$\quad C: \ \mathbf{r}(t) = t\mathbf{i} + \sqrt{4 - t^2}\,\mathbf{j}, \ \ -2 \le t \le 2$

$\quad \mathbf{F}(t) = 3t\mathbf{i} + 4\sqrt{4 - t^2}\,\mathbf{j}$

$\quad \mathbf{r}'(t) = \mathbf{i} - \dfrac{t}{\sqrt{4 - t^2}}\,\mathbf{j}$

$\quad \displaystyle\int_C \mathbf{F} \cdot d\mathbf{r} = \int_{-2}^2 (3t - 4t)\,dt = -\frac{t^2}{2}\bigg]_{-2}^2 = 0$

23. $\mathbf{F}(x, y, z) = x^2 y\mathbf{i} + (x - z)\mathbf{j} + xyz\mathbf{k}$

$\quad C: \ \mathbf{r}(t) = t\mathbf{i} + t^2\mathbf{j} + 2\mathbf{k}, \ \ 0 \le t \le 1$

$\quad \mathbf{F}(t) = t^4\mathbf{i} + (t - 2)\mathbf{j} + 2t^3\mathbf{k}$

$\quad \mathbf{r}'(t) = \mathbf{i} + 2t\mathbf{j}$

$\quad \displaystyle\int_C \mathbf{F} \cdot d\mathbf{r} = \int_0^1 [t^4 + 2t(t - 2)]\,dt = \left[\frac{t^5}{5} + \frac{2t^3}{3} - 2t^2 \right]_0^1 = -\frac{17}{15}$

24. $\mathbf{F}(x,\ y,\ z) = x^2\mathbf{i} + y^2\mathbf{j} + z^2\mathbf{k}$

   $C:\ \ \mathbf{r}(t) = \sin t\mathbf{i} + \cos t\mathbf{j} + t^2\mathbf{j},\ \ 0 \le t \le \dfrac{\pi}{2}$

   $\mathbf{F}(t) = \sin^2\mathbf{i} + \cos^2\mathbf{j} + t^4\mathbf{k}$

   $\mathbf{r}'(t) = \cos t\mathbf{i} - \sin t\mathbf{j} + 2t\mathbf{k}$

   $\displaystyle\int_C \mathbf{F} \cdot d\mathbf{r} = \int_0^{\pi/2} (\sin^2 t \cos t - \cos^2 t \sin t + 2t^5)\, dt = \left[\dfrac{\sin^3 t}{3} + \dfrac{\cos^3 t}{3} + \dfrac{t^6}{3}\right]_0^{\pi/2} = \dfrac{\pi^6}{192}$

25. $x = 2t,\ \ y = 10t,\ \ 0 \le t \le 1 \Rightarrow y = 5x$ or $x = \dfrac{y}{5},\ \ 0 \le y \le 10$

   $\displaystyle\int_C (x + 3y^2)\, dy = \int_0^{10} \left(\dfrac{y}{5} + 3y^2\right) dy = \left[\dfrac{y^2}{10} + y^3\right]_0^{10} = 1010$

26. $x = 2t,\ \ y = 10t,\ \ 0 \le t \le 1 \Rightarrow y = 5x,\ \ 0 \le x \le 2$

   $\displaystyle\int_C (x + 3y^2)\, dx = \int_0^2 (x + 75x^2)\, dx = \left[\dfrac{x^2}{2} + 25x^3\right]_0^2 = 202$

27. $x = 2t,\ \ y = 10t,\ \ 0 \le t \le 1 \Rightarrow x = \dfrac{y}{5},\ \ 0 \le y \le 10,\ \ dx = \dfrac{1}{5}\, dy$

   $\displaystyle\int_C xy\, dx + y\, dy = \int_0^{10} \left(\dfrac{y^2}{25} + y\right) dy = \left[\dfrac{y^3}{75} + \dfrac{y^2}{2}\right]_0^{10} = \dfrac{190}{3}$

   OR

   $y = 5x,\ \ dy = 5\, dx,\ \ 0 \le x \le 2$

   $\displaystyle\int_C xy\, dx + y\, dy = \int_0^2 (5x^2 + 25x)\, dx = \left[\dfrac{5x^3}{3} + \dfrac{25x^2}{2}\right]_0^2 = \dfrac{190}{3}$

28. $x = 2t,\ \ y = 10t,\ \ 0 \le t \le 1 \Rightarrow y = 5x,\ \ dy = 5\, dx,\ \ 0 \le x \le 2$

   $\displaystyle\int_C (y - 3x)\, dx + x^2\, dy = \int_0^2 (2x + 5x^2)\, dx = \left[x^2 + \dfrac{5x^3}{3}\right]_0^2 = \dfrac{52}{3}$

29. $\mathbf{r}(t) = t\mathbf{i},\ \ \ \ \ 0 \le t \le 5$

   $x(t) = t,\ \ \ \ \ y(t) = 0$

   $dx = dt,\ \ \ \ \ dy = 0$

   $\displaystyle\int_C (2x - y)\, dx + (x + 3y)\, dy = \int_0^5 2t\, dt = 25$

30. $\mathbf{r}(t) = t\mathbf{j},\ \ \ \ \ 0 \le t \le 2$

   $x(t) = 0,\ \ \ \ \ y(t) = t$

   $dx = 0,\ \ \ \ \ dy = dt$

   $\displaystyle\int_C (2x - y)\, dx + (x + 3y)\, dy = \int_0^2 3t\, dt = \dfrac{3}{2}t^2 \Big]_0^2 = 6$

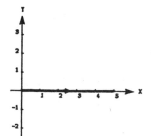

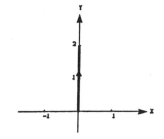

31. $\mathbf{r}(t) = \begin{cases} t\mathbf{i}, & 0 \le t \le 3 \\ 3\mathbf{i} + (t-3)\mathbf{j}, & 3 \le t \le 6 \end{cases}$

$C_1: \quad x(t) = t, \qquad y(t) = 0,$
$\qquad dx = dt, \qquad dy = 0$

$\displaystyle\int_{C_1} (2x - y)\,dx + (x + 3y)\,dy = \int_0^3 2t\,dt = 9$

$C_2: \quad x(t) = 3, \qquad y = t - 3$
$\qquad dx = 0, \qquad dy = dt$

$\displaystyle\int_{C_2} (2x - y)\,dx + (x + 3y)\,dy = \int_3^6 [3 + 3(t-3)]\,dt = \left[\frac{3t^2}{2} - 6t\right]_3^6 = \frac{45}{2}$

$\displaystyle\int_C (2x - y)\,dx + (x + 3y)\,dy = 9 + \frac{45}{2} = \frac{63}{2}$

32. $\mathbf{r}(t) = \begin{cases} -t\mathbf{j}, & 0 \le t \le 3 \\ (t-3)\mathbf{i} - 3\mathbf{j}, & 3 \le t \le 5 \end{cases}$

$C_1: \quad x(t) = 0, \qquad y(t) = -t$
$\qquad dx = 0, \qquad dy = -dt$

$\displaystyle\int_C (2x - y)\,dx + (x + 3y)\,dy = \int_0^3 t\,dt = \frac{9}{2}$

$C_2: \quad x(t) = t - 3, \qquad y(t) = -3$
$\qquad dx = dt, \qquad dy = 0$

$\displaystyle\int_{C_2} (2x - y)\,dx + (x + 3y)\,dy = \int_3^5 [2(t-3) + 3]\,dt = \left[(t-3)^2 + 3t\right]_3^5 = 10$

$\displaystyle\int_C (2x - y)\,dx + (x + 3y)\,dy = \frac{9}{2} + 10 = \frac{29}{2}$

33. $x(t) = t, \qquad y(t) = 2t^2, \qquad 0 \le t \le 2$
$dx = dt, \qquad dy = 4t\,dt$

$\displaystyle\int_C (2x - y)\,dx + (x + 3y)\,dy = \int_0^2 (2t - 2t^2)\,dt + (t + 6t^2)4t\,dt = \int_0^2 (24t^3 + 2t^2 + 2t)\,dt$

$\displaystyle\qquad\qquad = \left[6t^4 + \frac{2}{3}t^3 + t^2\right]_0^2 = \frac{316}{3}$

34. $x(t) = 4\sin t, \qquad y(t) = 3\cos t, \qquad 0 \le t \le \frac{\pi}{2}$
$dx = 4\cos t\,dt, \qquad dy = -3\sin t\,dt$

$\displaystyle\int_C (2x - y)\,dx + (x + 3y)\,dy = \int_0^{\pi/2} (8\sin t - 3\cos t)(4\cos t)\,dt + (4\sin t + 9\cos t)(-3\sin t)\,dt$

$\displaystyle\qquad\qquad = \int_0^{\pi/2} (5\sin t\cos t - 12\cos^2 t - 12\sin^2 t)\,dt$

$\displaystyle\qquad\qquad = \left(\frac{5}{2}\sin^2 t - 12t\right)\Big]_0^{\pi/2}$

$\displaystyle\qquad\qquad = \frac{5}{2} - 6\pi$

35. $\rho(x,\ y,\ z) = \dfrac{1}{2}(x^2 + y^2 + z^2)$

$\quad \mathbf{r}(t) = 3\cos t\mathbf{i} + 3\sin t\mathbf{j} + 2t\mathbf{k}, \quad 0 \le t \le 4\pi$

$\quad \mathbf{r}'(t) = -3\sin t\mathbf{i} + 3\cos t\mathbf{j} + 2\mathbf{k}$

$\quad \|\mathbf{r}'(t)\| = \sqrt{(-3\sin t)^2 + (3\cos t)^2 + (2)^2} = \sqrt{13}$

$\quad \text{mass} = \displaystyle\int_C \rho(x,\ y,\ z)\,ds$

$\qquad = \displaystyle\int_0^{4\pi} \frac{1}{2}[(3\cos t)^2 + (3\sin t)^2 + (2t)^2]\sqrt{13}\,dt$

$\qquad = \dfrac{\sqrt{13}}{2} \displaystyle\int_0^{4\pi} (9 + 4t^2)\,dt$

$\qquad = \dfrac{\sqrt{13}}{2}\left(9t + \dfrac{4}{3}t^3\right)\Big]_0^{4\pi}$

$\qquad = \dfrac{2\sqrt{13}\,\pi}{3}(27 + 64\pi^2) \approx 4973.8$

36. $\rho(x,\ y,\ z) = 2$

$\quad \mathbf{r}(t) = 3\cos t\mathbf{i} + 3\sin t\mathbf{j} + 2t\mathbf{k}, \quad 0 \le t \le 4\pi$

$\quad \mathbf{r}'(t) = -3\sin t\mathbf{i} + 3\cos t\mathbf{j} + 2\mathbf{k}$

$\quad \|\mathbf{r}'(t)\| = \sqrt{(-3\sin t)^2 + (3\cos t)^2 + (2)^2} = \sqrt{13}$

$\quad \text{mass} = \displaystyle\int_C \rho(x,\ y,\ z)\,ds = \int_0^{4\pi} 2\sqrt{13}\,dt = 8\sqrt{13}\,\pi$

37. $\mathbf{F}(x,\ y) = -x\mathbf{i} - 2y\mathbf{j}$

$\quad C:\ y = x^3$ from $(0, 0)$ to $(2, 8)$

$\quad \mathbf{r}(t) = t\mathbf{i} + t^3\mathbf{j}, \quad 0 \le t \le 2$

$\quad \mathbf{r}'(t) = \mathbf{i} + 3t^2\mathbf{j}$

$\quad \mathbf{F}(t) = -t\mathbf{i} - 2t^3\mathbf{j}$

$\quad \mathbf{F} \cdot \mathbf{r}' = -t - 6t^5$

$\quad \text{Work} = \displaystyle\int_C \mathbf{F} \cdot d\mathbf{r} = \int_0^2 (-t - 6t^5)\,dt = \left[-\frac{1}{2}t^2 - t^6\right]_0^2 = -66$

38. $\mathbf{F}(x,\ y) = x^2\mathbf{i} + xy\mathbf{j}$

$\quad C:\ x = \cos^3 t, \quad y = \sin^3 t$ from $(1, 0)$ to $(0, 1)$

$\quad \mathbf{r}(t) = \cos^3 t\mathbf{i} + \sin^3 t\mathbf{j}, \quad 0 \le t \le \dfrac{\pi}{2}$

$\quad \mathbf{r}'(t) = -3\cos^2 t\sin t\mathbf{i} + 3\sin^2 t\cos t\mathbf{j}$

$\quad \mathbf{F}(t) = \cos^6 t\mathbf{i} - \cos^3 t\sin^3 t\mathbf{j}$

$\quad \mathbf{F} \cdot \mathbf{r}' = -3\cos^8 t\sin t - 3\cos^4 t\sin^5 t$

$\qquad = -3\cos^4 t\sin t(\cos^4 t + \sin^4 t)$

$\qquad = -3\cos^4 t\sin t[\cos^4 t + (1 - \cos^2 t)^2]$

$\qquad = -3\cos^4 t\sin t(2\cos^4 t - 2\cos^2 t + 1)$

$\qquad = -6\cos^8 t\sin t + 6\cos^6 t\sin t - 3\cos^4 t\sin t$

$\quad \text{Work} = \displaystyle\int_C \mathbf{F} \cdot d\mathbf{r} = \int_0^{\pi/2} [-6\cos^8 t\sin t + 6\cos^6 t\sin t - 3\cos^4 t\sin t]\,dt$

$\qquad = \left[\dfrac{2\cos^9 t}{3} - \dfrac{6\cos^7 t}{7} + \dfrac{3\cos^5 t}{5}\right]_0^{\pi/2} = -\dfrac{43}{105}$

39. $\mathbf{F}(x,\ y) = 2x\mathbf{i} + y\mathbf{j}$

C : counterclockwise around the triangle whose vertices are $(0, 0)$, $(1, 0)$, $(1, 1)$

$$\mathbf{r}(t) = \begin{cases} t\mathbf{i}, & 0 \le t \le 1 \\ \mathbf{i} + (t-1)\mathbf{j}, & 1 \le t \le 2 \\ (3-t)\mathbf{i} + (3-t)\mathbf{j}, & 2 \le t \le 3 \end{cases}$$

On C_1 : $\mathbf{F}(t) = 2t\mathbf{i}$, $\mathbf{r}'(t) = \mathbf{i}$

$$\text{Work} = \int_{C_1} \mathbf{F} \cdot d\mathbf{r} = \int_0^1 2t\,dt = 1$$

On C_2 : $\mathbf{F}(t) = 2\mathbf{i} + (t-1)\mathbf{j}$, $\mathbf{r}'(t) = \mathbf{j}$

$$\text{Work} = \int_{C_2} \mathbf{F} \cdot d\mathbf{r} = \int_1^2 (t-1)\,dt = \frac{1}{2}$$

On C_3 : $\mathbf{F}(t) = 2(3-t)\mathbf{i} + (3-t)\mathbf{j}$, $\mathbf{r}'(t) = -\mathbf{i} - \mathbf{j}$

$$\text{Work} = \int_{C_3} \mathbf{F} \cdot d\mathbf{r} = \int_2^3 [-2(3-t) - (3-t)]\,dt = -\frac{3}{2}$$

$$\text{Total work} = \int_C \mathbf{F} \cdot d\mathbf{r} = 1 + \frac{1}{2} - \frac{3}{2} = 0$$

40. $\mathbf{F}(x,\ y) = -y\mathbf{i} - x\mathbf{j}$

C: counterclockwise along the semicircle $y = \sqrt{4 - x^2}$ from $(2, 0)$ to $(-2, 0)$

$\mathbf{r}(t) = 2\cos t\,\mathbf{i} + 2\sin t\,\mathbf{j}$, $0 \le t \le \pi$

$\mathbf{r}'(t) = -2\sin t\,\mathbf{i} + 2\cos t\,\mathbf{j}$

$\mathbf{F}(t) = -2\sin t\,\mathbf{i} - 2\cos t\,\mathbf{j}$

$\mathbf{F} \cdot \mathbf{r}' = 4\sin^2 t - 4\cos^2 t = -4\cos 2t$

$$\text{Work} = \int_C \mathbf{F} \cdot d\mathbf{r} = -4\int_0^\pi \cos 2t\,dt = \left. -2\sin 2t \right]_0^\pi = 0$$

41. $\mathbf{F}(x,\ y,\ z) = x\mathbf{i} + y\mathbf{j} - 5z\mathbf{k}$

C : $\mathbf{r}(t) = 2\cos t\,\mathbf{i} + 2\sin t\,\mathbf{j} + t\mathbf{k}$, $0 \le t \le 2\pi$

$\mathbf{r}'(t) = -2\sin t\,\mathbf{i} + 2\cos t\,\mathbf{j} + \mathbf{k}$

$\mathbf{F}(t) = 2\cos t\,\mathbf{i} + 2\sin t\,\mathbf{j} - 5t\mathbf{k}$

$\mathbf{F} \cdot \mathbf{r}' = -5t$

$$\text{Work} = \int_C \mathbf{F} \cdot d\mathbf{r} = \int_0^{2\pi} -5t\,dt = -10\pi^2$$

42. $\mathbf{F}(x,\ y,\ z) = yz\mathbf{i} + xz\mathbf{j} + xy\mathbf{k}$

C: line from $(0, 0, 0)$ to $(5, 3, 2)$

$\mathbf{r}(t) = 5t\mathbf{i} + 3t\mathbf{j} + 2t\mathbf{k}$, $0 \le t \le 1$

$\mathbf{r}'(t) = 5\mathbf{i} + 3\mathbf{j} + 2\mathbf{k}$

$\mathbf{F}(t) = 6t^2\mathbf{i} + 10t^2\mathbf{j} + 15t^2\mathbf{k}$

$\mathbf{F} \cdot \mathbf{r}' = 90t^2$

$$\text{Work} = \int_C \mathbf{F} \cdot d\mathbf{r} = \int_0^1 90t^2\,dt = 30$$

43. $\mathbf{F}(x,\ y) = y\mathbf{i} - x\mathbf{j}$

C : $\mathbf{r}(t) = t\mathbf{i} - 2t\mathbf{j}$

$\mathbf{r}'(t) = \mathbf{i} - 2\mathbf{j}$

$\mathbf{F}(t) = -2t\mathbf{i} - t\mathbf{j}$

$\mathbf{F} \cdot \mathbf{r}' = -2t + 2t = 0$

Thus, $\displaystyle\int_C \mathbf{F} \cdot d\mathbf{r} = 0$.

44. $\mathbf{F}(x,\ y) = -3y\mathbf{i} + x\mathbf{j}$

C : $\mathbf{r}(t) = t\mathbf{i} + t^3\mathbf{j}$

$\mathbf{r}'(t) = \mathbf{i} + 3t^2\mathbf{j}$

$\mathbf{F}(t) = -3t^3\mathbf{i} + t\mathbf{j}$

$\mathbf{F} \cdot \mathbf{r}' = -3t^3 + 3t^3 = 0$

Thus, $\displaystyle\int_C \mathbf{F} \cdot d\mathbf{r} = 0$.

45. $\mathbf{F}(x,\ y) = (x^3 - 2x^2)\mathbf{i} + \left(x - \dfrac{y}{2}\right)\mathbf{j}$

$\quad C:\ \mathbf{r}(t) = t\mathbf{i} + t^2\mathbf{j}$

$\qquad \mathbf{r}'(t) = \mathbf{i} + 2t\mathbf{j}$

$\qquad \mathbf{F}(t) = (t^3 - 2t^2)\mathbf{i} + \left(t - \dfrac{t^2}{2}\right)\mathbf{j}$

$\qquad \mathbf{F} \cdot \mathbf{r}' = (t^3 - 2t^2) + 2t\left(t - \dfrac{t^2}{2}\right) = 0$

Thus, $\displaystyle\int_C \mathbf{F} \cdot d\mathbf{r} = 0.$

46. $\mathbf{F}(x,\ y) = x\mathbf{i} + y\mathbf{j}$

$\quad C:\ \mathbf{r}(t) = 3\sin t\mathbf{i} + 3\cos t\mathbf{j}$

$\qquad \mathbf{r}'(t) = 3\cos t\mathbf{i} - 3\sin t\mathbf{j}$

$\qquad \mathbf{F}(t) = 3\sin t\mathbf{i} + 3\cos t\mathbf{j}$

$\qquad \mathbf{F} \cdot \mathbf{r}' = 9\sin t\cos t - 9\sin t\cos t = 0$

Thus, $\displaystyle\int_C \mathbf{F} \cdot d\mathbf{r} = 0.$

47. $f(x,\ y) = h$

$\quad C:$ line from $(0,\ 0)$ to $(3,\ 4)$

$\qquad \mathbf{r} = 3t\mathbf{i} + 4t\mathbf{j},\ \ 0 \le t \le 1$

$\qquad \mathbf{r}'(t) = 3\mathbf{i} + 4\mathbf{j}$

$\qquad \|\mathbf{r}'(t)\| = 5$

Lateral surface area $= \displaystyle\int_C f(x,\ y)\,ds = \int_0^1 5h\,dt = 5h$

48. $f(x,\ y) = y$

$\quad C:$ line from $(0,\ 0)$ to $(4,\ 4)$

$\qquad \mathbf{r}(t) = t\mathbf{i} + t\mathbf{j},\ \ 0 \le t \le 4$

$\qquad \mathbf{r}'(t) = \mathbf{i} + \mathbf{j}$

$\qquad \|\mathbf{r}'(t)\| = \sqrt{2}$

Lateral surface area $= \displaystyle\int_C f(x,\ y)\,ds$

$\qquad\qquad\qquad\qquad = \displaystyle\int_0^4 t(\sqrt{2})\,dt = 8\sqrt{2}$

49. $f(x,\ y) = xy$

$\quad C:\ x^2 + y^2 = 1$ from $(1,\ 0)$ to $(0,\ 1)$

$\qquad \mathbf{r}(t) = \cos t\mathbf{i} + \sin t\mathbf{j},\ \ 0 \le t \le \dfrac{\pi}{2}$

$\qquad \mathbf{r}'(t) = -\sin t\mathbf{i} + \cos t\mathbf{j}$

$\qquad \|\mathbf{r}'(t)\| = 1$

Lateral surface area $= \displaystyle\int_C f(x,\ y)\,ds = \int_0^{\pi/2} \cos t\sin t\,dt = \left.\dfrac{\sin^2 t}{2}\right]_0^{\pi/2} = \dfrac{1}{2}$

50. $f(x,\ y) = x + y$

$\quad C:\ x^2 + y^2 = 1$ from $(1,\ 0)$ to $(0,\ 1)$

$\qquad \mathbf{r}(t) = \cos t\mathbf{i} + \sin t\mathbf{j},\ \ 0 \le t \le \dfrac{\pi}{2}$

$\qquad \mathbf{r}'(t) = -\sin t\mathbf{i} + \cos t\mathbf{j}$

$\qquad \|\mathbf{r}'(t)\| = 1$

Lateral surface area $= \displaystyle\int_C f(x,\ y)\,ds = \int_0^{\pi/2} (\cos t + \sin t)\,dt = \Big[\sin t - \cos t\Big]_0^{\pi/2} = 2$

51. $f(x, y) = h$

$C: \quad y = 1 - x^2$ from $(1, 0)$ to $(0, 1)$

$\quad \mathbf{r}(t) = (1 - t)\mathbf{i} + [1 - (1 - t^2)]\mathbf{j}, \quad 0 \le t \le 1$

$\quad \mathbf{r}'(t) = -\mathbf{i} + 2(1 - t)\mathbf{j}$

$\quad \|\mathbf{r}'(t)\| = \sqrt{1 + 4(1 - t)^2}$

Lateral surface area $= \displaystyle\int_C f(x, y)\, ds$

$\qquad\qquad\qquad\quad = \displaystyle\int_0^1 h\sqrt{1 + 4(1 - t)^2}\, dt$

$\qquad\qquad\qquad\quad = -\dfrac{h}{4}\left[2(1 - t)\sqrt{1 + 4(1 - t)^2} + \ln\left|2(1 - t) + \sqrt{1 + 4(1 - t)^2}\right|\right]_0^1$

$\qquad\qquad\qquad\quad = \dfrac{h}{4}[2\sqrt{5} + \ln(2 + \sqrt{5})]$

52. $f(x, y) = y + 1$

$C: \quad y = 1 - x^2$ from $(1, 0)$ to $(0, 1)$

$\mathbf{r}(t) = (1 - t)\mathbf{i} + [1 - (1 - t)^2]\mathbf{j}, \quad 0 \le t \le 1$

$\mathbf{r}'(t) = -\mathbf{i} + 2(1 - t)\mathbf{j}$

$\|\mathbf{r}'(t)\| = \sqrt{1 + 4(1 - t)^2}$

Lateral surface area $= \displaystyle\int_C f(x, y)\, ds$

$\qquad\qquad\qquad\quad = \displaystyle\int_0^1 [2 - (1 - t)^2]\sqrt{1 + 4(1 - t)^2}\, dt$

$\qquad\qquad\qquad\quad = 2\displaystyle\int_0^1 \sqrt{1 + 4(1 - t)^2}\, dt - \int_0^1 (1 - t)^2\sqrt{1 + 4(1 - t)^2}\, dt$

$\qquad\qquad\qquad\quad = -\dfrac{1}{2}\left[2(1 - t)\sqrt{1 + 4(1 - t)^2} + \ln\left|2(1 - t) + \sqrt{1 + 4(1 - t)^2}\right|\right]_0^1$

$\qquad\qquad\qquad\qquad + \dfrac{1}{64}\left[2(1 - t)[2(4)(1 - t)^2 + 1]\sqrt{1 + 4(1 - t)^2} - \ln\left|2(1 - t) + \sqrt{1 + 4(1 - t)^2}\right|\right]_0^1$

$\qquad\qquad\qquad\quad = \dfrac{1}{2}[2\sqrt{5} + \ln(2 + \sqrt{5})] - \dfrac{1}{64}[18\sqrt{5} - \ln(2 + \sqrt{5})]$

$\qquad\qquad\qquad\quad = \dfrac{23}{32}\sqrt{5} + \dfrac{33}{64}\ln(2 + \sqrt{5})$

$\qquad\qquad\qquad\quad = \dfrac{1}{64}[46\sqrt{5} + 33\ln(2 + \sqrt{5})]$

53. $f(x, y) = xy$

$C: y = 1 - x^2$ from $(1, 0)$ to $(0, 1)$

Let $x = \cos t$, then

$$y = 1 - \cos^2 t = \sin^2 t$$

$$\mathbf{r}(t) = \cos t\mathbf{i} + \sin^2 t\mathbf{j}, \quad 0 \leq t \leq \frac{\pi}{2}$$

$$\mathbf{r}'(t) = -\sin t\mathbf{i} + 2\sin t \cos t\mathbf{j}$$

$$\|\mathbf{r}'(t)\| = \sqrt{\sin^2 t + 4\sin^2 t \cos^2 t} = \sin t\sqrt{1 + 4\cos^2 t}.$$

Lateral surface area $= \displaystyle\int_C f(x, y)\, ds = \int_0^{\pi/2} \cos t \sin^2 t\left(\sin t\sqrt{1 + 4\cos^2 t}\right) dt$

$$= \int_0^{\pi/2} \sin^2 t[(1 + 4\cos^2 t)^{1/2} \sin t \cos t]\, dt$$

Let $u = \sin^2 t$ and $dv = (1 + 4\cos^2 t)^{1/2} \sin t \cos t$, then $du = 2\sin t \cos t\, dt$ and $v = -\frac{1}{12}(1 + 4\cos^2 t)^{3/2}$.

$$\int_C f(x, y)\, ds = -\frac{1}{12}\sin^2 t(1 + 4\cos^2 t)^{3/2}\Big]_0^{\pi/2} + \frac{1}{6}\int_0^{\pi/2}(1 + 4\cos^2 t)^{3/2} \sin t \cos t\, dt$$

$$= \left[-\frac{1}{12}\sin^2 t(1 + 4\cos^2 t)^{3/2} - \frac{1}{120}(1 + 4\cos^2 t)^{5/2}\right]_0^{\pi/2}$$

$$= \left(-\frac{1}{12} - \frac{1}{120}\right) + \frac{1}{120}(5)^{5/2}$$

$$= \frac{1}{120}(25\sqrt{5} - 11)$$

54. $f(x, y) = x^2 - y^2 + 4$

$C: x^2 + y^2 = 4$

$\mathbf{r}(t) = 2\cos t\mathbf{i} + 2\sin t\mathbf{j}, \quad 0 \leq t \leq 2\pi$

$\mathbf{r}'(t) = -2\sin t\mathbf{i} + 2\cos t\mathbf{j}$

$\|\mathbf{r}'(t)\| = 2$

Lateral surface area $= \displaystyle\int_C f(x, y)\, ds = \int_0^{2\pi}(4\cos^2 t - 4\sin^2 t + 4)(2)\, dt$

$$= 8\int_0^{2\pi}(1 + \cos 2t)\, dt$$

$$= 8\left(t + \frac{1}{2}\sin 2t\right)\Big]_0^{2\pi} = 16\pi$$

55. $\mathbf{r}(t) = 3\sin t\mathbf{i} + 3\cos t\mathbf{j} + \frac{10}{2\pi}t\mathbf{k}, \quad 0 \leq t \leq 2\pi$

$\mathbf{F} = 150\mathbf{k}$

$d\mathbf{r} = \left(3\cos t\mathbf{i} - 3\sin t\mathbf{j} + \frac{10}{2\pi}\mathbf{k}\right) dt$

$\displaystyle\int_C \mathbf{F} \cdot d\mathbf{r} = \int_0^{2\pi} \frac{1500}{2\pi}\, dt = \frac{1500}{2\pi}t\Big]_0^{2\pi} = 1500 \text{ ft} \cdot \text{lb}$

56. $f(x, y) = 20 + \dfrac{1}{4}x$

$C: y = x^{3/2}, \quad 0 \le x \le 40$

$\mathbf{r}(t) = t\mathbf{i} + t^{3/2}\mathbf{j}, \quad 0 \le t \le 40$

$\mathbf{r}'(t) = \mathbf{i} + \dfrac{3}{2}t^{1/2}\mathbf{j}$

$\|\mathbf{r}'(t)\| = \sqrt{1 + (9/4)t}$

Lateral surface area $= \displaystyle\int_C f(x, y)\, ds = \int_0^{40} \left(20 + \dfrac{1}{4}t\right)\sqrt{1 + (9/4)t}\, dt$

Let $u = \sqrt{1 + (9/4)t}$, then $t = \dfrac{4}{9}(u^2 - 1)$ and $dt = \dfrac{8}{9}u\, du$.

$\displaystyle\int_0^{40} \left(20 + \dfrac{1}{4}t\right)\sqrt{1 + (9/4)t}\, dt = \int_1^{\sqrt{91}} \left[20 + \dfrac{1}{9}(u^2 - 1)\right](u)\left(\dfrac{8}{9}u\right) du$

$\displaystyle = \dfrac{8}{81}\int_1^{\sqrt{91}} (u^4 + 179u^2)\, du = \dfrac{8}{81}\left[\dfrac{u^5}{5} + \dfrac{179u^3}{3}\right]_1^{\sqrt{91}} = \dfrac{850,304\sqrt{91} - 7184}{1215}$

57. $\mathbf{F}(x, y, z) = x^2z\mathbf{i} + 6y\mathbf{j} + yz^2\mathbf{k}$

$\mathbf{r}(t) = t\mathbf{i} + t^2\mathbf{j} + \ln t\,\mathbf{k}, \quad 1 \le t \le 3$

$\mathbf{F}(t) = t^2 \ln t\,\mathbf{i} + 6t^2\mathbf{j} + t^2\ln^2 t\,\mathbf{k}$

$d\mathbf{r} = \left(\mathbf{i} + 2t\mathbf{j} + \dfrac{1}{t}\mathbf{k}\right) dt$

$\displaystyle\int_C \mathbf{F} \cdot d\mathbf{r} = \int_1^3 [t^2 \ln t + 12t^3 + t(\ln t)^2]\, dt \approx 249.49$

58. $\mathbf{F}(x, y, z) = \dfrac{x\mathbf{i} + y\mathbf{j} + z\mathbf{k}}{\sqrt{x^2 + y^2 + z^2}}$

$\mathbf{r}(t) = t\mathbf{i} + t\mathbf{j} + e^t\mathbf{k}, \quad 0 \le t \le 2$

$\mathbf{F}(t) = \dfrac{t\mathbf{i} + t\mathbf{j} + e^t\mathbf{k}}{\sqrt{2t^2 + e^{2t}}}$

$d\mathbf{r} = (\mathbf{i} + \mathbf{j} + e^t\mathbf{k})\, dt$

$\displaystyle\int_C \mathbf{F} \cdot d\mathbf{r} = \int_0^2 \dfrac{1}{\sqrt{2t^2 + e^{2t}}}(2t + e^{2t})\, dt \approx 6.91$

Section 16.3   Conservative Vector Fields and Independence of Path

1. $\mathbf{F}(x, y) = x^2\mathbf{i} + xy\mathbf{j}$

(a) $\mathbf{r}_1(t) = t\mathbf{i} + t^2\mathbf{j}, \quad 0 \le t \le 1$

$\mathbf{r}_1'(t) = \mathbf{i} + 2t\mathbf{j}$

$\mathbf{F}(t) = t^2\mathbf{i} + t^3\mathbf{j}$

$\displaystyle\int_C \mathbf{F} \cdot d\mathbf{r} = \int_0^1 (t^2 + 2t^4)\, dt = \dfrac{11}{15}$

(b) $\mathbf{r}_2(\theta) = \sin\theta\,\mathbf{i} + \sin^2\theta\,\mathbf{j}, \quad 0 \le \theta \le \dfrac{\pi}{2}$

$\mathbf{r}_2'(\theta) = \cos\theta\,\mathbf{i} + 2\sin\theta\cos\theta\,\mathbf{j}$

$\mathbf{F}(t) = \sin^2\theta\,\mathbf{i} + \sin^3\theta\,\mathbf{j}$

$\displaystyle\int_C \mathbf{F} \cdot d\mathbf{r} = \int_0^{\pi/2} (\sin^2\theta\cos\theta + 2\sin^4\theta\cos\theta)\, d\theta = \left[\dfrac{\sin^3\theta}{3} + \dfrac{2\sin^5\theta}{5}\right]_0^{\pi/2} = \dfrac{11}{15}$

2. $F(x, y) = (x^2 + y^2)i - xj$

(a) $r_1(t) = ti + \sqrt{t}\, j, \quad 0 \le t \le 4$

$r_1'(t) = i + \dfrac{1}{2\sqrt{t}}\, j$

$F(t) = (t^2 + t)i - t\, j$

$\displaystyle\int_C F \cdot dr = \int_0^4 \left(t^2 + t - \frac{1}{2}\sqrt{t}\right) dt$

$\displaystyle = \left[\frac{t^3}{3} + \frac{t^2}{2} - \frac{t^{3/2}}{3}\right]_0^4 = \frac{80}{3}$

(b) $r_2(w) = w^2 i + w j, \quad 0 \le w \le 2$

$r_2'(w) = 2wi + j$

$F(w) = (w^4 + w^2)i - w^2 j$

$\displaystyle\int_C F \cdot dr = \int_0^2 [2w(w^4 + w^2) - w^2]\, dw$

$\displaystyle = \left[\frac{w^6}{3} + \frac{w^4}{2} - \frac{w^3}{3}\right]_0^2 = \frac{80}{3}$

3. $F(x, y) = yi - xj$

(a) $r_1(\theta) = \sec\theta\, i + \tan\theta\, j, \quad 0 \le \theta \le \dfrac{\pi}{3}$

$r_1'(\theta) = \sec\theta \tan\theta\, i + \sec^2\theta\, j$

$F(\theta) = \tan\theta\, i - \sec\theta\, j$

$\displaystyle\int_C F \cdot dr = \int_0^{\pi/3} (\sec\theta \tan^2\theta - \sec^3\theta)\, d\theta = \int_0^{\pi/3} [\sec\theta(\sec^2\theta - 1) - \sec^3\theta]\, d\theta = -\int_0^{\pi/3} \sec\theta\, d\theta$

$\displaystyle = -\ln|\sec\theta + \tan\theta|\Big]_0^{\pi/3} = -\ln(2 + \sqrt{3}) \approx -1.317$

(b) $r_2(t) = \sqrt{t+1}\, i + \sqrt{t}\, j, \quad 0 \le t \le 3$

$r_2'(t) = \dfrac{1}{2\sqrt{t+1}}i + \dfrac{1}{2\sqrt{t}}j$

$F(t) = \sqrt{t}\, i - \sqrt{t+1}\, j$

$\displaystyle\int_C F \cdot dr = \int_0^3 \left[\frac{\sqrt{t}}{2\sqrt{t+1}} - \frac{\sqrt{t+1}}{2\sqrt{t}}\right] dt = -\frac{1}{2}\int_0^3 \frac{1}{\sqrt{t}\sqrt{t+1}}\, dt = -\frac{1}{2}\int_0^3 \frac{1}{\sqrt{t^2 + t + (1/4) - (1/4)}}\, dt$

$\displaystyle = -\frac{1}{2}\int_0^3 \frac{1}{\sqrt{[t + (1/2)]^2 - (1/4)}}\, dt = -\frac{1}{2}\ln\left|\left(t + \frac{1}{2}\right) + \sqrt{t^2 + t}\,\right|\Big]_0^3 = -\frac{1}{2}\left[\ln\left(\frac{7}{2} + 2\sqrt{3}\right) - \ln\left(\frac{1}{2}\right)\right]$

$\displaystyle = -\frac{1}{2}\ln(7 + 4\sqrt{3}) \approx -1.317$

4. $F(x, y) = yi - x^2 j$

(a) $r_1(t) = (2+t)i + (3-t)j, \quad 0 \le t \le 3$

$r_1'(t) = i - j$

$F(t) = (3-t)i - (2+t)^2 j$

$\displaystyle\int_C F \cdot dr = \int_0^3 [(3-t) + (2+t)^2]\, dt = \left[-\frac{(3-t)^2}{2} + \frac{(2+t)^3}{3}\right]_0^3 = \frac{87}{2}$

(b) $r_2(w) = (2 + \ln w)i + (3 - \ln w)j, \quad 1 \le w \le e^3$

$r_2'(w) = \dfrac{1}{w}i - \dfrac{1}{w}j$

$F(w) = (3 - \ln w)i - (2 + \ln w)^2 j$

$\displaystyle\int_C F \cdot dr = \int_1^{e^3} \left[(3 - \ln w)\left(\frac{1}{w}\right) + (2 + \ln w)^2\left(\frac{1}{w}\right)\right] dw = \left[-\frac{(3 - \ln w)^2}{2} + \frac{(2 + \ln w)^3}{3}\right]_1^{e^3} = \frac{87}{2}$

5. $F(x, y) = 2xy\mathbf{i} + x^2\mathbf{j}$

(a) $\mathbf{r}_1(t) = t\mathbf{i} + t^2\mathbf{j}, \quad 0 \le t \le 1$

$\mathbf{r}_1'(t) = \mathbf{i} + 2t\mathbf{j}$

$F(t) = 2t^3\mathbf{i} + t^2\mathbf{j}$

$\displaystyle\int_C F \cdot d\mathbf{r} = \int_0^1 4t^3 \, dt = 1$

(b) $\mathbf{r}_2(t) = t\mathbf{i} + t^3\mathbf{j}, \quad 0 \le t \le 1$

$\mathbf{r}_2'(t) = \mathbf{i} + 3t^2\mathbf{j}$

$F(t) = 2t^4\mathbf{i} + t^2\mathbf{j}$

$\displaystyle\int_C F \cdot d\mathbf{r} = \int_0^1 5t^4 \, dt = 1$

6. $F(x, y) = ye^{xy}\mathbf{i} + xe^{xy}\mathbf{j}$

(a) $\mathbf{r}_1(t) = t\mathbf{i} - \frac{3}{2}(t - 2)\mathbf{j}, \quad 0 \le t \le 2$

$\mathbf{r}_1'(t) = \mathbf{i} - \frac{3}{2}\mathbf{j}$

$F(t) = -\frac{3}{2}(t - 2)e^{-3t(t-2)/2}\mathbf{i} + te^{-3t(t-2)/2}\mathbf{j}$

$\displaystyle\int_C F \cdot d\mathbf{r} = \int_0^2 \left[-\frac{3}{2}(t - 2)e^{-3t(t-2)/2} - \frac{3}{2}te^{-3t(t-2)/2} \right] dt = \int_0^2 e^{-3t(t-2)/2}\left[-\frac{3}{2}(t-2) - \frac{3}{2}t \right] dt$

$$= e^{-3t(t-2)/2}\Big]_0^2 = 1 - 1 = 0$$

(b) Line segments from $(0, 3)$ to $(0, 0)$ and then from $(0, 0)$ to $(2, 0)$

$$\mathbf{r}_2(t) = \begin{cases} (3 - t)\mathbf{j}, & 0 \le t \le 3 \\ (t - 3)\mathbf{i}, & 3 \le t \le 5 \end{cases}$$

On C_1: $\mathbf{r}_2(t) = (3 - t)\mathbf{j}, \quad 0 \le t \le 3$

$\mathbf{r}_2'(t) = -\mathbf{j}$

$F(t) = (3 - t)\mathbf{i}$

$\displaystyle\int_{C_1} F \cdot d\mathbf{r} = 0$

On C_2: $\mathbf{r}_2(t) = (t - 3)\mathbf{i}, \quad 3 \le t \le 5$

$\mathbf{r}_2'(t) = \mathbf{i}$

$F(t) = (t - 3)\mathbf{j}$

$\displaystyle\int_{C_2} F \cdot d\mathbf{r} = 0$

Thus, $\displaystyle\int_C F \cdot d\mathbf{r} = \int_{C_1} F \cdot d\mathbf{r} + \int_{C_2} F \cdot d\mathbf{r} = 0 + 0 = 0.$

7. $F(x, y) = y\mathbf{i} - x\mathbf{j}$

(a) $\mathbf{r}_1(t) = t\mathbf{i} + t\mathbf{j}, \quad 0 \le t \le 1$

$\mathbf{r}_1'(t) = \mathbf{i} + \mathbf{j}$

$F(t) = t\mathbf{i} - t\mathbf{j}$

$\displaystyle\int_C F \cdot d\mathbf{r} = 0$

(c) $\mathbf{r}_3(t) = t\mathbf{i} + t^3\mathbf{j}, \quad 0 \le t \le 1$

$\mathbf{r}_3'(t) = \mathbf{i} + 3t^2\mathbf{j}$

$F(t) = t^3\mathbf{j} - t\mathbf{j}$

$\displaystyle\int_C F \cdot d\mathbf{r} = \int_0^1 -2t^3 \, dt = -\frac{1}{2}$

(b) $\mathbf{r}_2(t) = t\mathbf{i} + t^2\mathbf{j}, \quad 0 \le t \le 1$

$\mathbf{r}_2'(t) = \mathbf{i} + 2t\mathbf{j}$

$F(t) = t^2\mathbf{i} - t\mathbf{j}$

$\displaystyle\int_C F \cdot d\mathbf{r} = \int_0^1 -t^2 \, dt = -\frac{1}{3}$

8. $F(x, y) = xy^2 i + 2x^2 y j$

(a) $r_1(t) = ti + \dfrac{1}{t} j, \quad 1 \leq t \leq 3$

$r_1'(t) = i - \dfrac{1}{t^2} j$

$F(t) = \dfrac{1}{t} i + 2t j$

$\displaystyle \int_C F \cdot dr = \int_1^3 -\dfrac{1}{t} \, dt = -\ln|t| \Big]_1^3 = -\ln 3$

(b) $r_2(t) = (t + 1)i - \dfrac{1}{3}(t - 3) j, \quad 0 \leq t \leq 2$

$r_2'(t) = i - \dfrac{1}{3} j$

$F(t) = \dfrac{1}{9}(t + 1)(t - 3)^2 i - \dfrac{2}{3}(t + 1)^2(t - 3) j$

$\displaystyle \int_C F \cdot dr = \int_0^2 \left[\dfrac{1}{9}(t + 1)(t - 3)^2 + \dfrac{2}{9}(t + 1)^2(t - 3) \right] dt = \dfrac{1}{9} \int_0^2 (3t^3 - 7t^2 - 7t + 3) \, dt$

$$= \dfrac{1}{9} \left[\dfrac{3t^4}{4} - \dfrac{7t^3}{3} - \dfrac{7t^2}{2} + 3t \right]_0^2 = -\dfrac{44}{27}$$

9. $\displaystyle \int_C y^2 \, dx + 2xy \, dy$

Since $\dfrac{\partial M}{\partial y} = \dfrac{\partial N}{\partial x} = 2y$, $F(x, y) = y^2 i + 2xy j$ is conservative. The potential function is $f(x, y) = xy^2 + k$. Therefore, we can use the Fundamental Theorem of Line Integrals.

(a) $\displaystyle \int_C y^2 \, dx + 2xy \, dy = x^2 y \Big]_{(0,0)}^{(4,4)} = 64$

(b) $\displaystyle \int_C y^2 \, dx + 2xy \, dy = x^2 y \Big]_{(-1,0)}^{(1,0)} = 0$

(c) and (d) Since C is a closed curve, $\displaystyle \int_C y^2 \, dx + 2xy \, dy = 0$.

10. $\displaystyle \int_C (2x - 3y + 1) \, dx - (3x + y - 5) \, dy$

Since $\dfrac{\partial M}{\partial y} = \dfrac{\partial N}{\partial x} = -3$, $F(x, y) = (2x - 3y + 1)i - (3x + y - 5)j$ is conservative. The potential function is $f(x, y) = x^2 - 3xy - \dfrac{y^2}{2} + x + 5y + k$.

(a) and (d) Since C is a closed curve, $\displaystyle \int_C (2x - 3y + 1) \, dx - (3x + y - 5) \, dy = 0$.

(b) $\displaystyle \int_C (2x - 3y + 1) \, dx - (3x + y - 5) \, dy = \left[x^2 - 3xy - \dfrac{y^2}{2} + x + 5y \right]_{(0,-1)}^{(0,1)} = 10$

(c) $\displaystyle \int_C (2x - 3y + 1) \, dx - (3x + y - 5) \, dy = \left[x^2 - 3xy - \dfrac{y^2}{2} + x + 5y \right]_{(0,1)}^{(2,e^2)} = \dfrac{1}{2}(3 - 2e^2 - e^4)$

11. $\int_C 2xy\,dx + (x^2 + y^2)\,dy$

Since $\dfrac{\partial M}{\partial y} = \dfrac{\partial N}{\partial x} = 2x$, $\mathbf{F}(x,\,y) = 2xy\mathbf{i} + (x^2 + y^2)\mathbf{j}$ is conservative. The potential function is $f(x,\,y) = x^2 y + \dfrac{y^3}{3} + k$.

(a) $\int_C 2xy\,dx + (x^2 + y^2)\,dy = \left[x^2 y + \dfrac{y^3}{3}\right]_{(5,0)}^{(0,4)} = \dfrac{64}{3}$ (b) $\int_C 2xy\,dx + (x^2 + y^2)\,dy = \left[x^2 y + \dfrac{y^3}{3}\right]_{(2,0)}^{(0,4)} = \dfrac{64}{3}$

12. $\int_C (x^2 + y^2)\,dx + 2xy\,dy$

Since $\dfrac{\partial M}{\partial y} = \dfrac{\partial N}{\partial x} = 2y$, $\mathbf{F}(x,\,y) = (x^2 + y^2)\mathbf{i} + 2xy\mathbf{j}$ is conservative. The potential function is $f(x,\,y) = \dfrac{x^3}{3} + xy^2 + k$.

(a) $\int_C (x^2 + y^2)\,dx + 2xy\,dy = \left[\dfrac{x^3}{3} + xy^2\right]_{(0,0)}^{(8,4)} = \dfrac{896}{3}$ (b) $\int_C (x^2 + y^2)\,dx + 2xy\,dy = \left[\dfrac{x^3}{3} + xy^2\right]_{(2,0)}^{(0,2)} = -\dfrac{8}{3}$

13. $\mathbf{F}(x,\,y,\,z) = yz\mathbf{i} + xz\mathbf{j} + xy\mathbf{k}$

Since **curl $\mathbf{F} = \mathbf{0}$**, $\mathbf{F}(x,\,y,\,z)$ is conservative. The potential function is $f(x,\,y,\,z) = xyz + k$.

(a) $\mathbf{r}_1(t) = t\mathbf{i} + 2\mathbf{j} + t\mathbf{k}$, $0 \le t \le 4$

$\int_C \mathbf{F} \cdot d\mathbf{r} = xyz\Big]_{(0,2,0)}^{(4,2,4)} = 32$

(b) $\mathbf{r}_2(t) = t^2\mathbf{i} + t\mathbf{j} + t^2\mathbf{k}$, $0 \le t \le 2$

$\int_C \mathbf{F} \cdot d\mathbf{r} = xyz\Big]_{(0,0,0)}^{(4,2,4)} = 32$

14. $\mathbf{F}(x,\,y,\,z) = \mathbf{i} + z\mathbf{j} + y\mathbf{k}$

Since **curl $\mathbf{F} = \mathbf{0}$**, $\mathbf{F}(x,\,y,\,z)$ is conservative. The potential function is $f(x,\,y,\,z) = x + yz + k$.

(a) $\mathbf{r}_1(t) = \cos t\mathbf{i} + \sin t\mathbf{j} + t^2\mathbf{k}$, $0 \le t \le \pi$

$\int_C \mathbf{F} \cdot d\mathbf{r} = \left[x + yz\right]_{(1,0,0)}^{(-1,0,\pi^2)} = -2$

(b) $\mathbf{r}_2(t) = (1 - 2t)\mathbf{i} + \pi^2 t\mathbf{k}$, $0 \le t \le 1$

$\int_C \mathbf{F} \cdot d\mathbf{r} = \left[x + yz\right]_{(1,0,0)}^{(-1,0,\pi^2)} = -2$

15. $\mathbf{F}(x,\,y,\,z) = (2y + x)\mathbf{i} + (x^2 - z)\mathbf{j} + (2y - 4z)\mathbf{k}$

$\mathbf{F}(x,\,y,\,z)$ is not conservative.

(a) $\mathbf{r}_1(t) = t\mathbf{i} + t^2\mathbf{j} + \mathbf{k}$, $0 \le t \le 1$

$\mathbf{r}_1{}'(t) = \mathbf{i} + 2t\mathbf{j}$

$\mathbf{F}(t) = (2t^2 + t)\mathbf{i} + (t^2 - 1)\mathbf{j} + (2t^2 - 4)\mathbf{k}$

$\int_C \mathbf{F} \cdot d\mathbf{r} = \int_0^1 (2t^3 + 2t^2 - t)\,dt = \dfrac{2}{3}$

(b) $\mathbf{r}_2(t) = t\mathbf{i} + t\mathbf{j} + (2t - 1)^2\mathbf{k}$, $0 \le t \le 1$

$\mathbf{r}_2{}'(t) = \mathbf{i} + \mathbf{j} + 4(2t - 1)\mathbf{k}$

$\mathbf{F}(t) = 3t\mathbf{i} + [t^2 - (2t - 1)^2]\mathbf{j} + [2t - 4(2t - 1)^2]\mathbf{k}$

$\int_C \mathbf{F} \cdot d\mathbf{r} = \int_0^1 [3t + t^2 - (2t - 1)^2 + 8t(2t - 1) - 16(2t - 1)^3]\,dt$

$= \int_0^1 [17t^2 - 5t - (2t - 1)^2 - 16(2t - 1)^3]\,dt = \left[\dfrac{17t^3}{3} - \dfrac{5t^2}{2} - \dfrac{(2t - 1)^3}{6} - 2(2t - 1)^4\right]_0^1 = \dfrac{17}{6}$

16. $\mathbf{F}(x,\ y,\ z) = -y\mathbf{i} + x\mathbf{j} + 3xz^2\mathbf{k}$
$\mathbf{F}(x,\ y,\ z)$ is not conservative.

(a) $\mathbf{r}_1(t) = \cos t\mathbf{i} + \sin t\mathbf{j} + t\mathbf{k},\ \ 0 \le t \le \pi$
$\mathbf{r}_1{}'(t) = -\sin t\mathbf{i} + \cos t\mathbf{j} + \mathbf{k}$
$\mathbf{F}(t) = -\sin t\mathbf{i} + \cos t\mathbf{j} + 3t^2\cos t\mathbf{k}$

$$\int_C \mathbf{F} \cdot d\mathbf{r} = \int_0^\pi [\sin^2 t + \cos^2 t + 3t^2\cos t]\, dt$$

$$= \int_0^\pi [1 + 3t^2\cos t]\, dt = t\Big]_0^\pi + 3\Big[t^2\sin t\Big]_0^\pi - 2\int_0^\pi t\sin t\, dt = \Big[t + 3t^2\sin t - 6(\sin t - t\cos t)\Big]_0^\pi = -5\pi$$

(b) $\mathbf{r}_2(t) = (1 - 2t)\mathbf{i} + \pi t\mathbf{k},\ \ 0 \le t \le 1$
$\mathbf{r}_2{}'(t) = -2\mathbf{i} + \pi\mathbf{k}$
$\mathbf{F}(t) = (1 - 2t)\mathbf{j} + 3\pi^2 t^2(1 - 2t)\mathbf{k}$

$$\int_C \mathbf{F} \cdot d\mathbf{r} = \int_0^1 3\pi^3 t^2(1 - 2t)\, dt = 3\pi^3\int_0^1 (t^2 - 2t^3)\, dt = 3\pi^3\Big[\frac{t^3}{3} - \frac{t^4}{2}\Big]_0^1 = -\frac{\pi^3}{2}$$

17. $\mathbf{F}(x,\ y,\ z) = e^z(y\mathbf{i} + x\mathbf{j} + xy\mathbf{k})$
$\mathbf{F}(x,\ y,\ z)$ is conservative. The potential function is $f(x,\ y,\ z) = xye^z + k$.

(a) $\mathbf{r}_1(t) = 4\cos t\mathbf{i} + 4\sin t\mathbf{j} + 3\mathbf{k},\ \ 0 \le t \le \pi$

$$\int_C \mathbf{F} \cdot d\mathbf{r} = xye^z\Big]_{(4,0,3)}^{(-4,0,3)} = 0$$

(b) $\mathbf{r}_2(t) = (4 - 8t)\mathbf{i} + 3\mathbf{k},\ \ 0 \le t \le 1$

$$\int_C \mathbf{F} \cdot d\mathbf{r} = xye^z\Big]_{(4,0,3)}^{(-4,0,3)} = 0$$

18. $\mathbf{F}(x,\ y,\ z) = y\sin z\mathbf{i} + x\sin z\mathbf{j} + xy\cos z\mathbf{k}$
$\mathbf{F}(x,\ y,\ z)$ is conservative. The potential function is $f(x,\ y,\ z) = xy\sin z + k$.

(a) $\mathbf{r}_1(t) = t^2\mathbf{i} + t^2\mathbf{j},\ \ 0 \le t \le 2$

$$\int_C \mathbf{F} \cdot d\mathbf{r} = xy\sin z\Big]_{(0,0,0)}^{(4,4,0)} = 0$$

(b) $\mathbf{r}_2(t) = 4t\mathbf{i} + 4t\mathbf{j},\ \ 0 \le t \le 1$

$$\int_C \mathbf{F} \cdot d\mathbf{r} = xy\sin z\Big]_{(0,0,0)}^{(4,4,0)} = 0$$

19. $\displaystyle\int_C (y\mathbf{i} + x\mathbf{j}) \cdot d\mathbf{r} = xy\Big]_{(0,0)}^{(3,8)} = 24$

20. $\displaystyle\int_C [2(x + y)\mathbf{i} + 2(x + y)\mathbf{j}] \cdot d\mathbf{r} = (x + y)^2\Big]_{(-1,1)}^{(3,2)} = 25$

21. $\displaystyle\int_{(0,-\pi)}^{(3\pi/2,\pi/2)} \cos x\sin y\, dx + \sin x\cos y\, dy = \sin x\sin y\Big]_{(0,-\pi)}^{(3\pi/2,\pi/2)} = -1$

22. $\displaystyle\int_{(1,1)}^{(2\sqrt{3},2)} \frac{y\, dx - x\, dy}{x^2 + y^2} = \arctan\left(\frac{x}{y}\right)\Big]_{(1,1)}^{(2\sqrt{3},2)} = \frac{\pi}{3} - \frac{\pi}{4} = \frac{\pi}{12}$

23. $\displaystyle\int_C e^x\sin y\, dx + e^x\cos y\, dy = e^x\sin y\Big]_{(0,0)}^{(2\pi,0)} = 0$

24. $\displaystyle\int_C \frac{2x}{(x^2 + y^2)^2}\, dx + \frac{2y}{(x^2 + y^2)^2}\, dy = -\frac{1}{x^2 + y^2}\Big]_{(7,5)}^{(1,5)} = -\frac{1}{26} + \frac{1}{74} = \frac{-12}{481}$

25. $\int_C (z + 2y)\, dx + (2x - z)\, dy + (x - y)\, dz$

(a) $(xz + 2xy - yz)\Big]_{(0,0,0)}^{(1,1,1)} = 2$

(b) $(xz + 2xy - yz)\Big]_{(0,0,0)}^{(0,0,1)} + (xz + 2xy - yz)\Big]_{(0,0,1)}^{(1,1,1)} = 0 + 2 = 2$

(c) $(xz + 2xy - yz)\Big]_{(0,0,0)}^{(1,0,0)} + (xz + 2xy - yz)\Big]_{(1,0,0)}^{(1,1,0)} + (xz + 2xy - yz)\Big]_{(1,1,0)}^{(1,1,1)} = 0 + 2 + 2 - 2 = 2$

26. $\int_C zy\, dx + xz\, dy + xy\, dz$

(a) $xyz\Big]_{(0,0,0)}^{(1,1,1)} = 1$

(b) $xyz\Big]_{(0,0,0)}^{(0,0,1)} + xyz\Big]_{(0,0,1)}^{(1,1,1)} = 0 + 1 = 1$

(c) $xyz\Big]_{(0,0,0)}^{(1,0,0)} + xyz\Big]_{(1,0,0)}^{(1,1,0)} + xyz\Big]_{(1,1,0)}^{(1,1,1)} = 0 + 0 + 1 = 1$

27. $\int_{(0,0,0)}^{(\pi/2,3,4)} -\sin x\, dx + z\, dy + y\, dz = \Big[\cos x + yz\Big]_{(0,0,0)}^{(\pi/2,3,4)} = 11$

28. $\int_{(0,0,0)}^{(3,4,0)} 6x\, dx - 4z\, dy - (4y - 20z)\, dz = \Big[3x^2 - 4yz + 10z^2\Big]_{(0,0,0)}^{(3,4,0)} = 27$

29. $\mathbf{F}(x,\ y) = 9x^2y^2\mathbf{i} + (6x^3y - 1)\mathbf{j}$ is conservative.

Work $= \Big[3x^3y^2 - y\Big]_{(0,0)}^{(5,9)} = 30,366$

30. $\mathbf{F}(x,\ y) = \dfrac{2x}{y}\mathbf{i} - \dfrac{x^2}{y^2}\mathbf{j}$ is conservative.

Work $= \dfrac{x^2}{y}\Big]_{(-1,1)}^{(3,2)} = \dfrac{7}{2}$

31. $\mathbf{r}(t) = 2\cos 2\pi t\mathbf{i} + 2\sin 2\pi t\mathbf{j}$

$\mathbf{r}'(t) = -4\pi \sin 2\pi t\mathbf{i} + 4\pi \cos 2\pi t\mathbf{j}$

$\mathbf{a}(t) = -8\pi^2 \cos 2\pi t\mathbf{i} - 8\pi^2 \sin 2\pi t\mathbf{j}$

$\mathbf{F}(t) = m \cdot \mathbf{a}(t) = \dfrac{1}{32}\mathbf{a}(t) = -\dfrac{\pi^2}{4}(\cos 2\pi t\mathbf{i} + \sin 2\pi t\mathbf{j})$

$W = \int_C \mathbf{F} \cdot d\mathbf{r} = \int_C -\dfrac{\pi^2}{4}(\cos 2\pi t\mathbf{i} + \sin 2\pi t\mathbf{j}) \cdot 4\pi(-\sin 2\pi t\mathbf{i} + \cos 2\pi t\mathbf{j})\, dt = -\pi^3 \int_C 0\, dt = 0$

32. $\mathbf{F}(x,\ y,\ z) = a_1\mathbf{i} + a_2\mathbf{j} + a_3\mathbf{k}$

Since $\mathbf{F}(x,\ y,\ z)$ is conservative, the work done in moving a particle along any path from P to Q is

$f(x,\ y,\ z) = \Big[a_1 x + a_2 y + a_3 z\Big]_{P=(p_1,p_2,p_3)}^{Q=(q_1,q_2,q_3)}$

$= a_1(q_1 - p_1) + a_2(q_2 - p_2) + a_3(q_3 - p_3) = \mathbf{F} \cdot \overrightarrow{PQ}.$

33. Since the sum of the potential and kinetic energies remains constant from point to point, if the kinetic energy is decreasing at a rate of 10 units per minute, then the potential energy is increasing at a rate of 10 units per minute.

34. $\mathbf{F}(x, y) = \dfrac{y}{x^2 + y^2}\mathbf{i} - \dfrac{x}{x^2 + y^2}\mathbf{j}$

(a) $M = \dfrac{y}{x^2 + y^2}, \quad \dfrac{\partial M}{\partial y} = \dfrac{(x^2 + y^2)(1) - y(2y)}{(x^2 + y^2)^2} = \dfrac{x^2 - y^2}{(x^2 + y^2)^2}$

$N = -\dfrac{x}{x^2 + y^2}, \quad \dfrac{\partial N}{\partial x} = \dfrac{(x^2 + y^2)(-1) + x(2x)}{(x^2 + y^2)^2} = \dfrac{x^2 - y^2}{(x^2 + y^2)^2}$

Thus, $\dfrac{\partial N}{\partial x} = \dfrac{\partial M}{\partial y}$.

(b) $\mathbf{r}(t) = \cos t\,\mathbf{i} + \sin t\,\mathbf{j}, \quad 0 \le t \le \pi$

$\mathbf{F} = \sin t\,\mathbf{i} - \cos t\,\mathbf{j}$

$d\mathbf{r} = (-\sin t\,\mathbf{i} + \cos t\,\mathbf{j})\,dt$

$\displaystyle \int_C \mathbf{F} \cdot d\mathbf{r} = \int_0^{\pi} (-\sin^2 t - \cos^2 t)\,dt = -t \Big]_0^{\pi} = -\pi$

(c) $\mathbf{r}(t) = \cos t\,\mathbf{i} - \sin t\,\mathbf{j}, \quad 0 \le t \le \pi$

$\mathbf{F} = -\sin t\,\mathbf{i} - \cos t\,\mathbf{j}$

$d\mathbf{r} = (-\sin t\,\mathbf{i} - \cos t\,\mathbf{j})\,dt$

$\displaystyle \int_C \mathbf{F} \cdot d\mathbf{r} = \int_0^{\pi} (\sin^2 t + \cos^2 t)\,dt = t \Big]_0^{\pi} = \pi$

(d) $\mathbf{r}(t) = \cos t\,\mathbf{i} + \sin t\,\mathbf{j}, \quad 0 \le t \le 2\pi$

$\mathbf{F} = \sin t\,\mathbf{i} - \cos t\,\mathbf{j}$

$d\mathbf{r} = (-\sin t\,\mathbf{i} + \cos t\,\mathbf{j})\,dt$

$\displaystyle \int_C \mathbf{F} \cdot d\mathbf{r} = \int_0^{2\pi} (-\sin^2 t - \cos^2 t)\,dt = -t \Big]_0^{2\pi} = -2\pi$

This does not contradict Theorem 16.7 since $\mathbf{F}$ is not continuous at $(0, 0)$ in R enclosed by curve C.

Section 16.4   Green's Theorem

1. (a) $\mathbf{r}(t) = \begin{cases} t\mathbf{i}, & 0 \le t \le 4 \\ 4\mathbf{i} + (t-4)\mathbf{j}, & 4 \le t \le 8 \\ (12-t)\mathbf{i} + 4\mathbf{j}, & 8 \le t \le 12 \\ (16-t)\mathbf{j}, & 12 \le t \le 16 \end{cases}$

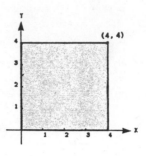

$$\int_C y^2 \, dx + x^2 \, dy$$

$$= \int_0^4 [0 \, dt + t^2(0)] + \int_4^8 [(t-4)^2(0) + 16 \, dt]$$

$$+ \int_8^{12} [16(-dt) + (12-t)^2(0)]$$

$$+ \int_{12}^{16} [(16-t)^2(0) + 0(-dt)]$$

$$= 0 + 64 - 64 + 0 = 0$$

(b) By Green's Theorem,

$$\iint_R \left(\frac{\partial N}{\partial x} - \frac{\partial M}{\partial y} \right) dA = \int_0^4 \int_0^4 (2x - 2y) \, dy \, dx = \int_0^4 (8x - 16) \, dx = 0.$$

2. (a) $\mathbf{r}(t) = \begin{cases} t\mathbf{i}, & 0 \le t \le 4 \\ 4\mathbf{i} + (t-4)\mathbf{j}, & 4 \le t \le 8 \\ (12-t)\mathbf{i} + (12-t)\mathbf{j}, & 8 \le t \le 12 \end{cases}$

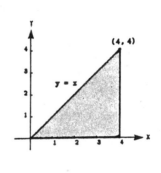

$$\int_C y^2 \, dx + x^2 \, dy = \int_0^4 [0 \, dt + t^2(0)] + \int_4^8 [(t-4)^2(0) + 16 \, dt]$$

$$+ \int_8^{12} [(12-t)^2(-dt) + (12-t)^2(-dt)]$$

$$= 0 + 64 - \frac{128}{3} = \frac{64}{3}$$

(b) By Green's Theorem,

$$\iint_R \left(\frac{\partial N}{\partial x} - \frac{\partial M}{\partial y} \right) dA = \int_0^4 \int_0^x (2x - 2y) \, dy \, dx = \int_0^4 x^2 \, dx = \frac{64}{3}.$$

3. (a) $\mathbf{r}(t) = \begin{cases} t\mathbf{i} + \dfrac{t^2}{4}\mathbf{j}, & 0 \le t \le 4 \\ (8-t)\mathbf{i} + (8-t)\mathbf{j}, & 4 \le t \le 8 \end{cases}$

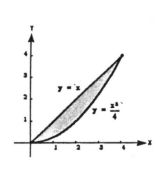

$$\int_C y^2 \, dx + x^2 \, dy = \int_0^4 \left[\frac{t^4}{16}(dt) + t^2 \left(\frac{t}{2} \, dt \right) \right] + \int_4^8 [(8-t)^2(-dt) + (8-t)^2(-dt)]$$

$$= \int_0^4 \left[\frac{t^4}{16} + \frac{t^3}{2} \right] dt + \int_4^8 -2(8-t)^2 \, dt = \frac{32}{15}$$

(b) By Green's Theorem,

$$\iint_R \left(\frac{\partial N}{\partial x} - \frac{\partial M}{\partial y} \right) dA = \int_0^4 \int_{x^2/4}^x (2x - 2y) \, dy \, dx = \int_0^4 \left(x^2 - \frac{x^3}{2} + \frac{x^4}{16} \right) dx = \frac{32}{15}.$$

4. (a) $\mathbf{r}(t) = \cos t\mathbf{i} + \sin t\mathbf{j}, \quad 0 \le t \le 2\pi$

$$\int_C y^2\,dx + x^2\,dy = \int_0^{2\pi} [\sin^2 t(-\sin t\,dt) + \cos^2 t(\cos t\,dt)]$$

$$= \int_0^{2\pi} (\cos^3 t - \sin^3 t)\,dt$$

$$= \int_0^{2\pi} [\cos t(1 - \sin^2 t) - \sin t(1 - \cos^2 t)]\,dt$$

$$= \left[\sin t - \frac{\sin^3 t}{3} + \cos t - \frac{\cos^3 t}{3}\right]_0^{2\pi} = 0$$

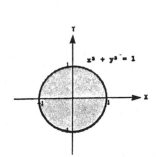

(b) By Green's Theorem,

$$\iint_R \left(\frac{\partial N}{\partial x} - \frac{\partial M}{\partial y}\right) dA = \int_{-1}^{1}\int_{-\sqrt{1-x^2}}^{\sqrt{1-x^2}} (2x - 2y)\,dy\,dx$$

$$= \int_0^{2\pi}\int_0^1 (2r\cos\theta - 2r\sin\theta)r\,dr\,d\theta$$

$$= \frac{2}{3}\int_0^{2\pi} (\cos\theta - \sin\theta)\,d\theta$$

$$= \frac{2}{3}(0) = 0.$$

5. $\displaystyle\int_C (y - x)\,dx + (2x - y)\,dy = \int_0^2\int_{x^2-x}^{x} dy\,dx$

$$= \int_0^2 (2x - x^2)\,dx$$

$$= \frac{4}{3}$$

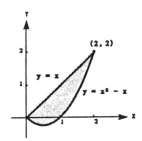

6. Since C is an ellipse with $a = 2$ and $b = 1$, then R is an ellipse of area $\pi ab = 2\pi$. Thus, Green's Theorem yields

$$\int_C (y - x)\,dx + (2x - y)\,dy = \iint_R 1\,dA = \text{ area of ellipse } = 2\pi.$$

7. From the accompanying figure, we see that R is the shaded region. Thus, Green's Theorem yields

$$\int_C (y - x)\,dx + (2x - y)\,dy = \iint_R 1\,dA$$

$$= \text{ area of } R$$

$$= 6(10) - 2(2)$$

$$= 56.$$

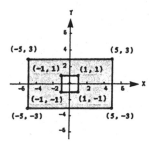

8. R is the shaded region of the accompanying figure. Thus, Green's Theorem yields

$$\int_C (y - x)\,dx + (2x - y)\,dy = \iint_R 1\,dA$$

$$= \text{area between the circles}$$

$$= 16\pi - \pi$$

$$= 15\pi.$$

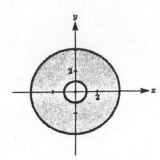

9. Since the curves $y = 0$ and $y = 4 - x^2$ intersect at $(-2,\ 0)$ and $(2,\ 0)$, Green's Theorem yields

$$\int_C 2xy\,dx + (x + y)\,dy = \iint_R (1 - 2x)\,dA = \int_{-2}^2 \int_0^{4-x^2} (1 - 2x)\,dy\,dx$$

$$= \int_{-2}^2 \Big[y - 2xy\Big]_0^{4-x^2}\,dx$$

$$= \int_{-2}^2 (4 - 8x - x^2 + 2x^3)\,dx$$

$$= \left[4x - 4x^2 - \frac{x^3}{3} + \frac{x^4}{2}\right]_{-2}^2$$

$$= -\frac{8}{3} - \frac{8}{3} + 16 = \frac{32}{3}.$$

10. The given curves intersect at $(0,\ 0)$ and $(4,\ 2)$. Thus, Green's Theorem yields

$$\int_C y^2\,dx + xy\,dy = \iint_R (y - 2y)\,dA$$

$$= \int_0^4 \int_0^{\sqrt{x}} -y\,dy\,dx = \int_0^4 \frac{-y^2}{2}\Big]_0^{\sqrt{x}}\,dx = \int_0^4 \frac{-x}{2}\,dx = \frac{-x^2}{4}\Big]_0^4 = -4.$$

11. Since R is the interior of the circle $x^2 + y^2 = a^2$, Green's Theorem yields

$$\int_C (x^2 - y^2)\,dx + 2xy\,dy = \iint_R (2y + 2y)\,dA$$

$$= \int_{-a}^a \int_{-\sqrt{a^2-x^2}}^{\sqrt{a^2-x^2}} 4y\,dy\,dx = 4\int_{-a}^a 0\,dx = 0.$$

12. In this case, let $y = r\sin\theta$, $x = r\cos\theta$. Then $dA = r\,dr\,d\theta$ and Green's Theorem yields

$$\int_C (x^2 - y^2)\,dx + 2xy\,dy = \iint_R 4y\,dA = 4\int_0^{2\pi} \int_0^{1+\cos\theta} r\sin\theta\, r\,dr\,d\theta$$

$$= 4\int_0^{2\pi} \int_0^{1+\cos\theta} r^2 \sin\theta\,dr\,d\theta$$

$$= \frac{4}{3}\int_0^{2\pi} \sin\theta(1 + \cos\theta)^3\,d\theta$$

$$= -\frac{(1 + \cos\theta)^4}{3}\Big]_0^{2\pi} = 0.$$

13. Since $\dfrac{\partial M}{\partial y} = \dfrac{2x}{x^2 + y^2} = \dfrac{\partial N}{\partial x}$,

we have path independence and

$$\int\int_R \left(\frac{\partial N}{\partial x} - \frac{\partial M}{\partial y} \right) dA = 0.$$

14. Since $\dfrac{\partial M}{\partial y} = 2e^x \cos 2y = \dfrac{\partial N}{\partial x}$

we have

$$\int\int_R \left(\frac{\partial N}{\partial x} - \frac{\partial M}{\partial y} \right) dA = 0.$$

15. By Green's Theorem,

$$\int_C \sin x \cos y \, dx + (xy + \cos x \sin y) \, dy = \int\int_R [(y - \sin x \sin y) - (-\sin x \sin y)] \, dA$$

$$= \int_0^1 \int_x^{\sqrt{x}} y \, dy \, dx = \frac{1}{2} \int_0^1 (x - x^2) \, dx = \frac{1}{2} \left[\frac{x^2}{2} - \frac{x^3}{3} \right]_0^1 = \frac{1}{12}.$$

16. By Green's Theorem,

$$\int_C (e^{-x^2/2} - y) \, dx + (e^{-y^2/2} + x) \, dy = \int\int_R 2 \, dA = 2(\text{area of } R) = 2[\pi(5)^2 - \pi(2)(1)] = 46\pi.$$

17. By Green's Theorem,

$$\int_C xy \, dx + (x + y) \, dy = \int\int_R (1 - x) \, dA$$

$$= \int_0^{2\pi} \int_1^3 (1 - r\cos\theta) r \, dr \, d\theta = \int_0^{2\pi} \left(4 - \frac{26}{3}\cos\theta \right) d\theta = 8\pi.$$

18. By Green's Theorem,

$$\int_C 3x^2 e^y \, dx + e^y \, dy = \int\int_R -3x^2 e^y \, dA$$

$$= \int_1^2 \int_{-2}^2 -3x^2 e^y \, dy \, dx + \int_{-1}^1 \int_1^2 -3x^2 e^y \, dy \, dx$$

$$+ \int_{-2}^{-1} \int_{-2}^2 -3x^2 e^y \, dy \, dx + \int_{-1}^1 \int_{-2}^{-1} -3x^2 e^y \, dy \, dx$$

$$= -7(e^2 - e^{-2}) - 2(e^2 - e) - 7(e^2 - e^{-2}) - 2(e^{-1} - e^{-2})$$

$$= -16e^2 + 16e^{-2} + 2e - 2e^{-1}.$$

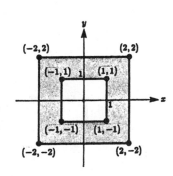

19. $\mathbf{F}(x, \, y) = xy\mathbf{i} + (x + y)\mathbf{j}$

$C: \ x^2 + y^2 = 4$

$$\text{Work} = \int_C xy \, dx + (x + y) \, dy = \int\int_R (1 - x) \, dA = \int_0^{2\pi} \int_0^2 (1 - r\cos\theta) r \, dr \, d\theta = \int_0^{2\pi} \left(2 - \frac{8}{3}\cos\theta \right) d\theta = 4\pi.$$

20. $\mathbf{F}(x, \, y) = (e^x - 3y)\mathbf{i} + (e^y + 6x)\mathbf{j}$

$C: \ r = 2\cos\theta$

$$\text{Work} = \int_C (e^x - 3y) \, dx + (e^y + 6x) \, dy = \int\int_R 9 \, dA = 9\pi \text{ since } r = 2\cos\theta \text{ is a circle with a radius of one.}$$

21. $\mathbf{F}(x, \, y) = (x^{3/2} - 3y)\mathbf{i} + (6x + 5\sqrt{y})\mathbf{j}$

$C:$ boundary of the triangle with vertices $(0, \, 0)$, $(5, \, 0)$, $(0, \, 5)$

$$\text{Work} = \int_C (x^{3/2} - 3y) \, dx + (6x + 5\sqrt{y}) \, dy = \int\int_R 9 \, dA = 9(\tfrac{1}{2})(5)(5) = \frac{225}{2}$$

22. $\mathbf{F}(x, y) = (3x^2 + y)\mathbf{i} + 4xy^2\mathbf{j}$

C: boundary of the region bounded by the graphs of $y = \sqrt{x}$, $y = 0$, $x = 4$

$$\text{Work} = \int_C (3x^2 + y)\,dx + 4xy^2\,dy = \int_0^4 \int_0^{\sqrt{x}} (4y^2 - 1)\,dy\,dx = \int_0^4 \left(\tfrac{4}{3}x^{3/2} - x^{1/2}\right)dx = \tfrac{176}{15}$$

23. C: let $x = a\cos t$, $y = a\sin t$, $0 \le t \le 2$. By Theorem 16.7, we have

$$A = \frac{1}{2}\int_C x\,dy - y\,dx = \frac{1}{2}\int_0^{2\pi} [a\cos t(a\cos t) - a\sin t(-a\sin t)]\,dt = \frac{1}{2}\int_0^{2\pi} a^2\,dt = \frac{a^2}{2}t\Big]_0^{2\pi} = \pi a^2.$$

24. From the accompanying figure we see that for

$$C_1: \quad y = \frac{2}{3}x, \quad dy = \frac{2}{3}\,dx$$

$$C_2: \quad y = -\frac{1}{3}x + 3, \quad dy = -\frac{1}{3}\,dx$$

$$C_3: \quad x = 0, \quad dx = 0.$$

Thus, by Theorem 16.7, we have

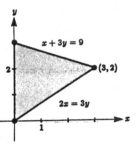

$$A = \frac{1}{2}\int_0^3 \left(\frac{2x}{3} - \frac{2x}{3}\right)dx + \frac{1}{2}\int_3^0 \left[\frac{-x}{3} - \left(\frac{-x}{3} + 3\right)\right]dx + \frac{1}{2}\int_3^0 0\,dy - y(0)$$

$$= \frac{1}{2}(0) + \frac{1}{2}\int_3^0 -3\,dx + \frac{1}{2}(0)$$

$$= \left[\frac{-3}{2}x\right]_3^0 = \frac{9}{2}.$$

25. From the accompanying figure we see that for

$$C_1: \quad y = 2x + 1, \quad dy = 2\,dx$$

$$C_2: \quad y = 4 - x^2, \quad dy = -2x\,dx.$$

Thus, by Theorem 16.7, we have

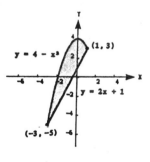

$$A = \frac{1}{2}\int_{-3}^1 [x(2) - (2x + 1)]\,dx + \frac{1}{2}\int_1^{-3} [x(-2x) - (4 - x^2)]\,dx$$

$$= \frac{1}{2}\int_{-3}^1 (-1)\,dx + \frac{1}{2}\int_1^{-3} (-x^2 - 4)\,dx$$

$$= \frac{1}{2}\int_{-3}^1 (-1)\,dx + \frac{1}{2}\int_{-3}^1 (x^2 + 4)\,dx$$

$$= \frac{1}{2}\int_{-3}^1 (3 + x^2)\,dx$$

$$= \frac{1}{2}\left[3x + \frac{x^3}{3}\right]_{-3}^1 = \frac{32}{3}.$$

26. Since the loop of the folium is formed on the interval $0 \le t < \infty$, $dx = \dfrac{3(1 - 2t^3)}{(t^3 + 1)^2}\,dt$ and $dy = \dfrac{3(2t - t^4)}{(t^3 + 1)^2}\,dt$, we have

$$A = \frac{1}{2}\int_0^\infty \left[\left(\frac{3t}{t^3 + 1}\right)\frac{3(2t - t^4)}{(t^3 + 1)^2} - \left(\frac{3t^2}{t^3 + 1}\right)\frac{3(1 - 2t^3)}{(t^3 + 1)^2}\right]dt$$

$$= \frac{9}{2}\int_0^\infty \frac{t^5 + t^2}{(t^3 + 1)^3}\,dt = \frac{9}{2}\int_0^\infty \frac{t^2(t^3 + 1)}{(t^3 + 1)^3}\,dt = \frac{3}{2}\int_0^\infty 3t^2(t^3 + 1)^{-2}\,dt = \frac{-3}{2(t^3 + 1)}\Bigg]_0^\infty = \frac{3}{2}.$$

27. For the moment about the x-axis, $M_x = \displaystyle\int\int_R y\,dA$, which by Green's Theorem implies that

$$M_x = \int_C 0\,dx - \frac{y^2}{2}\,dy = -\frac{1}{2}\int_C y^2\,dy \quad \text{and} \quad \overline{y} = \frac{M_x}{2A} = -\frac{1}{2A}\int_C y^2\,dy.$$

For the moment about the y-axis, $M_y = \displaystyle\int\int_R x\,dA$, which by Green's Theorem implies that

$$M_y = \int_C \frac{x^2}{2}\,dx - 0\,dy = \frac{1}{2}\int_C x^2\,dx \quad \text{and} \quad \overline{x} = \frac{M_y}{2A} = \frac{1}{2A}\int_C x^2\,dx.$$

28. By Theorem 16.9 and the fact that $x = r\cos\theta$, $y = r\sin\theta$, we have

$$A = \frac{1}{2}\int x\,dy - y\,dx = \frac{1}{2}\int (r\cos\theta)(r\cos\theta)\,d\theta - (r\sin\theta)(-r\sin\theta)\,d\theta = \frac{1}{2}\int_C r^2\,d\theta.$$

29. Since $A = \displaystyle\int_{-2}^{2}(4 - x^2)\,dx = \left[4x - \frac{x^3}{3}\right]_{-2}^{2} = \frac{32}{3}$, we have $dy = -2x\,dx$ and

$$\overline{x} = \frac{3}{64}\int_{2}^{-2} x^2(-2x\,dx) = \frac{3}{32}\left[-\frac{x^4}{4}\right]_{2}^{-2} = 0.$$

(Note the counterclockwise direction along C.)

$$\overline{y} = \frac{-3}{64}\int_{2}^{-2}(4 - x^2)^2\,dx = \frac{3}{64}\int_{-2}^{2}(16 - 8x^2 + x^4)\,dx = \frac{3}{64}\left[16x - \frac{8x^3}{3} + \frac{x^5}{5}\right]_{-2}^{2} = \frac{8}{5}.$$

30. Since $A = $ area of semicircle $= \dfrac{\pi a^2}{2}$, we have $\dfrac{1}{2A} = \dfrac{1}{\pi a^2}$.

Let $x = a\cos t$, $y = a\sin t$, $0 \le t \le \pi$, then

$$\overline{x} = \frac{1}{\pi a^2}\int_0^\pi a^2\cos^2 t(a\cos t)\,dt = \frac{a}{\pi}\int_0^\pi \cos^3 t\,dt = \frac{a}{\pi}\int_0^\pi (1 - \sin^2 t)\cos t\,dt = \frac{a}{\pi}\left[\sin t - \frac{\sin^3 t}{3}\right]_0^\pi = 0$$

$$\overline{y} = \frac{-1}{\pi a^2}\int_0^\pi a^2\sin^2 t(-a\sin t\,dt) = \frac{a}{\pi}\int_0^\pi \sin^3 t\,dt = \frac{a}{\pi}\left[-\cos t + \frac{\cos^3 t}{3}\right]_0^\pi = \frac{4a}{3\pi}.$$

31. Since $A = \displaystyle\int_0^1 (x - x^3)\,dx = \left[\frac{x^2}{2} - \frac{x^4}{4}\right]_0^1 = \frac{1}{4}$, we have $\dfrac{1}{2A} = 2$.

On C_1 we have $y = x^3$, $dy = 3x^2\,dx$ and on C_2 we have $y = x$, $dy = dx$. Thus,

$$\overline{x} = 2\int_C x^2\,dy = 2\int_{C_1} x^2(3x^2\,dx) + 2\int_{C_2} x^2\,dx = 6\int_0^1 x^4\,dx + 2\int_1^0 x^2\,dx = \frac{6}{5} - \frac{2}{3} = \frac{8}{15}$$

$$\overline{y} = -2\int_C y^2\,dx = -2\int_0^1 x^6\,dx - 2\int_1^0 x^2\,dx = -\frac{2}{7} + \frac{2}{3} = \frac{8}{21}.$$

32. Since $A = \frac{1}{2}(2a)(c) = ac$, we have $\frac{1}{2A} = \frac{1}{2ac}$,

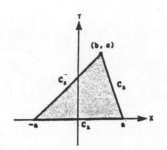

$$C_1 : \; y = 0, \quad dy = 0$$

$$C_2 : \; y = \frac{c}{b-a}(x-a), \quad dy = \frac{c}{b-a}\,dx$$

$$C_3 : \; y = \frac{c}{b+a}(x+a), \quad dy = \frac{c}{b+a}\,dx.$$

Thus,

$$\bar{x} = \frac{1}{2ac}\int_C x^2\,dy = \frac{1}{2ac}\left[\int_{-a}^{a}0 + \int_a^b x^2\frac{c}{b-a}\,dx + \int_b^{-a}x^2\frac{c}{b+a}\,dx\right] = \frac{1}{2ac}\left[0 + \frac{2abc}{3}\right] = \frac{b}{3}$$

$$\bar{y} = \frac{-1}{2ac}\int_C y^2\,dx = \frac{-1}{2ac}\left[0 + \int_a^b\left(\frac{c}{b-a}\right)^2(x-a)^2\,dx + \int_b^{-a}\left(\frac{c}{b+a}\right)^2(x+a)^2\,dx\right]$$

$$= \frac{-1}{2ac}\left[\frac{c^2(b-a)}{3} - \frac{c^2(b+a)}{3}\right] = \frac{c}{3}.$$

33. $A = \frac{1}{2}\int_0^{2\pi}a^2(1-\cos\theta)^2\,d\theta$

$$= \frac{a^2}{2}\int_0^{2\pi}\left(1 - 2\cos\theta + \frac{1}{2} + \frac{\cos 2\theta}{2}\right)d\theta = \frac{a^2}{2}\left[\frac{3\theta}{2} - 2\sin\theta + \frac{1}{2}\sin 2\theta\right]_0^{2\pi} = \frac{a^2}{2}(3\pi) = \frac{3\pi a^2}{2}$$

34. $A = \frac{1}{2}\int_0^{\pi}a^2\cos^2 3\theta\,d\theta = \frac{a^2}{2}\int_0^{\pi}\frac{1+\cos 6\theta}{2}\,d\theta = \frac{a^2}{4}\left[\theta + \frac{\sin 6\theta}{6}\right]_0^{\pi} = \frac{\pi a^2}{4}$

Note: In this case R is enclosed by $r = a\cos 3$ where $0 \le \theta \le \pi$.

35. In this case the inner loop has domain $\frac{2\pi}{3} \le \theta \le \frac{4\pi}{3}$. Thus,

$$A = \frac{1}{2}\int_{2\pi/3}^{4\pi/3}(1 + 4\cos\theta + 4\cos^2\theta)\,d\theta$$

$$= \frac{1}{2}\int_{2\pi/3}^{4\pi/3}(3 + 4\cos\theta + 2\cos 2\theta)\,d\theta = \frac{1}{2}\left[3\theta + 4\sin\theta + \sin 2\theta\right]_{2\pi/3}^{4\pi/3} = \pi - \frac{3\sqrt{3}}{2}.$$

36. In this case, $0 \le \theta \le 2\pi$ and we let $u = \frac{\sin\theta}{1+\cos\theta}$, $\cos\theta = \frac{1-u^2}{1+u^2}$, $d\theta = \frac{2\,du}{1+u^2}$. Now $u \Rightarrow \infty$ as $\theta \Rightarrow \pi$ and we have

$$A = 2\left(\frac{1}{2}\right)\int_0^{\pi}\frac{9}{(2-\cos\theta)^2}\,d\theta = 9\int_0^{\infty}\frac{\dfrac{2\,du}{1+u^2}}{4 - 4\left(\dfrac{1-u^2}{1+u^2}\right) + \dfrac{(1-u^2)^2}{(1+u^2)^2}}$$

$$= 18\int_0^{\infty}\frac{1+u^2}{(1+3u^2)^2}\,du$$

$$= 18\int_0^{\infty}\frac{1/3}{1+3u^2}\,du + 18\int_0^{\infty}\frac{2/3}{(1+3u^2)^2}\,du$$

$$= \frac{6}{\sqrt{3}}\arctan\sqrt{3}\,u\Big]_0^{\infty} + \frac{12}{\sqrt{3}}\left(\frac{1}{2}\right)\left[\frac{u}{1+3u^2} + \int\frac{\sqrt{3}}{1+3u^2}\,du\right]_0^{\infty}$$

$$= \frac{6}{\sqrt{3}}\left(\frac{\pi}{2}\right) + \frac{6}{\sqrt{3}}\left[\frac{u}{1+3u^2}\right]_0^{\infty} + \frac{6}{\sqrt{3}}\arctan\sqrt{3}\,u\Big]_0^{\infty}$$

$$= \frac{3\pi}{\sqrt{3}} + 0 + \frac{3\pi}{\sqrt{3}} = 2\sqrt{3}\,\pi.$$

37. $I = \displaystyle\int_C \frac{y\,dx - x\,dy}{x^2 + y^2}$

(a) Let $\mathbf{F} = \dfrac{y}{x^2 + y^2}\mathbf{i} - \dfrac{x}{x^2 + y^2}\mathbf{j}$.

$\mathbf{F}$ is conservative since $\dfrac{\partial N}{\partial x} = \dfrac{\partial M}{\partial y} = \dfrac{x^2 - y^2}{(x^2 + y^2)^2}$.

$\mathbf{F}$ is defined and has continuous first partials everywhere except at the origin.
If C is a circle (a closed path) that does not contain the origin, then

$$\int_C \mathbf{F} \cdot d\mathbf{r} = \int_C M\,dx + N\,dy = \int\int_R \left(\frac{\partial N}{\partial x} - \frac{\partial M}{\partial y}\right) dA = 0.$$

(b) Let $\mathbf{r} = a\cos t\,\mathbf{i} + a\sin t\,\mathbf{j}$, $0 \le t \le 2\pi$, a circle which contains the origin. Then,

$$\int_C M\,dx + N\,dy = \int_C \mathbf{F} \cdot d\mathbf{r} = \int_0^{2\pi}(-\sin^2 t - \cos^2 t)\,dt = -t\Big]_0^{2\pi} = -2\pi.$$

38. (a) Let C be the line segment joining $(x_1,\, y_1)$ and $(x_2,\, y_2)$.

$$y = \frac{y_2 - y_1}{x_2 - x_1}(x - x_1) + y_1$$

$$dy = \frac{y_2 - y_1}{x_2 - x_1}\,dx$$

$$\int_C -y\,dx + x\,dy = \int_{x_1}^{x_2}\left[-\frac{y_2 - y_1}{x_2 - x_1}(x - x_1) - y_1 + x\left(\frac{y_2 - y_1}{x_2 - x_1}\right)\right]dx$$

$$= \int_{x_1}^{x_2}\left[x_1\left(\frac{y_2 - y_1}{x_2 - x_1}\right) - y_1\right]dx$$

$$= \left[x_1\left(\frac{y_2 - y_1}{x_2 - x_1}\right) - y_1\right]x\;\Big]_{x_1}^{x_2}$$

$$= \left[x_1\left(\frac{y_2 - y_1}{x_2 - x_1}\right) - y_1\right](x_2 - x_1)$$

$$= x_1(y_2 - y_1) - y_1(x_2 - x_1)$$

$$= x_1 y_2 - x_2 y_1$$

(b) $\dfrac{1}{2}\displaystyle\int_C -y\,dx + x\,dy = \dfrac{1}{2}\int\int_R (1 - (-1))\,dA = \int\int_R dA$

Therefore, $\displaystyle\int\int_R dA = \frac{1}{2}\left[\int_{C_1} -y\,dx + x\,dy + \int_{C_2} -y\,dx + x\,dy + \cdots + \int_{C_n} -y\,dx + x\,dy\right]$

where C_1 is the line segment joining $(x_1,\, y_1)$ and $(x_2,\, y_2)$, C_2 is the line segment joining $(x_2,\, y_2)$ and $(x_3,\, y_3), \ldots$, and C_n is the line segment joining $(x_n,\, y_n)$ and $(x_1,\, y_1)$. Thus,

$$\int\int_R dA = \frac{1}{2}\left[(x_1 y_2 - x_2 y_1) + (x_2 y_3 - x_3 y_2) + \cdots + (x_{n-1}y_n - x_n y_{n-1}) + (x_n y_1 - x_1 y_n)\right].$$

39. Pentagon: $(0,\, 0)$, $(2,\, 0)$, $(3,\, 2)$, $(1,\, 4)$, $(-1,\, 1)$

$A = \dfrac{1}{2}[(0 - 0) + (4 - 0) + (12 - 2) + (1 + 4) + (0 - 0)] = \dfrac{19}{2}$

40. Hexagon: $(0,\, 0)$, $(2,\, 0)$, $(3,\, 2)$, $(2,\, 4)$, $(0,\, 3)$, $(-1,\, 1)$

$A = \dfrac{1}{2}[(0 - 0) + (4 - 0) + (12 - 4) + (6 - 0) + (0 + 3) + (0 - 0)] = \dfrac{21}{2}$

41. $\displaystyle\int_C f(x)\,dx + g(y)\,dy = \int\int_R \left[\frac{\partial}{\partial x}g(y) - \frac{\partial}{\partial y}f(x)\right]dA = \int\int_R (0 - 0)\,dA = 0$

42. $\mathbf{F} = M\mathbf{i} + N\mathbf{j}$

$\text{curl } \mathbf{F} = \left(\dfrac{\partial N}{\partial x} - \dfrac{\partial M}{\partial y} \right) \mathbf{k} = 0 \Rightarrow \dfrac{\partial N}{\partial x} = \dfrac{\partial M}{\partial y}$

$\displaystyle \int_C \mathbf{F} \cdot d\mathbf{r} = \int_C M\, dx + N\, dy = \iint_R \left(\dfrac{\partial N}{\partial x} - \dfrac{\partial N}{\partial y} \right) dA = \iint_R (0)\, dA = 0$

43. Since $\displaystyle \int_C \mathbf{F} \cdot \mathbf{N}\, ds = \iint_R \text{div } \mathbf{F}\, dA$, then

$\displaystyle \int_C f D_{\mathbf{N}} g\, ds = \int_C f\nabla g \cdot \mathbf{N}\, ds = \iint_R \text{div } (f\nabla g)\, dA = \iint_R (f \text{ div } (\nabla g) + \nabla f \cdot \nabla g)\, dA = \iint_R (f\nabla^2 g + \nabla f \cdot \nabla g)\, dA.$

44. $\displaystyle \int_C (f D_{\mathbf{N}} g - g D_{\mathbf{N}} f)\, ds = \int_C f D_{\mathbf{N}} g\, ds - \int_C g D_{\mathbf{N}} f\, ds$

$\displaystyle \qquad = \iint_R (f\nabla^2 g + \nabla f \cdot \nabla g)\, dA - \iint_R (g\nabla^2 f + \nabla g \cdot \nabla f)\, dA = \iint_R (f\nabla^2 g - g\nabla^2 f)\, dA$

45. $C : x^2 + y^2 = 4$

Let $x = 2\cos\theta$ and $y = 2\sin\theta$, $\ 0 \le \theta \le 2\pi$.

$\displaystyle \int_C xe^y\, dx + e^x\, dy = \int_0^{2\pi} [2\cos t\, e^{2\sin t}(-2\sin t) + e^{2\cos t}(2\cos t)]\, dt \approx 19.99$

$\displaystyle \iint_R \left(\dfrac{\partial N}{\partial x} - \dfrac{\partial M}{\partial y} \right) dA = \int_{-2}^{2} \int_{-\sqrt{4-x^2}}^{\sqrt{4-x^2}} (e^x - xe^y)\, dy\, dx = \int_{-2}^{2} \left[2\sqrt{4-x^2}\, e^x - xe^{\sqrt{4-x^2}} + xe^{-\sqrt{4-x^2}} \right] dx \approx 19.99$

46. C: boundary of the region lying between the graphs of $y = x$ and $y = x^3$

$\displaystyle \int_C xe^y\, dx + e^x\, dy = \int_0^1 (xe^{x^3} + 3x^2 e^x)\, dx + \int_1^0 (xe^x + e^x)\, dx \approx 2.936 - 2.718 \approx 0.22$

$\displaystyle \iint_R \left(\dfrac{\partial N}{\partial x} - \dfrac{\partial M}{\partial y} \right) dA = \int_0^1 \int_{x^3}^{x} (e^x - xe^y)\, dy\, dx = \int_0^1 (xe^{x^3} - x^3 e^x)\, dx \approx 0.22$

Section 16.5   Surface Integrals

1. $S : z = 4 - x, \quad 0 \le x \le 4, \quad 0 \le y \le 4, \quad \dfrac{\partial z}{\partial x} = -1, \quad \dfrac{\partial z}{\partial y} = 0$

$\displaystyle \iint_S (x - 2y + z)\, dS = \int_0^4 \int_0^4 (x - 2y + 4 - x)\sqrt{1 + (-1)^2 + (0)^2}\, dy\, dx = \sqrt{2} \int_0^4 \int_0^4 (4 - 2y)\, dy\, dx = 0$

2. $S : z = 10 - 2x + 2y, \quad 0 \le x \le 2, \quad 0 \le y \le 4, \quad \dfrac{\partial z}{\partial x} = -2, \quad \dfrac{\partial z}{\partial y} = 2$

$\displaystyle \iint_S (x - 2y + z)\, dS = \int_0^2 \int_0^4 (x - 2y + 10 - 2x + 2y)\sqrt{1 + (-2)^2 + (2)^2}\, dy\, dx = 3 \int_0^2 \int_0^4 (10 - x)\, dy\, dx = 216$

3. $S : z = 10, \quad x^2 + y^2 \le 1, \quad \dfrac{\partial z}{\partial x} = \dfrac{\partial z}{\partial y} = 0$

$$\iint_S (x - 2y + z)\, dS = \int_{-1}^{1} \int_{-\sqrt{1-x^2}}^{\sqrt{1-x^2}} (x - 2y + 10)\sqrt{1 + (0)^2 + (0)^2}\, dy\, dx$$

$$= \int_0^{2\pi} \int_0^1 (r\cos\theta - 2r\sin\theta + 10) r\, dr\, d\theta$$

$$= \int_0^{2\pi} \left(\frac{1}{3}\cos\theta - \frac{2}{3}\sin\theta + 5 \right) d\theta$$

$$= \left[\frac{1}{3}\sin\theta + \frac{2}{3}\cos\theta + 5\theta \right]_0^{2\pi}$$

$$= 10\pi$$

4. $S : z = \dfrac{2}{3}x^{3/2}, \quad 0 \le x \le 1, \quad 0 \le y \le x, \quad \dfrac{\partial z}{\partial x} = x^{1/2}, \quad \dfrac{\partial z}{\partial y} = 0$

$$\iint_S (x - 2y + z)\, dS = \int_0^1 \int_0^x \left(x - 2y + \frac{2}{3}x^{3/2} \right)\sqrt{1 + (x^{1/2})^2 + (0)^2}\, dy\, dx$$

$$= \int_0^1 \int_0^x \left(x - 2y + \frac{2}{3}x^{3/2} \right)\sqrt{1 + x}\, dy\, dx$$

$$= \frac{2}{3}\int_0^1 x^{5/2}\sqrt{x + 1}\, dx$$

$$= \frac{2}{3}\left[\frac{1}{4}x^{5/2}(1 + x)^{3/2} \right]_0^1 - \frac{5}{3}\int_0^1 x^{3/2}\sqrt{1 + x}\, dx$$

$$= \frac{1}{6}x^{5/2}(1 + x)^{3/2}\bigg]_0^1 - \frac{5}{12}\left(\frac{1}{3}\right)\left[x^{3/2}(1 + x)^{3/2} \right]_0^1 + \frac{5}{24}\int_0^1 x^{1/2}\sqrt{1 + x}\, dx$$

$$= \frac{\sqrt{2}}{3} - \frac{5\sqrt{2}}{18} + \frac{5}{24}\int_0^1 \sqrt{x + x^2}\, dx$$

$$= \frac{\sqrt{2}}{18} + \frac{5}{24}\int_0^1 \sqrt{\left(x + \tfrac{1}{2}\right)^2 - \tfrac{1}{4}}\, dx$$

$$= \frac{\sqrt{2}}{18} + \frac{5}{24}\left(\frac{1}{2}\right)\left[\left(x + \frac{1}{2}\right)\sqrt{x^2 + x} - \frac{1}{4}\ln\left|\left(x + \frac{1}{2}\right) + \sqrt{x^2 + x}\right| \right]_0^1$$

$$= \frac{\sqrt{2}}{18} + \frac{5}{48}\left[\frac{3}{2}\sqrt{2} - \frac{1}{4}\ln\left|\frac{3}{2} + \sqrt{2}\right| + \frac{1}{4}\ln\left|\frac{1}{2}\right| \right]$$

$$= \frac{\sqrt{2}}{18} + \frac{15\sqrt{2}}{96} + \frac{5}{192}\ln\left|\frac{1}{3 + 2\sqrt{2}}\right|$$

$$= \frac{61\sqrt{2}}{288} - \frac{5}{192}\ln\left|3 + 2\sqrt{2}\right|$$

5. $S: z = 6 - x - 2y,$ (first octant) $\dfrac{\partial z}{\partial x} = -1,$ $\dfrac{\partial z}{\partial y} = -2$

$$\iint_S xy\, dS = \int_0^6 \int_0^{3-(x/2)} xy\sqrt{1 + (-1)^2 + (-2)^2}\, dy\, dx$$

$$= \sqrt{6} \int_0^6 \left. \frac{xy^2}{2} \right]_0^{3-(x/2)} dx$$

$$= \frac{\sqrt{6}}{2} \int_0^6 x\left(9 - 3x + \frac{1}{4}x^2\right) dx$$

$$= \frac{\sqrt{6}}{2} \left. \left(\frac{9x^2}{2} - x^3 + \frac{x^4}{16}\right) \right]_0^6$$

$$= \frac{27\sqrt{6}}{2}$$

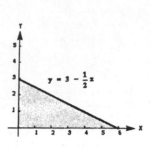

$y = 3 - \frac{1}{2}x$

6. $S: z = xy,$ $0 \le x \le 2,$ $0 \le y \le 2,$ $\dfrac{\partial z}{\partial x} = y,$ $\dfrac{\partial z}{\partial y} = x$

$$\iint_S xy\, dS = \int_0^2 \int_0^2 xy\sqrt{1 + x^2 + y^2}\, dy\, dx = \int_0^2 \left. \frac{x}{3}(1 + x^2 + y^2)^{3/2} \right]_0^2 dx$$

$$= \frac{1}{3} \int_0^2 \left[x(5 + x^2)^{3/2} - x(1 + x^2)^{3/2} \right] dx$$

$$= \frac{1}{3}\left(\frac{1}{5}\right) \left. \left[(5 + x^2)^{5/2} - (1 + x^2)^{5/2} \right] \right]_0^2$$

$$= \frac{1}{15}\left[(243 - 25\sqrt{5}) - (25\sqrt{5} - 1)\right]$$

$$= \frac{244 - 50\sqrt{5}}{15}$$

7. $S: z = 9 - x^2,$ $0 \le x \le 2,$ $0 \le y \le x,$ $\dfrac{\partial z}{\partial x} = -2x,$ $\dfrac{\partial z}{\partial y} = 0$

$$\iint xy\, dS = \int_0^2 \int_y^2 xy\sqrt{1 + 4x^2}\, dx\, dy = \int_0^2 \left. \frac{y}{12}(1 + 4x^2)^{3/2} \right]_y^2 dy$$

$$= \frac{1}{12} \int_0^2 \left[17\sqrt{17}\, y - (1 + 4y^2)^{3/2} y \right] dy$$

$$= \frac{1}{12} \left. \left[\frac{17\sqrt{17}\, y^2}{2} - \frac{1}{20}(1 + 4y^2)^{5/2} \right] \right]_0^2$$

$$= \frac{1}{12} \left[34\sqrt{17} - \frac{1}{20}(289\sqrt{17}) + \frac{1}{20} \right]$$

$$= \frac{391\sqrt{17} + 1}{240}$$

8. $S: z = h,$ $0 \le x \le 2,$ $0 \le y \le \sqrt{4 - x^2},$ $\dfrac{\partial z}{\partial x} = \dfrac{\partial z}{\partial y} = 0$

$$\iint_S dx\, dS = \int_0^2 \int_0^{\sqrt{4-x^2}} xy\, dy\, dx = \frac{1}{2} \int_0^2 x(4 - x^2)\, dx = \frac{1}{2} \left. \left[2x^2 - \frac{x^4}{4} \right] \right]_0^2 = 2$$

9. $f(x, y, z) = x^2 + y^2 + z^2$
$S : z = x + 2, \quad x^2 + y^2 \leq 1$

$$\iint_S f(x, y, z) \, dS = \int_{-1}^{1} \int_{-\sqrt{1-x^2}}^{\sqrt{1-x^2}} [x^2 + y^2 + (x + 2)^2] \sqrt{1 + (1)^2 + (0)^2} \, dy \, dx$$

$$= \sqrt{2} \int_0^{2\pi} \int_0^1 [r^2 + (r\cos\theta + 2)^2] r \, dr \, d\theta$$

$$= \sqrt{2} \int_0^{2\pi} \int_0^1 [r^2 + r^2\cos^2\theta + 4r\cos\theta + 4] r \, dr \, d\theta$$

$$= \sqrt{2} \int_0^{2\pi} \left[\frac{r^4}{4} + \frac{r^4}{4}\cos^2\theta + \frac{4r^3}{3}\cos\theta + 2r^2\right]_0^1 d\theta$$

$$= \sqrt{2} \int_0^{2\pi} \left[\frac{9}{4} + \left(\frac{1}{4}\right)\frac{1 + \cos 2\theta}{2} + \frac{4}{3}\cos\theta\right] d\theta$$

$$= \sqrt{2} \left[\frac{9}{4}\theta + \frac{1}{8}\left(\theta + \frac{1}{2}\sin 2\theta\right) + \frac{4}{3}\sin\theta\right]_0^{2\pi}$$

$$= \sqrt{2} \left[\frac{18\pi}{4} + \frac{\pi}{4}\right] = \frac{19\sqrt{2}\,\pi}{4}$$

10. $f(x, y, z) = \dfrac{xy}{z}$
$S : z = x^2 + y^2, \quad 4 \leq x^2 + y^2 \leq 16$

$$\iint_S f(x, y, z) \, dS = \iint_S \frac{xy}{x^2 + y^2} \sqrt{1 + 4x^2 + 4y^2} \, dy \, dx$$

$$= \int_0^{2\pi} \int_2^4 \frac{r^2 \sin\theta \cos\theta}{r^2} \sqrt{1 + 4r^2}\, r \, dr \, d\theta$$

$$= \int_0^{2\pi} \int_2^4 r\sqrt{1 + 4r^2} \sin\theta \cos\theta \, dr \, d\theta$$

$$= \int_0^{2\pi} \frac{1}{12}(1 + 4r^2)^{3/2} \Big]_2^4 \sin\theta \cos\theta \, d\theta$$

$$= \frac{65\sqrt{65} - 17\sqrt{17}}{12} \left(\frac{\sin^2\theta}{2}\right)\Big]_0^{2\pi} = 0$$

11. $f(x, y, z) = \sqrt{x^2 + y^2 + z^2}$
$S : z = \sqrt{x^2 + y^2}, \quad x^2 + y^2 \leq 4$

$$\iint_S f(x, y, z) \, dS = \int_{-2}^{2} \int_{-\sqrt{4-x^2}}^{\sqrt{4-x^2}} \sqrt{x^2 + y^2 + (\sqrt{x^2+y^2})^2} \sqrt{1 + \left(\frac{x}{\sqrt{x^2+y^2}}\right)^2 + \left(\frac{y}{\sqrt{x^2+y^2}}\right)^2} \, dy \, dx$$

$$= \sqrt{2} \int_{-2}^{2} \int_{-\sqrt{4-x^2}}^{\sqrt{4-x^2}} \sqrt{x^2 + y^2} \sqrt{\frac{x^2 + y^2 + x^2 + y^2}{x^2 + y^2}} \, dy \, dx$$

$$= 2 \int_{-2}^{2} \int_{-\sqrt{4-x^2}}^{\sqrt{4-x^2}} \sqrt{x^2 + y^2} \, dy \, dx$$

$$= 2 \int_0^{2\pi} \int_0^2 r^2 \, dr \, d\theta = 2 \int_0^{2\pi} \frac{r^3}{3}\Big]_0^2 d\theta = \frac{16}{3}\theta\Big]_0^{2\pi} = \frac{32\pi}{3}$$

12. $f(x, y, z) = \sqrt{x^2 + y^2 + z^2}$

$S: z = \sqrt{x^2 + y^2}, \quad (x-1)^2 + y^2 \le 1$

$$\iint_S f(x, y, z) \, dS = \iint_S \sqrt{x^2 + y^2 + (\sqrt{x^2 + y^2})^2} \sqrt{1 + \left(\frac{x}{\sqrt{x^2 + y^2}}\right)^2 + \left(\frac{y}{\sqrt{x^2 + y^2}}\right)^2} \, dy \, dx$$

$$= \iint_S \sqrt{2(x^2 + y^2)} \sqrt{\frac{2(x^2 + y^2)}{x^2 + y^2}} \, dy \, dx$$

$$= 2 \iint_S \sqrt{x^2 + y^2} \, dy \, dx$$

$$= 2 \int_0^\pi \int_0^{2\cos\theta} r^2 \, dr \, d\theta$$

$$= \frac{16}{3} \int_0^\pi \cos^3 \theta \, d\theta$$

$$= \frac{16}{3} \int_0^\pi (1 - \sin^2 \theta) \cos \theta \, d\theta$$

$$= \frac{16}{3} \left(\sin \theta - \frac{\sin^3 \theta}{3} \right) \Big]_0^\pi = 0$$

13. $f(x, y, z) = x^2 + y^2 + z^2$

$S: x^2 + y^2 = 9, \quad 0 \le x \le 3, \quad 0 \le y \le 3, \quad 0 \le z \le 9$

Project the solid onto the yz-plane; $x = \sqrt{9 - y^2}, \quad 0 \le y \le 3, \quad 0 \le z \le 9$.

$$\iint_S f(x, y, z) \, dS = \int_0^3 \int_0^9 [(9 - y^2) + y^2 + z^2] \sqrt{1 + \left(\frac{-y}{\sqrt{9 - y^2}}\right)^2 + (0)^2} \, dz \, dy$$

$$= \int_0^3 \int_0^9 (9 + z^2) \frac{3}{\sqrt{9 - y^2}} \, dz \, dy$$

$$= \int_0^3 \frac{3}{\sqrt{9 - y^2}} \left(9z + \frac{z^3}{3} \right) \Big]_0^9 \, dy$$

$$= 324 \int_0^3 \frac{3}{\sqrt{9 - y^2}} \, dy$$

$$= 972 \arcsin \left(\frac{y}{3} \right) \Big]_0^3$$

$$= 972 \left(\frac{\pi}{2} - 0 \right) = 486\pi$$

14. $f(x, y, z) = x^2 + y^2 + z^2$

$S : x^2 + y^2 = 9, \quad 0 \le x \le 3, \quad 0 \le z \le x$

Project the solid onto the xz-plane; $y = \sqrt{9 - x^2}$.

$$\iint_S f(x, y, z)\, dS = \int_0^3 \int_0^x [x^2 + (9 - x^2) + z^2]\sqrt{1 + \left(\frac{-x}{\sqrt{9 - x^2}}\right)^2 + (0)^2}\, dz\, dx$$

$$= \int_0^3 \int_0^x (9 + z^2)\frac{3}{\sqrt{9 - x^2}}\, dz\, dx$$

$$= \int_0^3 \frac{3}{\sqrt{9 - x^2}}\left(9z + \frac{z^3}{3}\right)\Big]_0^x dx$$

$$= \int_0^3 \frac{3}{\sqrt{9 - x^2}}\left(9x - \frac{x^3}{3}\right) dx$$

$$= \int_0^3 27x(9 - x^2)^{-1/2}\, dx - \int_0^3 x^3(9 - x^2)^{-1/2}\, dx$$

Let $u = x^2, \quad dv = x(9 - x^2)^{-1/2}\, dx$, then $du = 2x\, dx, \quad v = -\sqrt{9 - x^2}$.

$$= -27\sqrt{9 - x^2}\,\Big]_0^3 - \left[\left[-x^2\sqrt{9 - x^2}\,\right]_0^3 + \int_0^3 2x\sqrt{9 - x^2}\, dx\right]$$

$$= 81 + \frac{2}{3}(9 - x^2)^{3/2}\Big]_0^3$$

$$= 81 - 18 = 63$$

15. $\mathbf{F}(x, y, z) = 3z\mathbf{i} - 4\mathbf{j} + y\mathbf{k}$

$S : x + y + z = 1 \quad$ (first octant)

$G(x, y, z) = x + y + z - 1$

$\nabla G(x, y, z) = \mathbf{i} + \mathbf{j} + \mathbf{k}$

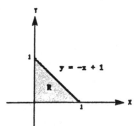

$$\iint_S \mathbf{F} \cdot \mathbf{N}\, dS = \iint_R \mathbf{F} \cdot \nabla G\, dA = \int_0^1 \int_0^{1-x} (3z - 4 + y)\, dy\, dx$$

$$= \int_0^1 \int_0^{1-x} [3(1 - x - y) - 4 + y]\, dy\, dx$$

$$= \int_0^1 \int_0^{1-x} (-1 - 3x - 2y)\, dy\, dx$$

$$= \int_0^1 (-y - 3xy - y^2)\Big]_0^{1-x} dx$$

$$= -\int_0^1 [(1 - x) + 3x(1 - x) + (1 - x)^2]\, dx$$

$$= -\int_0^1 (2 - 2x^2)\, dx = -\frac{4}{3}$$

16. $\mathbf{F}(x, y, z) = x\mathbf{i} + y\mathbf{j}$

$S : 2x + 3y + z = 6$ (first octant)

$G(x, y, z) = 2x + 3y + z - 6$

$\nabla \mathbf{G}(x, y, z) = 2\mathbf{i} + 3\mathbf{j} + \mathbf{k}$

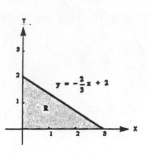

$$\iint_S \mathbf{F} \cdot \mathbf{N} \, dS = \iint_R \mathbf{F} \cdot \nabla \mathbf{G} \, dA = \int_0^3 \int_0^{-(2x/3)+2} (2x + 3y) \, dy \, dx$$

$$= \int_0^3 \left[-\frac{4}{3}x^2 + 4x + \frac{3}{2}\left(-\frac{2}{3}x + 2\right)^2 \right] dx$$

$$= \left[-\frac{4}{9}x^3 + 2x^2 - \frac{3}{4}\left(-\frac{2}{3}x + 2\right)^3 \right]_0^3 = 12$$

17. $\mathbf{F}(x, y, z) = x\mathbf{i} + y\mathbf{j} + z\mathbf{k}$

$S : z = 9 - x^2 - y^2, \quad 0 \leq z$

$G(x, y, z) = x^2 + y^2 + z - 9$

$\nabla \mathbf{G}(x, y, z) = 2x\mathbf{i} + 2y\mathbf{j} + \mathbf{k}$

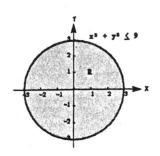

$$\iint_S \mathbf{F} \cdot \mathbf{N} \, dS = \iint_R \mathbf{F} \cdot \nabla \mathbf{G} \, dA = \iint_R (2x^2 + 2y^2 + z) \, dA$$

$$= \iint_R [2x^2 + 2y^2 + (9 - x^2 - y^2)] \, dA$$

$$= \iint_R (x^2 + y^2 + 9) \, dA$$

$$= \int_0^{2\pi} \int_0^3 (r^2 + 9)r \, dr \, d\theta$$

$$= \int_0^{2\pi} \left(\frac{r^4}{4} + \frac{9r^2}{2}\right)\Big]_0^3 \, d\theta = \frac{243\pi}{2}$$

18. $\mathbf{F}(x, y, z) = x\mathbf{i} + y\mathbf{j} + z\mathbf{k}$

$S : x^2 + y^2 + z^2 = 16$ (first octant)

$z = \sqrt{16 - x^2 - y^2}$

$G(x, y, z) = z - \sqrt{16 - x^2 - y^2}$

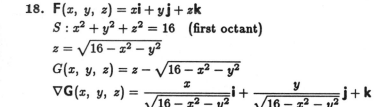

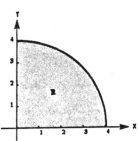

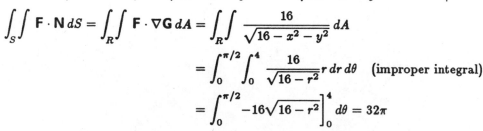

$$\iint_S \mathbf{F} \cdot \mathbf{N} \, dS = \iint_R \mathbf{F} \cdot \nabla \mathbf{G} \, dA = \iint_R \frac{16}{\sqrt{16 - x^2 - y^2}} \, dA$$

$$= \int_0^{\pi/2} \int_0^4 \frac{16}{\sqrt{16 - r^2}} r \, dr \, d\theta \quad \text{(improper integral)}$$

$$= \int_0^{\pi/2} -16\sqrt{16 - r^2}\Big]_0^4 \, d\theta = 32\pi$$

19. $\mathbf{F}(x,\ y,\ z) = 4\mathbf{i} - 3\mathbf{j} + 5\mathbf{k}$

$S : z = x^2 + y^2, \quad x^2 + y^2 \le 4$

$G(x,\ y,\ z) = -x^2 - y^2 + z$

$\nabla G(x,\ y,\ z) = -2x\mathbf{i} - 2y\mathbf{j} + \mathbf{k}$

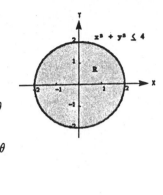

$$\iint_S \mathbf{F} \cdot \mathbf{N} \, dS = \iint_R \mathbf{F} \cdot \nabla G \, dA = \iint_R (-8x + 6y + 5) \, dA$$

$$= \int_0^{2\pi} \int_0^2 [-8r\cos\theta + 6r\sin\theta + 5] r \, dr \, d\theta$$

$$= \int_0^{2\pi} \left[-\frac{8}{3} r^3 \cos\theta + 2r^3 \sin\theta + \frac{5}{2} r^2 \right]_0^2 d\theta$$

$$= \int_0^{2\pi} \left[-\frac{64}{3} \cos\theta + 16 \sin\theta + 10 \right] d\theta$$

$$= \left[-\frac{64}{3} \sin\theta - 16 \cos\theta + 10\theta \right]_0^{2\pi} = 20\pi$$

20. $\mathbf{F}(x,\ y,\ z) = x\mathbf{i} + y\mathbf{j} - 2z\mathbf{k}$

$S : z = \sqrt{a^2 - x^2 - y^2}$

$G(x,\ y,\ z) = z - \sqrt{a^2 - x^2 - y^2}$

$\nabla G(x,\ y,\ z) = \dfrac{x}{\sqrt{a^2 - x^2 - y^2}}\mathbf{i} + \dfrac{y}{\sqrt{a^2 - x^2 - y^2}}\mathbf{j} + \mathbf{k}$

$\mathbf{F} \cdot \nabla G = \dfrac{x^2}{\sqrt{a^2 - x^2 - y^2}} + \dfrac{y^2}{\sqrt{a^2 - x^2 - y^2}} - 2\sqrt{a^2 - x^2 - y^2}$

$\qquad = \dfrac{3x^2 + 3y^2 - 2a^2}{\sqrt{a^2 - x^2 - y^2}}$

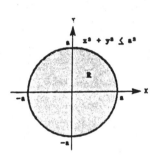

$$\iint_S \mathbf{F} \cdot \mathbf{N} \, dS = \iint_R \mathbf{F} \cdot \nabla G \, dA$$

$$= \iint_R \frac{3x^2 + 3y^2 - 2a^2}{\sqrt{a^2 - x^2 - y^2}} \, dA$$

$$= \int_0^{2\pi} \int_0^a \frac{3r^2 - 2a^2}{\sqrt{a^2 - r^2}} r \, dr \, d\theta$$

$$= 3 \int_0^{2\pi} \int_0^a \frac{r^3}{\sqrt{a^2 - r^2}} \, dr \, d\theta - 2a^2 \int_0^{2\pi} \int_0^a \frac{r}{\sqrt{a^2 - r^2}} \, dr \, d\theta$$

$$= 3 \left[\int_0^{2\pi} \left[-r^2 \sqrt{a^2 - r^2} - \frac{2}{3}(a^2 - r^2)^{3/2} \right]_0^a d\theta \right] - 2a^2 \int_0^{2\pi} \left[-\sqrt{a^2 - r^2} \right]_0^a d\theta$$

$$= 3 \int_0^{2\pi} \frac{2}{3} a^3 \, d\theta - 2a^2 \int_0^{2\pi} a \, d\theta = 0$$

21. $\mathbf{F}(x,\ y,\ z) = 4xy\mathbf{i} + z^2\mathbf{j} + yz\mathbf{k}$

S: unit cube bounded by $x = 0,\quad x = 1,\quad y = 0,\quad y = 1,\quad z = 0,\quad z = 1$

S_1 : The top of the cube

$\mathbf{N} = \mathbf{k},\quad z = 1$

$$\int_{S_1}\!\!\int \mathbf{F} \cdot \mathbf{N}\,dS = \int_0^1\!\int_0^1 y(1)\,dy\,dx = \frac{1}{2}$$

S_2 : The bottom of the cube

$\mathbf{N} = -\mathbf{k},\quad z = 0$

$$\int_{S_2}\!\!\int \mathbf{F} \cdot \mathbf{N}\,dS = \int_0^1\!\int_0^1 -y(0)\,dy\,dx = 0$$

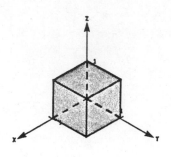

S_3 : The front of the cube

$\mathbf{N} = \mathbf{i},\quad x = 1$

$$\int_{S_3}\!\!\int \mathbf{F} \cdot \mathbf{N}\,dS = \int_0^1\!\int_0^1 4(1)y\,dy\,dz = 2$$

S_5 : The right side of the cube

$\mathbf{N} = \mathbf{j},\quad y = 1$

$$\int_{S_5}\!\!\int \mathbf{F} \cdot \mathbf{N}\,dS = \int_0^1\!\int_0^1 z^2\,dz\,dx = \frac{1}{3}$$

S_4 : The back of the cube

$\mathbf{N} = -\mathbf{i},\quad x = 0$

$$\int_{S_4}\!\!\int \mathbf{F} \cdot \mathbf{N}\,dS = \int_0^1\!\int_0^1 -4(0)y\,dy\,dz = 0$$

S_6 : The left side of the cube

$\mathbf{N} = -\mathbf{j},\quad y = 0$

$$\int_{S_6}\!\!\int \mathbf{F} \cdot \mathbf{N}\,dS = \int_0^1\!\int_0^1 -z^2\,dz\,dx = -\frac{1}{3}$$

$$\int_S\!\!\int \mathbf{F} \cdot \mathbf{N}\,dS = \frac{1}{2} + 0 + 2 + \frac{1}{3} - \frac{1}{3} = \frac{5}{2}$$

22. $\mathbf{F}(x,\ y,\ z) = (x+y)\mathbf{i} + y\mathbf{j} + z\mathbf{k}$

$S:\ z = 1 - x^2 - y^2,\quad z = 0$

$G(x,\ y,\ z) = z + x^2 + y^2 - 1$

$\nabla G(x,\ y,\ z) = 2x\mathbf{i} + 2y\mathbf{j} + \mathbf{k}$

$\mathbf{F} \cdot \nabla G = 2x(x+y) + 2y(y) + (1 - x^2 - y^2) = x^2 + 2xy + y^2 + 1$

$$\int_S\!\!\int \mathbf{F} \cdot \mathbf{N}\,dS = \int_R\!\!\int \mathbf{F} \cdot \nabla G\,dA = \int_R\!\!\int (x^2 + 2xy + y^2 + 1)\,dA$$

$$= \int_0^{2\pi}\!\int_0^1 (r^2 + 2r^2 \cos\theta \sin\theta + 1)r\,dr\,d\theta$$

$$= \int_0^{2\pi} \left(\frac{3}{4} + \frac{1}{2}\sin\theta\cos\theta\right) d\theta$$

$$= \left[\frac{3}{4}\theta + \frac{\sin^2\theta}{4}\right]_0^{2\pi} = \frac{3\pi}{2}$$

The flux across the bottom $z = 0$ is zero.

23. $S: 2x + 3y + 6z = 12$ (first octant) $\Rightarrow z = 2 - \dfrac{1}{3}x - \dfrac{1}{2}y$

$\rho(x, y, z) = x^2 + y^2$

$m = \displaystyle\int\!\!\int_R (x^2 + y^2)\sqrt{1 + (-1/3)^2 + (-1/2)^2}\, dA$

$\quad = \dfrac{7}{6}\displaystyle\int_0^6 \int_0^{4-(2x/3)} (x^2 + y^2)\, dy\, dx$

$\quad = \dfrac{7}{6}\displaystyle\int_0^6 \left[x^2\left(4 - \dfrac{2}{3}x\right) + \dfrac{1}{3}\left(4 - \dfrac{2}{3}x\right)^3 \right] dx$

$\quad = \dfrac{7}{6}\left[\dfrac{4}{3}x^3 - \dfrac{1}{6}x^4 - \dfrac{1}{8}\left(4 - \dfrac{2}{3}x\right)^4 \right]_0^6$

$\quad = \dfrac{364}{3}$

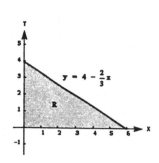

24. $S: z = \sqrt{a^2 - x^2 - y^2}$

$\rho(x, y, z) = kz$

$m = \displaystyle\int\!\!\int_S kz\, dS$

$\quad = \displaystyle\int\!\!\int_R k\sqrt{a^2 - x^2 - y^2}\sqrt{1 + \left(\dfrac{-x}{\sqrt{a^2 - x^2 - y^2}}\right)^2 + \left(\dfrac{-y}{\sqrt{a^2 - x^2 - y^2}}\right)^2}\, dA$

$\quad = \displaystyle\int\!\!\int_R k\sqrt{a^2 - x^2 - y^2}\left(\dfrac{a}{\sqrt{a^2 - x^2 - y^2}}\right) dA$

$\quad = \displaystyle\int\!\!\int_R ka\, dA$

$\quad = ka\displaystyle\int\!\!\int_R dA$

$\quad = ka(\pi a^2) = ka^3\pi$

25. $z = \sqrt{x^2 + y^2}, \;\; 0 \le z \le a$

$m = \displaystyle\int\!\!\int_S k\, dS = k\displaystyle\int\!\!\int_R \sqrt{1 + \left(\dfrac{x}{\sqrt{x^2 + y^2}}\right)^2 + \left(\dfrac{y}{\sqrt{x^2 + y^2}}\right)^2}\, dA = k\displaystyle\int\!\!\int_R \sqrt{2}\, dA = \sqrt{2}\, k\pi a^2$

$I_z = \displaystyle\int\!\!\int_S k(x^2 + y^2)\, dS$

$\quad = \displaystyle\int\!\!\int_R k(x^2 + y^2)\sqrt{2}\, dA = \sqrt{2}\, k\displaystyle\int_0^{2\pi}\int_0^a r^3\, dr\, d\theta = \dfrac{\sqrt{2}\, ka^4}{4}(2\pi) = \dfrac{\sqrt{2}\, k\pi a^4}{2} = \dfrac{a^2}{2}(\sqrt{2}\, k\pi a^2) = \dfrac{a^2 m}{2}$

26. $x^2 + y^2 + z^2 = a^2$

$z = \pm\sqrt{a^2 - x^2 - y^2}$

$$m = 2\iint_S k\,dS = 2k\iint_R \sqrt{1 + \left(\frac{-x}{\sqrt{a^2 - x^2 - y^2}}\right)^2 + \left(\frac{-y}{\sqrt{a^2 - x^2 - y^2}}\right)^2}\,dA$$

$$= 2k\iint_R \frac{a}{\sqrt{a^2 - x^2 - y^2}}\,dA$$

$$= 2ka\int_0^{2\pi}\int_0^a \frac{r}{\sqrt{a^2 - r^2}}\,dr\,d\theta$$

$$= 2ka\left[-\sqrt{a^2 - r^2}\right]_0^a (2\pi) = 4\pi ka^2$$

$$I_z = 2\iint_S k(x^2 + y^2)\,dS = 2k\iint_R (x^2 + y^2)\frac{a}{\sqrt{a^2 - x^2 - y^2}}\,dA$$

$$= 2ka\int_0^{2\pi}\int_0^a \frac{r^3}{\sqrt{a^2 - r^2}}\,dr\,d\theta \quad \text{(use integration by parts)}$$

$$= 2ka\left[-r^2\sqrt{a^2 - r^2} - \frac{2}{3}(a^2 - r^2)^{3/2}\right]_0^a (2\pi)$$

$$= 2ka\left(\frac{2}{3}a^3\right)(2\pi)$$

$$= \frac{2}{3}a^2(4\pi ka^2) = \frac{2}{3}a^2 m$$

Let $u = r^2$, $dv = r(a^2 - r^2)^{-1/2}\,dr$, $du = 2r\,dr$, $v = -\sqrt{a^2 - r^2}$.

27. $x^2 + y^2 = a^2$, $0 \le z \le h$

$\rho(x, y, z) = 1$

$y = \pm\sqrt{a^2 - x^2}$

Project the solid onto the xz-plane.

$$I_z = 4\iint_S (x^2 + y^2)(1)\,dS = 4\int_0^h\int_0^a [x^2 + (a^2 - x^2)]\sqrt{1 + \left(\frac{-x}{\sqrt{a^2 - x^2}}\right)^2 + (0)^2}\,dx\,dz$$

$$= 4a^3\int_0^h\int_0^a \frac{1}{\sqrt{a^2 - x^2}}\,dx\,dz$$

$$= 4a^3\int_0^h \arcsin\frac{x}{a}\Big]_0^a\,dz$$

$$= 4a^3\left(\frac{\pi}{2}\right)(h)$$

$$= 2\pi a^3 h$$

28. $z = x^2 + y^2, \quad 0 \leq z \leq h$

Project the solid onto the xy-plane.

$$I_z = \int\!\!\int_S (x^2 + y^2)(1)\, dS = \int_{-\sqrt{h}}^{\sqrt{h}} \int_{-\sqrt{h-x^2}}^{\sqrt{h-x^2}} (x^2 + y^2)\sqrt{1 + 4x^2 + 4y^2}\, dy\, dx$$

$$= \int_0^{2\pi} \int_0^{\sqrt{h}} r^2 \sqrt{1 + 4r^2}\, r\, dr\, d\theta$$

$$= 2\pi \left[\frac{h}{12}(1 + 4h)^{3/2} - \frac{1}{120}(1 + 4h)^{5/2} \right] + \frac{2\pi}{120}$$

$$= \frac{(1 + 4h)^{3/2}\pi}{60}[10h - (1 + 4h)] + \frac{\pi}{60}$$

$$= \frac{\pi}{60}[(1 + 4h)^{3/2}(6h - 1) + 1]$$

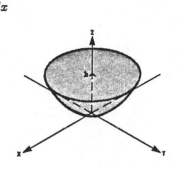

29. $S : z = 16 - x^2 - y^2, \quad z \geq 0$

$\mathbf{F}(x, y, z) = 0.5z\mathbf{k}$

$$\int\!\!\int_S \rho \mathbf{F} \cdot \mathbf{N}\, dS = \int\!\!\int_R \rho \mathbf{F} \cdot (-g_x(x, y)\mathbf{i} - g_y(x, y)\mathbf{j} + \mathbf{k})\, dA$$

$$= \int\!\!\int_R 0.5\rho z\mathbf{k} \cdot (2x\mathbf{i} + 2y\mathbf{j} + \mathbf{k})\, dA$$

$$= \int\!\!\int_R 0.5\rho z\, dA$$

$$= \int\!\!\int_R 0.5\rho(16 - x^2 - y^2)\, dA$$

$$= 0.5\rho \int_0^{2\pi} \int_0^4 (16 - r^2)r\, dr\, d\theta$$

$$= 0.5\rho \int_0^{2\pi} 64\, d\theta = 64\pi\rho$$

30. $S : z = \sqrt{16 - x^2 - y^2}$

$\mathbf{F}(x, y, z) = 0.5z\mathbf{k}$

$$\int\!\!\int_S \rho \mathbf{F} \cdot \mathbf{N}\, dS = \int\!\!\int_R \rho \mathbf{F} \cdot (-g_x(x, y)\mathbf{i} - g_y(x, y)\mathbf{j} + \mathbf{k})\, dA$$

$$= \int\!\!\int_R 0.5\rho z\mathbf{k} \cdot \left[\frac{x}{\sqrt{16 - x^2 - y^2}}\mathbf{i} + \frac{y}{\sqrt{16 - x^2 - y^2}}\mathbf{j} + \mathbf{k} \right] dA$$

$$= \int\!\!\int_R 0.5\rho z\, dA$$

$$= \int\!\!\int_R 0.5\rho\sqrt{16 - x^2 - y^2}\, dA$$

$$= 0.5\rho \int_0^{2\pi} \int_0^4 \sqrt{16 - r^2}\, r\, dr\, d\theta$$

$$= 0.5\rho \int_0^{2\pi} \frac{64}{3}\, d\theta$$

$$= \frac{64\pi\rho}{3}$$

31. This surface is not orientable since a normal vector at a point P on the surface will point in the opposite direction if it is moved around the mobius strip one time.

32. Orientable

33. $S: z = 10 - x^2 - y^2$, $0 \le x \le 2$, $0 \le y \le 2$

$$\iint_S (x^2 - 2xy)\, dS = \int_0^2 \int_0^2 (x^2 - 2xy)\sqrt{1 + 4x^2 + 4y^2}\, dy\, dx \approx -11.47$$

34. $S: z = \cos x$, $0 \le x \le \dfrac{\pi}{2}$, $0 \le y \le x$

$$\iint_S (x^2 - 2xy)\, dS = \int_0^{\pi/2} \int_0^x (x^2 - 2xy)\sqrt{1 + \sin^2 x}\, dy\, dx = \int_0^{\pi/2} (x^3 - x^3)\sqrt{1 + \sin^2 x}\, dx = 0$$

35. $\mathbf{E} = yz\mathbf{i} + xz\,\mathbf{j} + xy\mathbf{k}$

$S: z = \sqrt{1 - x^2 - y^2}$

$$\iint_S \mathbf{E} \cdot \mathbf{N}\, dS = \iint_R \mathbf{E} \cdot (-g_x(x,\ y)\mathbf{i} - g_y(x,\ y)\,\mathbf{j} + \mathbf{k})\, dA$$

$$= \iint_R (yz\mathbf{i} + xz\,\mathbf{j} + xy\mathbf{k}) \cdot \left(\frac{x}{\sqrt{1 - x^2 - y^2}}\mathbf{i} + \frac{y}{\sqrt{1 - x^2 - y^2}}\mathbf{j} + \mathbf{k} \right) dA$$

$$= \iint_R \left(\frac{2xyz}{\sqrt{1 - x^2 - y^2}} + xy \right) dA = \iint_R 3xy\, dA = \int_{-1}^1 \int_{-\sqrt{1-x^2}}^{\sqrt{1-x^2}} 3xy\, dy\, dx = 0$$

Section 16.6   Divergence Theorem

1. (a) There are six surfaces to the cube, each with $dS = \sqrt{1}\, dA$.

$z = 0$, $\mathbf{N} = -\mathbf{k}$, $\mathbf{F} \cdot \mathbf{N} = -z^2$, $\displaystyle\int_{S_1}\!\!\int 0\, dA = 0$

$z = a$, $\mathbf{N} = \mathbf{k}$, $\mathbf{F} \cdot \mathbf{N} = z^2$, $\displaystyle\int_{S_2}\!\!\int a^2\, dA = \int_0^a \int_0^a a^2\, dx\, dy = a^4$

$x = 0$, $\mathbf{N} = -\mathbf{i}$, $\mathbf{F} \cdot \mathbf{N} = -2x$, $\displaystyle\int_{S_3}\!\!\int 0\, dA = 0$

$x = a$, $\mathbf{N} = \mathbf{i}$, $\mathbf{F} \cdot \mathbf{N} = 2x$, $\displaystyle\int_{S_4}\!\!\int 2a\, dy\, dz = \int_0^a \int_0^a 2a\, dy\, dz = 2a^3$

$y = 0$, $\mathbf{N} = -\mathbf{j}$, $\mathbf{F} \cdot \mathbf{N} = 2y$, $\displaystyle\int_{S_5}\!\!\int 0\, dA = 0$

$y = a$, $\mathbf{N} = \mathbf{j}$, $\mathbf{F} \cdot \mathbf{N} = -2y$, $\displaystyle\int_{S_6}\!\!\int -2a\, dA = \int_0^a \int_0^a -2a\, dz\, dx = -2a^3$

Therefore, $\displaystyle\iint_S \mathbf{F} \cdot \mathbf{N}\, dS = a^4 + 2a^3 - 2a^3 = a^4$.

(b) Since div $\mathbf{F} = 2z$, the Divergence Theorem yields

$$\iiint_Q \operatorname{div} \mathbf{F}\, dV = \int_0^a \int_0^a \int_0^a 2z\, dz\, dy\, dx = \int_0^a \int_0^a a^2\, dy\, dx = a^4.$$

2. (a) There are three surfaces to the cylinder.

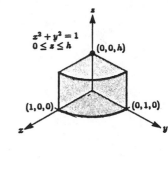

$z = 0$, $\quad \mathbf{N} = -\mathbf{k}$, $\quad \mathbf{F} \cdot \mathbf{N} = -z^2$

$$\int_{S_1}\!\!\int 0 \, dS = 0$$

$z = h$, $\quad \mathbf{N} = \mathbf{k}$, $\quad \mathbf{F} \cdot \mathbf{N} = z^2$

$$\int_{S_2}\!\!\int h^2 \, dS = h^2(\text{area of circle}) = \pi h^2$$

$x^2 + y^2 = 1$, $\quad \mathbf{N} = \dfrac{2x\mathbf{i} + 2y\mathbf{j}}{2\sqrt{x^2 + y^2}} = x\mathbf{i} + y\mathbf{j}$, $\quad \mathbf{F} \cdot \mathbf{N} = 2x^2 - 2y^2$, $\quad dS = 2 \, dA$

$$\int_{S_3}\!\!\int \mathbf{F} \cdot \mathbf{N} \, dS = \int_{-1}^{1}\int_{-\sqrt{1-x^2}}^{\sqrt{1-x^2}} (4x^2 - 4y^2) \, dy \, dx$$

$$= \int_0^{2\pi}\int_0^1 (4\cos^2\theta - 4\sin^2\theta) r \, dr \, d\theta$$

$$= 4\int_0^{2\pi}\int_0^1 (\cos 2\theta) r \, dr \, d\theta = 2\int_0^{2\pi} \cos 2\theta, \; d\theta = \sin 2\theta \Big]_0^{2\pi} = 0$$

Therefore, $\displaystyle\int_S\!\!\int \mathbf{F} \cdot \mathbf{N} \, dS = 0 + \pi h^2 + 0 = \pi h^2.$

(b) Since div $\mathbf{F} = 2z$, we have

$$\iiint_Q 2z \, dV = \int_{-1}^{1}\int_{-\sqrt{1-x^2}}^{\sqrt{1-x^2}}\int_0^h 2z \, dz \, dy \, dx = h^2\int_{-1}^{1}\int_{-\sqrt{1-x^2}}^{\sqrt{1-x^2}} dy \, dx = h^2(\text{area of circle}) = \pi h^2.$$

3. (a) There are four surfaces to this solid.

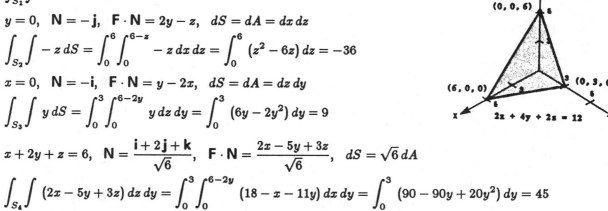

$z = 0$, $\quad \mathbf{N} = -\mathbf{k}$, $\quad \mathbf{F} \cdot \mathbf{N} = -z$

$$\int_{S_1}\!\!\int 0 \, dS = 0$$

$y = 0$, $\quad \mathbf{N} = -\mathbf{j}$, $\quad \mathbf{F} \cdot \mathbf{N} = 2y - z$, $\quad dS = dA = dx \, dz$

$$\int_{S_2}\!\!\int -z \, dS = \int_0^6\int_0^{6-z} -z \, dx \, dz = \int_0^6 (z^2 - 6z) \, dz = -36$$

$x = 0$, $\quad \mathbf{N} = -\mathbf{i}$, $\quad \mathbf{F} \cdot \mathbf{N} = y - 2x$, $\quad dS = dA = dz \, dy$

$$\int_{S_3}\!\!\int y \, dS = \int_0^3\int_0^{6-2y} y \, dz \, dy = \int_0^3 (6y - 2y^2) \, dy = 9$$

$x + 2y + z = 6$, $\quad \mathbf{N} = \dfrac{\mathbf{i} + 2\mathbf{j} + \mathbf{k}}{\sqrt{6}}$, $\quad \mathbf{F} \cdot \mathbf{N} = \dfrac{2x - 5y + 3z}{\sqrt{6}}$, $\quad dS = \sqrt{6} \, dA$

$$\int_{S_4}\!\!\int (2x - 5y + 3z) \, dz \, dy = \int_0^3\int_0^{6-2y} (18 - x - 11y) \, dx \, dy = \int_0^3 (90 - 90y + 20y^2) \, dy = 45$$

Therefore, $\displaystyle\int_S\!\!\int \mathbf{F} \cdot \mathbf{N} \, dS = 0 - 36 + 9 + 45 = 18.$

(b) Since div $\mathbf{F} = 1$, we have

$$\iiint_Q dV = (\text{volume of solid}) = \frac{1}{3}(\text{area of base}) \times (\text{height}) = \frac{1}{3}(9)(6) = 18$$

4. $\mathbf{F}(x,\ y,\ z) = xy\mathbf{i} + z\mathbf{j} + (x+y)\mathbf{k}$

S: surface bounded by the planes $y = 4$, $z = 4 - x$ and the coordinate planes

(a) There are five surfaces to this solid.

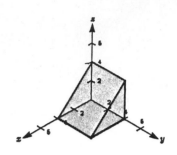

$z = 0$, $\mathbf{N} = -\mathbf{k}$, $\mathbf{F} \cdot \mathbf{N} = -(x+y)$

$$\int_{S_1}\!\!\int -(x+y)\,dS = \int_0^4 \int_0^4 -(x+y)\,dy\,dx = -\int_0^4 (4x+8)\,dx = -64$$

$y = 0$, $\mathbf{N} = -\mathbf{j}$, $\mathbf{F} \cdot \mathbf{N} = -z$

$$\int_{S_2}\!\!\int -z\,dS = \int_0^4 \int_0^{4-x} -z\,dz\,dx = -\int_0^4 \frac{(4-x)^2}{2}\,dx = -\frac{32}{3}$$

$y = 4$, $\mathbf{N} = \mathbf{j}$, $\mathbf{F} \cdot \mathbf{N} = z$

$$\int_{S_3}\!\!\int z\,dS = \int_0^4 \int_0^{4-x} z\,dz\,dx = \int_0^4 \frac{(4-x)^2}{2}\,dx = \frac{32}{3}$$

$x = 0$, $\mathbf{N} = -\mathbf{i}$, $\mathbf{F} \cdot \mathbf{N} = -xy$

$$\int_{S_4}\!\!\int -xy\,dS = \int_0^4 \int_0^4 0\,dS = 0$$

$x + z = 4$, $\mathbf{N} = \dfrac{\mathbf{i}+\mathbf{k}}{\sqrt{2}}$, $\mathbf{F} \cdot \mathbf{N} = \dfrac{1}{\sqrt{2}}[xy + x + y]$, $dS = \sqrt{2}\,dA$

$$\int_{S_5}\!\!\int \frac{1}{\sqrt{2}}[xy + x + y]\sqrt{2}\,dA = \int_0^4 \int_0^4 (xy + x + y)\,dy\,dx = 128$$

Therefore, $\displaystyle\int_S\!\!\int \mathbf{F} \cdot \mathbf{N}\,dS = -64 - \frac{32}{3} + \frac{32}{3} + 0 + 128 = 64.$

(b) Since div $\mathbf{F} = y$, we have $\displaystyle\int\!\!\int\!\!\int_Q \text{div } \mathbf{F}\,dV = \int_0^4 \int_0^4 \int_0^{4-x} y\,dz\,dy\,dx = 64.$

5. Since div $\mathbf{F} = 2x + 2y + 2z$, we have

$$\int\!\!\int\!\!\int_Q \text{div } \mathbf{F}\,dV = \int_0^a \int_0^a \int_0^a (2x + 2y + 2z)\,dz\,dy\,dx$$

$$= \int_0^a \int_0^a (2ax + 2ay + a^2)\,dy\,dx = \int_0^a (2a^2 x + 2a^3)\,dx = \left[a^2 x^2 + 2a^3 x\right]_0^a = 3a^4.$$

6. Since div $\mathbf{F} = 2xz - 1 + xy$

$$\int\!\!\int\!\!\int_Q \text{div } \mathbf{F}\,dV = \int_0^a \int_0^a \int_0^a (2xz - 1 + xy)\,dz\,dy\,dx$$

$$= \int_0^a \int_0^a (a^2 x - a + axy)\,dy\,dx$$

$$= \int_0^a \left(a^3 x - a^2 + \frac{a^2}{2}x\right)dx = \frac{a^5}{2} - a^3 + \frac{a^4}{4} = \frac{a^3}{4}(2a^2 + a - 4).$$

7. Since div $\mathbf{F} = 2x - 2x + 2xyz = 2xyz$

$$\iiint_Q \text{div } \mathbf{F} \, dV = \iiint_Q 2xyz \, dV$$

$$= \int_0^a \int_0^{2\pi} \int_0^{\pi/2} 2(\rho \sin \phi \cos \theta)(\rho \sin \phi \sin \theta)(\rho \cos \phi)\rho^2 \sin \phi \, d\phi \, d\theta \, d\rho$$

$$= \int_0^a \int_0^{2\pi} \int_0^{\pi/2} 2\rho^5 (\sin \theta \cos \theta)(\sin^3 \phi \cos \phi) \, d\phi \, d\theta \, d\rho$$

$$= \int_0^a \int_0^{2\pi} \frac{1}{2}\rho^5 \sin \theta \cos \theta \, d\theta \, d\rho$$

$$= \int_0^a \left(\frac{\rho^5}{2}\right) \frac{\sin^2 \theta}{2}\bigg]_0^{2\pi} d\rho = 0.$$

8. Since div $\mathbf{F} = y + z - y = z$, we have

$$\iiint_Q \text{div } \mathbf{F} \, dV = \int_{-a}^a \int_{-\sqrt{a^2-x^2}}^{\sqrt{a^2-x^2}} \int_0^{\sqrt{a^2-x^2-y^2}} z \, dz \, dy \, dx$$

$$= \int_0^{2\pi} \int_0^a \int_0^{\sqrt{a^2-r^2}} zr \, dz \, dr \, d\theta$$

$$= \int_0^{2\pi} \int_0^a \left[\frac{a^2 r}{2} - \frac{r^3}{2}\right] dr \, d\theta = \int_0^{2\pi} \left[\frac{a^2 r^2}{4} - \frac{r^4}{8}\right]_0^a d\theta = \int_0^{2\pi} \frac{a^4}{8} \, d\theta = \frac{\pi a^4}{4}.$$

9. Since div $\mathbf{F} = 3$, we have $\displaystyle\iiint_Q 3 \, dV = 3(\text{volume of sphere}) = 3\left[\frac{4}{3}\pi(2)^3\right] = 32\pi.$

10. Since div $\mathbf{F} = xz$, we have

$$\iiint_Q xz \, dV = \int_0^4 \int_{-3}^3 \int_{-\sqrt{9-y^2}}^{\sqrt{9-y^2}} xz \, dx \, dy \, dz = \int_0^4 \int_{-3}^3 \frac{z}{2}(0) \, dy \, dz = 0.$$

11. Since div $\mathbf{F} = 1 + 2y - 1 = 2y$, we have

$$\iiint_Q 2y \, dV = \int_0^4 \int_{-3}^3 \int_{-\sqrt{9-y^2}}^{\sqrt{9-y^2}} 2y \, dx \, dy \, dz = \int_0^4 \int_{-3}^3 4y\sqrt{9-y^2} \, dy \, dz = \int_0^4 -\frac{4}{3}(9-y^2)^{3/2}\bigg]_{-3}^3 dz = 0.$$

12. Since div $\mathbf{F} = y^2 + x^2 + e^z$, we have

$$\iiint\limits_{Q} (x^2 + y^2 + e^z)\, dV = \int_0^2 \int_{-\sqrt{4-x^2}}^{\sqrt{4-x^2}} \int_{\sqrt{x^2+y^2}}^4 (x^2 + y^2 + e^z)\, dz\, dy\, dx$$

$$= \int_0^{2\pi} \int_0^2 \int_r^4 (r^2 + e^z) r\, dz\, dr\, d\theta$$

$$= \int_0^{2\pi} \int_0^2 (4r^3 + re^4 - r^4 - re^r)\, dr\, d\theta$$

$$= \int_0^{2\pi} \left[r^4 + \frac{r^2}{2}e^4 - \frac{r^5}{5} - re^r + e^r \right]_0^2 d\theta$$

$$= \left[16 + 2e^4 - \frac{32}{5} - 2e^2 + e^2 - 1 \right](2\pi)$$

$$= \left[\frac{43}{5} + 2e^4 - e^2 \right](2\pi)$$

$$= \frac{2\pi}{5}(10e^4 - 5e^2 + 43).$$

13. Since div $\mathbf{F} = 3x^2 + x^2 + 0 = 4x^2$, we have

$$\iiint\limits_{Q} 4x^2\, dV = \int_0^6 \int_0^4 \int_0^{4-y} 4x^2\, dz\, dy\, dx = \int_0^6 \int_0^4 4x^2(4-y)\, dy\, dx = \int_0^6 32x^2\, dx = 2304.$$

14. Since div $\mathbf{F} = e^z + e^z + e^z = 3e^z$, we have

$$\iiint\limits_{Q} 3e^z\, dV = \int_0^6 \int_0^4 \int_0^{4-y} 3e^z\, dz\, dy\, dx = \int_0^6 \int_0^4 3[e^{4-y} - 1]\, dy\, dx = \int_0^6 3(e^4 - 5)\, dx = 18(e^4 - 5).$$

15. Using the triple integral to find volume, we need $\mathbf{F}$ so that div $\mathbf{F} = \dfrac{\partial M}{\partial x} + \dfrac{\partial N}{\partial y} + \dfrac{\partial P}{\partial z} = 1$. Hence, we could have $\mathbf{F} = x\mathbf{i}$, $\mathbf{F} = y\mathbf{j}$, or $\mathbf{F} = z\mathbf{k}$.

For $dA = dy\, dz$ consider $\mathbf{F} = x\mathbf{i}$, $x = f(y,\, z)$, then $\mathbf{N} = \dfrac{\mathbf{i} + f_y\mathbf{j} + f_z\mathbf{k}}{\sqrt{1 + f_y{}^2 + f_z{}^2}}$ and $dS = \sqrt{1 + f_y{}^2 + f_z{}^2}\, dy\, dz$.

For $dA = dz\, dx$ consider $\mathbf{F} = y\mathbf{j}$, $y = f(x,\, z)$, then $\mathbf{N} = \dfrac{f_x\mathbf{i} + \mathbf{j} + f_z\mathbf{k}}{\sqrt{1 + f_x{}^2 + f_z{}^2}}$ and $dS = \sqrt{1 + f_x{}^2 + f_z{}^2}\, dz\, dx$.

For $dA = dx\, dy$ consider $\mathbf{F} = z\mathbf{k}$, $z = f(x,\, y)$, then $\mathbf{N} = \dfrac{f_x\mathbf{i} + f_y\mathbf{j} + \mathbf{k}}{\sqrt{1 + f_x{}^2 + f_y{}^2}}$ and $dS = \sqrt{1 + f_x{}^2 + f_y{}^2}\, dx\, dy$.

Correspondingly, we then have $V = \iint\limits_{S} \mathbf{F} \cdot \mathbf{N}\, dS = \iint\limits_{S} x\, dy\, dz = \iint\limits_{S} y\, dz\, dx = \iint\limits_{S} z\, dx\, dy$.

16. $\displaystyle\int_0^a \int_0^a x\, dy\, dz = \int_0^a \int_0^a a\, dy\, dz = \int_0^a a^2\, dz = a^3$

Similarly, $\displaystyle\int_0^a \int_0^a y\, dz\, dx = \int_0^a \int_0^a z\, dx\, dy = a^3$.

17. Using the Divergence Theorem, we have

$$\iint_S \text{curl } \mathbf{F} \cdot \mathbf{N} \, dS = \iiint_Q \text{div (curl } \mathbf{F}) \, dV$$

$$\text{curl } \mathbf{F}(x, y, z) = \begin{vmatrix} \mathbf{i} & \mathbf{j} & \mathbf{k} \\ \dfrac{\partial}{\partial x} & \dfrac{\partial}{\partial y} & \dfrac{\partial}{\partial z} \\ 4xy + z^2 & 2x^2 + 6yz & 2xz \end{vmatrix} = -6y\mathbf{i} - (2z - 2z)\mathbf{j} + (4x - 4x)\mathbf{k} = -6y\mathbf{i}$$

$$\text{div (curl } \mathbf{F}) = 0.$$

Therefore, $\displaystyle\iiint_Q \text{div (curl } \mathbf{F}) \, dV = 0.$

18. Using the Divergence Theorem, we have

$$\iint_S \text{curl } \mathbf{F} \cdot \mathbf{N} \, dS = \iiint_Q \text{div (curl } \mathbf{F}) \, dV$$

$$\text{curl } \mathbf{F}(x, y, z) = \begin{vmatrix} \mathbf{i} & \mathbf{j} & \mathbf{k} \\ \dfrac{\partial}{\partial x} & \dfrac{\partial}{\partial y} & \dfrac{\partial}{\partial z} \\ xy \cos z & yz \sin x & xyz \end{vmatrix} = (xz - y \sin x)\mathbf{i} - (yz + xy \sin z)\mathbf{j} + (yz \cos x - x \cos z)\mathbf{k}.$$

Now, div **curl** $\mathbf{F}(x, y, z) = (z - y \cos x) - (z + x \sin z) + (y \cos x + x \sin z) = 0.$

Therefore, $\displaystyle\iint_S \text{curl } \mathbf{F} \cdot \mathbf{N} \, dS = \iiint_Q \text{div (curl } \mathbf{F}) \, dV = 0.$

19. Using the Divergence Theorem, we have $\displaystyle\iint_S \text{curl } \mathbf{F} \cdot \mathbf{N} \, dS = \iiint_Q \text{div (curl } \mathbf{F}) \, dV.$ Let

$$\mathbf{F}(x, y, z) = M\mathbf{i} + N\mathbf{j} + P\mathbf{k}$$

$$\text{curl } \mathbf{F} = \left(\frac{\partial P}{\partial y} - \frac{\partial N}{\partial z} \right)\mathbf{i} - \left(\frac{\partial P}{\partial x} - \frac{\partial M}{\partial z} \right)\mathbf{j} + \left(\frac{\partial N}{\partial x} - \frac{\partial M}{\partial y} \right)\mathbf{k}$$

$$\text{div (curl } \mathbf{F}) = \frac{\partial^2 P}{\partial x \partial y} - \frac{\partial^2 N}{\partial x \partial z} - \frac{\partial^2 P}{\partial y \partial x} + \frac{\partial^2 M}{\partial y \partial z} + \frac{\partial^2 N}{\partial z \partial x} - \frac{\partial^2 M}{\partial z \partial y} = 0.$$

Therefore, $\displaystyle\iint_S \text{curl } \mathbf{F} \cdot \mathbf{N} \, dS = \iiint_Q 0 \, dV = 0.$

20. If $\mathbf{F}(x, y, z) = a_1\mathbf{i} + a_2\mathbf{j} + a_3\mathbf{k}$, then div $\mathbf{F} = 0.$

Therefore, $\displaystyle\iint_S \mathbf{F} \cdot \mathbf{N} \, dS = \iiint_Q \text{div } \mathbf{F} \, dV = \iiint_Q 0 \, dV = 0.$

21. If $\mathbf{F}(x, y, z) = x\mathbf{i} + y\mathbf{j} + z\mathbf{k}$, then div $\mathbf{F} = 3.$

$$\iint_S \mathbf{F} \cdot \mathbf{N} \, dS = \iiint_Q \text{div } \mathbf{F} \, dV = \iiint_Q 3 \, dV = 3V.$$

22. If $\mathbf{F}(x, y, z) = x\mathbf{i} + y\mathbf{j} + z\mathbf{k}$, then div $\mathbf{F} = 3.$

$$\frac{1}{\|\mathbf{F}\|} \iint_S \mathbf{F} \cdot \mathbf{N} \, dS = \frac{1}{\|\mathbf{F}\|} \iiint_Q \text{div } \mathbf{F} \, dV = \frac{1}{\|\mathbf{F}\|} \iiint_Q 3 \, dV = \frac{3}{\|\mathbf{F}\|} \iiint_Q dV$$

23. $\displaystyle\iint_S f D_N g\, dS = \iint_S f\nabla g \cdot \mathbf{N}\, dS$

$$= \iiint_Q \text{div } (f\nabla g)\, dV = \iiint_Q (f \text{ div } \nabla g + \nabla f \cdot \nabla g)\, dV = \iiint_Q (f\nabla^2 g + \nabla f \cdot \nabla g)\, dV$$

24. $\displaystyle\iint_S (f D_N g - g D_N f)\, dS = \iint_S f D_N g\, dS - \iint_S g D_N f\, dS$

$$= \iiint_Q (f\nabla^2 g + \nabla f \cdot \nabla g)\, dV - \iiint_Q (g\nabla^2 f + \nabla g \cdot \nabla f)\, dV$$

$$= \iiint_Q (f\nabla^2 g - g\nabla^2 f)\, dV$$

Section 16.7   Stoke's Theorem

1. $\mathbf{F}(x,\, y,\, z) = (2y - z)\mathbf{i} + xyz\,\mathbf{j} + e^z\mathbf{k}$

$$\text{curl } \mathbf{F} = \begin{vmatrix} \mathbf{i} & \mathbf{j} & \mathbf{k} \\ \dfrac{\partial}{\partial x} & \dfrac{\partial}{\partial y} & \dfrac{\partial}{\partial z} \\ 2y - z & xyz & e^z \end{vmatrix}$$

$$= -xy\mathbf{i} - \mathbf{j} + (yz - 2)\mathbf{k}$$

2. $\mathbf{F}(x,\, y,\, z) = z^2\mathbf{i} + y^2\mathbf{j} + x^2\mathbf{k}$

$$\text{curl } \mathbf{F} = \begin{vmatrix} \mathbf{i} & \mathbf{j} & \mathbf{k} \\ \dfrac{\partial}{\partial x} & \dfrac{\partial}{\partial y} & \dfrac{\partial}{\partial z} \\ z^2 & y^2 & x^2 \end{vmatrix} = (2z - 2x)\mathbf{j}$$

3. $\mathbf{F}(x,\, y,\, z) = 2z\mathbf{i} - 4x^2\mathbf{j} + \arctan x\mathbf{k}$

$$\text{curl } \mathbf{F} = \begin{vmatrix} \mathbf{i} & \mathbf{j} & \mathbf{k} \\ \dfrac{\partial}{\partial x} & \dfrac{\partial}{\partial y} & \dfrac{\partial}{\partial z} \\ 2z & -4x^2 & \arctan x \end{vmatrix} = \left(2 - \dfrac{1}{1 + x^2}\right)\mathbf{j} - 8x\mathbf{k}$$

4. $\mathbf{F}(x,\, y,\, z) = x\sin y\mathbf{i} - y\cos x\mathbf{j} + yz^2\mathbf{k}$

$$\text{curl } \mathbf{F} = \begin{vmatrix} \mathbf{i} & \mathbf{j} & \mathbf{k} \\ \dfrac{\partial}{\partial x} & \dfrac{\partial}{\partial y} & \dfrac{\partial}{\partial z} \\ x\sin y & -y\cos x & yz^2 \end{vmatrix} = z^2\mathbf{i} + (y\sin x - x\cos y)\mathbf{k}$$

5. $\mathbf{F}(x,\, y,\, z) = e^{x^2 + y^2}\mathbf{i} + e^{y^2 + z^2}\mathbf{j} + xyz\mathbf{k}$

$$\text{curl } \mathbf{F} = \begin{vmatrix} \mathbf{i} & \mathbf{j} & \mathbf{k} \\ \dfrac{\partial}{\partial x} & \dfrac{\partial}{\partial y} & \dfrac{\partial}{\partial z} \\ e^{x^2 + y^2} & e^{y^2 + z^2} & xyz \end{vmatrix} = (xz - 2ze^{y^2 + z^2})\mathbf{i} - yz\mathbf{j} - 2ye^{x^2 + y^2}\mathbf{k} = z(x - 2e^{y^2 + z^2})\mathbf{i} - yz\mathbf{j} - 2ye^{x^2 + y^2}\mathbf{k}$$

6. $\mathbf{F}(x,\, y,\, z) = \arcsin y\mathbf{i} + \sqrt{1 - x^2}\,\mathbf{j} + y^2\mathbf{k}$

$$\text{curl } \mathbf{F} = \begin{vmatrix} \mathbf{i} & \mathbf{j} & \mathbf{k} \\ \dfrac{\partial}{\partial x} & \dfrac{\partial}{\partial y} & \dfrac{\partial}{\partial z} \\ \arcsin y & \sqrt{1 - x^2} & y^2 \end{vmatrix} = 2y\mathbf{i} + \left[\dfrac{-x}{\sqrt{1 - x^2}} - \dfrac{1}{\sqrt{1 - y^2}}\right]\mathbf{k} = 2y\mathbf{i} - \left[\dfrac{x}{\sqrt{1 - x^2}} + \dfrac{1}{\sqrt{1 - y^2}}\right]\mathbf{k}$$

7. In this case, $M = -y + z$, $N = x - z$, $P = x - y$ and C is the circle $x^2 + y^2 = 1$, $z = 0$, $dz = 0$.

(a) $\displaystyle\int_C \mathbf{F} \cdot d\mathbf{r} = \int_C -y\,dx + x\,dy$

Letting $x = \cos t$, $y = \sin t$, we have $dx = -\sin t\,dt$, $dy = \cos t\,dt$ and

$$\int_C -y\,dx + x\,dy = \int_0^{2\pi} (\sin^2 t + \cos^2 t)\,dt = 2\pi.$$

(b) Consider $F(x, y, z) = x^2 + y^2 + z^2 - 1$. Then $\mathbf{N} = \dfrac{\nabla F}{\|\nabla F\|} = \dfrac{2x\mathbf{i} + 2y\mathbf{j} + 2z\mathbf{k}}{2\sqrt{x^2 + y^2 + z^2}} = x\mathbf{i} + y\mathbf{j} + z\mathbf{k}.$

Since $z^2 = 1 - x^2 - y^2$, $z_x = \dfrac{-2x}{2z} = \dfrac{-x}{z}$, and $z_y = \dfrac{-y}{z}$, $dS = \sqrt{1 + \dfrac{x^2}{z^2} + \dfrac{y^2}{z^2}}\,dA = \dfrac{1}{z}\,dA.$

Now, since **curl F** $= 2\mathbf{k}$, we have

$$\iint_S (\text{curl }\mathbf{F}) \cdot \mathbf{N}\,dS = \iint_R 2z\left(\frac{1}{z}\right) dA = \iint_R 2\,dA = 2(\text{area of circle of radius } 1) = 2\pi.$$

8. In this case C is the circle $x^2 + y^2 = 4$, $z = 0$, $dz = 0$.

(a) $\displaystyle\int_C \mathbf{F} \cdot d\mathbf{r} = \int_C -y\,dx + x\,dy$

Let $x = 2\cos t$, $y = 2\sin t$, then $dx = -2\sin t\,dt$, $dy = 2\cos t\,dt$, and $\displaystyle\int_C -y\,dx + x\,dy = \int_0^{2\pi} 4\,dt = 8\pi.$

(b) $F(x, y, z) = z + x^2 + y^2 - 4$, $\mathbf{N} = \dfrac{\nabla F}{\|\nabla F\|} = \dfrac{2x\mathbf{i} + 2y\mathbf{j} + \mathbf{k}}{\sqrt{1 + 4x^2 + 4y^2}}$, $dS = \sqrt{1 + 4x^2 + 4y^2}\,dA$

curl F $= 2\mathbf{k}$, therefore

$$\iint_S (\text{curl }\mathbf{F}) \cdot \mathbf{N}\,dS = \iint_R 2\,dA = \int_{-2}^2 \int_{-\sqrt{4-x^2}}^{\sqrt{4-x^2}} 2\,dy\,dx = 2\int_{-2}^2 2\sqrt{4 - x^2}\,dx$$

$$= 4\int_{-2}^2 \sqrt{4 - x^2}\,dx = 2\left[x\sqrt{4 - x^2} + 4\arcsin\frac{x}{2}\right]_{-2}^2 = 8\pi.$$

9. (a) From the accompanying figure we see that for

$C_1:\ z = 0,\ dz = 0$

$C_2:\ x = 0,\ dx = 0$

$C_3:\ y = 0,\ dy = 0.$

Hence, $\displaystyle\int_C \mathbf{F} \cdot d\mathbf{r} = \int_C xyz\,dx + y\,dy + z\,dz$

$$= \int_{C_1} y\,dy + \int_{C_2} y\,dy + z\,dz + \int_{C_3} z\,dz$$

$$= \int_0^3 y\,dy + \int_3^0 y\,dy + \int_0^6 z\,dz + \int_6^0 z\,dz = 0.$$

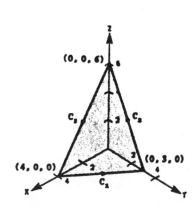

(b) **curl F** $= xy\mathbf{j} - xz\mathbf{k}$

Considering $F(x, y, z) = 3x + 4y + 2z - 12$, then $\mathbf{N} = \dfrac{\nabla F}{\|\nabla F\|} = \dfrac{3\mathbf{i} + 4\mathbf{j} + 2\mathbf{k}}{\sqrt{29}}$ and $dS = \sqrt{29}\,dA$. Thus,

$$\iint_S (\text{curl }\mathbf{F}) \cdot \mathbf{N}\,dS = \iint_R (4xy - 2xz)\,dy\,dx = \int_0^4 \int_0^{(-3x+12)/4} \left[4xy - 2x\left(6 - 2y - \frac{3}{2}x\right)\right] dy\,dx$$

$$= \int_0^4 \int_0^{(12-3x)/4} (8xy + 3x^2 - 12x)\,dy\,dx = \int_0^4 0\,dx = 0.$$

10. (a) From the accompanying figure we see that for

$C_1 : \ x = 0, \ \ z = 0, \ \ dx = dz = 0$

$C_2 : \ z = x^2, \ \ y = a, \ \ dy = 0, \ \ dz = 2x\,dx$

$C_3 : \ x = a, \ \ z = a, \ \ dx = dz = 0$

$C_4 : \ z = x^2, \ \ y = 0, \ \ dy = 0, \ \ dz = 2x\,dx.$

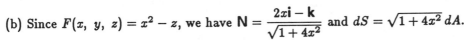

Oriented Surface S
$x^2 - z = 0$

Hence, $\displaystyle\int_C \mathbf{F} \cdot d\mathbf{r} = \int_C z^2\,dx + x^2\,dy + y^2\,dz$

$$= \int_{C_1} 0 + \int_{C_2} x^4\,dx + a^2(2x\,dx) + \int_{C_3} a^2\,dy + \int_{C_4} x^4\,dx$$

$$= \int_0^a x^4\,dx + \int_0^a 2a^2x\,dx + \int_a^0 a^2\,dy + \int_a^0 x^4\,dx = a^2x^2\Big]_0^a + a^2y\Big]_a^0 = a^4 - a^3 = a^3(a-1).$$

(b) Since $F(x, \ y, \ z) = x^2 - z$, we have $\mathbf{N} = \dfrac{2x\mathbf{i} - \mathbf{k}}{\sqrt{1 + 4x^2}}$ and $dS = \sqrt{1 + 4x^2}\,dA.$

Furthermore, $\mathbf{curl\ F} = 2y\mathbf{i} + 2z\mathbf{j} + 2x\mathbf{k}$. Therefore,

$$\iint_S (\mathbf{curl\ F}) \cdot \mathbf{N}\,dS = \int_0^a \int_0^a (4xy - 2x)\,dy\,dx = \int_0^a (2a^2x - 2ax)\,dx = \Big[a^2x^2 - ax^2\Big]_0^a = a^3(a-1).$$

11. Let $A = (0, \ 0, \ 0)$, $B = (1, \ 1, \ 1)$ and $C = (0, \ 2, \ 0)$. Then $\mathbf{U} = \overrightarrow{AB} = \mathbf{i} + \mathbf{j} + \mathbf{k}$ and $\mathbf{V} = \overrightarrow{AC} = 2\mathbf{j}$. Thus,

$\mathbf{N} = \dfrac{\mathbf{U} \times \mathbf{V}}{\|\mathbf{U} \times \mathbf{V}\|} = \dfrac{-2\mathbf{i} + 2\mathbf{k}}{2\sqrt{2}} = \dfrac{-\mathbf{i} + \mathbf{k}}{\sqrt{2}}$. Surface S has direction numbers $-1, \ 0, \ 1$, with equation $z - x = 0$ and

$dS = \sqrt{2}\,dA$. Since $\mathbf{curl\ F} = -3\mathbf{i} + \mathbf{j} - 2\mathbf{k}$, we have

$$\iint_S (\mathbf{curl\ F}) \cdot \mathbf{N}\,dS = \iint_R \frac{1}{\sqrt{2}}(\sqrt{2})\,dA = \iint_R dA = \text{(area of triangle with } a = 1, \ b = 2) = 1.$$

12. Let $A = (0, \ 0, \ 0)$, $B = (1, \ 1, \ 1)$, and $C = (0, \ 0, \ 2)$. Then $\mathbf{U} = \overrightarrow{AB} = \mathbf{i} + \mathbf{j} + \mathbf{k}$, $\mathbf{V} = \overrightarrow{AC} = 2\mathbf{k}$, and

$\mathbf{N} = \dfrac{\mathbf{U} \times \mathbf{V}}{\|\mathbf{U} \times \mathbf{V}\|} = \dfrac{2\mathbf{i} - 2\mathbf{j}}{2\sqrt{2}} = \dfrac{\mathbf{i} - \mathbf{j}}{\sqrt{2}}$. Hence, $F(x, \ y, \ z) = x - y$ and $dS = \sqrt{2}\,dA$. Since $\mathbf{curl\ F} = \dfrac{2x}{x^2 + y^2}\mathbf{k}$, we

have

$$\iint_S (\mathbf{curl\ F}) \cdot \mathbf{N}\,dS = \iint_R 0\,dS = 0.$$

13. $\mathbf{F}(x, \ y, \ z) = z^2\mathbf{i} + x^2\mathbf{j} + y^2\mathbf{k}$, $S : \ z = 4 - x^2 - y^2$, $0 \le z$

$\mathbf{curl\ F} = \begin{vmatrix} \mathbf{i} & \mathbf{j} & \mathbf{k} \\ \dfrac{\partial}{\partial x} & \dfrac{\partial}{\partial y} & \dfrac{\partial}{\partial z} \\ z^2 & x^2 & y^2 \end{vmatrix} = 2y\mathbf{i} + 2z\mathbf{j} + 2x\mathbf{k}$

$G(x, \ y, \ z) = x^2 + y^2 + z - 4$

$\nabla G(x, \ y, \ z) = 2x\mathbf{i} + 2y\mathbf{j} + \mathbf{k}$

$$\iint_S (\mathbf{curl\ F}) \cdot \mathbf{N}\,dS = \iint_R (4xy + 4yz + 2x)\,dA = \int_{-2}^2 \int_{-\sqrt{4-x^2}}^{\sqrt{4-x^2}} [4xy + 4y(4 - x^2 - y^2) + 2x]\,dy\,dx$$

$$= \int_{-2}^2 \int_{-\sqrt{4-x^2}}^{\sqrt{4-x^2}} [4xy + 16y - 4x^2y - 4y^3 + 2x]\,dy\,dx$$

$$= \int_{-2}^2 4x\sqrt{4 - x^2}\,dx = 0$$

14. $F(x,\ y,\ z) = 4xz\mathbf{i} + y\mathbf{j} + 4xy\mathbf{k}, \quad S: \ z = 4 - x^2 - y^2, \quad 0 \le z$

$$\text{curl } F = \begin{vmatrix} \mathbf{i} & \mathbf{j} & \mathbf{k} \\ \dfrac{\partial}{\partial x} & \dfrac{\partial}{\partial y} & \dfrac{\partial}{\partial z} \\ 4xz & y & 4xy \end{vmatrix} = 4x\mathbf{i} + (4x - 4y)\mathbf{j}$$

$G(x,\ y,\ z) = x^2 + y^2 + z - 4$

$\nabla G(x,\ y,\ z) = 2x\mathbf{i} + 2y\mathbf{j} + \mathbf{k}$

$$\iint_S (\text{curl } F) \cdot N\, dS = \iint_R \left[8x^2 + 2y(4x - 4y) \right] dA$$

$$= \int_{-2}^{2} \int_{-\sqrt{4-x^2}}^{\sqrt{4-x^2}} \left[8x^2 + 8xy - 8y^2 \right] dy\, dx$$

$$= \int_{-2}^{2} \left[8x^2 y + 4xy^2 - \frac{8}{3} y^3 \right]_{-\sqrt{4-x^2}}^{\sqrt{4-x^2}} dx$$

$$= \int_{-2}^{2} \left[16x^2 \sqrt{4 - x^2} - \frac{16}{3}(4 - x^2)\sqrt{4 - x^2} \right] dx$$

$$= \int_{-2}^{2} \left[\frac{64}{3} x^2 \sqrt{4 - x^2} - \frac{64}{3}\sqrt{4 - x^2} \right] dx$$

$$= \frac{64}{3} \int_{-2}^{2} \left[x^2 \sqrt{4 - x^2} - \sqrt{4 - x^2} \right] dx$$

$$= \frac{64}{3} \left[\frac{1}{8}\left(x(2x^2 - 4)\sqrt{4 - x^2} + 16 \arcsin \frac{x}{2} \right) - \frac{1}{2}\left(x\sqrt{4 - x^2} + 4 \arcsin \frac{x}{2} \right) \right]_{-2}^{2}$$

$$= \frac{64}{3} \left[\frac{1}{8}(8\pi) - \frac{1}{2}(2\pi) - \frac{1}{8}(-8\pi) + \frac{1}{2}(-2\pi) \right] = 0$$

15. $F(x,\ y,\ z) = z^2\mathbf{i} + y\mathbf{j} + xz\mathbf{k}, \quad S: \ z = \sqrt{4 - x^2 - y^2}$

$$\text{curl } F = \begin{vmatrix} \mathbf{i} & \mathbf{j} & \mathbf{k} \\ \dfrac{\partial}{\partial x} & \dfrac{\partial}{\partial y} & \dfrac{\partial}{\partial z} \\ z^2 & y & xz \end{vmatrix} = z\mathbf{j}$$

$G(x,\ y,\ z) = z - \sqrt{4 - x^2 - y^2}$

$$\nabla G(x,\ y,\ z) = \frac{x}{\sqrt{4 - x^2 - y^2}}\mathbf{i} + \frac{y}{\sqrt{4 - x^2 - y^2}}\mathbf{j} + \mathbf{k}$$

$$\iint_S (\text{curl } F) \cdot F\, dS = \iint_R \frac{yz}{\sqrt{4 - x^2 - y^2}}\, dA = \iint_R \frac{y\sqrt{4 - x^2 - y^2}}{\sqrt{4 - x^2 - y^2}}\, dA = \int_{-2}^{2} \int_{-\sqrt{4-x^2}}^{\sqrt{4-x^2}} y\, dy\, dx = 0$$

16. $\mathbf{F}(x,\ y,\ z) = x^2\mathbf{i} + z^2\mathbf{j} - xyz\mathbf{k},\quad S:\ z = \sqrt{4 - x^2 - y^2}$

$$\text{curl } \mathbf{F} = \begin{vmatrix} \mathbf{i} & \mathbf{j} & \mathbf{k} \\ \dfrac{\partial}{\partial x} & \dfrac{\partial}{\partial y} & \dfrac{\partial}{\partial z} \\ x^2 & z^2 & -xyz \end{vmatrix} = (-xz - 2z)\mathbf{i} + yz\mathbf{j}$$

$$G(x,\ y,\ z) = z - \sqrt{4 - x^2 - y^2}$$

$$\nabla G(x,\ y,\ z) = \frac{x}{\sqrt{4 - x^2 - y^2}}\mathbf{i} + \frac{y}{\sqrt{4 - x^2 - y^2}}\mathbf{j} + \mathbf{k}$$

$$\iint_S (\text{curl }\mathbf{F}) \cdot \mathbf{N}\, dS = \iint_R \left[\frac{-z(x+2)x}{\sqrt{4 - x^2 - y^2}} + \frac{y^2 z}{\sqrt{4 - x^2 - y^2}} \right] dA$$

$$= \iint_R [-x(x+2) + y^2]\, dA = \int_{-2}^{2}\int_{-\sqrt{4-x^2}}^{\sqrt{4-x^2}} (-x^2 - 2x + y^2)\, dy\, dx$$

$$= \int_{-2}^{2}\left[-x^2 y - 2xy + \frac{y^3}{3} \right]_{-\sqrt{4-x^2}}^{\sqrt{4-x^2}} dx$$

$$= \int_{-2}^{2}\left[-2x^2\sqrt{4 - x^2} - 4x\sqrt{4 - x^2} + \frac{2}{3}(4 - x^2)\sqrt{4 - x^2} \right] dx$$

$$= \int_{-2}^{2}\left[-\frac{8}{3}x^2\sqrt{4 - x^2} - 4x\sqrt{4 - x^2} + \frac{8}{3}\sqrt{4 - x^2} \right] dx$$

$$= \left[-\frac{8}{3}\left(\frac{1}{8}\right)\left[x(2x^2 - 4)\sqrt{4 - x^2} + 16\arcsin\frac{x}{2} \right] \right.$$

$$\left. + \frac{4}{3}(4 - x^2)^{3/2} + \frac{8}{3}\left(\frac{1}{2}\right)\left[x\sqrt{4 - x^2} + 4\arcsin\frac{x}{2} \right] \right]_{-2}^{2}$$

$$= \left[\left(-\frac{1}{3}\right)(8\pi) + \frac{4}{3}(2\pi) + \frac{1}{3}(-8\pi) - \frac{4}{3}(-2\pi) \right] = 0$$

17. $\mathbf{F}(x,\ y,\ z) = -\ln\sqrt{x^2 + y^2}\,\mathbf{i} + \arctan\dfrac{x}{y}\,\mathbf{j} + \mathbf{k}$

$$\text{curl } \mathbf{F} = \begin{vmatrix} \mathbf{i} & \mathbf{j} & \mathbf{k} \\ \dfrac{\partial}{\partial x} & \dfrac{\partial}{\partial y} & \dfrac{\partial}{\partial z} \\ -\dfrac{1}{2}\ln(x^2 + y^2) & \arctan\dfrac{x}{y} & 1 \end{vmatrix} = \left[\frac{(1/y)}{1 + (x^2/y^2)} + \frac{y}{x^2 + y^2} \right]\mathbf{k} = \left[\frac{2y}{x^2 + y^2} \right]\mathbf{k}$$

$S:\ z = 9 - 2x - 3y$ over one petal of $r = 2\sin 2\theta$ in the first octant.

$$G(x,\ y,\ z) = 2x + 3y + z - 9$$

$$\nabla G(x,\ y,\ z) = 2\mathbf{i} + 3\mathbf{j} + \mathbf{k}$$

$$\iint_S (\text{curl }\mathbf{F}) \cdot \mathbf{N}\, dS = \iint_R \frac{2y}{x^2 + y^2}\, dA = \int_0^{\pi/2}\int_0^{2\sin 2\theta} \frac{2r\sin\theta}{r^2}\, r\, dr\, d\theta$$

$$= \int_0^{\pi/2}\int_0^{4\sin\theta\cos\theta} 2\sin\theta\, dr\, d\theta$$

$$= \int_0^{\pi/2} 8\sin^2\theta\cos\theta\, d\theta = \frac{8\sin^3\theta}{3}\Big]_0^{\pi/2} = \frac{8}{3}$$

18. $\mathbf{F}(x,\ y,\ z) = yz\mathbf{i} + (2 - 3y)\mathbf{j} + (x^2 + y^2)\mathbf{k}$

$$\text{curl } \mathbf{F} = \begin{vmatrix} \mathbf{i} & \mathbf{j} & \mathbf{k} \\ \dfrac{\partial}{\partial x} & \dfrac{\partial}{\partial y} & \dfrac{\partial}{\partial z} \\ yz & 2 - 3y & x^2 + y^2 \end{vmatrix} = 2y\mathbf{i} + (y - 2x)\mathbf{j} - z\mathbf{k}$$

S: the first octant portion of $x^2 + z^2 = 16$ over $x^2 + y^2 = 16$

$G(x,\ y,\ z) = z - \sqrt{16 - x^2}$

$\nabla G(x,\ y,\ z) = \dfrac{x}{\sqrt{16 - x^2}}\mathbf{i} + \mathbf{k}$

$$\iint_S (\text{curl } \mathbf{F}) \cdot \mathbf{N}\, dS = \iint_R \left[\frac{2xy}{\sqrt{16 - x^2}} - z \right] dA = \iint_R \left[\frac{2xy}{\sqrt{16 - x^2}} - \sqrt{16 - x^2} \right] dA$$

$$= \int_0^4 \int_0^{\sqrt{16 - x^2}} \left[\frac{2xy}{\sqrt{16 - x^2}} - \sqrt{16 - x^2} \right] dy\, dx = \int_0^4 \left[\frac{x}{\sqrt{16 - x^2}}y^2 - \sqrt{16 - x^2}\, y \right]_0^{\sqrt{16 - x^2}} dx$$

$$= \int_0^4 [x\sqrt{16 - x^2} - (16 - x^2)]\, dx = \left[-\frac{1}{3}(16 - x^2)^{3/2} - 16x + \frac{x^3}{3} \right]_0^4$$

$$= \left(-64 + \frac{64}{3} \right) - \left(-\frac{64}{3} \right) = -\frac{64}{3}$$

19. From Exercise 10, we have $\mathbf{N} = \dfrac{2x\mathbf{i} - \mathbf{k}}{\sqrt{1 + 4x^2}}$ and $dS = \sqrt{1 + 4x^2}\, dA$. Since $\text{curl } \mathbf{F} = xy\mathbf{j} - xz\mathbf{k}$, we have

$$\iint_S (\text{curl } \mathbf{F}) \cdot \mathbf{N}\, dS = \iint_R xz\, dA = \int_0^a \int_0^a x^3\, dy\, dx = \int_0^a ax^3\, dx = \frac{ax^4}{4} \bigg]_0^a = \frac{a^5}{4}.$$

20. $\mathbf{F}(x,\ y,\ z) = xyz\mathbf{i} + y\mathbf{j} + z\mathbf{k}$

$$\text{curl } \mathbf{F} = \begin{vmatrix} \mathbf{i} & \mathbf{j} & \mathbf{k} \\ \dfrac{\partial}{\partial x} & \dfrac{\partial}{\partial y} & \dfrac{\partial}{\partial z} \\ xyz & y & z \end{vmatrix} = xy\mathbf{j} - xz\mathbf{k}$$

S: the first octant portion of $z = x^2$ over $x^2 + y^2 = a^2$.

From Exercise 10, we have $\mathbf{N} = \dfrac{2x\mathbf{i} - \mathbf{k}}{\sqrt{1 + 4x^2}}$ and $dS = \sqrt{1 + 4x^2}\, dA$.

$$\iint_S (\text{curl } \mathbf{F}) \cdot \mathbf{N}\, dS = \iint_R xz\, dA = \iint_R x^3\, dA = \int_0^a \int_0^{\sqrt{a^2 - x^2}} x^3\, dy\, dx = \int_0^a x^3 \sqrt{a^2 - x^2}\, dx$$

$$= \left[-\frac{1}{3}x^2(a^2 - x^2)^{3/2} - \frac{2}{15}(a^2 - x^2)^{5/2} \right]_0^a = \frac{2}{15}a^5$$

21. $\mathbf{F}(x,\ y,\ z) = \mathbf{i} + \mathbf{j} - 2\mathbf{k}$

$$\text{curl } \mathbf{F} = \begin{vmatrix} \mathbf{i} & \mathbf{j} & \mathbf{k} \\ \dfrac{\partial}{\partial x} & \dfrac{\partial}{\partial y} & \dfrac{\partial}{\partial z} \\ 1 & 1 & -2 \end{vmatrix} = \mathbf{0}$$

Letting $\mathbf{N} = \mathbf{K}$, we have $\displaystyle\iint_S (\text{curl } \mathbf{F}) \cdot \mathbf{N}\, dS = 0.$

22. $F(x, y, z) = -y\mathbf{i} + x\mathbf{j}$

$S: x^2 + y^2 = 1$

$$\text{curl } F = \begin{vmatrix} \mathbf{i} & \mathbf{j} & \mathbf{k} \\ \dfrac{\partial}{\partial x} & \dfrac{\partial}{\partial y} & \dfrac{\partial}{\partial z} \\ -y & x & 0 \end{vmatrix} = 2\mathbf{k}$$

Letting $\mathbf{N} = \mathbf{k}$, we have $\displaystyle\iint_S (\text{curl } F) \cdot \mathbf{N}\, dS = \iint_R 2\, dA = \int_0^{2\pi} \int_0^1 2r\, dr\, d\theta = 2\pi.$

23. (a) $\displaystyle\int_C f\nabla g \cdot d\mathbf{r} = \iint_S \text{curl } [f\nabla g] \cdot \mathbf{N}\, dS$ Stoke's Theorem

$$f\nabla g = f\frac{\partial g}{\partial x}\mathbf{i} + f\frac{\partial g}{\partial y}\mathbf{j} + f\frac{\partial g}{\partial z}\mathbf{k}$$

$$\text{curl } (f\nabla g) = \begin{vmatrix} \mathbf{i} & \mathbf{j} & \mathbf{k} \\ \dfrac{\partial}{\partial x} & \dfrac{\partial}{\partial y} & \dfrac{\partial}{\partial z} \\ f\left(\dfrac{\partial g}{\partial x}\right) & f\left(\dfrac{\partial g}{\partial y}\right) & f\left(\dfrac{\partial g}{\partial z}\right) \end{vmatrix}$$

$$= \left[\left[f\left(\frac{\partial^2 g}{\partial y \partial z}\right) + \left(\frac{\partial f}{\partial y}\right)\left(\frac{\partial g}{\partial z}\right)\right] - \left[f\left(\frac{\partial^2 g}{\partial z \partial y}\right) + \left(\frac{\partial f}{\partial z}\right)\left(\frac{\partial g}{\partial y}\right)\right]\right]\mathbf{i}$$

$$- \left[\left[f\left(\frac{\partial^2 g}{\partial x \partial z}\right) + \left(\frac{\partial f}{\partial x}\right)\left(\frac{\partial g}{\partial z}\right)\right] - \left[f\left(\frac{\partial^2 g}{\partial z \partial x}\right) + \left(\frac{\partial f}{\partial z}\right)\left(\frac{\partial g}{\partial x}\right)\right]\right]\mathbf{j}$$

$$+ \left[\left[f\left(\frac{\partial^2 g}{\partial x \partial y}\right) + \left(\frac{\partial f}{\partial x}\right)\left(\frac{\partial g}{\partial y}\right)\right] - \left[f\left(\frac{\partial^2 g}{\partial y \partial x}\right) + \left(\frac{\partial f}{\partial y}\right)\left(\frac{\partial g}{\partial x}\right)\right]\right]\mathbf{k}$$

$$= \left[\left(\frac{\partial f}{\partial y}\right)\left(\frac{\partial g}{\partial z}\right) - \left(\frac{\partial f}{\partial z}\right)\left(\frac{\partial g}{\partial y}\right)\right]\mathbf{i} - \left[\left(\frac{\partial f}{\partial x}\right)\left(\frac{\partial g}{\partial z}\right) - \left(\frac{\partial f}{\partial z}\right)\left(\frac{\partial g}{\partial x}\right)\right]\mathbf{j}$$

$$+ \left[\left(\frac{\partial f}{\partial x}\right)\left(\frac{\partial g}{\partial y}\right) - \left(\frac{\partial f}{\partial y}\right)\left(\frac{\partial g}{\partial x}\right)\right]\mathbf{k}$$

$$= \begin{vmatrix} \mathbf{i} & \mathbf{j} & \mathbf{k} \\ \dfrac{\partial f}{\partial x} & \dfrac{\partial f}{\partial y} & \dfrac{\partial f}{\partial z} \\ \dfrac{\partial g}{\partial x} & \dfrac{\partial g}{\partial y} & \dfrac{\partial g}{\partial z} \end{vmatrix} = \nabla f \times \nabla g$$

Therefore, $\displaystyle\int_C f\nabla g \cdot d\mathbf{r} = \iint_S \text{curl } [f\nabla g] \cdot \mathbf{N}\, dS = \iint_S [\nabla f \times \nabla g] \cdot \mathbf{N}\, dS.$

(b) $\displaystyle\int_C (f\nabla f) \cdot d\mathbf{r} = \iint_S (\nabla f \times \nabla f) \cdot \mathbf{N}\, dS$ using part (a)

$\qquad\qquad = 0$ since $\nabla f \times \nabla f = 0$

(c) $\displaystyle\int_C (f\nabla g + g\nabla f) \cdot d\mathbf{r} = \int_C (f\nabla g) \cdot d\mathbf{r} + \int_C (g\nabla f) \cdot d\mathbf{r}$

$$= \iint_S (\nabla f \times \nabla g) \cdot \mathbf{N}\, dS + \iint_S (\nabla g \times \nabla f) \cdot \mathbf{N}\, dS \quad \text{using part (a)}$$

$$= \iint_S (\nabla f \times \nabla g) \cdot \mathbf{N}\, dS + \iint_S -(\nabla f \times \nabla g) \cdot \mathbf{N}\, dS = 0$$

24. $f(x, y, z) = xyz, \quad g(x, y, z) = z, \quad S: z = \sqrt{4 - x^2 - y^2}$

(a) $\nabla g(x, y, z) = \mathbf{k}$

$f(x, y, z)\nabla g(x, y, z) = xyz\mathbf{k}$

$\mathbf{r}(t) = 2\cos t\mathbf{i} + 2\sin t\mathbf{j} + 0\mathbf{k}, \quad 0 \le t \le 2\pi$

$\int_C [f(x, y, z)\nabla g(x, y, z)] \cdot d\mathbf{r} = 0$

(b) $\nabla f(x, y, z) = yz\mathbf{i} + xz\mathbf{j} + xy\mathbf{k}$

$\nabla g(x, y, z) = \mathbf{k}$

$\nabla f \times \nabla g = \begin{vmatrix} \mathbf{i} & \mathbf{j} & \mathbf{k} \\ yz & xz & xy \\ 0 & 0 & 1 \end{vmatrix} = xz\mathbf{i} - yz\mathbf{j}$

$\mathbf{N} = \dfrac{x}{\sqrt{4 - x^2 - y^2}}\mathbf{i} + \dfrac{y}{\sqrt{4 - x^2 - y^2}}\mathbf{j} + \mathbf{k}$

$dS = \sqrt{1 + \left(\dfrac{-x}{\sqrt{4 - x^2 - y^2}}\right)^2 + \left(\dfrac{-y}{\sqrt{4 - x^2 - y^2}}\right)^2}\, dA = \dfrac{2}{\sqrt{4 - x^2 - y^2}}\, dA$

$\displaystyle\iint_S [\nabla f(x, y, z) \times \nabla g(x, y, z)] \cdot \mathbf{N}\, dS = \iint_S \left[\dfrac{x^2 z}{\sqrt{4 - x^2 - y^2}} - \dfrac{y^2 z}{\sqrt{4 - x^2 - y^2}}\right]\dfrac{2}{\sqrt{4 - x^2 - y^2}}\, dA$

$\displaystyle\qquad\qquad = \iint_S \dfrac{2(x^2 - y^2)}{\sqrt{4 - x^2 - y^2}}\, dA$

$\displaystyle\qquad\qquad = \int_0^2 \int_0^{2\pi} \dfrac{2r^2(\cos^2\theta - \sin^2\theta)}{\sqrt{4 - r^2}} r\, d\theta\, dr$

$\displaystyle\qquad\qquad = \int_0^2 \dfrac{2r^3}{\sqrt{4 - r^2}}\left(\dfrac{1}{2}\sin 2\theta\right)\Big]_0^{2\pi}\, dr = 0$

Chapter 16 Review Exercises

1. $f(x, y, z) = 8x^2 + xy + z^2$

$\mathbf{F}(x, y, z) = (16x + y)\mathbf{i} + x\mathbf{j} + 2z\mathbf{k}$

2. $f(x, y, z) = x^2 e^{yz}$

$\mathbf{F}(x, y, z) = 2xe^{yz}\mathbf{i} + x^2 ze^{yz}\mathbf{j} + x^2 ye^{yz}\mathbf{k}$

$\qquad\qquad = xe^{yz}(2\mathbf{i} + xz\mathbf{j} + xy\mathbf{k})$

3. Since $\dfrac{\partial M}{\partial y} = \dfrac{-1}{y^2} \ne \dfrac{\partial N}{\partial x}$, $\mathbf{F}$ is not conservative.

4. Since $\dfrac{\partial M}{\partial y} = -\dfrac{1}{x^2} = \dfrac{\partial N}{\partial x}$, $\mathbf{F}$ is conservative. From $M = \dfrac{\partial U}{\partial x} = -\dfrac{y}{x^2}$ and $N = \dfrac{\partial U}{\partial y} = \dfrac{1}{x}$, partial integration yields $U = \dfrac{y}{x} + h(y)$ and $U = \dfrac{y}{x} + g(x)$ which suggests that $U(x, y) = \dfrac{y}{x} + C$.

5. Since $\dfrac{\partial M}{\partial y} = 12xy = \dfrac{\partial N}{\partial x}$, $\mathbf{F}$ is conservative. From $M = \dfrac{\partial U}{\partial x} = 6xy^2 - 3x^2$ and $N = \dfrac{\partial U}{\partial y} = 6x^2 y + 3y^2 - 7$, partial integration yields $U = 3x^2 y^2 - x^3 + h(y)$ and $U = 3x^2 y^2 + y^3 - 7y + g(x)$ which suggests $h(y) = y^3 - 7y$, $g(x) = -x^3$, and $U(x, y) = 3x^2 y^2 - x^3 + y^3 - 7y + C$.

6. Since $\dfrac{\partial M}{\partial y} = -6y^2 \sin 2x = \dfrac{\partial N}{\partial x}$,   **F** is conservative.

From $M = \dfrac{\partial U}{\partial x} = -2y^3 \sin 2x$ and $N = \dfrac{\partial U}{\partial y} = 3y^2(1 + \cos 2x)$, we obtain $U = y^3 \cos 2x + h(y)$ and $U = y^3(1 + \cos 2x) + g(x)$ which suggests that $h(y) = y^3$, $g(x) = C$, and $U(x,\ y) = y^3(1 + \cos 2x) + C$.

7. Since $\dfrac{\partial M}{\partial y} = 4x = \dfrac{\partial N}{\partial x}$, $\dfrac{\partial M}{\partial z} = 1 \neq \dfrac{\partial P}{\partial x}$,   **F** is not conservative.

8. Since $\dfrac{\partial M}{\partial y} = 4x = \dfrac{\partial N}{\partial x}$, $\dfrac{\partial M}{\partial z} = 2z = \dfrac{\partial P}{\partial x}$, $\dfrac{\partial N}{\partial z} = 6y \neq \dfrac{\partial P}{\partial y}$,   **F** is not conservative.

9. Since $\dfrac{\partial M}{\partial y} = \dfrac{-1}{y^2 z} = \dfrac{\partial N}{\partial x}$, $\dfrac{\partial M}{\partial z} = \dfrac{-1}{yz^2} = \dfrac{\partial P}{\partial x}$, $\dfrac{\partial N}{\partial z} = \dfrac{x}{y^2 z^2} = \dfrac{\partial P}{\partial y}$,   **F** is conservative.

From $M = \dfrac{\partial U}{\partial x} = \dfrac{1}{yz}$, $N = \dfrac{\partial U}{\partial y} = \dfrac{-x}{y^2 z}$, $P = \dfrac{\partial U}{\partial z} = \dfrac{-x}{yz^2}$, we obtain $U = \dfrac{x}{yz} + f(y,\ z)$, $U = \dfrac{x}{yz} + g(x,\ z)$, $U = \dfrac{x}{yz} + h(x,\ y)$ which suggests that $f(y,\ z) = C_1$, $g(x,\ z) = C_2$, $h(x,\ y) = C_3$. Thus, $U(x,\ y,\ z) = \dfrac{x}{yz} + C$.

10. Since $\dfrac{\partial M}{\partial y} = \sin z = \dfrac{\partial N}{\partial x}$, $\dfrac{\partial M}{\partial z} = y \cos z \neq \dfrac{\partial P}{\partial x}$,   **F** is not conservative.

11. Since $\mathbf{F} = x^2\mathbf{i} + y^2\mathbf{j} + z^2\mathbf{k}$
    (a) div $\mathbf{F} = 2x + 2y + 2z$
    (b) **curl F** $= \left(\dfrac{\partial P}{\partial y} - \dfrac{\partial N}{\partial z}\right)\mathbf{i} - \left(\dfrac{\partial P}{\partial x} - \dfrac{\partial M}{\partial z}\right)\mathbf{j} + \left(\dfrac{\partial N}{\partial x} - \dfrac{\partial M}{\partial y}\right)\mathbf{k} = 0\mathbf{i} - 0\mathbf{j} + 0\mathbf{k} = \mathbf{0}$

12. Since $\mathbf{F} = xy^2\mathbf{j} - zx^2\mathbf{k}$
    (a) div $\mathbf{F} = 2xy - x^2$
    (b) **curl F** $= 2xz\mathbf{j} + y^2\mathbf{k}$

13. Since $\mathbf{F} = (\cos y + y \cos x)\mathbf{i} + (\sin x - x \sin y)\mathbf{j} + xyz\mathbf{k}$
    (a) div $\mathbf{F} = -y \sin x - x \cos y + xy$
    (b) **curl F** $= xz\mathbf{i} - yz\mathbf{j} + (\cos x - \sin y + \sin y - \cos x)\mathbf{k} = xz\mathbf{i} - yz\mathbf{j}$

14. Since $\mathbf{F} = (3x - y)\mathbf{i} + (y - 2z)\mathbf{j} + (z - 3x)\mathbf{k}$
    (a) div $\mathbf{F} = 3 + 1 + 1 = 5$
    (b) **curl F** $= 2\mathbf{i} + 3\mathbf{j} + \mathbf{k}$

15. Since $\mathbf{F} = \arcsin x\,\mathbf{i} + xy^2\mathbf{j} + yz^2\mathbf{k}$
    (a) div $\mathbf{F} = \dfrac{1}{\sqrt{1 - x^2}} + 2xy + 2yz$
    (b) **curl F** $= z^2\mathbf{i} + y^2\mathbf{k}$

16. Since $\mathbf{F} = (x^2 - y)\mathbf{i} - (x + \sin^2 y)\mathbf{j}$
    (a) div $\mathbf{F} = 2x - 2\sin y \cos y$
    (b) **curl F** $= \mathbf{0}$

17. Since $\mathbf{F} = \ln(x^2 + y^2)\mathbf{i} + \ln(x^2 + y^2)\mathbf{j} + z\mathbf{k}$

(a) div $\mathbf{F} = \dfrac{2x}{x^2 + y^2} + \dfrac{2y}{x^2 + y^2} + 1$

$\qquad = \dfrac{2x + 2y}{x^2 + y^2} + 1$

(b) curl $\mathbf{F} = \dfrac{2x - 2y}{x^2 + y^2}\mathbf{k}$

18. Since $\mathbf{F} = \dfrac{z}{x}\mathbf{i} + \dfrac{z}{y}\mathbf{j} + z^2\mathbf{k}$

(a) div $\mathbf{F} = -\dfrac{z}{x^2} - \dfrac{z}{y^2} + 2z = z\left(2 - \dfrac{1}{x^2} - \dfrac{1}{y^2}\right)$

(b) curl $\mathbf{F} = -\dfrac{1}{y}\mathbf{i} + \dfrac{1}{x}\mathbf{j}$

19. (a) Let $x = t,\ \ y = t,\ \ -1 \le t \le 2$, then $ds = \sqrt{2}\,dt$.

$$\int_C (x^2 + y^2)\,ds = \int_{-1}^{2} 2t^2 \sqrt{2}\,dt = 2\sqrt{2}\left(\frac{t^3}{3}\right)\Big]_{-1}^{2} = 6\sqrt{2}$$

(b) Let $x = 4\cos t,\ \ y = 4\sin t,\ \ 0 \le t \le 2\pi$, then $ds = 4\,dt$.

$$\int_C (x^2 + y^2)\,ds = \int_0^{2\pi} 16(4\,dt) = 128\pi$$

20. (a) Let $x = 5t,\ \ y = 4t,\ \ 0 \le t \le 1$, then $ds = \sqrt{41}\,dt$.

$$\int_C xy\,ds = \int_0^1 20t^2\sqrt{41}\,dt = \frac{20\sqrt{41}}{3}$$

(b) $C_1:\ x = t,\ \ y = 0,\ \ 0 \le t \le 4,\ \ ds = dt$

$\quad C_2:\ x = 4 - 4t,\ \ y = 2t,\ \ 0 \le t \le 1,\ \ ds = 2\sqrt{5}\,dt$

$\quad C_3:\ x = 0,\ \ y = 2 - t,\ \ 0 \le t \le 2,\ \ ds = dt$

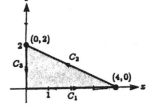

Therefore, $\displaystyle\int_C xy\,ds = \int_0^4 0\,dt + \int_0^1 (8t - 8t^2)2\sqrt{5}\,dt + \int_0^2 0\,dt$

$$= 16\sqrt{5}\left[\frac{t^2}{2} - \frac{t^3}{3}\right]_0^1 = \frac{8\sqrt{5}}{3}$$

21. $x = \cos t + t\sin t,\ \ y = \sin t - t\cos t,\ \ 0 \le t \le 2\pi,\ \ \dfrac{dx}{dt} = t\cos t,\ \ \dfrac{dy}{dt} = t\sin t$

$$\int_C (x^2 + y^2)\,ds = \int_0^{2\pi} [(\cos t + t\sin t)^2 + (\sin t - t\cos t)^2]\sqrt{t^2\cos^2 t + t^2\sin^2 t}\,dt$$

$$= \int_0^{2\pi} (1 + t^2)t\,dt$$

$$= \left[\frac{t^2}{2} + \frac{t^4}{4}\right]_0^{2\pi}$$

$$= 2\pi^2 + 4\pi^4 = 2\pi^2(1 + 2\pi^2)$$

22. $x = t - \sin t$, $y = 1 - \cos t$, $0 \le t \le 2\pi$, $\dfrac{dx}{dt} = 1 - \cos t$, $\dfrac{dy}{dt} = \sin t$

$$\int_C x\,ds = \int_0^{2\pi} (t - \sin t)\sqrt{(1 - \cos t)^2 + (\sin t)^2}\,dt = \int_0^{2\pi} (t - \sin t)\sqrt{2 - 2\cos t}\,dt$$

$$= \sqrt{2}\int_0^{2\pi} [t\sqrt{1 - \cos t} - \sin t\sqrt{1 - \cos t}\,]\,dt$$

$$= \sqrt{2}\left[-\frac{2}{3}(1 - \cos t)^{3/2}\right]_0^{2\pi} + \sqrt{2}\int_0^{2\pi} t\sqrt{1 - \cos t}\,dt$$

$$= \sqrt{2}\int_0^{2\pi} t\sqrt{1 - \cos t}\,dt$$

$$= \sqrt{2}\int_0^{2\pi} \frac{t\sin t}{\sqrt{1 + \cos t}}\,dt$$

$$= \sqrt{2}\left[-2t\sqrt{1 + \cos t}\right]_0^{2\pi} + 2\int_0^{2\pi} \sqrt{1 + \cos t}\,dt\Big]$$

$$= \sqrt{2}\left[-4\sqrt{2}\,\pi + 2\int_0^{2\pi} \frac{\sin t}{\sqrt{1 - \cos t}}\,dt\right]$$

$$= -8\pi + 2\sqrt{2}\left[2\sqrt{1 - \cos t}\right]_0^{2\pi} = -8\pi$$

23. (a) Let $x = 2t$, $y = -3t$, $0 \le t \le 1$

$$\int_C (2x - y)\,dx + (x + 3y)\,dy = \int_0^1 [7t(2) + (-7t)(-3)]\,dt = \int_0^1 35t\,dt = \frac{35}{2}$$

(b) $x = 3\cos t$, $y = 3\sin t$, $dx = -3\sin t\,dt$, $dy = 3\cos t\,dt$, $0 \le t \le 2\pi$

$$\int_C (2x - y)\,dx + (x + 3y)\,dy = \int_0^{2\pi} (9 + 9\sin t\cos t)\,dt = 18\pi$$

24. $x = \cos t + t\sin t$, $y = \sin t - t\sin t$, $0 \le t \le \dfrac{\pi}{2}$, $dx = t\cos t\,dt$, $dy = (\cos t - t\cos t - \sin t)\,dt$

$$\int_C (2x - y)\,dx + (x + 3y)\,dy = \int_0^{\pi/2} [\sin t\cos t(5t^2 - 6t + 2) + \cos^2 t(t + 1) + \sin^2 t(2t - 3)]\,dt \approx 1.01$$

25. $d\mathbf{r} = (2t\mathbf{i} + 3t^2\,\mathbf{j})\,dt$

$\mathbf{F} = t^5\mathbf{i} + t^4\mathbf{j}$, $0 \le t \le 1$

$$\int_C \mathbf{F}\cdot d\mathbf{r} = \int_0^1 5t^6\,dt = \frac{5}{7}$$

26. $d\mathbf{r} = [(-4\sin t)\mathbf{i} + 3\cos t\,\mathbf{j}]\,dt$

$\mathbf{F} = (4\cos t - 3\sin t)\mathbf{i} + (4\cos t + 3\sin t)\,\mathbf{j}$, $0 \le t \le 2\pi$

$$\int_C \mathbf{F}\cdot d\mathbf{r} = \int_0^{2\pi} (12 - 7\sin t\cos t)\,dt = \left[12t - \frac{7\sin^2 t}{2}\right]_0^{2\pi} = 24\pi$$

27. $d\mathbf{r} = [(-2\sin t)\mathbf{i} + (2\cos t)\,\mathbf{j} + \mathbf{k}]\,dt$

$\mathbf{F} = (2\cos t)\mathbf{i} + (2\sin t)\,\mathbf{j} + t\mathbf{k}$, $0 \le t \le 2\pi$

$$\int_C \mathbf{F}\cdot d\mathbf{r} = \int_0^{2\pi} t\,dt = 2\pi^2$$

28. $x = 2 - t, \quad y = 2 - t, \quad z = \sqrt{4t - t^2}, \quad 0 \le t \le 2$

$$d\mathbf{r} = \left[-\mathbf{i} - \mathbf{j} + \frac{2 - t}{\sqrt{4t - t^2}} \mathbf{k} \right] dt$$

$$\mathbf{F} = (4 - 2t - \sqrt{4t - t^2})\mathbf{i} + (\sqrt{4t - t^2} - 2 + t)\mathbf{j} + 0\mathbf{k}$$

$$\int_C \mathbf{F} \cdot d\mathbf{r} = \int_0^2 (t - 2)\, dt = \left[\frac{t^2}{2} - 2t \right]_0^2 = -2$$

29. Let $x = t, \quad y = -t, \quad z = 2t^2, \quad -2 \le t \le 2, \quad d\mathbf{r} = [\mathbf{i} - \mathbf{j} + 4t\mathbf{k}]\, dt.$

$$\mathbf{F} = (-t - 2t^2)\mathbf{i} + (2t^2 - t)\mathbf{j} + (2t)\mathbf{k}$$

$$\int_C \mathbf{F} \cdot d\mathbf{r} = \int_{-2}^2 4t^2\, dt = \frac{4t^3}{3} \bigg]_{-2}^2 = \frac{64}{3}$$

30. Let $x = 2\sin t, \quad y = -2\cos t, \quad z = 4\sin^2 t, \quad 0 \le t \le \pi.$

$$d\mathbf{r} = [(2\cos t)\mathbf{i} + (2\sin t)\mathbf{j} + (8\sin t\cos t)\mathbf{k}]\, dt$$

$$\mathbf{F} = 0\mathbf{i} + 4\mathbf{j} + (2\sin t)\mathbf{k}$$

$$\int_C \mathbf{F} \cdot d\mathbf{r} = \int_0^\pi (8\sin t + 16\sin^2 t\cos t)\, dt = \left[-8\cos t + \frac{16}{3}\sin^3 t \right]_0^\pi = 16$$

31. $\displaystyle\int_{(0,0,0)}^{(1,4,3)} 2xyz\, dx + x^2 z\, dy + x^2 y\, dz = x^2 yz \bigg]_{(0,0,0)}^{(1,4,3)} = 12$

32. $\displaystyle\int_{(0,0,1)}^{(4,4,4)} y\, dx + x\, dy + \frac{1}{z}\, dz = \left[xy + \ln|z| \right]_{(0,0,1)}^{(4,4,4)} = 16 + \ln 4$

33. $\displaystyle\int_C y\, dx + 2x\, dy = \int_0^2 \int_0^2 (2 - 1)\, dy\, dx = \int_0^2 2\, dx = 4$

34. $\displaystyle\int_C xy\, dx + (x^2 + y^2)\, dy = \int_0^2 \int_0^2 (2x - x)\, dy\, dx = \int_0^2 2x\, dx = 4$

35. $\displaystyle\int_C xy^2\, dx + x^2 y\, dy = \iint_R (2xy - 2xy)\, dA = 0$

36. $\displaystyle\int_C (x^2 - y^2)\, dx + 2xy\, dy = \int_{-a}^a \int_{-\sqrt{a^2 - x^2}}^{\sqrt{a^2 - x^2}} 4y\, dy\, dx = \int_{-a}^a 0\, dx = 0$

37. $\displaystyle\int_C xy\, dx + x^2\, dy = \int_0^1 \int_{x^2}^x x\, dy\, dx = \int_0^1 (x^2 - x^3)\, dx = \frac{1}{12}$

38. $\displaystyle\int_C y^2\, dx + x^{2/3}\, dy = \int_{-1}^1 \int_{-(1-x^{2/3})^{3/2}}^{(1-x^{2/3})^{3/2}} \left(\frac{2}{3} x^{-1/3} - 2y \right) dy\, dx$

$$= \int_{-1}^1 \left[\frac{2y}{3x^{1/3}} - y^2 \right]_{-(1-x^{2/3})^{3/2}}^{(1-x^{2/3})^{3/2}} dx$$

$$= \frac{4}{3} \int_{-1}^1 \frac{(1 - x^{2/3})^{3/2}}{x^{1/3}}\, dx$$

$$= -\frac{4}{3} \left(-\frac{3}{2} \right) \left(\frac{2}{5} \right) (1 - x^{2/3})^{5/2} \bigg]_{-1}^1 = \frac{4}{5}[0] = 0$$

39. $\mathbf{F}(x,\ y,\ z) = x^2\mathbf{i} + xy\mathbf{j} + z\mathbf{k}$

Q: solid region bounded by the coordinate planes and the plane $2x + 3y + 4z = 12$

(a) There are four surfaces for this solid.

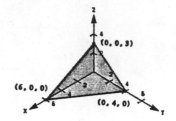

$$z = 0, \qquad \mathbf{N} = -\mathbf{k}, \qquad \mathbf{F} \cdot \mathbf{N} = -z, \qquad \int_{S_1}\int 0\,dS = 0$$

$$y = 0, \qquad \mathbf{N} = -\mathbf{j}, \qquad \mathbf{F} \cdot \mathbf{N} = -xy, \qquad \int_{S_2}\int 0\,dS = 0$$

$$x = 0, \qquad \mathbf{N} = -\mathbf{i}, \qquad \mathbf{F} \cdot \mathbf{N} = -x^2, \qquad \int_{S_3}\int 0\,dS = 0$$

$$2x + 3y + 4z = 12, \quad \mathbf{N} = \frac{2\mathbf{i} + 3\mathbf{j} + 4\mathbf{k}}{\sqrt{29}}, \quad dS = \sqrt{1 + (1/4) + (9/16)}\,dA = \frac{\sqrt{29}}{4}\,dA$$

$$\int_{S_4}\int \mathbf{F} \cdot \mathbf{N}\,dS = \frac{1}{4}\int_{R}\int (2x^2 + 3xy + 4z)\,dA$$

$$= \frac{1}{4}\int_0^6 \int_0^{4-(2x/3)} (2x^2 + 3xy + 12 - 2x - 3y)\,dy\,dx$$

$$= \frac{1}{4}\int_0^6 \left[2x^2\left(\frac{12-2x}{3}\right) + \frac{3x}{2}\left(\frac{12-2x}{3}\right)^2 + 12\left(\frac{12-2x}{3}\right) - 2x\left(\frac{12-2x}{3}\right) - \frac{3}{2}\left(\frac{12-2x}{3}\right)^2 \right] dx$$

$$= \frac{1}{6}\int_0^6 (-x^3 + x^2 + 24x + 36)\,dx$$

$$= \frac{1}{6}\left[-\frac{x^4}{4} + \frac{x^3}{3} + 12x^2 + 36x \right]_0^6 = 66$$

(b) Since div $\mathbf{F} = 2x + x + 1 = 3x + 1$, the Divergence Theorem yields

$$\iiint_Q \text{div } \mathbf{F}\,dV = \int_0^6 \int_0^{(12-2x)/3} \int_0^{(12-2x-3y)/4} (3x+1)\,dz\,dy\,dx$$

$$= \int_0^6 \int_0^{(12-2x)/3} (3x+1)\left(\frac{12-2x-3y}{4}\right) dy\,dx$$

$$= \frac{1}{4}\int_0^6 (3x+1)\left[12y - 2xy - \frac{3}{2}y^2 \right]_0^{(12-2x)/3} dx$$

$$= \frac{1}{4}\int_0^6 (3x+1)\left[4(12-2x) - 2x\left(\frac{12-2x}{3}\right) - \frac{3}{2}\left(\frac{12-2x}{3}\right)^2 \right] dx$$

$$= \frac{1}{4}\int_0^6 \frac{2}{3}(3x^3 - 35x^2 + 96x + 36)\,dx$$

$$= \frac{1}{6}\left[\frac{3x^4}{4} - \frac{35x^3}{3} + 48x^2 + 36x \right]_0^6 = 66.$$

40. $\mathbf{F}(x,\ y,\ z) = x\mathbf{i} + y\mathbf{j} + z\mathbf{k}$

Q: solid region bounded by the coordinate planes and the plane $2x + 3y + 4z = 12$

(a) There are four surfaces for this solid.

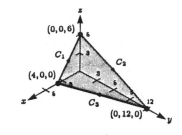

$$z = 0, \qquad \mathbf{N} = -\mathbf{k}, \qquad \mathbf{F} \cdot \mathbf{N} = -z, \qquad \int_{S_1}\!\!\int 0\,dS = 0$$

$$y = 0, \qquad \mathbf{N} = -\mathbf{j}, \qquad \mathbf{F} \cdot \mathbf{N} = -y, \qquad \int_{S_2}\!\!\int 0\,dS = 0$$

$$x = 0, \qquad \mathbf{N} = -\mathbf{i}, \qquad \mathbf{F} \cdot \mathbf{N} = -x, \qquad \int_{S_3}\!\!\int 0\,dS = 0$$

$$2x + 3y + 4z = 12, \quad \mathbf{N} = \frac{2\mathbf{i} + 3\mathbf{j} + 4\mathbf{k}}{\sqrt{29}}, \quad dS = \sqrt{1 + (1/4) + (9/16)}\,dA = \frac{\sqrt{29}}{4}\,dA$$

$$\int_{S_4}\!\!\int \mathbf{N} \cdot \mathbf{F}\,dS = \frac{1}{4}\int_R\!\!\int (2x + 3y + 4z)\,dy\,dx$$

$$= \frac{1}{4}\int_0^6 \int_0^{(12-2x)/3} 12\,dy\,dx = 3\int_0^6 \left(4 - \frac{2x}{3}\right)dx = 3\left[4x - \frac{x^2}{3}\right]_0^6 = 36$$

(b) Since div $\mathbf{F} = 3$, the Divergence Theorem yields

$$\iiint_Q \text{div }\mathbf{F}\,dV = \iiint_Q 3\,dV = 3(\text{volume of solid}) = 3\left[\frac{1}{3}(\text{area of base})(\text{height})\right] = \frac{1}{2}(6)(4)(3) = 36.$$

41. $\mathbf{F}(x,\ y,\ z) = (\cos y + y\cos x)\mathbf{i} + (\sin x - x\sin y)\mathbf{j} + xyz\mathbf{k}$

S: portion of $z = y^2$ over the square in the xy-plane with vertices $(0,\ 0)$, $(a,\ 0)$, $(a,\ a)$, $(0,\ a)$

(a) Using the line integral we have

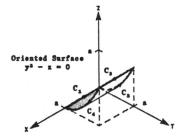

$C_1 : y = 0, \quad dy = 0$

$C_2 : x = 0, \quad dx = 0, \quad z = y^2, \quad dz = 2y\,dy$

$C_3 : y = a, \quad dy = 0, \quad z = a^2, \quad dz = 0$

$C_4 : x = a, \quad dx = 0, \quad z = y^2, \quad dz = 2y\,dy$

$$\int_C \mathbf{F} \cdot d\mathbf{r} = \int_C (\cos y + y\cos x)\,dx + (\sin x - x\sin y)\,dy + xyz\,dz$$

$$= \int_{C_1} dx + \int_{C_2} 0 + \int_{C_3} (\cos a + a\cos x)\,dx + \int_{C_4} (\sin a - a\sin y)\,dy + ay^3(2y\,dy)$$

$$= \int_a^0 dx + \int_0^a (\cos a + a\cos x)\,dx + \int_a^0 (\sin a - a\sin y)\,dy + \int_a^0 2ay^4\,dy$$

$$= -a + \Big[x\cos a + a\sin x\Big]_0^a + \Big[y\sin a + a\cos y\Big]_a^0 + 2a\frac{y^5}{5}\Big]_a^0$$

$$= -a + a\cos a + a\sin a + a - a\sin a - a\cos a - \frac{2a^6}{5} = -\frac{2a^6}{5}$$

(b) Considering $f(x,\ y,\ z) = y^2 - z$, we have $\mathbf{N} = \dfrac{\nabla f}{\|\nabla f\|} = \dfrac{2y\mathbf{j} - \mathbf{k}}{\sqrt{1 + 4y^2}}$, $dS = \sqrt{1 + 4y^2}\,dA$, and

curl F $= xz\mathbf{i} - yz\mathbf{j}$. Hence,

$$\int_S\!\!\int (\text{curl }\mathbf{F}) \cdot \mathbf{N}\,dS = \int_0^a \int_0^a -2y^2z\,dy\,dx = \int_0^a \int_0^a -2y^4\,dy\,dx = \int_0^a -\frac{2a^5}{5}\,dx = -\frac{2a^6}{5}.$$

42. $\mathbf{F}(x,\ y,\ z) = (x-z)\mathbf{i} + (y-z)\mathbf{j} + x^2\mathbf{k}$

S: first octant portion of the plane $3x + y + 2z = 12$

(a) $C_1: \ y = 0, \quad dy = 0, \quad z = \dfrac{12 - 3x}{2}, \quad dz = -\dfrac{3}{2}\,dx$

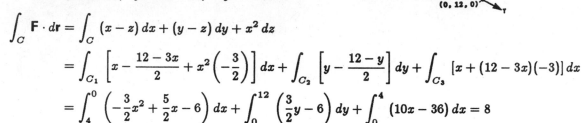

$\quad C_2: \ x = 0, \quad dx = 0, \quad z = \dfrac{12 - y}{2}, \quad dz = -\dfrac{1}{2}\,dy$

$\quad C_3: \ z = 0, \quad dz = 0, \quad y = 12 - 3x, \quad dy = -3\,dx$

$$\int_C \mathbf{F} \cdot d\mathbf{r} = \int_C (x - z)\,dx + (y - z)\,dy + x^2\,dz$$

$$= \int_{C_1}\left[x - \frac{12 - 3x}{2} + x^2\left(-\frac{3}{2}\right)\right] dx + \int_{C_2}\left[y - \frac{12 - y}{2}\right] dy + \int_{C_3}\left[x + (12 - 3x)(-3)\right] dx$$

$$= \int_4^0 \left(-\frac{3}{2}x^2 + \frac{5}{2}x - 6\right) dx + \int_0^{12}\left(\frac{3}{2}y - 6\right) dy + \int_0^4 (10x - 36)\,dx = 8$$

(b) $G(x,\ y,\ z) = \dfrac{12 - 3x - y}{2} - z$

$\quad \nabla G(x,\ y,\ z) = -\dfrac{3}{2}\mathbf{i} - \dfrac{1}{2}\mathbf{j} - \mathbf{k}$

curl $\mathbf{F} = \mathbf{i} - (2x + 1)\mathbf{j}$

$$\iint_S (\textbf{curl }\mathbf{F}) \cdot \mathbf{N}\,dS = \int_0^4 \int_0^{12 - 3x} (x - 1)\,dy\,dx = \int_0^4 (-3x^2 + 15x - 12)\,dx = 8$$

43. In this case $x = \cos t, \quad y = \sin t, \quad 0 \le t \le 2\pi, \quad \mathbf{F} = (-\sin t)\mathbf{i} + (\cos t)\mathbf{j}$, and $d\mathbf{r} = [(-\sin t)\mathbf{i} + (\cos t)\mathbf{j}]\,dt$. Hence,

$$\int_C \mathbf{F} \cdot d\mathbf{r} = \int_0^{2\pi} 1\,dt = 2\pi.$$

Since $M(x,\ y)$ and $N(x,\ y)$ are not defined at the point $(0,\ 0)$ in region R enclosed by curve C, the condition of a zero value for the integral is not guaranteed.

CHAPTER 17
Differential Equations

Section 17.1   Definitions and Basic Concepts

		Type	Order
1.	$\dfrac{dy}{dx} + 3xy = x^2$	Ordinary	1
2.	$y'' + 2y' + y = 1$	Ordinary	2
3.	$\dfrac{d^2x}{dt^2} + 2\dfrac{dx}{dt} - 4x = e^t$	Ordinary	2
4.	$\dfrac{d^2u}{dt^2} + \dfrac{du}{dt} = \sec t$	Ordinary	2
5.	$y^{(4)} + 3(y')^2 - 4y = 0$	Ordinary	4
6.	$x^2 y'' + 3xy' = 0$	Ordinary	2
7.	$(y'')^2 + 3y' - 4y = 0$	Ordinary	2
8.	$\dfrac{\partial u}{\partial t} = C^2 \dfrac{\partial^2 u}{\partial x^2}$	Partial	2
9.	$\dfrac{\partial u}{\partial t} + \dfrac{\partial u}{\partial y} = 2u$	Partial	1
10.	$\dfrac{\partial^2 u}{\partial x \partial y} = \dfrac{\partial u}{\partial y}$	Partial	2
11.	$\dfrac{d^2 y}{dx^2} = \sqrt{1 + \left(\dfrac{dy}{dx}\right)^2}$	Ordinary	2
12.	$\sqrt{\dfrac{d^2 y}{dx^2}} = \dfrac{dy}{dx}$	Ordinary	2

13. Differential equation: $y' = 4y$

Solution: $y = Ce^{4x}$

Check: $y' = 4Ce^{4x} = 4y$

14. Differential equation: $y' = \dfrac{2xy}{x^2 - y^2}$

Solution: $x^2 + y^2 = Cy$

Check: $2x + 2yy' = Cy'$

$$y' = \frac{-2x}{(2y - C)}$$

$$y' = \frac{-2xy}{2y^2 - Cy} = \frac{-2xy}{2y^2 - (x^2 + y^2)} = \frac{-2xy}{y^2 - x^2} = \frac{2xy}{x^2 - y^2}$$

15. Differential equation: $y'' + y = 0$

Solution: $y = C_1 \cos x + C_2 \sin x$

Check: $\quad y' = -C_1 \sin x + C_2 \cos x$

$\qquad y'' = -C_1 \cos x - C_2 \sin x$

$\quad y'' + y = -C_1 \cos x - C_2 \sin x + C_1 \cos x + C_2 \sin x = 0$

16. Differential equation: $y'' + 2y' + 2y = 0$

Solution: $y = C_1 e^{-x} \cos x + C_2 e^{-x} \sin x$

Check: $\qquad y' = -(C_1 + C_2)e^{-x} \sin x + (-C_1 + C_2)e^{-x} \cos x$

$\qquad y'' = 2C_1 e^{-x} \sin x - 2C_2 e^{-x} \cos x$

$y'' + 2y' + 2y = (2C_1 - 2C_1 - 2C_2 + 2C_2)e^{-x} \sin x + (-2C_2 - 2C_1 + 2C_2 + 2C_1)e^{-x} \cos x = 0$

17. Differential equation: $b^2 \dfrac{\partial u}{\partial t} = \dfrac{\partial^2 u}{\partial x^2}$

Solution: $u = e^{-t} \sin bx$

Check: $\quad \dfrac{\partial u}{\partial t} = -e^{-t} \sin bx$

$\qquad \dfrac{\partial u}{\partial x} = be^{-t} \cos bx$

$\qquad \dfrac{\partial^2 u}{\partial x^2} = -b^2 e^{-t} \sin bx$

$\quad b^2 \dfrac{\partial u}{\partial t} = -b^2 e^{-t} \sin bx = \dfrac{\partial^2 u}{\partial x^2}$

18. Differential equation: $\dfrac{\partial^2 u}{\partial x^2} + \dfrac{\partial^2 u}{\partial y^2} = 0$

Solution: $u = \dfrac{y}{x^2 + y^2}$

Check: $\qquad \dfrac{\partial u}{\partial x} = \dfrac{-2xy}{(x^2 + y^2)^2}$

$\qquad \dfrac{\partial^2 u}{\partial x^2} = -\dfrac{2y(y^2 - 3x^2)}{(x^2 + y^2)^3}$

$\qquad \dfrac{\partial u}{\partial y} = \dfrac{x^2 - y^2}{(x^2 + y^2)^2}$

$\qquad \dfrac{\partial^2 u}{\partial y^2} = \dfrac{2y(y^2 - 3x^2)}{(x^2 + y^2)^3}$

$\qquad \dfrac{\partial^2 u}{\partial x^2} + \dfrac{\partial^2 u}{\partial y^2} = 0$

In Exercises 19–24, the differential equation is $y^{(4)} - 16y = 0$.

19. $\qquad y = 3 \cos x$

$\qquad y^{(4)} = 3 \cos x$

$\quad y^{(4)} - 16y = -45 \cos x \neq 0$

20. $\qquad y = 3 \cos 2x$

$\qquad y^{(4)} = 48 \cos 2x$

$\quad y^{(4)} - 16y = 48 \cos 2x - 48 \cos 2x = 0$

21. $\qquad y = e^{-2x}$

$\qquad y^{(4)} = 16e^{-2x}$

$\quad y^{(4)} - 16y = 16e^{-2x} - 16e^{-2x} = 0$

22. $\qquad y = 5 \ln x$

$\qquad y^{(4)} = -\dfrac{30}{x^4}$

$\quad y^{(4)} - 16y = -\dfrac{30}{x^4} - 80 \ln x \neq 0$

23. $\qquad y = C_1 e^{2x} + C_2 e^{-2x} + C_3 \sin 2x + C_4 \cos 2x$

$\qquad y^{(4)} = 16C_1 e^{2x} + 16C_2 e^{-2x} + 16C_3 \cos 2x + 16C_4 \sin 2x$

$\quad y^{(4)} - 16y = 0$

24. $\qquad y = 5e^{-2x} + 3 \cos 2x$

$\qquad y^{(4)} = 80e^{-2x} + 48 \cos 2x$

$\quad y^{(4)} - 16y = \left(80e^{-2x} + 48 \cos 2x\right) - \left(80e^{-2x} + 48 \cos 2x\right) = 0$

In Exercises 25–30, the differential equation is $x\dfrac{\partial u}{\partial x} - y\dfrac{\partial u}{\partial y} = 0$.

25.
$$u = e^{x+y}$$
$$\frac{\partial u}{\partial x} = e^{x+y}$$
$$\frac{\partial u}{\partial y} = e^{x+y}$$
$$x\frac{\partial u}{\partial x} - y\frac{\partial u}{\partial y} \neq 0$$

26.
$$u = 5$$
$$\frac{\partial u}{\partial x} = 0$$
$$\frac{\partial u}{\partial y} = 0$$
$$x\frac{\partial u}{\partial x} - y\frac{\partial u}{\partial y} = 0$$

27.
$$u = x^2 y^2$$
$$\frac{\partial u}{\partial x} = 2xy^2$$
$$\frac{\partial u}{\partial y} = 2x^2 y$$
$$x\frac{\partial u}{\partial x} - y\frac{\partial u}{\partial y} = 2x^2 y^2 - 2x^2 y^2 = 0$$

28.
$$u = \sin xy$$
$$\frac{\partial u}{\partial x} = y \cos xy$$
$$\frac{\partial u}{\partial y} = x \cos xy$$
$$x\frac{\partial u}{\partial x} - y\frac{\partial u}{\partial y} = xy \cos xy - xy \cos xy = 0$$

29.
$$u = (xy)^n$$
$$\frac{\partial u}{\partial x} = ny(xy)^{n-1}$$
$$\frac{\partial u}{\partial y} = nx(xy)^{n-1}$$
$$x\frac{\partial u}{\partial x} - y\frac{\partial u}{\partial y} = nxy(xy)^{n-1} - nxy(xy)^{n-1} = 0$$

30.
$$u = x^2 + y^2$$
$$\frac{\partial u}{\partial x} = 2x$$
$$\frac{\partial u}{\partial y} = 2y$$
$$x\frac{\partial u}{\partial x} - y\frac{\partial u}{\partial y} = 2x^2 - 2y^2 \neq 0$$

31. $y^2 = Cx^3$ passes through $(4, 4)$
$16 = C(64) \Rightarrow C = \frac{1}{4}$
Particular solution: $y^2 = \frac{1}{4}x^3$ or $4y^2 = x^3$

32. $2x^2 - y^2 = C$ passes through $(3, 4)$
$2(9) - 16 = C \Rightarrow C = 2$
Particular solution: $2x^2 - y^2 = 2$

33. Differential equation: $y' + 2y = 0$
General solution: $y = Ce^{-2x}$
Initial condition: $y(0) = 3$, $\;3 = Ce^0 = C$
Particular solution: $y = 3e^{-2x}$

34. Differential equation: $2x + 3yy' = 0$
General solution: $2x^2 + 3y^2 = C$
Initial condition: $y(1) = 2$, $\;2(1) + 3(4) = 14 = C$
Particular solution: $2x^2 + 3y^2 = 14$

35. Differential equation: $y'' + 9y = 0$
General solution: $y = C_1 \sin 3x + C_2 \cos 3x$
Initial conditions: $y\left(\dfrac{\pi}{6}\right) = 2, \quad y'\left(\dfrac{\pi}{6}\right) = 1$
$$2 = C_1 \sin\left(\frac{\pi}{2}\right) + C_2 \cos\left(\frac{\pi}{2}\right) \Rightarrow C_1 = 2$$
$$y' = 3C_1 \cos 3x - 3C_2 \sin 3x$$
$$1 = 3C_1 \cos\left(\frac{\pi}{2}\right) - 3C_2 \sin\left(\frac{\pi}{2}\right)$$
$$= -3C_2 \Rightarrow C_2 = -\frac{1}{3}$$
Particular solution: $y = 2 \sin 3x - \dfrac{1}{3} \cos 3x$

36. Differential equation: $xy'' + y' = 0$
General solution: $y = C_1 + C_2 \ln x$
Initial conditions: $y(2) = 0, \quad y'(2) = \dfrac{1}{2}$
$$0 = C_1 + C_2 \ln 2$$
$$y' = \frac{C_2}{x}$$
$$\frac{1}{2} = \frac{C_2}{2} \Rightarrow C_2 = 1, \quad C_1 = -\ln 2$$
Particular solution: $y = -\ln 2 + \ln x$

37. Differential equation: $x^2y'' - 3xy' + 3y = 0$
General solution: $y = C_1x + C_2x^3$
Initial conditions: $y(2) = 0$, $y'(2) = 4$

$$0 = 2C_1 + 8C_2$$
$$y' = C_1 + 3C_2x^2$$
$$4 = C_1 + 12C_2$$

$$\left.\begin{array}{l} C_1 + 4C_2 = 0 \\ C_1 + 12C_2 = 4 \end{array}\right\} \quad C_2 = \tfrac{1}{2}, \quad C_1 = -2$$

Particular solution: $y = -2x + \tfrac{1}{2}x^3$

38. Differential equation: $9y'' - 12y' + 4y = 0$
General solution: $y = e^{2x/3}(C_1 + C_2x)$
Initial conditions: $y(0) = 4$, $y(3) = 0$

$$0 = e^2(C_1 + 3C_2)$$
$$4 = (1)(C_1 + 0) \Rightarrow C_1 = 4$$
$$0 = e^2(4 + 3C_2) \Rightarrow C_2 = -\tfrac{4}{3}$$

Particular solution: $y = e^{2x/3}\left(4 - \tfrac{4}{3}x\right)$

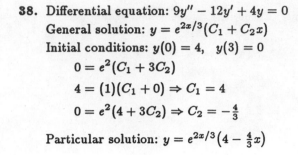

39. Differential equation: $4yy' - x = 0$
General solution: $4y^2 - x^2 = C$
Particular solutions: $C = 0$, Two intersecting lines
$\qquad\qquad\qquad\quad C = \pm 1$, $C = \pm 4$, Hyperbolas

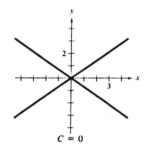

$C = 0$

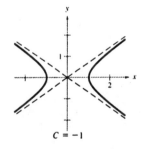

$C = -1$

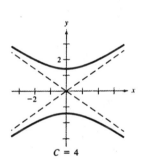

$C = 4$

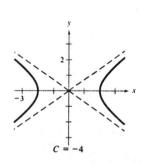

$C = -4$

40. Differential equation: $yy' + x = 0$
General solution: $x^2 + y^2 = C$
Particular solutions: $C = 0$, Point
$\qquad C = 1, \quad C = 4,$ Circles

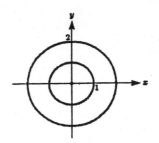

41. $\dfrac{dy}{dx} = 3x^2$

$\qquad y = \displaystyle\int 3x^2 \, dx = x^3 + C$

42. $\dfrac{dy}{dx} = \dfrac{1}{1+x^2}$

$\qquad y = \displaystyle\int \dfrac{1}{1+x^2} \, dx = \arctan x + C$

43. $\dfrac{dy}{dx} = \dfrac{x-2}{x} = 1 - \dfrac{2}{x}$

$\qquad y = \displaystyle\int \left[1 - \dfrac{2}{x} \right] dx$

$\qquad\quad = x - 2\ln|x| + C = x - \ln x^2 + C$

44. $\dfrac{dy}{dx} = x \cos x$

$\qquad y = \displaystyle\int x \cos x \, dx = x \sin x + \cos x + C$

45. $\dfrac{dy}{dx} = e^x \sin 2x$

$\qquad y = \displaystyle\int e^x (\sin 2x) \, dx = \dfrac{e^x}{5}[\sin 2x - 2\cos 2x] + C$

46. $\dfrac{dy}{dx} = \tan^2 x = \sec^2 x - 1$

$\qquad y = \displaystyle\int (\sec^2 x - 1) \, dx = \tan x - x + C$

47. $\dfrac{dy}{dx} = x\sqrt{x-3}$

$\qquad y = \displaystyle\int x\sqrt{x-3}\,dx = \int (u^2+3)(u)(2u)\,du = 2\int (u^4 + 3u^2)\,du = 2\left(\dfrac{u^5}{5} + u^3 \right) + C$

$\qquad\qquad\qquad\qquad = \dfrac{2}{5}(x-3)^{5/2} + 2(x-3)^{3/2} + C$

Let $u = \sqrt{x-3}$, then $x = u^2 + 3$ and $dx = 2u\,du$.

48. $\dfrac{dy}{dx} = xe^x$

$\qquad y = \displaystyle\int xe^x \, dx = xe^x - e^x + C$

49. (a) $N = L - Ce^{-kt}$, $L = 750$

When $t = 0$, $N = 100$.

$$100 = 750 - Ce^{-k(0)}$$

$$-650 = -C(1) \Rightarrow C = 650$$

$$N = 750 - 650e^{-kt}$$

When $t = 2$, $N = 160$.

$$160 = 750 - 650e^{-k(2)}$$

$$\frac{590}{650} = e^{-2k}$$

$$\ln\left(\frac{59}{65}\right) = -2k$$

$$-\frac{1}{2}\ln\left(\frac{59}{65}\right) = k$$

$$k = \frac{1}{2}\ln\left(\frac{65}{59}\right)$$

$$N = 750 - 650e^{-(1/2)\ln(65/59)t}$$

$$\approx 750 - 650e^{-0.0484t}$$

(b) $$\frac{dN}{dt} = -650\left[-\frac{1}{2}\ln\left(\frac{65}{59}\right)\right]e^{-(1/2)\ln(65/59)t} = 325\ln\left(\frac{65}{59}\right)e^{-(1/2)\ln(65/59)t}$$

$$k(L - N) = k\left[750 - \left(750 - 650e^{-(1/2)\ln(65/59)t}\right)\right]$$

$$= \frac{1}{2}\ln\left(\frac{65}{59}\right)\left[650e^{-(1/2)\ln(65/59)t}\right] = 325\ln\left(\frac{65}{59}\right)e^{-(1/2)\ln(65/59)t}$$

Therefore, $\dfrac{dN}{dt} = k(L - N)$.

(c)

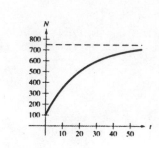

50. (a) $A = Ce^{kt}$

$$\frac{dA}{dt} = Cke^{kt} = k(Ce^{kt}) = kA$$

(b) $A = Ce^{kt}$

When $t = 0$, $A = \$1000$.

$$1000 = Ce^{k(0)} \Rightarrow C = 1000$$

$$A = 1000e^{kt}$$

When $t = 10$, $A = \$3320.12$.

$$3320.12 = 1000e^{k(10)}$$

$$3.32012 = e^{10k}$$

$$\ln 3.32012 = 10k$$

$$k = \frac{\ln 3.32012}{10} \approx 0.12$$

Particular solution: $A = 1000e^{0.12t}$

Section 17.2 Separation of Variables in First-Order Equations

1. $\dfrac{dy}{dx} = \dfrac{x}{y}$

$$\int y\, dy = \int x\, dx$$

$$\frac{y^2}{2} = \frac{x^2}{2} + C_1$$

$$y^2 - x^2 = C$$

2. $\dfrac{dy}{dx} = \dfrac{x^2 + 2}{3y^2}$

$$\int 3y^2\, dy = \int (x^2 + 2)\, dx$$

$$y^3 = \frac{x^3}{3} + 2x + C$$

3. $(2 + x)y' = 3y$

$$\int \frac{dy}{y} = \int \frac{3}{2 + x}\, dx$$

$$\ln y = \ln(2 + x)^3 + \ln C$$

$$y = C(x + 2)^3$$

4. $xy' = y$

$$\int \frac{dy}{y} = \int \frac{dx}{x}$$

$$\ln y = \ln x + \ln C = \ln Cx$$

$$y = Cx$$

5. $yy' = \sin x$

$$\int y\, dy = \int \sin x\, dx$$

$$\frac{y^2}{2} = -\cos x + C_1$$

$$y^2 = -2\cos x + C$$

6. $\sqrt{1 - 4x^2}\, y' = 1$

$$\int dy = \int \frac{1}{\sqrt{1 - 4x^2}}\, dx$$

$$y = \frac{1}{2}\arcsin 2x + C$$

7. $yy' - e^x = 0$

$$\int y\, dy = \int e^x\, dx$$

$$\frac{y^2}{2} = e^x + C_1$$

$$y^2 = 2e^x + C$$

Initial condition: $y(0) = 4$, $16 = 2 + C$, $C = 14$
Particular solution: $y^2 = 2e^x + 14$

8. $\sqrt{x} + \sqrt{y}\, y' = 0$

$$\int y^{1/2}\, dy = -\int x^{1/2}\, dx$$

$$\tfrac{2}{3}y^{3/2} = -\tfrac{2}{3}x^{3/2} + C_1$$

$$y^{3/2} + x^{3/2} = C$$

Initial condition:
$$y(1) = 4, \quad (4)^{3/2} + (1)^{3/2} = 8 + 1 = 9 = C$$
Particular solution: $y^{3/2} + x^{3/2} = 9$

9. $y(x + 1) + y' = 0$

$$\int \frac{dy}{y} = -\int (x + 1)\, dx$$

$$\ln y = -\frac{(x + 1)^2}{2} + C_1$$

$$y = Ce^{-[(x+1)^2]/2}$$

Initial condition: $y(-2) = 1$, $1 = Ce^{-1/2}$, $C = e^{1/2}$
Particular solution: $y = e^{[1-(x+1)^2]/2} = e^{-(x^2+2x)/2}$

10. $xyy' - \ln x = 0$

$$\int y\, dy = \int \frac{\ln x}{x}\, dx$$

$$\frac{y^2}{2} = \frac{\ln^2 x}{2} + C_1$$

$$y^2 = \ln^2 x + C$$

Initial condition: $y(1) = 0$, $0 = \ln^2(1) + C = C$
Particular solution: $y^2 = \ln^2 x$

11. $(1+x^2)y' - (1+y^2) = 0$

$$\int \frac{dy}{1+y^2} = \int \frac{dx}{1+x^2}$$

$$\arctan y = \arctan x + C_1$$

$$C_1 = \arctan y - \arctan x$$

$$C = \tan C_1 = \tan(\arctan y - \arctan x)$$

$$= \frac{\tan(\arctan y) - \tan(\arctan x)}{1 + \tan(\arctan y)\tan(\arctan x)}$$

$$C = \frac{y-x}{1+xy}$$

Initial condition: $y(0) = \sqrt{3}$, $C = \dfrac{\sqrt{3} - 0}{1 + 0} = \sqrt{3}$

Particular solution: $\dfrac{y-x}{1+xy} = \sqrt{3}$, $y = \dfrac{x+\sqrt{3}}{1-\sqrt{3}\,x}$

12. $\sqrt{1-x^2}\,y' - \sqrt{1-y^2} = 0$

$$\int \frac{dy}{\sqrt{1-y^2}} = \int \frac{dx}{\sqrt{1-x^2}}$$

$$\arcsin y = \arcsin x + C$$

Initial condition: $y(0) = 1$, $\dfrac{\pi}{2} = 0 + C = C$

Particular solution:

$$\arcsin y = \arcsin x + \frac{\pi}{2}$$

$$y = \sin\left(\arcsin x + \frac{\pi}{2}\right)$$

$$y = x\cos\frac{\pi}{2} + [\cos(\arcsin x)]\sin\frac{\pi}{2}$$

$$y = \cos(\arcsin x) = \sqrt{1-x^2}$$

13. $dP - kP\,dt = 0$

$$\int \frac{dP}{P} = k\int dt$$

$$\ln P = kt + C_1$$

$$P = Ce^{kt}$$

Initial condition: $P(0) = P_0$, $P_0 = Ce^0 = C$
Particular solution: $P = P_0 e^{kt}$

14. $dT + k(T - 70)\,dt = 0$

$$\int \frac{dT}{T-70} = -k\int dt$$

$$\ln(T - 70) = -kt + C_1$$

$$T - 70 = Ce^{-kt}$$

Initial condition:

$$T(0) = 140, \quad 140 - 70 = 70 = Ce^0 = C$$

Particular solution:

$$T - 70 = 70e^{-kt}, \quad T = 70(1 + e^{-kt})$$

15. $f(x,\ y) = x^3 - 4xy^2 + y^3$

$f(tx,\ ty) = t^3x^3 - 4txt^2y^2 + t^3y^3$

$\qquad = t^3(x^3 - 4xy^2 + y^3)$

Homogeneous of degree 3

16. $f(x,\ y) = \dfrac{xy}{\sqrt{x^2 + y^2}}$

$f(tx,\ ty) = \dfrac{txty}{\sqrt{t^2x^2 + t^2y^2}}$

$\qquad = \dfrac{t^2xy}{t\sqrt{x^2+y^2}} = t\dfrac{xy}{\sqrt{x^2+y^2}}$

Homogeneous of degree 1

17. $f(x,\ y) = 2\ln xy$

$f(tx,\ ty) = 2\ln txty$

$\qquad = 2\ln t^2xy = 2(\ln t^2 + \ln xy)$

Not homogeneous

18. $f(x,\ y) = \tan(x+y)$

$f(tx,\ ty) = \tan(tx + ty) = \tan[t(x+y)]$

Not homogeneous

19. $f(x,\ y) = 2\ln\dfrac{x}{y}$

$f(tx,\ ty) = 2\ln\dfrac{tx}{ty} = 2\ln\dfrac{x}{y}$

Homogeneous of degree 0

20. $f(x,\ y) = \tan\dfrac{y}{x}$

$f(tx,\ ty) = \tan\dfrac{ty}{tx} = \tan\dfrac{y}{x}$

Homogeneous of degree 0

21.
$$y' = \frac{x+y}{2x}$$
$$v + x\frac{dv}{dx} = \frac{x+vx}{2x}$$
$$2\int \frac{dv}{1-v} = \int \frac{dx}{x}$$
$$-\ln(1-v)^2 = \ln x + \ln C = \ln Cx$$
$$\frac{1}{(1-v)^2} = Cx$$
$$\frac{1}{[1-(y/x)]^2} = Cx$$
$$\frac{x^2}{(x-y)^2} = Cx$$
$$x = C(x-y)^2$$

22.
$$y' = \frac{2x+y}{y}$$
$$v + x\frac{dv}{dx} = \frac{2x+xv}{xv}$$
$$\int \frac{v}{v^2-v-2}\,dv = -\int \frac{dx}{x}$$
$$\frac{2}{3}\ln(v-2) + \frac{1}{3}\ln(v+1) = -\ln x + \ln C_1 = \ln \frac{C_1}{x}$$
$$(v-2)^2(v+1) = \frac{C}{x^3}$$
$$\left(\frac{y}{x}-2\right)^2\left(\frac{y}{x}+1\right) = \frac{C}{x^3}$$
$$(y-2x)^2(y+x) = C$$

23.
$$y' = \frac{x-y}{x+y}$$
$$v + x\frac{dv}{dx} = \frac{x-xv}{x+xv}$$
$$\int \frac{v+1}{v^2+2v-1}\,dv = -\int \frac{dx}{x}$$
$$\frac{1}{2}\ln(v^2+2v-1) = -\ln x + \ln C_1 = \ln \frac{C_1}{x}$$
$$v^2+2v-1 = \frac{C}{x^2}$$
$$\left(\frac{y^2}{x^2}+2\frac{y}{x}-1\right) = \frac{C}{x^2}$$
$$y^2+2xy-x^2 = C$$

24.
$$y' = \frac{x^2+y^2}{2xy}$$
$$v + x\frac{dv}{dx} = \frac{x^2+v^2x^2}{2x^2v}$$
$$\int \frac{2v}{v^2-1}\,dv = -\int \frac{dx}{x}$$
$$\ln(v^2-1) = -\ln x + \ln C = \ln \frac{C}{x}$$
$$v^2-1 = \frac{C}{x}$$
$$\frac{y^2}{x^2}-1 = \frac{C}{x}$$
$$y^2-x^2 = Cx$$

25.
$$y' = \frac{xy}{x^2-y^2}$$
$$v + x\frac{dv}{dx} = \frac{x^2v}{x^2-x^2v^2}$$
$$\int \frac{1-v^2}{v^3}\,dv = \int \frac{dx}{x}$$
$$-\frac{1}{2v^2} - \ln v = \ln x + \ln C_1 = \ln C_1 x$$
$$\frac{-1}{2v^2} = \ln C_1 xv$$
$$\frac{-x^2}{2y^2} = \ln C_1 y$$
$$y = Ce^{-x^2/2y^2}$$

26.
$$y' = \frac{3x+2y}{x}$$
$$v + x\frac{dv}{dx} = \frac{3x+2vx}{x} = 3+2v$$
$$\int \frac{dv}{v+3} = \int \frac{dx}{x}$$
$$\ln(v+3) = \ln x + \ln C$$
$$v+3 = Cx$$
$$\frac{y}{x}+3 = Cx$$
$$y = Cx^2 - 3x$$

27.
$$x\,dy - (2xe^{-y/x} + y)\,dx = 0$$
$$x(v\,dx + x\,dv) - (2xe^{-v} + vx)\,dx = 0$$
$$\int e^v\,dv = \int \frac{2}{x}\,dx$$
$$e^v = \ln C_1 x^2$$
$$e^{y/x} = \ln C_1 + \ln x^2$$
$$e^{y/x} = C + \ln x^2$$

Initial condition: $y(1) = 0, \quad 1 = C$
Particular solution: $e^{y/x} = 1 + \ln x^2$

28.
$$-y^2\,dx + x(x + y)\,dy = 0$$
$$-x^2 v^2\,dx + (x^2 + x^2 v)(v\,dx + x\,dv) = 0$$
$$\int \frac{1+v}{v}\,dv = -\int \frac{dx}{x}$$
$$v + \ln v = -\ln x + \ln C_1 = \ln \frac{C_1}{x}$$
$$v = \ln \frac{C_1}{xv}$$
$$\frac{C_1}{vx} = e^v$$
$$\frac{C_1}{y} = e^{y/x}$$
$$y = Ce^{-y/x}$$

Initial condition: $y(1) = 1, \quad 1 = Ce^{-1}, \quad C = e$
Particular solution: $y = e^{1-y/x}$

29.
$$\left(x\sec\frac{y}{x} + y\right)dx - x\,dy = 0$$
$$(x\sec v + xv)\,dx - x(v\,dx + x\,dv) = 0$$
$$\int \cos v\,dv = \int \frac{dx}{x}$$
$$\sin v = \ln x + \ln C_1$$
$$x = Ce^{\sin v} = Ce^{\sin(y/x)}$$

Initial condition: $y(1) = 0, \quad 1 = Ce^0 = C$
Particular solution: $x = e^{\sin(y/x)}$

30.
$$(y - \sqrt{x^2 - y^2})\,dx - x\,dy = 0$$
$$(xv - \sqrt{x^2 - x^2 v^2})\,dx - x(v\,dx + x\,dv) = 0$$
$$\int \frac{dv}{\sqrt{1 - v^2}} = -\int \frac{dx}{x}$$
$$\arcsin v = -\ln x + \ln C_1$$
$$\frac{C_1}{x} = e^{\arcsin v} = e^{\arcsin(y/x)}$$
$$x = Ce^{-\arcsin(y/x)}$$

Initial condition: $y(1) = 0, \quad 1 = Ce^0 = C$
Particular solution: $x = e^{-\arcsin(y/x)}$

31. Given family (circles): $\quad x^2 + y^2 = C$
$$2x + 2yy' = 0$$
$$y' = -\frac{x}{y}$$

Orthogonal trajectory (lines): $\quad y' = \dfrac{y}{x}$
$$\int \frac{dy}{y} = \int \frac{dx}{x}$$
$$\ln y = \ln x + \ln K$$
$$y = Kx$$

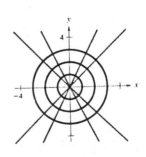

32. Given family (hyperbolas): $2x^2 - y^2 = C$

$$4x - 2yy' = 0$$

$$y' = \frac{2x}{y}$$

Orthogonal trajectory (hyperbolas): $\qquad y' = -\frac{y}{2x}$

$$2\int \frac{dy}{y} = -\int \frac{dx}{x}$$

$$2\ln y = -\ln x + \ln K$$

$$y^2 = \frac{K}{x}$$

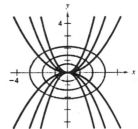

33. Given family (parabolas): $x^2 = Cy$

$$2x = Cy'$$

$$y' = \frac{2x}{C} = \frac{2x}{x^2/y} = \frac{2y}{x}$$

Orthogonal trajectory (ellipses): $\qquad y' = -\frac{x}{2y}$

$$2\int y\,dy = -\int x\,dx$$

$$y^2 = -\frac{x^2}{2} + K_1$$

$$x^2 + 2y^2 = K$$

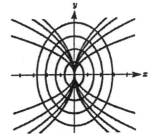

34. Given family (parabolas): $y^2 = 2Cx$

$$2yy' = 2C$$

$$y' = \frac{C}{y} = \frac{y^2}{2x}\left(\frac{1}{y}\right) = \frac{y}{2x}$$

Orthogonal trajectory (ellipses): $\qquad y' = -\frac{2x}{y}$

$$\int y\,dy = -\int 2x\,dx$$

$$\frac{y^2}{2} = -x^2 + K_1$$

$$2x^2 + y^2 = K$$

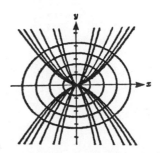

35. Given family (parabolas): $y^2 = Cx^3$

$$2yy' = 3Cx^2$$

$$y' = \frac{3Cx^2}{2y} = \frac{3x^2}{2y}\left(\frac{y^2}{x^3}\right) = \frac{3y}{2x}$$

Orthogonal trajectory (ellipses): $\qquad y' = -\frac{2x}{3y}$

$$3\int y\,dy = -2\int x\,dx$$

$$\frac{3y^2}{2} = -x^2 + K_1$$

$$3y^2 + 2x^2 = K$$

36. Given family (exponential functions): $y = Ce^x$

$$y' = Ce^x = y$$

Orthogonal trajectory (parabolas): $y' = -\dfrac{1}{y}$

$$\int y \, dy = -\int dx$$

$$\frac{y^2}{2} = -x + K_1$$

$$y^2 = -2x + K$$

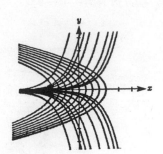

37. $\dfrac{dy}{dx} = \dfrac{-9x}{16y}$

$$\int 16y \, dy = -\int 9x \, dx$$

$$8y^2 = \frac{-9}{2}x^2 + C$$

Initial condition:

$$y(1) = 1, \quad 8 = -\frac{9}{2} + C, \quad C = \frac{25}{2}$$

Particular solution:

$$8y^2 = \frac{-9}{2}x^2 + \frac{25}{2}, \quad 16y^2 + 9x^2 = 25$$

38. $\dfrac{dy}{dx} = \dfrac{2y}{3x}$

$$\int \frac{3}{y} \, dy = \int \frac{2}{x} \, dx$$

$$\ln y^3 = \ln x^2 + \ln C$$

$$y^3 = Cx^2$$

Initial condition: $y(8) = 2, \quad 2^3 = C(8^2), \quad C = \dfrac{1}{8}$

Particular solution: $8y^3 = x^2, \quad y = \dfrac{1}{2}x^{2/3}$

39. (a) $\dfrac{dA}{dt} = kA, \quad \int \dfrac{dA}{A} = \int k \, dt, \quad \ln A = kt + C_1, \quad A = Pe^{kt}$

  (b) $P = 1000, \quad R = 0.11, \quad A = 1000e^{0.11(10)} \approx \$3,004.17$

  (c) $2000 = 1000e^{0.11t}, \quad 0.11t = \ln 2, \quad t = \dfrac{1}{0.11}\ln 2 \approx 6.3$ years

40. $\dfrac{dP}{dt} = kP, \quad \int \dfrac{dP}{P} = \int k \, dt, \quad \ln P = kt + C_1, \quad P = Ce^{kt}$

Initial conditions: $P(2) = 180$

$$P(4) = 300$$

$$180 = Ce^{2k}$$

$$300 = Ce^{4k} = C(e^{2k})^2$$

$$300 = C\left(\frac{180}{C}\right)^2 = \frac{32400}{C}$$

$$C = 108$$

$$\frac{180}{108} = e^{2k}$$

$$k = \frac{1}{2}\ln\frac{180}{108} = \ln\sqrt{5/3}$$

Particular solution: $P = 108e^{t \ln \sqrt{5/3}} \approx 108e^{0.2554t}$

41. $\dfrac{dy}{dt} = ky, \quad y = Ce^{kt}$

Initial conditions:

$$y(0) = y_0$$

$$y(1600) = \dfrac{y_0}{2}$$

$$C = y_0$$

$$\dfrac{y_0}{2} = y_0 e^{1600k}$$

$$k = \dfrac{\ln(1/2)}{1600}$$

Particular solution: $y = y_0 e^{-t(\ln 2)/1600}$

When $t = 25$, $y = y_0 e^{-(\ln 2)/64} \approx 0.989 y_0$,

$y = 98.9\%$ of y_0.

42. $\dfrac{dy}{dt} = ky, \quad y = Ce^{kt}$

Initial conditions: $y(0) = 20$

$$y(1) = 16$$

$$20 = Ce^0 = C$$

$$16 = 20e^k$$

$$k = \ln \dfrac{4}{5}$$

Particular solution: $y = 20e^{t \ln(4/5)}$

When 75% has been changed,

$$5 = 20e^{t \ln(4/5)}$$

$$\dfrac{1}{4} = e^{t \ln(4/5)}$$

$$t = \dfrac{\ln(1/4)}{\ln(4/5)} \approx 6.2 \text{ hours.}$$

43. $\dfrac{dT}{dt} = k(T - 70)$

$$\int \dfrac{dT}{T - 70} = \int k \, dt$$

$$T - 70 = Ce^{kt}$$

Initial conditions:

$$T(0) = 350$$

$$T(45) = 150$$

$$350 - 70 = 280 = Ce^0 = C$$

$$150 - 70 = 80 = 280e^{45k}$$

$$k = \dfrac{\ln(2/7)}{45}$$

Particular solution: $T = 70 + 280e^{t[\ln(2/7)]/45}$

When $T = 80$,

$$80 = 70 + 280e^{t[\ln(2/7)]/45}$$

$$\dfrac{t \ln(2/7)}{45} = \ln\left(\dfrac{1}{28}\right)$$

$$t = \dfrac{45 \ln(1/28)}{\ln(2/7)} \approx 119.7 \text{ minutes.}$$

44. $F = mv\dfrac{dv}{ds} = \dfrac{w}{32} v \dfrac{dv}{ds}$

$$\dfrac{640}{32} \dfrac{dv}{dt} = 60 - 3v$$

$$20 \int \dfrac{dv}{20 - v} = 3 \int dt$$

$$-20 \ln(20 - v) = 3t + C_1$$

$$20 - v = Ce^{-3t/20}$$

$$v = 20 - Ce^{-3t/20}$$

Initial condition: $v(0) = 0$, $\quad 0 = 20 - C$, $\quad C = 20$

Particular solution: $v = 20(1 - e^{-3t/20})$

Limiting speed:

$$\lim_{t \to \infty} 20(1 - e^{-3t/20}) = 20(1 - 0) = 20 \text{ ft/sec}$$

Section 17.3   Exact First-Order Equations

1. $(2x - 3y)\,dx + (2y - 3x)\,dy = 0$

$\dfrac{\partial M}{\partial y} = -3 = \dfrac{\partial N}{\partial x}$

$U(x,\ y) = x^2 - 3xy + f(y)$

$f'(y) = 2y$

$f(y) = y^2 + C_1$

$x^2 - 3xy + y^2 = C$

2. $ye^x\,dx + e^x\,dy = 0$

$\dfrac{\partial M}{\partial y} = e^x = \dfrac{\partial N}{\partial x}$

$U(x,\ y) = ye^x + g(x)$

$g'(x) = 0$

$g(x) = C_1$

$ye^x = C$

3. $(3y^2 + 10xy^2)\,dx + (6xy - 2 + 10x^2 y)\,dy = 0$

$\dfrac{\partial M}{\partial y} = 6y + 20xy = \dfrac{\partial N}{\partial x}$

$U(x,\ y) = 3y^2 x + 5x^2 y^2 + f(y)$

$f'(y) = -2$

$f(y) = -2y + C_1$

$3y^2 x + 5x^2 y^2 - 2y = C$

4. $2\cos(2x - y)\,dx - \cos(2x - y)\,dy = 0$

$\dfrac{\partial M}{\partial y} = 2\sin(2x - y) = \dfrac{\partial N}{\partial x}$

$U(x,\ y) = \sin(2x - y) + f(y)$

$f'(y) = 0$

$f(y) = C_1$

$\sin(2x - y) = C$

5. $(4x^3 - 6xy^2)\,dx + (4y^3 - 6xy)\,dy = 0$

$\dfrac{\partial M}{\partial y} = -12xy$

$\dfrac{\partial N}{\partial x} = -6y$

Not exact

6. $2y^2 e^{xy^2}\,dx + 2xye^{xy^2}\,dy = 0$

$\dfrac{\partial M}{\partial y} = 4(xy^3 + y)e^{xy^2}$

$\dfrac{\partial N}{\partial x} = 2(xy^3 + y)e^{xy^2}$

Not exact

7. $\dfrac{-y}{x^2 + y^2}\,dx + \dfrac{x}{x^2 + y^2}\,dy = 0$

$\dfrac{\partial M}{\partial y} = \dfrac{y^2 - x^2}{(x^2 + y^2)^2} = \dfrac{\partial N}{\partial x}$

$U(x,\ y) = -\arctan\dfrac{x}{y} + f(y)$

$f'(y) = 0$

$f(y) = C_1$

$\arctan\dfrac{x}{y} = C$

8. $xe^{-(x^2 + y^2)}\,dx + ye^{-(x^2 + y^2)}\,dy = 0$

$\dfrac{\partial M}{\partial y} = -2xye^{-(x^2 + y^2)} = \dfrac{\partial N}{\partial x}$

$U(x,\ y) = -\dfrac{1}{2}e^{-(x^2 + y^2)} + f(y)$

$f'(y) = 0$

$f(y) = C_1$

$C = e^{-(x^2 + y^2)}$

9. $\left(\dfrac{y}{x - y}\right)^2 dx + \left(\dfrac{x}{x - y}\right)^2 dy = 0$

$\dfrac{\partial M}{\partial y} = \dfrac{2xy}{(x - y)^3}$

$\dfrac{\partial N}{\partial x} = \dfrac{-2xy}{(x - y)^3}$

Not exact

10. $(ye^y \cos xy)\,dx + e^y(x\cos xy + \sin xy)\,dy = 0$

$\dfrac{\partial M}{\partial y} = -xye^y \sin xy + (y + 1)e^y \cos xy = \dfrac{\partial N}{\partial x}$

$U(x,\ y) = e^y \sin xy + f(y)$

$f'(y) = 0$

$f(y) = C_1$

$e^y \sin xy = C$

11. $\dfrac{y}{x-1}\,dx + [\ln(x-1) + 2y]\,dy = 0$

$\dfrac{\partial M}{\partial y} = \dfrac{1}{x-1} = \dfrac{\partial N}{\partial x}$

$U(x,\ y) = y\ln(x-1) + f(y)$

$f'(y) = 2y$

$f(y) = y^2 + C_1$

$y\ln(x-1) + y^2 = C$

Initial condition: $y(2) = 4,\quad 4\ln(1) + 16 = C,\quad C = 16$

Particular solution: $y\ln(x-1) + y^2 = 16$

12. $\dfrac{x}{\sqrt{x^2+y^2}}\,dx + \dfrac{y}{\sqrt{x^2+y^2}}\,dy = 0$

$\dfrac{\partial M}{\partial y} = \dfrac{-xy}{\sqrt{(x^2+y^2)^3}} = \dfrac{\partial N}{\partial x}$

$U(x,\ y) = \sqrt{x^2+y^2} + f(y)$

$f'(y) = 0$

$f(y) = C_1$

$\sqrt{x^2+y^2} = C$

Initial condition: $y(4) = 3,\quad \sqrt{16+9} = \sqrt{25} = 5 = C$

Particular solution: $\sqrt{x^2+y^2} = 5$

13. $\dfrac{x}{x^2+y^2}\,dx + \dfrac{y}{x^2+y^2}\,dy = 0$

$\dfrac{\partial M}{\partial y} = \dfrac{-2xy}{(x^2+y^2)^2} = \dfrac{\partial N}{\partial x}$

$U(x,\ y) = \dfrac{1}{2}\ln(x^2+y^2) + f(y)$

$f'(y) = 0$

$f(y) = C_1$

$\ln(x^2+y^2) = C_2$

$x^2 + y^2 = C$

Initial condition: $y(0) = 4,\quad 16 = C$

Particular solution: $x^2 + y^2 = 16$

14. $(e^{3x}\sin 3y)\,dx + (e^{3x}\cos 3y)\,dy = 0$

$\dfrac{\partial M}{\partial y} = 3e^{3x}\cos 3y = \dfrac{\partial N}{\partial x}$

$U(x,\ y) = \dfrac{1}{3}e^{3x}\sin 3y + f(y)$

$f'(y) = 0$

$f(y) = C_1$

$e^{3x}\sin 3y = C$

Initial condition: $y(0) = \pi,\quad C = 0$

Particular solution: $e^{3x}\sin 3y = 0$

15. $(2x\tan y + 5)\,dx + (x^2\sec^2 y)\,dy = 0$

$\dfrac{\partial M}{\partial y} = 2x\sec^2 y = \dfrac{\partial N}{\partial x}$

$U(x,\ y) = x^2\tan y + 5x + f(y)$

$f'(y) = 0$

$f(y) = C_1$

$x^2\tan y + 5x = C$

Initial condition: $y(0) = 0,\quad C = 0$

Particular solution: $x^2\tan y + 5x = 0$

16. $(x^2+y^2)\,dx + 2xy\,dy = 0$

$\dfrac{\partial M}{\partial y} = 2y = \dfrac{\partial N}{\partial x}$

$U(x,\ y) = \dfrac{x^3}{3} + xy^2 + f(y)$

$f'(y) = 0$

$f(y) = C_1$

$x^3 + 3xy^2 = C$

Initial condition: $y(3) = 1,\quad C = 36$

Particular solution: $x^3 + 3xy^2 = 36$

17. $y\,dx - (x + 6y^2)\,dy = 0$

$\dfrac{(\partial N/\partial x) - (\partial M/\partial y)}{M} = -\dfrac{2}{y} = k(y)$

Integrating factor: $e^{\int k(y)\,dy} = e^{\ln y^{-2}} = \dfrac{1}{y^2}$

Exact equation: $\dfrac{1}{y}\,dx - \left(\dfrac{x}{y^2} + 6\right)dy = 0$

$U(x,\ y) = \dfrac{x}{y} + f(y)$

$f'(y) = -6$

$f(y) = -6y + C_1$

$\dfrac{x}{y} - 6y = C$

18. $(2x^3 + y)\,dx - x\,dy = 0$

$\dfrac{(\partial M/\partial y) - (\partial N/\partial x)}{N} = -\dfrac{2}{x} = h(x)$

Integrating factor: $e^{\int h(x)\,dx} = e^{\ln x^{-2}} = \dfrac{1}{x^2}$

Exact equation: $\left(2x + \dfrac{y}{x^2}\right)dx - \dfrac{1}{x}\,dy = 0$

$U(x,\ y) = x^2 - \dfrac{y}{x} + f(y)$

$f'(y) = 0$

$f(y) = C_1$

$x^2 - \dfrac{y}{x} = C$

19. $(5x^2 - y)\,dx + x\,dy = 0$

$\dfrac{(\partial M/\partial y) - (\partial N/\partial x)}{N} = \dfrac{-2}{x} = h(x)$

Integrating factor: $e^{\int h(x)\,dx} = e^{\ln x^{-2}} = \dfrac{1}{x^2}$

Exact equation: $\left(5 - \dfrac{y}{x^2}\right) dx + \dfrac{1}{x}\,dy = 0$

$U(x,\ y) = 5x + \dfrac{y}{x} + f(y)$

$f'(y) = 0$

$f(y) = C_1$

$5x + \dfrac{y}{x} = C$

20. $(5x^2 - y^2)\,dx + 2y\,dy = 0$

$\dfrac{(\partial M/\partial y) - (\partial N/\partial x)}{N} = -1 = h(x)$

Integrating factor: $e^{\int h(x)\,dx} = e^{-x}$

Exact equation: $(5x^2 - y^2)e^{-x}\,dx + 2ye^{-x}\,dy = 0$

$U(x,\ y) = -5x^2 e^{-x} - 10xe^{-x} - 10e^{-x} + y^2 e^{-x} + f(y)$

$f'(y) = 0$

$f(y) = C_1$

$y^2 e^{-x} - 5x^2 e^{-x} - 10xe^{-x} - 10e^{-x} = C$

21. $(x + y)\,dx + (\tan x)\,dy = 0$

$\dfrac{(\partial M/\partial y) - (\partial N/\partial x)}{N} = -\tan x = h(x)$

Integrating factor: $e^{\int h(x)\,dx} = e^{\ln \cos x} = \cos x$

Exact equation: $(x + y)\cos x\,dx + \sin x\,dy = 0$

$U(x,\ y) = x\sin x + \cos x + y\sin x + f(y)$

$f'(y) = 0$

$f(y) = C_1$

$x\sin x + \cos x + y\sin x = C$

22. $(2x^2 y - 1)\,dx + x^3\,dy = 0$

$\dfrac{(\partial M/\partial y) - (\partial N/\partial x)}{N} = -\dfrac{1}{x} = h(x)$

Integrating factor: $e^{\int h(x)\,dx} = e^{\ln(1/x)} = \dfrac{1}{x}$

Exact equation: $\left(2xy - \dfrac{1}{x}\right) dx + x^2\,dy = 0$

$U(x,\ y) = x^2 y - \ln|x| + f(y)$

$f'(y) = 0$

$f(y) = C_1$

$x^2 y - \ln|x| = C$

23. $y^2\,dx + (xy - 1)\,dy = 0$

$\dfrac{(\partial N/\partial x) - (\partial M/\partial y)}{M} = -\dfrac{1}{y} = k(y)$

Integrating factor: $e^{\int k(y)\,dy} = e^{\ln(1/y)} = \dfrac{1}{y}$

Exact equation: $y\,dx + \left(x - \dfrac{1}{y}\right) dy = 0$

$U(x,\ y) = xy + f(y)$

$f'(y) = -\dfrac{1}{y}$

$f(y) = -\ln|y| + C_1$

$xy - \ln|y| = C$

24. $(x^2 + 2x + y)\,dx + 2\,dy = 0$

$\dfrac{(\partial M/\partial y) - (\partial N/\partial x)}{N} = \dfrac{1}{2} = h(x)$

Integrating factor: $e^{\int h(x)\,dx} = e^{x/2}$

Exact equation: $(x^2 + 2x + y)e^{x/2}\,dx + 2e^{x/2}\,dy = 0$

$U(x,\ y) = 2(x^2 - 2x + 4 + y)e^{x/2} + f(y)$

$f'(y) = 0$

$f(y) = C_1$

$(x^2 - 2x + 4 + y)e^{x/2} = C$

25. $2y \, dx + (x - \sin \sqrt{y}) \, dy = 0$

$\dfrac{(\partial N / \partial x) - (\partial M / \partial y)}{M} = \dfrac{-1}{2y} = k(y)$

Integrating factor: $e^{\int k(y) \, dy} = e^{\ln(1/\sqrt{y})} = \dfrac{1}{\sqrt{y}}$

Exact equation: $2\sqrt{y} \, dx + \left(\dfrac{x}{\sqrt{y}} - \dfrac{\sin \sqrt{y}}{\sqrt{y}} \right) dy = 0$

$U(x, \, y) = 2\sqrt{y} \, x + f(y)$

$f'(y) = -\dfrac{\sin \sqrt{y}}{\sqrt{y}}$

$f(y) = 2 \cos \sqrt{y} + C_1$

$\sqrt{y} \, x + \cos \sqrt{y} = C$

26. $(-2y^3 + 1) \, dx + (3xy^2 + x^3) \, dy = 0$

$\dfrac{(\partial M / \partial y) - (\partial N / \partial x)}{N} = \dfrac{-3}{x} = h(x)$

Integrating factor: $e^{\int h(x) \, dx} = e^{\ln(1/x^3)} = \dfrac{1}{x^3}$

Exact equation: $\left(\dfrac{-2y^3}{x^3} + \dfrac{1}{x^3} \right) dx + \left(\dfrac{3y^2}{x^2} + 1 \right) dy = 0$

$U(x, \, y) = \dfrac{y^3}{x^2} - \dfrac{1}{2x^2} + f(y)$

$f'(y) = 1$

$f(y) = y + C_1$

$\dfrac{y^3}{x^2} - \dfrac{1}{2x^2} + y = C$

27. $(4x^2 y + 2y^2) \, dx + (3x^3 + 4xy) \, dy = 0$

Integrating factor: xy^2

Exact equation:

$\quad (4x^3 y^3 + 2xy^4) \, dx + (3x^4 y^2 + 4x^2 y^3) \, dy = 0$

$U(x, \, y) = x^4 y^3 + x^2 y^4 + f(y)$

$f'(y) = 0$

$f(y) = C_1$

$x^4 y^3 + x^2 y^4 = C$

28. $(3y^2 + 5x^2 y) \, dx + (3xy + 2x^3) \, dy = 0$

Integrating factor: $x^2 y$

Exact equation:

$\quad (3x^2 y^3 + 5x^4 y^2) \, dx + (3x^3 y^2 + 2x^5 y) \, dy = 0$

$U(x, \, y) = x^3 y^3 + x^5 y^2 + f(y)$

$f'(y) = 0$

$f(y) = C_1$

$x^3 y^3 + x^5 y^2 = C$

29. $(-y^5 + x^2 y) \, dx + (2xy^4 - 2x^3) \, dy = 0$

Integrating factor: $x^{-2} y^{-3}$

Exact equation: $\left(-\dfrac{y^2}{x^2} + \dfrac{1}{y^2} \right) dx + \left(2\dfrac{y}{x} - 2\dfrac{x}{y^3} \right) dy = 0$

$U(x, \, y) = \dfrac{y^2}{x} + \dfrac{x}{y^2} + f(y)$

$f'(y) = 0$

$f(y) = C_1$

$\dfrac{y^2}{x} + \dfrac{x}{y^2} = C$

30. $-y^3 \, dx + (xy^2 - x^2) \, dy = 0$

Integrating factor: $x^{-2} y^{-2}$

Exact equation: $\dfrac{-y}{x^2} \, dx + \left(\dfrac{1}{x} - \dfrac{1}{y^2} \right) dy = 0$

$U(x, \, y) = \dfrac{y}{x} + f(y)$

$f'(y) = \dfrac{-1}{y^2}$

$f(y) = \dfrac{1}{y} + C_1$

$\dfrac{y}{x} + \dfrac{1}{y} = C$

31. $y \, dx - x \, dy = 0$

(a) $\dfrac{1}{x^2}$, $\quad \dfrac{y}{x^2} \, dx - \dfrac{1}{x} \, dy = 0$, $\quad \dfrac{\partial M}{\partial y} = \dfrac{1}{x^2} = \dfrac{\partial N}{\partial x}$

(b) $\dfrac{1}{y^2}$, $\quad \dfrac{1}{y} \, dx - \dfrac{x}{y^2} \, dy = 0$, $\quad \dfrac{\partial M}{\partial y} = \dfrac{-1}{y^2} = \dfrac{\partial N}{\partial x}$

(c) $\dfrac{1}{xy}$, $\quad \dfrac{1}{x} \, dx - \dfrac{1}{y} \, dy = 0$, $\quad \dfrac{\partial M}{\partial y} = 0 = \dfrac{\partial N}{\partial x}$

(d) $\dfrac{1}{x^2 + y^2}$, $\quad \dfrac{y}{x^2 + y^2} \, dx - \dfrac{x}{x^2 + y^2} \, dy = 0$, $\quad \dfrac{\partial M}{\partial y} = \dfrac{x^2 - y^2}{(x^2 + y^2)^2} = \dfrac{\partial N}{\partial x}$

32. $(axy^2 + by)\,dx + (bx^2y + ax)\,dy = 0$

Exact equation: $\dfrac{\partial M}{\partial y} = 2axy + b,\quad \dfrac{\partial N}{\partial x} = 2bxy + a,\quad \dfrac{\partial M}{\partial y} = \dfrac{\partial M}{\partial x}$ only if $a = b$

Integrating factor: $x^m y^n$

$(ax^{m+1}y^{n+2} + bx^m y^{n+1})\,dx + (bx^{m+2}y^{n+1} + ax^{m+1}y^n)\,dy = 0$

$\left.\begin{array}{l} \dfrac{\partial M}{\partial y} = a(n+2)x^{m+1}y^{n+1} + b(n+1)x^m y^n \\[2mm] \dfrac{\partial N}{\partial x} = b(m+2)x^{m+1}y^{n+1} + a(m+1)x^m y^n \end{array}\right\} \quad \begin{array}{l} a(n+2) = b(m+2) \\[2mm] b(n+1) = a(m+1) \end{array}$

$\left.\begin{array}{l} an - bm = 2(b - a) \\[2mm] bn - am = a - b \end{array}\right\} \quad \begin{array}{l} abn - b^2 m = 2b(b - a) \\[2mm] abn - a^2 m = a(a - b) \\ \hline \end{array}$

$$(a^2 - b^2)m = -(2b + a)(a - b)$$
$$m = -\frac{2b + a}{a + b}$$

$$bn - a\left(-\frac{2b + a}{a + b}\right) = a - b$$

$$bn = \frac{-2ab - a^2 + a^2 - b^2}{a + b} = \frac{-b(2a + b)}{a + b}$$

$$n = -\frac{2a + b}{a + b}$$

33. $\mathbf{F}(x,\,y) = \dfrac{y}{\sqrt{x^2 + y^2}}\,\mathbf{i} - \dfrac{x}{\sqrt{x^2 + y^2}}\,\mathbf{j}$

$\dfrac{dy}{dx} = -\dfrac{x}{y}$

$y\,dy + x\,dx = 0$

$y^2 + x^2 = C^2$

Family of circles

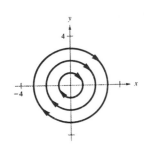

34. $\mathbf{F}(x,\,y) = \dfrac{x}{\sqrt{x^2 + y^2}}\,\mathbf{i} - \dfrac{y}{\sqrt{x^2 + y^2}}\,\mathbf{j}$

$\dfrac{dy}{dx} = -\dfrac{y}{x}$

$x\,dy + y\,dx = 0$

$xy = C$

Family of hyperbolas

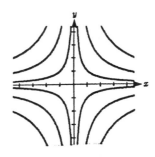

35. $\mathbf{F}(x,\,y) = 4x^2y\,\mathbf{i} - \left(2xy^2 + \dfrac{x}{y^2}\right)\mathbf{j}$

$\dfrac{dy}{dx} = \dfrac{-y}{2x} - \dfrac{1}{4xy^3}$

$\dfrac{8y^3}{2y^4 + 1}\,dy = -\dfrac{2}{x}\,dx$

$\ln(2y^4 + 1) = \ln\left(\dfrac{1}{x^2}\right) + \ln C$

$2y^4 + 1 = \dfrac{C}{x^2}$

$2x^2y^4 + x^2 = C$

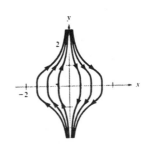

36. $\mathbf{F}(x,\ y) = (1+x^2)\,\mathbf{i} - 2xy\,\mathbf{j}$

$$\frac{dy}{dx} = \frac{-2xy}{1+x^2}$$

$$\frac{1}{y}\,dy = -\frac{2x}{1+x^2}\,dx$$

$$\ln y = \ln\left(\frac{1}{1+x^2}\right) + \ln C$$

$$y = \frac{C}{1+x^2}$$

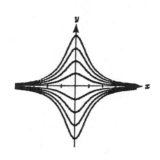

37. $\dfrac{dy}{dx} = \dfrac{y-x}{3y-x}$

$(x-y)\,dx + (3y-x)\,dy = 0$

$\dfrac{\partial M}{\partial y} = -1 = \dfrac{\partial N}{\partial x}$

$U(x,\ y) = \dfrac{x^2}{2} - xy + f(y)$

$f'(y) = 3y$

$f(y) = \dfrac{3y^2}{2} + C_1$

$x^2 - 2xy + 3y^2 = C$

Initial condition: $y(2) = 1,\ \ 4 - 4 + 3 = C,\ C = 3$

Particular solution: $x^2 - 2xy + 3y^2 = 3$

38. $\dfrac{dy}{dx} = \dfrac{-2xy}{x^2 + y^2}$

$2xy\,dx + (x^2 + y^2)\,dy = 0$

$\dfrac{\partial M}{\partial y} = 2x = \dfrac{\partial N}{\partial x}$

$U(x,\ y) = x^2 y + f(y)$

$f'(y) = y^2$

$f(y) = \dfrac{y^3}{3} + C_1$

$3x^2 y + y^3 = C$

Initial condition: $y(0) = 2,\ \ 8 = C$

Particular solution: $3x^2 y + y^3 = 8$

39. $E(x) = \dfrac{20x - y}{2y - 10x} = \dfrac{x}{y}\dfrac{dy}{dx}$

$(20xy - y^2)\,dx + (10x^2 - 2xy)\,dy = 0$

$\dfrac{\partial M}{\partial y} = 20x - 2y = \dfrac{\partial N}{\partial x}$

$U(x,\ y) = 10x^2 y - xy^2 + f(y)$

$f'(y) = 0$

$f(y) = C_1$

$10x^2 y - xy^2 = K$

Initial condition: $C(100) = 500,\ \ 100 \le x,\ \ K = 25{,}000{,}000$

$$10x^2 y - xy^2 = 25{,}000{,}000$$

$xy^2 - 10x^2 y + 25{,}000{,}000 = 0 \quad \text{Quadratic Formula}$

$$y = \frac{10x^2 + \sqrt{100x^4 - 4x(25{,}000{,}000)}}{2x} = \frac{5\left(x^2 + \sqrt{x^4 - 1{,}000{,}000x}\right)}{x}$$

Section 17.4   First-Order Linear Differential Equations

1. $\dfrac{dy}{dx} + \left(\dfrac{1}{x}\right)y = 3x + 4$

Integrating factor: $e^{\int (1/x)\,dx} = e^{\ln x} = x$

$xy = \displaystyle\int x(3x + 4)\,dx = x^3 + 2x^2 + C$

$y = x^2 + 2x + \dfrac{C}{x}$

2. $\dfrac{dy}{dx} + \left(\dfrac{2}{x}\right)y = 3x + 1$

Integrating factor: $e^{\int (2/x)\,dx} = e^{\ln x^2} = x^2$

$x^2 y = \displaystyle\int x^2(3x + 1)\,dx = \dfrac{3}{4}x^4 + \dfrac{1}{3}x^3 + C$

$y = \dfrac{3x^2}{4} + \dfrac{x}{3} + \dfrac{C}{x^2}$

3. $\dfrac{dy}{dx} = e^x - y, \quad \dfrac{dy}{dx} + y = e^x$

Integrating factor: $e^{\int dx} = e^x$

$ye^x = \displaystyle\int e^{2x}\,dx = \dfrac{1}{2}e^{2x} + C$

$y = \dfrac{1}{2}e^x + Ce^{-x}$

4. $y' + 2y = \sin x$

Integrating factor: $e^{\int 2\,dx} = e^{2x}$

$ye^{2x} = \displaystyle\int e^{2x}\sin x\,dx = \tfrac{1}{5}e^{2x}(2\sin x - \cos x) + C$

$y = \tfrac{1}{5}(2\sin x - \cos x) + Ce^{-2x}$

5. $y' - y = \cos x$

Integrating factor: $e^{\int -1\,dx} = e^{-x}$

$ye^{-x} = \displaystyle\int e^{-x}\cos x\,dx = \tfrac{1}{2}e^{-x}(-\cos x + \sin x) + C$

$y = \tfrac{1}{2}(\sin x - \cos x) + Ce^x$

6. $y' + 2xy = 2x$

Integrating factor: $e^{\int 2x\,dx} = e^{x^2}$

$ye^{x^2} = \displaystyle\int 2xe^{x^2}\,dx = e^{x^2} + C$

$y = 1 + Ce^{-x^2}$

7. $(3y + \sin 2x)\,dx - dy = 0$

$y' - 3y = \sin 2x$

Integrating factor: $e^{\int -3\,dx} = e^{-3x}$

$ye^{-3x} = \displaystyle\int e^{-3x}\sin 2x\,dx$

$\qquad = \tfrac{1}{13}e^{-3x}(-3\sin 2x - 2\cos 2x) + C$

$y = -\tfrac{1}{13}(3\sin 2x + 2\cos 2x) + Ce^{3x}$

8. $[(y - 1)\sin x]\,dx - dy = 0$

$y' - (\sin x)y = -\sin x$

Integrating factor: $e^{\int -\sin x\,dx} = e^{\cos x}$

$ye^{\cos x} = \displaystyle\int -\sin xe^{\cos x}\,dx = e^{\cos x} + C$

$y = 1 + Ce^{-\cos x}$

9. $(x - 1)y' + y = x^2 - 1$

$y' + \left(\dfrac{1}{x - 1}\right)y = x + 1$

Integrating factor: $e^{\int [1/(x-1)]\,dx} = e^{\ln|x-1|} = x - 1$

$y(x - 1) = \displaystyle\int (x^2 - 1)\,dx = \dfrac{1}{3}x^3 - x + C_1$

$y = \dfrac{x^3 - 3x + C}{3(x - 1)}$

10. $y' + 5y = e^{5x}$

Integrating factor: $e^{\int 5\,dx} = e^{5x}$

$ye^{5x} = \displaystyle\int e^{10x}\,dx = \dfrac{1}{10}e^{10x} + C$

$y = \dfrac{1}{10}e^{5x} + Ce^{-5x}$

11. $y' \cos^2 x + y - 1 = 0$

$y' + (\sec^2 x)y = \sec^2 x$

Integrating factor: $e^{\int \sec^2 x \, dx} = e^{\tan x}$

$ye^{\tan x} = \int \sec^2 x e^{\tan x} \, dx = e^{\tan x} + C$

$y = 1 + Ce^{-\tan x}$

Initial condition: $y(0) = 5$, $C = 4$

Particular solution: $y = 1 + 4e^{-\tan x}$

12. $x^3 y' + 2y = e^{1/x^2}$

$y' + \left(\dfrac{2}{x^3}\right)y = \dfrac{1}{x^3}e^{1/x^2}$

Integrating factor: $e^{\int (2/x^3) \, dx} = e^{-(1/x^2)}$

$ye^{-1/x^2} = \int \dfrac{1}{x^3} \, dx = -\dfrac{1}{2x^2} + C_1$

$y = e^{1/x^2}\left(\dfrac{Cx^2 - 1}{2x^2}\right)$

Initial condition: $y(1) = e$, $C = 3$

Particular solution: $y = e^{1/x^2}\left(\dfrac{3x^2 - 1}{2x^2}\right)$

13. $y' + y \tan x = \sec x + \cos x$

Integrating factor: $e^{\int \tan x \, dx} = e^{\ln|\sec x|} = \sec x$

$y \sec x = \int \sec x(\sec x + \cos x) \, dx = \tan x + x + C$

$y = \sin x + x \cos x + C \cos x$

Initial condition: $y(0) = 1$, $1 = C$

Particular solution: $y = \sin x + (x + 1)\cos x$

14. $y' + y \sec x = \sec x$

Integrating factor: $e^{\int \sec x \, dx} = e^{\ln|\sec x + \tan x|} = \sec x + \tan x$

$y(\sec x + \tan x) = \int (\sec x + \tan x)\sec x \, dx = \sec x + \tan x + C$

$y = 1 + \dfrac{C}{\sec x + \tan x}$

Initial condition: $y(0) = 4$, $4 = 1 + \dfrac{C}{1 + 0}$, $C = 3$

Particular solution: $y = 1 + \dfrac{3}{\sec x + \tan x} = 1 + \dfrac{3 \cos x}{1 + \sin x}$

15. $y' + \left(\dfrac{1}{x}\right)y = 0$

Integrating factor: $e^{\int (1/x) \, dx} = e^{\ln|x|} = x$, $xy = C$

Separation of variables:

$\dfrac{dy}{dx} = -\dfrac{y}{x}$

$\int \dfrac{1}{y} \, dy = \int -\dfrac{1}{x} \, dx$

$\ln y = -\ln x + \ln C$

$\ln xy = \ln C$

$xy = C$

Initial condition: $y(2) = 2$, $C = 4$

Particular solution: $xy = 4$

16. $y' + (2x - 1)y = 0$

Integrating factor: $e^{\int (2x-1)\,dx} = e^{x^2-x}$

$ye^{x^2-x} = C$

$y = Ce^{x-x^2}$

Separation of variables: $\displaystyle\int \frac{1}{y}\,dy = \int (1 - 2x)\,dx$

$$\ln y + \ln C_1 = x - x^2$$

$$yC_1 = e^{x-x^2}$$

$$y = Ce^{x-x^2}$$

Initial condition: $y(1) = 2, \quad 2 = C$

Particular solution: $y = 2e^{x-x^2}$

17. $y' + 3x^2 y = x^2 y^3$

$n = 3, \quad Q = x^2, \quad P = 3x^2$

$\displaystyle y^{-2} e^{\int (-2)3x^2\,dx} = \int (-2)x^2 e^{\int (-2)3x^2\,dx}\,dx$

$\displaystyle y^{-2} e^{-2x^3} = -\int 2x^2 e^{-2x^3}\,dx$

$\displaystyle y^{-2} e^{-2x^3} = \frac{1}{3} e^{-2x^3} + C$

$\displaystyle y^{-2} = \frac{1}{3} + Ce^{2x^3}$

$\displaystyle \frac{1}{y^2} = Ce^{2x^3} + \frac{1}{3}$

18. $y' + 2xy = xy^2$

$n = 2, \quad Q = x, \quad P = 2x, \quad e^{\int -2x\,dx} = e^{-x^2}$

$\displaystyle y^{-1} e^{-x^2} = \int -xe^{-x^2}\,dx = \frac{1}{2} e^{-x^2} + C_1$

$\displaystyle \frac{1}{y} = \frac{1 + Ce^{x^2}}{2}$

$\displaystyle y = \frac{2}{1 + Ce^{x^2}}$

19. $y' + \left(\dfrac{1}{x}\right)y = xy^2$

$n = 2, \quad Q = x, \quad P = x^{-1}$

$e^{\int -(1/x)\,dx} = e^{-\ln|x|} = x^{-1}$

$\displaystyle y^{-1} x^{-1} = \int -x(x^{-1})\,dx = -x + C$

$\displaystyle \frac{1}{y} = -x^2 + Cx$

$\displaystyle y = \frac{1}{Cx - x^2}$

20. $y' + \left(\dfrac{1}{x}\right)y = x\sqrt{y}$

$n = \dfrac{1}{2}, \quad Q = x, \quad P = x^{-1}$

$e^{\int (1/2)(1/x)\,dx} = e^{(1/2)\ln x} = \sqrt{x}$

$\displaystyle y^{1/2} x^{1/2} = \int \frac{1}{2} x^{1/2}(x)\,dx$

$\displaystyle = \frac{1}{5} x^{5/2} + C_1 = \frac{x^{5/2} + C}{5}$

$\displaystyle y = \frac{(x^{5/2} + C)^2}{25x}$

21. $y' - y = x^3 \sqrt[3]{y}$

$n = \frac{1}{3}, \quad Q = x^3, \quad P = -1$

$e^{\int -(2/3)\,dx} = e^{-(2/3)x}$

$\displaystyle y^{2/3} e^{-(2/3)x} = \int \frac{2}{3} x^3 e^{-(2/3)x}\,dx$

$\displaystyle y^{2/3} = -\frac{1}{4}(4x^3 + 18x^2 + 54x + 81) + Ce^{2x/3}$

22. $yy' - 2y^2 = e^x$

$y' - 2y = e^x y^{-1}$

$n = -1, \quad Q = e^x, \quad P = -2$

$e^{\int 2(-2)\,dx} = e^{-4x}$

$\displaystyle y^2 e^{-4x} = \int 2e^{-4x} e^x\,dx = -\frac{2}{3} e^{-3x} + C$

$\displaystyle y^2 = -\frac{2}{3} e^x + Ce^{4x}$

23. $L\dfrac{dI}{dt} + RI = E_0, \quad I' + \dfrac{R}{L}I = \dfrac{E_0}{L}$

Integrating factor: $e^{\int (R/L)\, dt} = e^{Rt/L}$

$Ie^{Rt/L} = \displaystyle\int \dfrac{E_0}{L}e^{Rt/L}\, dt = \dfrac{E_0}{R}e^{Rt/L} + C$

$I = \dfrac{E_0}{R} + Ce^{-Rt/L}$

24. $I(0) = 0, \quad E_0 = 110$ volts

$R = 550$ ohms, $\quad L = 4$ henrys

$0 = \dfrac{110}{550} + Ce^0, \quad C = -\dfrac{1}{5}$

$y = \dfrac{E_0}{R} - \dfrac{1}{5}e^{-Rt/L} = \dfrac{1}{5}\left(1 - e^{-137.5t}\right)$

$\displaystyle\lim_{t\to\infty} \dfrac{1}{5}\left(1 - e^{-137.5t}\right) = \dfrac{1}{5}$ amp

$(0.90)\left(\dfrac{1}{5}\right) = 0.18$

$0.18 = \dfrac{1}{5}\left(1 - e^{-137.5t}\right)$

$0.90 = 1 - e^{-137.5t}$

$-137.5t = \ln(0.10)$

$t = \dfrac{\ln(0.10)}{-137.5} \approx 0.0167$

25. $\dfrac{dP}{dt} = kP + N, \quad N$ constant

$\dfrac{dP}{kP + N} = dt$

$\displaystyle\int \dfrac{1}{kP + N}\, dP = \int dt$

$\dfrac{1}{k}\ln(kP + N) = t + C_1$

$\ln(kP + N) = kt + C_2$

$kP + N = e^{kt + C_2}$

$P = \dfrac{C_3 e^{kt} - N}{k}$

$P = Ce^{kt} - \dfrac{N}{k}$

When $t = 0, \quad P = P_0$

$P_0 = C - \dfrac{N}{k} \Rightarrow C = P_0 + \dfrac{N}{k}$

$P = \left(P_0 + \dfrac{N}{k}\right)e^{kt} - \dfrac{N}{k}$

26. $\dfrac{dA}{dt} = rA + P$

$\dfrac{dA}{rA + P} = dt$

$\displaystyle\int \dfrac{dA}{rA + P} = \int dt$

$\dfrac{1}{r}\ln(rA + P) = t + C_1$

$\ln(rA + P) = rt + C_2$

$rA + P = e^{rt + C_2}$

$A = \dfrac{C_3 e^{rt} - P}{r}$

$A = Ce^{rt} - \dfrac{P}{r}$

When $t = 0, \quad A = 0,$

$0 = C - \dfrac{P}{r} \Rightarrow C = \dfrac{P}{r}$

$A = \dfrac{P}{r}\left(e^{rt} - 1\right)$

27. (a) $A = \dfrac{P}{r}\left(e^{rt} - 1\right)$

$A = \dfrac{100,000}{0.12}\left(e^{0.12(5)} - 1\right) \approx \$685,099.00$

(b) $A = \dfrac{250,000}{0.15}\left(e^{0.15(10)} - 1\right) \approx \$5,802,815.12$

28. $800,000 = \dfrac{75,000}{0.13}\left(e^{0.13t} - 1\right)$

$2.386666667 = e^{0.13t}$

$t = \dfrac{\ln 2.386666667}{0.13}$

$t \approx 6.69$ years

29. (a) $\dfrac{dQ}{dt} = q - kQ$, q constant

(b) $Q' + kQ = q$

Let $P(t) = k$, $Q(t) = q$, then the integrating factor is $u(t) = e^{kt}$.

$$Q = e^{-kt} \int q e^{kt} \, dt = e^{-kt}\left(\frac{q}{k}e^{kt} + C\right) = \frac{q}{k} + Ce^{-kt}$$

When $t = 0$, $Q = Q_0$, $Q_0 = \dfrac{q}{k} + C \Rightarrow C = Q_0 - \dfrac{q}{k}$

$$Q = \frac{q}{k} + \left(Q_0 - \frac{q}{k}\right)e^{-kt}$$

(c) $\displaystyle\lim_{t\to\infty} Q = \dfrac{q}{k}$

30. (a) $\dfrac{dN}{dt} = k(30 - N)$

(b) $N' + kN = 30k$

Let $P(t) = k$, $Q(t) = 30k$, then the integrating factor is $u(t) = e^{kt}$.

$$N = e^{-kt} \int 30k e^{kt} \, dt = e^{-kt}(30e^{kt} + C) = 30 + Ce^{-kt}$$

When $t = 1$, $N = 10$ and   when $t = 20$, $N = 19$

$\qquad 10 = 30 + Ce^{-k} \qquad\qquad\qquad\qquad 19 = 30 + Ce^{-20k}$

$\qquad C = -20e^{k} \qquad\qquad\qquad\qquad\qquad C = -11e^{20k}$

$$-20e^{k} = -11e^{20k}$$

$$\frac{20}{11} = e^{19k}$$

$$k = \frac{\ln(20/11)}{19} \approx 0.0315$$

$$C = -20e^{[\ln(20/11)]/19} \approx -20.6393$$

$$N = 30 - 20.6393e^{-0.0315t}$$

31. Let Q be the number of pounds of concentrate in the solution at any time t. Since the number of gallons of solution in the tank at any time t is $v_0 + (r_1 - r_2)t$ and since the tank loses r_2 gallons of solution per minute, it must lose concentrate at the rate

$$\left[\frac{Q}{v_0 + (r_1 - r_2)t}\right]r_2.$$

The solution gains concentrate at the rate $r_1 q_1$. Therefore, the net rate of change is

$$\frac{dQ}{dt} = q_1 r_1 - \left[\frac{Q}{v_0 + (r_1 - r_2)t}\right]r_2 \quad \text{or} \quad \frac{dQ}{dt} + \frac{r_2 Q}{v_0 + (r_1 - r_2)t} = q_1 r_1.$$

32. (a) $Q' + \dfrac{r_2 Q}{v_0 + (r_1 - r_2)t} = q_1 r_1$

$Q(0) = q_0, \quad q_0 = 25, \quad q_1 = 0, \quad v_0 = 200, \quad r_1 = 10, \quad r_2 = 10, \quad Q' + \dfrac{1}{20}Q = 0$

$\displaystyle\int \frac{1}{Q}\,dQ = \int -\frac{1}{20}\,dt$

$\ln Q = -\dfrac{1}{20}t + \ln C_1$

$Q = Ce^{-(1/20)t}$

Initial condition: $Q(0) = 25, \quad C = 25$

Particular solution: $Q = 25e^{-(1/20)t}$

(b) $\qquad 15 = 25e^{-(1/20)t}$

$\ln\left(\dfrac{3}{5}\right) = -\dfrac{1}{20}t$

$t = -20\ln\left(\dfrac{3}{5}\right) \approx 10.2 \text{ min}$

33. (a) $Q' + \dfrac{r_2 Q}{v_0 + (r_1 - r_2)t} = q_1 r_1$

$Q(0) = q_0, \quad q_0 = 25, \quad q_1 = 0.05, \quad v_0 = 200, \quad r_1 = 10, \quad r_2 = 10, \quad Q' + \dfrac{1}{20}Q = 0.5$

Integrating factor: $e^{\int (1/20)\,dt} = e^{(1/20)t}$

$Qe^{(1/20)t} = \displaystyle\int 0.5e^{(1/20)t}\,dt = 10e^{(1/20)t} + C$

$Q = 10 + Ce^{-(1/20)t}$

Initial condition: $Q(0) = 25, \quad 25 = 10 + C, \quad C = 15$

Particular solution: $Q = 10 + 15e^{-(1/20)t}$

(b) $\qquad 15 = 10 + 15e^{-(1/20)t}$

$\ln\left(\dfrac{1}{3}\right) = -\dfrac{1}{20}t$

$t = -20\ln\left(\dfrac{1}{3}\right) \approx 21.97 \text{ min}$

34. (a) The volume of the solution in the tank is given by $v_0 + (r_1 - r_2)t$. Therefore, $100 + (5 - 3)t = 200$ or $t = 50$ minutes.

(b) $Q' + \dfrac{r_2 Q}{v_0 + (r_1 - r_2)t} = q_1 r_1$

$Q(0) = q_0, \quad q_0 = 0, \quad q_1 = 0.5, \quad v_0 = 100, \quad r_1 = 5, \quad r_2 = 3, \quad Q' + \dfrac{3}{100 + 2t}Q = 2.5$

Integrating factor: $e^{\int [3/(100+2t)]\,dt} = (50 + t)^{3/2}$

$Q(50 + t)^{3/2} = \displaystyle\int 2.5(50 + t)^{3/2}\,dt = (50 + t)^{5/2} + C$

$Q = (50 + t) + C(50 + t)^{-3/2}$

Initial condition: $Q(0) = 0, \quad 0 = 50 + C(50^{-3/2}), \quad C = -50^{5/2}$

Particular solution: $\qquad Q = (50 + t) - 50^{-5/2}(50 + t)^{-3/2}$

$$Q(50) = 100 - 50^{5/2}(100)^{-3/2} = 100 - \frac{25}{\sqrt{2}} \approx 82.32 \text{ lbs}$$

35. $e^{2x+y} \, dx - e^{x-y} \, dy = 0$

Separation of variables:

$$e^{2x} e^y \, dx = e^x e^{-y} \, dy$$

$$\int e^x \, dx = \int e^{-2y} \, dy$$

$$e^x = -\tfrac{1}{2} e^{-2y} + C_1$$

$$2e^x + e^{-2y} = C$$

36. $(x+1) \, dx - (y^2 + 2y) \, dy = 0$

Separation of variables:

$$\int (x+1) \, dx = \int (y^2 + 2y) \, dy$$

$$\tfrac{1}{2}x^2 + x = \tfrac{1}{3}y^3 + y^2 + C_1$$

$$3x^2 + 6x - 2y^3 - 6y^2 = C$$

37. $(1+y^2) \, dx + (2xy + y + 2) \, dy = 0$

Exact: $\dfrac{\partial M}{\partial y} = 2y = \dfrac{\partial N}{\partial x}$

$$U(x, \, y) = \int (1+y^2) \, dx = x + xy^2 + f(y)$$

$$U_y(x, \, y) = 2xy + f'(y) = 2xy + y + 2 \text{ or } f'(y) = y + 2$$

$$f(y) = \frac{1}{2}y^2 + 2y + C_1$$

$$U(x, \, y) = x + xy^2 + \frac{1}{2}y^2 + 2y + C_1$$

$$x + xy^2 + \frac{1}{2}y^2 + 2y = C$$

38. $(1 + 2e^{2x+y}) \, dx + e^{2x+y} \, dy = 0$

Exact: $\dfrac{\partial M}{\partial y} = 2e^{2x+y} = \dfrac{\partial N}{\partial x}$

$$U(x, \, y) = \int e^{2x+y} \, dy = e^{2x+y} + g(x)$$

$$U_x(x, \, y) = 2e^{2x+y} + g'(x) = 1 + 2e^{2x+y}$$

$$g'(x) = 1 \text{ or } g(x) = x + C_1$$

$$U(x, \, y) = e^{2x+y} + x + C_1 \text{ or } x + e^{2x+y} = C$$

39. $(y \cos x - \cos x) \, dx + dy = 0$

Separation of variables:

$$\int \cos x \, dx = \int \frac{-1}{y-1} \, dy$$

$$\sin x = -\ln(y-1) + \ln C$$

$$\ln(y-1) = -\sin x + \ln C$$

$$y = Ce^{-\sin x} + 1$$

40. $(x+1) \, dy + (y - e^x) \, dx = 0$

Exact: $\dfrac{\partial M}{\partial y} = 1 = \dfrac{\partial N}{\partial x}$

$$U(x, \, y) = \int (x+1) \, dy = xy + y + g(x)$$

$$U_x(x, \, y) = y + g'(x) = y - e^x \text{ or } g'(x) = -e^x$$

$$g(x) = -e^x + C_1$$

$$U(x, \, y) = xy + y - e^x + C_1$$

$$xy + y - e^x = C$$

41. $2xy \, dx + (x^2 + \cos y) \, dy = 0$

Exact: $\dfrac{\partial M}{\partial y} = 2x = \dfrac{\partial N}{\partial x}$

$$U(x, \, y) = \int 2xy \, dx = x^2 y + f(y)$$

$$U_y(x, \, y) = x^2 + f'(y) = x^2 + \cos y$$

$$\text{or } f'(y) = \cos y$$

$$f(y) = \sin y + C_1$$

$$U(x, \, y) = x^2 y + \sin y + C_1$$

$$x^2 y + \sin y = C$$

42. $y' = 2x\sqrt{1 - y^2}$

Separation of variables:

$$\int \frac{1}{\sqrt{1-y^2}} \, dy = \int 2x \, dx$$

$$\arcsin y = x^2 + C$$

$$y = \sin(x^2 + C)$$

43. $(3y^2 + 4xy)\,dx + (2xy + x^2)\,dy = 0$

Homogeneous: $y = vx$, $dy = v\,dx + x\,dv$

$(3v^2x^2 + 4vx^2)\,dx + (2vx^2 + x^2)(v\,dx + x\,dv) = 0$

$$\int \frac{5}{x}\,dx + \int \left(\frac{2v+1}{v^2+v}\right)dv = 0$$

$$\ln x^5 + \ln|v^2 + v| = \ln C$$

$$x^5(v^2 + v) = C$$

$$x^3y^2 + x^4y = C$$

44. $(x + y)\,dx - x\,dy = 0$

Linear: $y' - \dfrac{1}{x}y = 1$

Integrating factor: $e^{\int -(1/x)\,dx} = e^{\ln|x^{-1}|} = \dfrac{1}{x}$

$$y\frac{1}{x} = \int \frac{1}{x}\,dx = \ln|x| + C$$

$$y = x(\ln|x| + C)$$

45. $(2y - e^x)\,dx + x\,dy = 0$

Linear: $y' + \left(\dfrac{2}{x}\right)y = \dfrac{1}{x}e^x$

Integrating factor: $e^{\int (2/x)\,dx} = e^{\ln x^2} = x^2$

$$yx^2 = \int x^2 \frac{1}{x}e^x\,dx = e^x(x - 1) + C$$

$$y = \frac{e^x}{x^2}(x - 1) + \frac{C}{x^2}$$

46. $(y^2 + xy)\,dx - x^2\,dy = 0$

Homogeneous: $y = vx$, $dy = v\,dx + x\,dv$

$(v^2x^2 + vx^2)\,dx - x^2(v\,dx + x\,dv) = 0$

$$v^2\,dx - x\,dv = 0$$

$$\int \frac{1}{x}\,dx = \int \frac{1}{v^2}\,dv$$

$$\ln x = -\frac{1}{v} + C$$

$$y = \frac{x}{C - \ln|x|}$$

47. $(x^2y^4 - 1)\,dx + x^3y^3\,dy = 0$

$$y' + \left(\frac{1}{x}\right)y = x^{-3}y^{-3}$$

Bernoulli: $n = -3$, $Q = x^{-3}$, $P = x^{-1}$, $e^{\int (4/x)\,dx} = e^{\ln x^4} = x^4$

$$y^4x^4 = \int 4(x^{-3})(x^4)\,dx = 2x^2 + C$$

$$x^4y^4 - 2x^2 = C$$

48. $y\,dx + (3x + 4y)\,dy = 0$

Homogeneous: $x = vy$, $dx = v\,dy + y\,dv$

$y(v\,dy + y\,dv) + (3vy + 4y)\,dy = 0$

$$\int \frac{1}{v+1}\,dv = \int -\frac{4}{y}\,dy$$

$$\ln|v + 1| = -\ln y^4 + \ln C$$

$$y^4(v + 1) = C$$

$$y^3(x + y) = C$$

49. $3y\,dx - (x^2 + 3x + y^2)\,dy = 0$

Multiplying by the integrating factor, $\dfrac{1}{x^2 + y^2}$, and regrouping, we have

$$3\left[\frac{y\,dx - x\,dy}{x^2 + y^2}\right] - dy = 0$$

$$\int 3d\left[\arctan\frac{x}{y}\right] - \int dy = 0$$

$$3\arctan\frac{x}{y} - y = C.$$

50. $x\,dx + (y + e^y)(x^2 + 1)\,dy = 0$

Separation of variables:

$$\int \frac{x}{x^2 + 1}\,dx = \int -(y + e^y)\,dy$$

$$\frac{1}{2}\ln(x^2 + 1) = -\frac{1}{2}y^2 - e^y + C_1$$

$$\ln(x^2 + 1) + y^2 + 2e^y = C$$

Section 17.5   Second-Order Homogeneous Linear Equations

1. $y'' - y' = 0$
Characteristic equation: $m^2 - m = 0$
Roots: $m = 0,\ 1$
$y = C_1 + C_2 e^x$

2. $y'' + 2y' = 0$
Characteristic equation: $m^2 + 2m = 0$
Roots: $m = 0,\ -2$
$y = C_1 + C_2 e^{-2x}$

3. $y'' - y' - 6y = 0$
Characteristic equation: $m^2 - m - 6 = 0$
Roots: $m = 3,\ -2$
$y = C_1 e^{3x} + C_2 e^{-2x}$

4. $y'' + 6y' + 5y = 0$
Characteristic equation: $m^2 + 6m + 5 = 0$
Roots: $m = -1,\ -5$
$y = C_1 e^{-x} + C_2 e^{-5x}$

5. $2y'' + 3y' - 2y = 0$
Characteristic equation: $2m^2 + 3m - 2 = 0$
Roots: $m = \frac{1}{2},\ -2$
$y = C_1 e^{(1/2)x} + C_2 e^{-2x}$

6. $16y'' - 16y' + 3y = 0$
Characteristic equation: $16m^2 - 16m + 3 = 0$
Roots: $m = \frac{1}{4},\ \frac{3}{4}$
$y = C_1 e^{(1/4)x} + C_2 e^{(3/4)x}$

7. $y'' + 6y' + 9y = 0$
Characteristic equation: $m^2 + 6m + 9 = 0$
Roots: $m = -3,\ -3$
$y = C_1 e^{-3x} + C_2 x e^{-3x}$

8. $y'' - 10y' + 25y = 0$
Characteristic equation: $m^2 - 10m + 25 = 0$
Roots: $m = 5,\ 5$
$y = C_1 e^{5x} + C_2 x e^{5x}$

9. $16y'' - 8y' + y = 0$
Characteristic equation: $16m^2 - 8m + 1 = 0$
Roots: $m = \frac{1}{4},\ \frac{1}{4}$
$y = C_1 e^{(1/4)x} + C_2 x e^{(1/4)x}$

10. $9y'' - 12y' + 4y = 0$
Characteristic equation: $9m^2 - 12m + 4 = 0$
Roots: $m = \frac{2}{3},\ \frac{2}{3}$
$y = C_1 e^{(2/3)x} + C_2 x e^{(2/3)x}$

11. $y'' + y = 0$
Characteristic equation: $m^2 + 1 = 0$
Roots: $m = -i,\ i$
$y = C_1 \cos x + C_2 \sin x$

12. $y'' + 4y = 0$
Characteristic equation: $m^2 + 4 = 0$
Roots: $m = -2i,\ 2i$
$y = C_1 \cos 2x + C_2 \sin 2x$

13. $y'' - 9y = 0$
Characteristic equation: $m^2 - 9 = 0$
Roots: $m = -3,\ 3$
$y = C_1 e^{3x} + C_2 e^{-3x}$

14. $y'' - 2y = 0$
Characteristic equation: $m^2 - 2 = 0$
Roots: $m = -\sqrt{2},\ \sqrt{2}$
$y = C_1 e^{\sqrt{2}x} + C_2^{-\sqrt{2}x}$

15. $y'' - 2y' + 4y = 0$
Characteristic equation: $m^2 - 2m + 4 = 0$
Roots: $m = 1 - \sqrt{3}i,\ 1 + \sqrt{3}i$
$y = e^x(C_1 \cos \sqrt{3}\,x + C_2 \sin \sqrt{3}\,x)$

16. $y'' - 4y' + 21y = 0$
Characteristic equation: $m^2 - 4m + 21 = 0$
Roots: $m = 2 - \sqrt{17}i,\ 2 + \sqrt{17}i$
$y = e^{2x}(C_1 \cos \sqrt{17}\,x + C_2 \sin \sqrt{17}\,x)$

17. $y'' - 3y' + y = 0$
Characteristic equation: $m^2 - 3m + 1 = 0$
Roots: $m = \dfrac{3 - \sqrt{5}}{2},\ \dfrac{3 + \sqrt{5}}{2}$
$y = C_1 e^{[(3+\sqrt{5})/2]x} + C_2 e^{[(3-\sqrt{5})/2]x}$

18. $3y'' + 4y' - y = 0$
Characteristic equation: $3m^2 + 4m - 1 = 0$
Roots: $m = \dfrac{-2 - \sqrt{7}}{3},\ \dfrac{-2 + \sqrt{7}}{3}$
$y = C_1 e^{[(-2+\sqrt{7})/3]x} + C_2 e^{[(-2-\sqrt{7})/3]x}$

19. $9y'' - 12y' + 11y = 0$

Characteristic equation: $9m^2 - 12m + 11 = 0$

Roots: $m = \dfrac{2 + \sqrt{7}\,i}{3},\ \dfrac{2 - \sqrt{7}\,i}{3}$

$y = e^{(2/3)x}\left[C_1 \cos\left(\dfrac{\sqrt{7}}{3}x\right) + C_2 \sin\left(\dfrac{\sqrt{7}}{3}x\right)\right]$

20. $2y'' - 6y' + 7y = 0$

Characteristic equation: $2m^2 - 6m + 7 = 0$

Roots: $m = \dfrac{3 + \sqrt{5}\,i}{2},\ \dfrac{3 - \sqrt{5}\,i}{2}$

$y = e^{(3/2)x}\left[C_1 \cos\left(\dfrac{\sqrt{5}}{2}x\right) + C_2 \sin\left(\dfrac{\sqrt{5}}{2}x\right)\right]$

21. $y^{(4)} - y = 0$

Characteristic equation: $m^4 - 1 = 0$

Roots: $m = -1,\ 1,\ -i,\ i$

$y = C_1 e^x + C_2 e^{-x} + C_3 \cos x + C_4 \sin x$

22. $y^{(4)} - y'' = 0$

Characteristic equation: $m^4 - m^2 = 0$

Roots: $m = 0,\ 0,\ -1,\ 1$

$y = C_1 + C_2 x + C_3 e^x + C_4 e^{-x}$

23. $y''' - 6y'' + 11y' - 6y = 0$

Characteristic equation: $m^3 - 6m^2 + 11m - 6 = 0$

Roots: $m = 1,\ 2,\ 3$

$y = C_1 e^x + C_2 e^{2x} + C_3 e^{3x}$

24. $y''' - y'' - y' + y = 0$

Characteristic equation: $m^3 - m^2 - m + 1 = 0$

Roots: $m = -1,\ 1,\ 1$

$y = C_1 e^x + C_2 x e^x + C_3 e^{-x}$

25. $y''' - 3y'' + 7y' - 5y = 0$

Characteristic equation: $m^3 - 3m^2 + 7m - 5 = 0$

Roots: $m = 1,\ 1 - 2i,\ 1 + 2i$

$y = C_1 e^x + e^x(C_2 \cos 2x + C_3 \sin 2x)$

26. $y''' - 3y'' + 3y' - y = 0$

Characteristic equation: $m^3 - 3m^2 + 3m - 1 = 0$

Roots: $m = 1,\ 1,\ 1$

$y = C_1 e^x + C_2 x e^x + C_3 x^2 e^x$

27. $y'' - y' - 30y = 0,\ \ y(0) = 1,\ \ y'(0) = -4$

Characteristic equation: $m^2 - m - 30 = 0$

Roots: $m = 6,\ -5$

$y = C_1 e^{6x} + C_2 e^{-5x},\ \ y' = 6C_1 e^{6x} - 5C_2 e^{-5x}$

Initial conditions: $y(0) = 1,\ \ y'(0) = -4,\ \ 1 = C_1 + C_2,\ \ -4 = 6C_1 - 5C_2$

Solving simultaneously: $C_1 = \frac{1}{11},\ \ C_2 = \frac{10}{11}$

Particular solution: $y = \frac{1}{11}(e^{6x} + 10e^{-5x})$

28. $y'' + 2y' + 3y = 0,\ \ y(0) = 2,\ \ y'(0) = 1$

Characteristic equation: $m^2 + 2m + 3 = 0$

Roots: $m = -1 + \sqrt{2}\,i,\ -1 - \sqrt{2}\,i$

$y = e^{-x}(C_1 \cos \sqrt{2}\,x + C_2 \sin \sqrt{2}\,x)$

$y' = e^{-x}(-\sqrt{2}\,C_1 \sin \sqrt{2}\,x + \sqrt{2}\,C_2 \cos \sqrt{2}\,x) - e^{-x}(C_1 \cos \sqrt{2}\,x + C_2 \sin \sqrt{2}\,x)$

Initial conditions: $y(0) = 2,\ \ y'(0) = 1,\ \ 2 = C_1,\ \ 1 = \sqrt{2}\,C_2 - C_1,\ \ C_2 = \dfrac{3}{\sqrt{2}}$

Particular solution: $y = e^{-x}\left(2\cos\sqrt{2}\,x + \dfrac{3}{\sqrt{2}}\sin\sqrt{2}\,x\right)$

29. By Hooke's Law, $F = kx$

$$k = \frac{F}{x} = \frac{32}{2/3} = 48.$$

Also, $F = ma$, and $m = \dfrac{F}{a} = \dfrac{32}{32} = 1.$

Therefore, $y = \dfrac{1}{2}\cos(4\sqrt{3}\,t).$

30. By Hooke'e Law, $F = kx$

$$k = \frac{F}{x} = \frac{32}{2/3} = 48.$$

Also, $F = ma$, and $m = \dfrac{F}{a} = \dfrac{32}{32} = 1.$

Therefore, $y = -\dfrac{2}{3}\cos(4\sqrt{3}\,t).$

31. $y = C_1 \cos(\sqrt{k/m}\, t) + C_2 \sin(\sqrt{k/m}\, t)$, $\quad \sqrt{k/m} = \sqrt{48} = 4\sqrt{3}$

Initial conditions: $y(0) = \dfrac{2}{3}$, $\quad y'(0) = -\dfrac{1}{2}$

$$y = C_1 \cos(4\sqrt{3}\, t) + C_2 \sin(4\sqrt{3}\, t)$$

$$y(0) = C_1 = \frac{2}{3}$$

$$y'(t) = -4\sqrt{3}\, C_1 \sin(4\sqrt{3}\, t) + 4\sqrt{3}\, C_2 \cos(4\sqrt{3}\, t)$$

$$y'(0) = 4\sqrt{3}\, C_2 = -\frac{1}{2} \Rightarrow C_2 = -\frac{1}{8\sqrt{3}} = -\frac{\sqrt{3}}{24}$$

$$y(t) = \frac{2}{3} \cos(4\sqrt{3}\, t) - \frac{\sqrt{3}}{24} \sin(4\sqrt{3}\, t)$$

32. $y = C_1 \cos(4\sqrt{3}\, t) + C_2 \sin(4\sqrt{3}\, t)$

Initial conditions: $y(0) = -\dfrac{1}{2}$, $\quad y'(0) = \dfrac{1}{2}$

$$y(0) = C_1 = -\frac{1}{2}$$

$$y'(t) = -4\sqrt{3}\, C_1 \sin(4\sqrt{3}\, t) + 4\sqrt{3}\, C_2 \cos(4\sqrt{3}\, t)$$

$$y'(0) = 4\sqrt{3}\, C_2 = \frac{1}{2} \Rightarrow C_2 = \frac{1}{8\sqrt{3}}$$

$$y(t) = -\frac{1}{2} \cos(4\sqrt{3}\, t) + \frac{1}{8\sqrt{3}} \sin(4\sqrt{3}\, t)$$

33. By Hooke's Law, $32 = k(2/3)$ so that $k = 48$. Moreover, since the weight w is given by mg, it follows that $m = w/g = 32/32 = 1$. Also, the damping force is given by $(-1/8)(dy/dt)$. Thus, the differential equation for the oscillations of the weight is

$$m\left(\frac{d^2 y}{dt^2}\right) = -\frac{1}{8}\left(\frac{dy}{dt}\right) - 48y$$

$$m\left(\frac{d^2 y}{dt^2}\right) + \frac{1}{8}\left(\frac{dy}{dt}\right) + 48y = 0.$$

In this case the characteristic equation is $8m^2 + m + 384 = 0$ with complex roots $m = (-1/16) \pm (\sqrt{12287}/16)i$. Therefore, the general solution is

$$y(t) = e^{-t/16}\left(C_1 \cos \frac{\sqrt{12287}\, t}{16} + C_2 \sin \frac{\sqrt{12287}\, t}{16}\right).$$

Using the initial conditions, we have

$$y(0) = C_1 = \frac{1}{2}$$

$$y'(t) = e^{-t/16}\left[\left(-\frac{\sqrt{12287}}{16}C_1 - \frac{C_2}{16}\right)\sin\frac{\sqrt{12287}\, t}{16} + \left(\frac{\sqrt{12287}}{16}C_2 - \frac{C_1}{16}\right)\cos\frac{\sqrt{12287}\, t}{16}\right]$$

$$y'(0) = \frac{\sqrt{12287}}{16}C_2 - \frac{C_1}{16} = 0 \Rightarrow C_2 = \frac{\sqrt{12287}}{24574}$$

and the particular solution is

$$y(t) = \frac{e^{-t/16}}{2}\left(\cos\frac{\sqrt{12287}\, t}{16} + \frac{\sqrt{12287}}{12287}\sin\frac{\sqrt{12287}\, t}{16}\right).$$

34. By Hooke's Law, $32 = k(2/3)$ so $k = 48$. Also, $m = w/g = 32/32 = 1$. The damping force is given by $(-1/4)(dy/dt)$. Thus,

$$m\left(\frac{d^2y}{dt^2}\right) = -\frac{1}{4}\left(\frac{dy}{dt}\right) - 48y$$

$$m\left(\frac{d^2y}{dt^2}\right) + \frac{1}{4}\left(\frac{dy}{dt}\right) + 48y = 0.$$

The characteristic equation is $4m^2 + m + 192 = 0$ with complex roots $m = (-1/8) \pm (\sqrt{3071}/8)i$. Therefore, the general solution is

$$y(t) = e^{-t/8}\left(C_1 \cos\frac{\sqrt{3071}\,t}{8} + C_2 \sin\frac{\sqrt{3071}\,t}{8}\right).$$

Using the initial conditions $y(0) = C_1 = 1/2$,

$$y'(t) = e^{-t/8}\left[\left(-\frac{\sqrt{3071}}{8}C_1 - \frac{C_2}{8}\right)\sin\frac{\sqrt{3071}\,t}{8} + \left(\frac{\sqrt{3071}\,C_2}{8} - \frac{C_1}{8}\right)\cos\frac{\sqrt{3071}\,t}{8}\right]$$

$$y'(0) = \frac{\sqrt{3071}}{8}C_2 - \frac{C_1}{8} = 0 \Rightarrow C_2 = \frac{\sqrt{3071}}{6142}$$

and the particular solution is

$$y(t) = \frac{e^{-t/8}}{2}\left[\cos\frac{\sqrt{3071}\,t}{8} + \frac{\sqrt{3071}}{3071}\sin\frac{\sqrt{3071}\,t}{8}\right].$$

35. Since $m = -a/2$ is a double root of the characteristic equation, we have $\left(m + \frac{a}{2}\right)^2 = m^2 + am + \frac{a^2}{4} = 0$

and the differential equation is $y'' + ay' + \frac{a^2}{4}y = 0$.

The solution is

$$y = (C_1 + C_2x)e^{-(a/2)x}$$

$$y' = \left(-\frac{C_1a}{2} + C_2 - \frac{C_2a}{2}x\right)e^{-(a/2)x}$$

$$y'' = \left(\frac{C_1a^2}{4} - aC_2 + \frac{C_2a^2}{4}x\right)e^{-(a/2)x}$$

$$y'' + ay' + \frac{a^2}{4}y = \left(\frac{C_1a^2}{4} - C_2a + \frac{C_2a^2}{4}x\right) + \left(-\frac{C_1a^2}{2} + C_2a - \frac{C_2a^2}{2}x\right)e^{-(a/2)x}$$

$$+ \left(\frac{C_1a^2}{4} + \frac{C_2a^2}{4}x\right)e^{-(a/2)x} = 0.$$

36. Since $m = \alpha \pm \beta i$ are roots to the characteristic equation, we have

$$[m - (\alpha + \beta i)][m - (\alpha - \beta i)] = m^2 - 2\alpha m + (\alpha^2 + \beta^2) = 0$$

and the differential equation is $y'' - 2\alpha y' + (\alpha^2 + \beta^2)y = 0$. (**Note:** $i^2 = -1$.) The solution is

$$y = e^{\alpha x}(C_1 \cos\beta x + C_2 \sin\beta x)$$

$$y' = e^{\alpha x}[(C_1\alpha + C_2\beta)\cos\beta x + (C_2\alpha - C_1\beta)\sin\beta x]$$

$$y'' = e^{\alpha x}[(C_1\alpha^2 - C_1\beta^2 + 2C_2\alpha\beta)\cos\beta x + (C_2\alpha^2 - C_2\beta^2 - 2C_1\alpha\beta)\sin\beta x]$$

$$-2\alpha y' = e^{\alpha x}[(-2C_1\alpha^2 - 2C_2\alpha\beta)\cos\beta x + (-2C_2\alpha^2 + 2C_1\alpha\beta)\sin\beta x]$$

$$(\alpha^2 + \beta^2)y = e^{\alpha x}[(C_1\alpha^2 + C_1\beta^2)\cos\beta x + (C_2\alpha^2 + C_2\beta^2)\sin\beta x]$$

Therefore, $y'' - 2\alpha y' + (\alpha^2 + \beta^2)y = 0$.

37. $y_1 = e^{ax}, \quad y_2 = e^{bx}, \quad a \neq b$

$$W(y_1, \ y_2) = \begin{vmatrix} e^{ax} & e^{bx} \\ ae^{ax} & be^{bx} \end{vmatrix}$$

$$= (b-a)e^{ax+bx} \neq 0 \text{ for any value of } x.$$

38. $y_1 = e^{ax}, \quad y_2 = xe^{ax}$

$$W(y_1, \ y_2) = \begin{vmatrix} e^{ax} & xe^{ax} \\ ae^{ax} & e^{ax} + axe^{ax} \end{vmatrix}$$

$$= e^{2ax} \neq 0 \text{ for any value of } x.$$

39. $y_1 = e^{ax} \sin bx, \quad y_2 = e^{ax} \cos bx, \quad b \neq 0$

$$W(y_1, \ y_2) = \begin{vmatrix} e^{ax} \sin bx & e^{ax} \cos bx \\ ae^{ax} \sin bx + be^{ax} \cos bx & ae^{ax} \cos bx - be^{ax} \sin bx \end{vmatrix}$$

$$= -be^{2ax} \sin^2 bx - be^{2ax} \cos^2 bx$$

$$= -be^{2ax} \neq 0 \text{ for any value of } x.$$

40. $y_1 = x, \quad y_2 = x^2$

$$W(y_1, \ y_2) = \begin{vmatrix} x & x^2 \\ 1 & 2x \end{vmatrix} = x^2 \neq 0 \text{ for } x \neq 0.$$

Section 17.6   Second-Order Nonhomogeneous Linear Equations

1. $y'' - 3y' + 2y = 2x$

$y'' - 3y' + 2y = 0$

$m^2 - 3m + 2 = 0$ when $m = 1, \ 2.$

$$y_h = C_1 e^x + C_2 e^{2x}$$

$$y_p = A_0 + A_1 x$$

$$y_p{}' = A_1$$

$$y_p{}'' = 0$$

$$y_p{}'' - 3y_p{}' + 2y_p = (2A_0 - 3A_1) + 2A_1 x = 2x$$

$$\left. \begin{array}{l} 2A_0 - 3A_1 = 0 \\ 2A_1 = 2 \end{array} \right\} A_1 = 1, \quad A_0 = \tfrac{3}{2}$$

$$y = C_1 e^x + C_2 e^{2x} + x + \tfrac{3}{2}$$

2. $y'' - 2y' - 3y = x^2 - 1$

$y'' - 2y' - 3y = 0$

$m^2 - 2m - 3 = 0$ when $m = -1, \ 3.$

$$y_h = C_1 e^{-x} + C_2 e^{3x}$$

$$y_p = A_0 + A_1 x + A_2 x^2$$

$$y_p{}' = A_1 + 2A_2 x$$

$$y_p{}'' = 2A_2$$

$$y_p{}'' - 2y_p{}' - 3y_p = (-3A_2)x^2 + (-3A_1 - 4A_2)x + (-3A_0 - 2A_1 + 2A_2) = x^2 - 1$$

$$\left. \begin{array}{l} -3A_2 = 1 \\ -3A_1 - 4A_2 = 0 \\ -3A_0 - 2A_1 + 2A_2 = N \end{array} \right\} A_0 = -\tfrac{5}{27}, \quad A_1 = \tfrac{4}{9}, \quad A_2 = -\tfrac{1}{3}$$

$$y = C_1 e^{-x} + C_2 e^{3x} - \tfrac{1}{3}x^2 + \tfrac{4}{9}x - \tfrac{5}{27}$$

3. $y'' + y = x^3$, $y(0) = 1$, $y'(0) = 0$

$y'' + y = 0$

$m^2 + 1 = 0$ when $m = i$, $-i$.

$\qquad y_h = C_1 \cos x + C_2 \sin x$

$\qquad y_p = A_0 + A_1 x + A_2 x^2 + A_3 x^3$

$\qquad y_p{'} = A_1 + 2A_2 x + 3A_3 x^2$

$\qquad y_p{''} = 2A_2 + 6A_3 x$

$y_p{''} + y_p = A_3 x^3 + A_2 x^2 + (A_1 + 6A_3)x + (A_0 + 2A_2) = x^3$ OR $A_3 = 1$, $A_2 = 0$, $A_1 = -6$, $A_0 = 0$

$\qquad y = C_1 \cos x + C_2 \sin x + x^3 - 6x$

$\qquad y' = -C_1 \sin x + C_2 \cos x + 3x^2 - 6$

Initial conditions: $y(0) = 1$, $y'(0) = 0$, $1 = C_1$, $0 = C_2 - 6$, $C_2 = 6$

$y = \cos x + 6 \sin x + x^3 - 6x$

4. $y'' + 4y = 4$, $y(0) = 1$, $y'(0) = 6$

$y'' + 4y = 0$

$m^2 + 4 = 0$ when $m = 2i$, $-2i$.

$\qquad y_h = C_1 \cos 2x + C_2 \sin 2x$

$\qquad y_p = A_0$

$\qquad y_p{''} = 0$

$y_p{''} + 4y_p = 4A_0 = 4$ or $A_0 = 1$

$\qquad y = C_1 \cos 2x + C_2 \sin 2x + 1$

$\qquad y' = -2C_1 \sin 2x + 2C_2 \cos 2x$

Initial conditions: $y(0) = 1$, $y'(0) = 6$, $1 = C_1 + 1$, $C_1 = 0$, $6 = 2C_2$, $C_2 = 3$

Particular solution: $y = 3 \sin 2x + 1$

5. $y'' + 2y' = 2e^x$

$y'' + 2y' = 0$

$m^2 + 2m = 0$ when $m = 0$, -2.

$\qquad y_h = C_1 + C_2 e^{-2x}$

$\qquad y_p = Ae^x = y_p{'} = y_p{''}$

$y_p{''} + 2y_p{'} = 3Ae^x = 2e^x$ or $A = \frac{2}{3}$

$\qquad y = C_1 + C_2 e^{-2x} + \frac{2}{3} e^x$

6. $y'' - 9y = 5e^{3x}$

$y'' - 9y = 0$

$m^2 - 9 = 0$ when $m = -3$, 3.

$\qquad y_h = C_1 e^{-3x} + C_2 e^{3x}$

$\qquad y_p = Axe^{3x}$

$\qquad y_p{'} = Ae^{3x}(3x + 1)$

$\qquad y_p{''} = Ae^{3x}(9x + 6)$

$y_p{''} - 9y_p = 6Ae^{3x} = 5e^{3x}$ or $A = \frac{5}{6}$

$\qquad y = C_1 e^{-3x} + \left(C_2 + \frac{5}{6} x \right) e^{3x}$

7. $y'' - 10y' + 25y = 5 + 6e^x$

$y'' - 10y' + 25y = 0$

$m^2 - 10m + 25 = 0$ when $m = 5$, 5.

$\qquad y_h = C_1 e^{5x} + C_2 x e^{5x}$

$\qquad y_p = A_0 + A_1 e^x$

$\qquad y_p{'} = y_p{''} = A_1 e^x$

$y_p{''} - 10y_p{'} + 25y_p = 25A_0 + 16A_1 e^x = 5 + 6e^x$ or $A_0 = \frac{1}{5}$, $A_1 = \frac{3}{8}$

$\qquad y = (C_1 + C_2 x)e^{5x} + \frac{3}{8} + \frac{1}{5}$

8. $16y'' - 8y' + y = 4(x + e^x)$

$16y'' - 8y' + y = 0$

$16m^2 - 8m + 1 = 0$ when $m = \frac{1}{4}, \frac{1}{4}$.

$$y_h = (C_1 + C_2 x)e^{(1/4)x}$$

$$y_p = A_0 + A_1 x + A_2 e^x$$

$$y_p' = A_1 + A_2 e^x$$

$$y_p'' = A_2 e^x$$

$16y_p'' - 8y_p' + y_p = (A_0 - 8A_1) + A_1 x + 9A_2 e^x = 4x + 4e^x$ or $A_2 = \frac{4}{9}$, $A_1 = 4$, $A_0 = 32$

$$y = (C_1 + C_2 x)e^{(1/4)x} + 32 + 4x + \frac{4}{9}e^x$$

9. $y'' + y' = 2\sin x$, $y(0) = 0$, $y'(0) = -3$

$y'' + y' = 0$

$m^2 + m = 0$ when $m = 0, -1$.

$$y_h = C_1 + C_2 e^{-x}$$

$$y_p = A\cos x + B\sin x$$

$$y_p' = -A\sin x + B\cos x$$

$$y_p'' = -A\cos x - B\sin x$$

$$y_p'' + y_p' = (-A + B)\cos x + (-A - B)\sin x = 2\sin x$$

$\left.\begin{array}{l} -A + B = 0 \\ -A - B = 2 \end{array}\right\}$ $A = -1$, $B = -1$

$$y = C_1 + C_2 e^x - (\cos x + \sin x)$$

$$y' = -C_2 e^{-x} - (-\sin x + \cos x)$$

Initial conditions: $y(0) = 0$, $y'(0) = -3$, $0 = C_1 + C_2 - 1$, $-3 = -C_2 - 1$, $C_2 = 2$, $C_1 = -1$

Particular solution: $y = -1 + 2e^{-x} - (\cos x + \sin x)$

10. $y'' + y' - 2y = 3\cos 2x$, $y(0) = -1$, $y'(0) = 2$

$y'' + y' - 2y = 0$

$m^2 + m - 2 = 0$ when $m = 1, -2$.

$$y_h = C_1 e^x + C_2 e^{-2x}$$

$$y_p = A\cos 2x + B\sin 2x$$

$$y_p' = -2A\sin 2x + 2B\cos 2x$$

$$y_p'' = -4A\cos 2x - 4B\sin 2x$$

$$y_p'' + y_p' - 2y_p = (-6A + 2B)\cos 2x + (-2A - 6B)\sin 2x = 3\cos 2x$$

$\left.\begin{array}{l} -6A + 2B = 3 \\ -2A - 6B = 0 \end{array}\right\}$ $A = -\frac{9}{20}$, $B = \frac{3}{20}$

$$y = C_1 e^x + C_2 e^{-2x} - \frac{9}{20}\cos 2x + \frac{3}{20}\sin 2x$$

$$y' = C_1 e^x - 2C_2 e^{-2x} + \frac{9}{10}\sin 2x + \frac{3}{10}\cos 2x$$

Initial conditions: $y(0) = -1$, $y'(0) = 2$, $-1 = C_1 + C_2 - \frac{9}{20}$, $2 = C_1 - 2C_2 + \frac{3}{10}$

$\left.\begin{array}{l} C_1 + C_2 = \frac{-11}{20} \\ C_1 - 2C_2 = \frac{17}{10} \end{array}\right\}$ $C_1 = \frac{1}{5}$, $C_2 = -\frac{3}{4}$

Particular solution: $y = \frac{1}{20}(4e^x - 15e^{-2x} - 9\cos 2x + 3\sin 2x)$

11. $y'' + 9y = \sin 3x$

$y'' + 9y = 0$

$m^2 + 9 = 0$ when $m = -3i,\ 3i$.

$$y_h = C_1 \cos 3x + C_2 \sin 3x$$

$$y_p = A_0 \sin 3x + A_1 x \sin 3x + A_2 \cos 3x + A_3 x \cos 3x$$

$$y_p{}'' = (-9A_0 - 6A_3)\sin 3x - 9A_1 x \sin 3x + (6A_1 - 9A_2)\cos 3x - 9A_3 x \cos 3x$$

$$y_p{}'' + 9y_p = -6A_3 \sin 3x + 6A_1 \cos 3x = \sin 3x, \quad A_1 = 0, \quad A_3 = -\tfrac{1}{6}$$

$$y = \left(C_1 - \tfrac{1}{6}x\right)\cos 3x + C_2 \sin 3x$$

12. $y'' + 4y' + 5y = \sin x + \cos x$

$y'' + 4y' + 5y = 0$

$m^2 + 4m + 5 = 0$ when $m = -2 - i,\ -2 + i$.

$$y_h = e^{-2x}(C_1 \cos x + C_2 \sin x)$$

$$y_p = A \cos x + B \sin x$$

$$y_p{}' = -A \sin x + B \cos x$$

$$y_p{}'' = -A \cos x - B \sin x$$

$$y_p{}'' + 4y_p{}' + 5y_p = (4A + 4B)\cos x + (-4A + 4B)\sin x = \sin x + \cos x$$

$$\left.\begin{array}{r} 4A + 4B = 1 \\ -4A + 4B = 1 \end{array}\right\} A = 0, \quad B = \tfrac{1}{4}$$

$$y = e^{-2x}(C_1 \cos x + C_2 \sin x) + \tfrac{1}{4}\sin x$$

13. $y''' - 3y' + 2y = 2e^{-2x}$

$y''' - 3y' + 2y = 0$

$m^3 - 3m + 2 = 0$ when $m = 1,\ 1,\ -2$.

$$y_h = C_1 e^x + C_2 x e^x + C_3 e^{-2x}$$

$$y_p = A_0 e^{-2x} + A_1 x e^{-2x}$$

$$y_p{}' = (-2A_0 + A_1)e^{-2x} - 2A_1 x e^{-2x}$$

$$y_p{}'' = (4A_0 - 4A_1)e^{-2x} + 4A_1 x e^{-2x}$$

$$y_p{}''' = (-8A_0 + 12A_1)e^{-2x} - 8A_1 x e^{-2x}$$

$$y_p{}''' - 3y_p{}' + 2y_p = 9A_1 e^{-2x} = 2e^{-2x} \text{ or } A_1 = \tfrac{2}{9}$$

$$y = C_1 e^x + C_2 x e^x + \left(C_3 + \tfrac{2}{9}x\right)e^{-2x}$$

14. $y''' - y'' = 4x^2$, $y(0) = 1$, $y'(0) = 1$, $y''(0) = 1$

$y''' - y'' = 0$

$m^3 - m^2 = 0$ when $m = 0,\ 0,\ 1$.

$$y_h = C_1 + C_2 x + C_3 e^x$$

$$y_p = A_0 x^2 + A_1 x^3 + A_2 x^4$$

$$y_p{}' = 2A_0 x + 3A_1 x^2 + 4A_2 x^3$$

$$y_p{}'' = 2A_0 + 6A_1 x + 12A_2 x^2$$

$$y_p{}''' = 6A_1 + 24A_2 x$$

$$y_p{}''' - y_p{}'' = (-2A_0 + 6A_1) + (-6A_1 + 24A_2)x - 12A_2 x^2 = 4x^2 \text{ or } A_0 = -4,\quad A_1 = -\tfrac{4}{3},\quad A_2 = -\tfrac{1}{3}$$

$$y = C_1 + C_2 x + C_3 e^x - 4x^2 - \tfrac{4}{3}x^3 - \tfrac{1}{3}x^4$$

$$y' = C_2 + C_3 e^x - 8x - 4x^2 - \tfrac{4}{3}x^3$$

$$y'' = C_3 e^x - 8 - 8x - 4x^2$$

Initial conditions:

$\quad y(0) = 1$, $y'(0) = 1$, $y''(0) = 1$, $1 = C_1 + C_3$, $1 = C_2 + C_3$, $1 = C_3 - 8$, $C_1 = -8$, $C_2 = -8$, $C_3 = 9$

Particular solution: $y = -8 - 8x - 4x^2 - \tfrac{4}{3}x^3 - \tfrac{1}{3}x^4 + 9e^x$

15. $y' - 4y = xe^x - xe^{4x}$, $y(0) = \tfrac{1}{3}$

$y' - 4y = 0$

$m - 4 = 0$ when $m = 4$.

$$y_h = Ce^{4x}$$

$$y_p = (A_0 + A_1 x)e^x + (A_2 x + A_3 x^2)e^{4x}$$

$$y_p{}' = (A_0 + A_1 x)e^x + A_1 e^x + 4(A_2 x + A_3 x^2)e^{4x} + (A_2 + 2A_3 x)e^{4x}$$

$$y_p{}' - 4y_p = (-3A_0 - 3A_1 x)e^x + A_1 e^x + A_2 e^{4x} + 2A_3 x e^{4x} = xe^x - xe^{4x}$$

$$A_0 = -\tfrac{1}{9},\quad A_1 = -\tfrac{1}{3},\quad A_2 = 0,\quad A_3 = -\tfrac{1}{2}$$

$$y = \left(C - \tfrac{1}{2}x^2\right)e^{4x} - \tfrac{1}{9}(1 + 3x)e^x$$

Initial condition: $y(0) = \tfrac{1}{3}$, $\tfrac{1}{3} = C - \tfrac{1}{9}$, $C = \tfrac{4}{9}$

Particular solution: $y = \left(\tfrac{4}{9} - \tfrac{1}{2}x^2\right)e^{4x} - \tfrac{1}{9}(1 + 3x)e^x$

16. $y' + 2y = \sin x$, $y\left(\dfrac{\pi}{2}\right) = \dfrac{2}{5}$

$y' + 2y = 0$

$m + 2 = 0$ when $m = -2$.

$$y_h = Ce^{-2x}$$

$$y_p = A\cos x + B\sin x$$

$$y_p{}' = -A\sin x + B\cos x$$

$$y_p{}' + 2y_p = (-A\sin x + B\cos x) + 2(A\cos x + B\sin x) = (2B - A)\sin x + (2A + B)\cos x = \sin x$$

$$2B - A = 1,\quad 2A + B = 0 \Rightarrow B = \frac{2}{5},\quad A = -\frac{1}{5}$$

$$y = y_h + y_p = Ce^{-2x} - \frac{1}{5}\cos x + \frac{2}{5}\sin x$$

Initial condition: $y\left(\dfrac{\pi}{2}\right) = \dfrac{2}{5}$, $\dfrac{2}{5} = Ce^{-\pi} + \dfrac{2}{5}$, $C = 0$

Particular solution: $y = \dfrac{2}{5}\sin x - \dfrac{1}{5}\cos x$

17. $y'' + y = \sec x$

$y'' + y = 0$

$m^2 + 1 = 0$ when $m = -i, \ i$.

$y_h = C_1 \cos x + C_2 \sin x$

$y_p = v_1 \cos x + v_2 \sin x$

$\qquad v_1{}' \cos x + v_2{}' \sin x = 0$

$\qquad v_1{}'(-\sin x) + v_2{}'(\cos x) = \sec x$

$$v_1{}' = \frac{\begin{vmatrix} 0 & \sin x \\ \sec x & \cos x \end{vmatrix}}{\begin{vmatrix} \cos x & \sin x \\ -\sin x & \cos x \end{vmatrix}} = -\tan x$$

$$v_1 = \int -\tan x \, dx = \ln|\cos x|$$

$$v_2{}' = \frac{\begin{vmatrix} \cos x & 0 \\ -\sin x & \sec x \end{vmatrix}}{\begin{vmatrix} \cos x & \sin x \\ -\sin x & \cos x \end{vmatrix}} = 1$$

$$v_2 = \int dx = x$$

$$y = (C_1 + \ln|\cos x|)\cos x + (C_2 + x)\sin x$$

18. $y'' + y = \sec x \tan x$

$y'' + y = 0$

$m^2 + 1 = 0$ when $m = \pm i$

$y_h = C_1 \cos x + C_2 \sin x$

$y_p = v_1 \cos x + v_2 \sin x$

$\qquad v_1{}' \cos x + v_2{}' \sin x = 0$

$\qquad v_1{}'(-\sin x) + v_2{}' \cos x = \sec x \tan x$

$$v_1{}' = \frac{\begin{vmatrix} 0 & \sin x \\ \sec x \tan x & \cos x \end{vmatrix}}{\begin{vmatrix} \cos x & \sin x \\ -\sin x & \cos x \end{vmatrix}} = -\tan^2 x$$

$$v_1 = \int -\tan^2 x \, dx = -\int (\sec^2 x - 1) \, dx$$

$$= -\tan x + x$$

$$v_2{}' = \frac{\begin{vmatrix} \cos x & 0 \\ -\sin x & \sec x \tan x \end{vmatrix}}{\begin{vmatrix} \cos x & \sin x \\ -\sin x & \cos x \end{vmatrix}} = \tan x$$

$$v_2 = \int \tan x \, dx = -\ln|\cos x| = \ln|\sec x|$$

$y = y_h + y_p$

$\quad = C_1 \cos x + C_2 \sin x + (x - \tan x)\cos x$

$\qquad + \ln|\sec x| \sin x$

$\quad = (C_1 + x - \tan x)\cos x + (C_2 + \ln|\sec x|)\sin x$

19. $y'' + 4y = \csc 2x$

$y'' + 4y = 0$

$m^2 + 4 = 0$ when $m = -2i, \ 2i$.

$y_h = C_1 \cos 2x + C_2 \sin 2x$

$y_p = v_1 \cos 2x + v_2 \sin 2x = 0$

$\qquad v_1{}' \cos 2x + v_2{}' \sin 2x = 0$

$\qquad v_1{}'(-2\sin 2x) + v_2{}'(2\cos 2x) = \csc 2x$

$$v_1{}' = \frac{\begin{vmatrix} 0 & \sin 2x \\ \csc 2x & 2\cos 2x \end{vmatrix}}{\begin{vmatrix} \cos 2x & \sin 2x \\ -2\sin 2x & 2\cos 2x \end{vmatrix}} = -\frac{1}{2}$$

$$v_1 = \int -\frac{1}{2} \, dx = -\frac{1}{2}x$$

$$v_2{}' = \frac{\begin{vmatrix} \cos 2x & 0 \\ -2\sin 2x & \csc 2x \end{vmatrix}}{\begin{vmatrix} \cos 2x & \sin 2x \\ -2\sin 2x & 2\cos 2x \end{vmatrix}} = \frac{1}{2}\cot 2x$$

$$v_2 = \int \frac{1}{2}\cot 2x \, dx = \frac{1}{4}\ln|\sin 2x|$$

$$y = \left(C_1 - \frac{1}{2}x\right)\cos 2x + \left(C_2 + \frac{1}{4}\ln|\sin 2x|\right)\sin 2x$$

20. $y'' - 4y' + 4y = x^2 e^{2x}$

$y'' - 4y' + 4y = 0$

$m^2 - 4m + 4 = 0$ when $m = 2, \ 2$.

$y_h = (C_1 + C_2 x)e^{2x}$

$y_p = (v_1 + v_2 x)e^{2x}$

$\qquad v_1{}' e^{2x} + v_2{}' x e^{2x} = 0$

$\qquad v_1{}'(2e^{2x}) + v_2{}'(2x + 1)e^{2x} = x^2 e^{2x}$

$$v_1{}' = \frac{\begin{vmatrix} 0 & xe^{2x} \\ x^2 e^{2x} & (2x+1)e^{2x} \end{vmatrix}}{\begin{vmatrix} e^{2x} & xe^{2x} \\ 2e^{2x} & (2x+1)e^{2x} \end{vmatrix}} = \frac{-x^3 e^{4x}}{e^{4x}} = -x^3$$

$$v_1 = \int -x^3 \, dx = -\frac{1}{4}x^4$$

$$v_2{}' = \frac{\begin{vmatrix} e^{2x} & 0 \\ 2e^{2x} & x^2 e^{2x} \end{vmatrix}}{e^{4x}} = \frac{x^2 e^{4x}}{e^{4x}} = x^2$$

$$v_2 = \int x^2 \, dx = \frac{1}{3}x^3$$

$$y = \left(C_1 + C_2 x + \frac{1}{12}x^4\right)e^{2x}$$

21. $y'' - 2y' + y = e^x \ln x$

$y'' - 2y' + y = 0$

$m^2 - 2m + 1 = 0$ when $m = 1, \ 1.$

$y_h = (C_1 + C_2 x)e^x$

$y_p = (v_1 + v_2 x)e^x$

$\qquad v_1{}'e^x + v_2 x e^x = 0$

$\qquad v_1{}'e^x + v_2{}'(x+1)e^x = e^x \ln x$

$\qquad v_1{}' = -x \ln x$

$\qquad v_1 = \int -x \ln x \, dx = -\dfrac{x^2}{2} \ln x + \dfrac{x^2}{4}$

$\qquad v_2{}' = \ln x$

$\qquad v_2 = \int \ln x \, dx = x \ln x - x$

$y = (C_1 + C_2 x)e^x + \dfrac{x^2 e^x}{4}(\ln x^2 - 3)$

22. $y'' - 4y' + 4y = \dfrac{e^{2x}}{x}$

$y'' - 4y' + 4y = 0$

$m^2 - 4m + 4 = 0$ when $m = 2, \ 2.$

$y_h = (C_1 + C_2 x)e^{2x}$

$y_p = (v_1 + v_2 x)e^{2x}$

$\qquad v_1{}'e^{2x} + v_2{}' x e^{2x} = 0$

$\qquad v_1{}'e^{2x}(2) + v_2{}'(2x+1)e^{2x} = \dfrac{e^{2x}}{x}$

$\qquad v_1{}' = -1$

$\qquad v_1 = \int -1 \, dx = -x$

$\qquad v_2{}' = \dfrac{1}{x}$

$\qquad v_2 = \int \dfrac{1}{x} \, dx = \ln|x|$

$y = (C_1 + C_2 x - x + x \ln|x|)e^{2x}$

23. $q'' + 10q' + 25q = 6 \sin 5t, \quad q(0) = 0, \quad q'(0) = 0$

$m^2 + 10m + 25 = 0$ when $m = -5, \ -5.$

$\qquad q_h = (C_1 + C_2 t)e^{-5t}$

$\qquad q_p = A \cos 5t + B \sin 5t$

$\qquad q_p{}' = -5A \sin 5t + 5B \cos 5t$

$\qquad q_p{}'' = -25A \cos 5t - 25B \sin 5t$

$q_p{}'' + 10q_p{}' + 25q_p = 50B \cos 5t - 50A \sin 5t = 6 \sin 5t, \quad A = -\tfrac{3}{25}, \quad B = 0$

$\qquad q = (C_1 + C_2 t)e^{-5t} - \tfrac{3}{25} \cos 5t$

Initial conditions: $q(0) = 0, \quad q'(0) = 0, \quad C_1 - \tfrac{3}{25} = 0, \quad -5C_1 + C_2 = 0, \quad C_1 = \tfrac{3}{25}, \quad C_2 = \tfrac{3}{5}$

Particular solution: $q = \tfrac{3}{25}(e^{-5t} + 5te^{-5t} - \cos 5t)$

24. $q'' + 20q' + 50q = 10 \sin 5t$

$m^2 + 20m + 50 = 0$ when $m = -10 \pm 5\sqrt{2}.$

$\qquad q_h = C_1 e^{(-10+5\sqrt{2})t} + C_2 e^{(-10-5\sqrt{2})t}$

$\qquad q_p = A \cos 5t + B \sin 5t$

$\qquad q_p{}' = 5B \cos 5t - 5A \sin 5t$

$\qquad q_p{}'' = -25A \cos 5t - 25B \sin 5t$

$q_p{}'' + 20q_p{}' + 50q_p = (25A + 100B) \cos 5t + (25B - 100A) \sin 5t = 10 \sin 5t$

$\left.\begin{array}{l} 25A + 100B = 0 \\ 25B - 100A = 10 \end{array}\right\} B = \tfrac{2}{85}, \quad A = -\tfrac{8}{85}$

$q = C_1 e^{(-10+5\sqrt{2})t} + C_2 e^{(-10-5\sqrt{2})t} - \tfrac{8}{85} \cos 5t + \tfrac{2}{85} \sin 5t$

Initial conditions: $q(0) = 0, \quad q'(0) = 0, \quad C_1 + [C_2 - (8/85)] = 0,$

$\qquad (-10 + 5\sqrt{2})C_1 + (-10 - 5\sqrt{2})C_2 + (2/17) = 0,$

$\qquad C_1 = (8 + 7\sqrt{2})/(170), \quad C_2 = (8 - 7\sqrt{2})/(170)$

Particular solution: $q = [(8 + 7\sqrt{2})/(170)]e^{(-10+5\sqrt{2})t} + [(8 - 7\sqrt{2})/(170)]e^{(-10-5\sqrt{2})t}$

$\qquad\qquad - (8/85) \cos 5t + (2/85) \sin 5t$

25. $\frac{24}{32}y'' + 48y = \frac{24}{32}(48\sin 4t)$, $y(0) = \frac{1}{4}$, $y'(0) = 0$

$\frac{24}{32}m^2 + 48 = 0$ when $m = \pm 8i$.

$$y_h = C_1\cos 8t + C_2\sin 8t$$

$$y_p = A\sin 4t + B\cos 4t$$

$$y_p' = 4A\cos 4t - 4B\sin 4t$$

$$y_p'' = -16A\sin 4t - 16B\cos 4t$$

$$\frac{24}{32}y_p'' + 48y_p = 36A\sin 4t + 36B\cos 4t = \frac{24}{32}(48\sin 4t), \quad B = 0, \quad A = 1$$

$$y = y_h + y_p = C_1\cos 8t + C_2\sin 8t + \sin 4t$$

Initial conditions: $y(0) = \frac{1}{4}$, $y'(0) = 0$, $\frac{1}{4} = C_1$, $0 = 8C_2 + 4 \Rightarrow C_2 = -\frac{1}{2}$

Particular solution: $y = \frac{1}{4}\cos 8t - \frac{1}{2}\sin 8t + \sin 4t$

26. $\frac{2}{32}y'' + 4y = \frac{2}{32}(4\sin 8t)$, $y(0) = \frac{1}{4}$, $y'(0) = 0$

$\frac{2}{32}m^2 + 4 = 0$ when $m = \pm 8i$.

$$y_h = C_1\cos 8t + C_2\sin 8t$$

$$y_p = At\sin 8t + Bt\cos 8t$$

$$y_p'' = (-64At - 16B)\sin 8t + (16A - 64Bt)\cos 8t$$

$$\frac{2}{32}y_p'' + 4y_p = -B\sin 8t + A\cos 8t = \frac{2}{32}(4\sin 8t), \quad A = 0, \quad B = -\frac{1}{4}$$

$$y = C_1\cos 8t + C_2\sin 8t - \frac{1}{4}t\cos 8t$$

Initial conditions: $y(0) = \frac{1}{4}$, $y'(0) = 0$, $\frac{1}{4} = C_1$, $0 = 8C_2 - \frac{1}{4} \Rightarrow C_2 = \frac{1}{32}$

Particular solution: $y = \frac{1}{4}\cos 8t + \frac{1}{32}\sin 8t - \frac{1}{4}t\cos 8t$

27. $\frac{2}{32}y'' + y' + 4y = \frac{2}{32}(4\sin 8t)$, $y(0) = \frac{1}{4}$, $y'(0) = -3$

$\frac{1}{16}m^2 + m + 4 = 0$ when $m = -8, -8$.

$$y_h = (C_1 + C_2t)e^{-8t}$$

$$y_p = A\sin 8t + B\cos 8t$$

$$y_p' = 8A\cos 8t - 8B\sin 8t$$

$$y_p'' = -64A\sin 8t - 64B\cos 8t$$

$$\frac{2}{32}y_p'' + y_p' + 4y_p = -8B\sin 8t + 8A\cos 8t = \frac{2}{32}(4\sin 8t)$$

$$-8B = \frac{1}{4} \Rightarrow B = -\frac{1}{32}, \quad 8A = 0 \Rightarrow A = 0$$

$$y = y_h + y_p = (C_1 + C_2t)e^{-8t} - \frac{1}{32}\cos 8t$$

Initial conditions: $y(0) = \frac{1}{4}$, $y'(0) = -3$, $\frac{1}{4} = C_1 - \frac{1}{32} \Rightarrow C_1 = \frac{9}{32}$, $-3 = -8C_1 + C_2 \Rightarrow C_2 = -\frac{3}{4}$

Particular solution: $y = \left(\frac{9}{32} - \frac{3}{4}t\right)e^{-8t} - \frac{1}{32}\cos 8t$

28. $\frac{4}{32}y'' + \frac{1}{2}y' + \frac{25}{2}y = 0$, $y(0) = \frac{1}{2}$, $y'(0) = -4$

$\frac{1}{8}m^2 + \frac{1}{2}m + \frac{25}{2} = 0$

$m^2 + 4m + 100 = 0$ when $m = -2 \pm 4\sqrt{6}\,i$.

$y = C_1 e^{-2t}\cos(4\sqrt{6}\,t) + C_2 e^{-2t}\sin(4\sqrt{6}\,t)$

Initial conditions: $y(0) = \frac{1}{2}$, $y'(0) = -4$, $\frac{1}{2} = C_1$, $-4 = -2C_1 + 4\sqrt{6}\,C_2$, $C_2 = -\frac{3}{4\sqrt{6}} = -\frac{\sqrt{6}}{8}$

Particular solution: $y = \frac{1}{2}e^{-2t}\cos(4\sqrt{6}\,t) - \frac{\sqrt{6}}{8}e^{-2t}\sin(4\sqrt{6}\,t)$

29. In Exercise 25,

$$y_h = \frac{1}{4}\cos 8t - \frac{1}{2}\sin 8t = \frac{\sqrt{5}}{4}\sin\left[8t + \arctan\left(-\frac{1}{2}\right)\right] = \frac{\sqrt{5}}{4}\sin\left(8t - \arctan\frac{1}{2}\right) \approx \frac{\sqrt{5}}{4}\sin(8t - 0.4636).$$

30. $\displaystyle\lim_{t\to\infty}\left[\frac{1}{2}e^{-2t}\cos(4\sqrt{6}\,t) - \frac{\sqrt{6}}{8}e^{-2t}\sin(4\sqrt{6}\,t)\right] = 0$

Thus, when there is no external acceleration imposed on the system, ultimately the retarding force has no effect.

31. $-5y'' - 8y' = 160$

$-5m^2 - 8m = 0$ when $m = 0, -\frac{8}{5}$.

$\qquad y_h = C_1 + C_2 e^{-1.6t}$

$\qquad y_p = At + B$

$\qquad y_p{}' = A$

$\qquad y_p{}'' = 0$

$-5y'' - 8y' = -8A = 160 \Rightarrow A = -20$

$\qquad y = C_1 + C_2 e^{-1.6t} - 20t$

Initial conditions:

$\qquad y(0) = 2000$, $y'(0) = -100$, $2000 = C_1 + C_2$, $-100 = -1.6C_2 - 20$, $C_2 = 50 \Rightarrow C_1 = 1950$

Particular solution: $y = 1950 + 50e^{-1.6t} - 20t$

32. $-6y'' - 9y' = 192$

$-6m^2 - 9m = 0$ when $m = 0, -1.5$.

$\qquad y_h = C_1 + C_2 e^{-1.5t}$

$\qquad y_p = At + B$

$-6y'' - 9y' = -9A = 192 \Rightarrow A = -\frac{64}{3}$

$\qquad y = C_1 + C_2 e^{-1.5t} - \frac{64}{3}t$

Initial conditions:

$\qquad y(0) = 2000$, $y'(0) = -100$, $2000 = C_1 + C_2$, $-100 = -1.5C_2 - \frac{64}{3}$, $C_2 = \frac{472}{9} \Rightarrow C_1 = \frac{17528}{9}$

$y = \frac{17528}{9} + \frac{472}{9}e^{-1.5t} - \frac{64}{3}t$

Section 17.7   Series Solutions of Differential Equations

1. $y' - y = 0$

Letting $y = \sum\limits_{n=0}^{\infty} a_n x^n$,

$$y' - y = \sum_{n=0}^{\infty} n a_n x^{n-1} - \sum_{n=0}^{\infty} a_n x^n$$

$$= \sum_{n=-1}^{\infty} (n+1) a_{n+1} x^n - \sum_{n=0}^{\infty} a_n x^n = 0$$

$$(n+1) a_{n+1} = a_n$$

$$a_{n+1} = \frac{a_n}{n+1}$$

$$a_1 = a_0, \ a_2 = \frac{a_1}{2} = \frac{a_0}{2}, \ a_3 = \frac{a_2}{3} = \frac{a_0}{1 \cdot 2 \cdot 3}, \ \ldots, \ a_n = \frac{a_0}{n!}$$

$$y = \sum_{n=0}^{\infty} \frac{a_0}{n!} x^n = a_0 e^x.$$

2. $y' - ky = 0$

Letting $y = \sum\limits_{n=0}^{\infty} a_n x^n$,

$$y' - ky = \sum_{n=0}^{\infty} n a_n x^{n-1} - k \sum_{n=0}^{\infty} a_n x^n$$

$$= \sum_{n=-1}^{\infty} (n+1) a_{n+1} x^n - \sum_{n=0}^{\infty} k a_n x^n = 0$$

$$(n+1) a_{n+1} = k a_n$$

$$a_{n+1} = \frac{k a_n}{n+1}$$

$$a_1 = k a_0, \ a_2 = \frac{k a_1}{2} = \frac{k^2 a_0}{2}, \ a_3 = \frac{k a_2}{3} = \frac{k^3 a_0}{1 \cdot 2 \cdot 3}, \ \ldots, \ a_n = \frac{k^n}{n!} a_0$$

$$y = \sum_{n=0}^{\infty} \frac{k^n}{n!} a_0 x^n = a_0 \sum_{n=0}^{\infty} \frac{(kx)^n}{n!} = a_0 e^{kx}.$$

3. $y'' - 9y = 0$. Letting $y = \sum_{n=0}^{\infty} a_n x^n$,

$$y'' - 9y = \sum_{n=0}^{\infty} n(n-1)a_n x^{n-2} - 9\sum_{n=0}^{\infty} a_n x^n = \sum_{n=-2}^{\infty} (n+2)(n+1)a_{n+2}x^n - \sum_{n=0}^{\infty} 9a_n x^n = 0$$

$$(n+2)(n+1)a_{n+2} = 9a_n$$

$$a_{n+2} = \frac{9a_n}{(n+2)(n+1)}$$

$$a_0 = a_0 \qquad\qquad\qquad a_1 = a_1$$

$$a_2 = \frac{9a_0}{2} \qquad\qquad\qquad a_3 = \frac{9a_1}{3 \cdot 2}$$

$$a_4 = \frac{9a_2}{4 \cdot 3} = \frac{9^2 a_0}{4 \cdot 3 \cdot 2 \cdot 1} \qquad\qquad a_5 = \frac{9a_3}{5 \cdot 4} = \frac{9^2 a_1}{5 \cdot 4 \cdot 3 \cdot 2 \cdot 1}$$

$$\vdots \qquad\qquad\qquad\qquad \vdots$$

$$a_{2n} = \frac{9^n a_0}{(2n)!} \qquad\qquad\qquad a_{2n+1} = \frac{9^n a_1}{(2n+1)!}$$

$$y = \sum_{n=0}^{\infty} \frac{9^n a_0}{(2n)!}x^{2n} + \sum_{n=0}^{\infty} \frac{9^n a_1}{(2n+1)!}x^{2n+1} = a_0\sum_{n=0}^{\infty} \frac{(3x)^{2n}}{(2n)!} + \frac{a_1}{3}\sum_{n=0}^{\infty} \frac{(3x)^{2n+1}}{(2n+1)!}$$

$$= C_0\sum_{n=0}^{\infty} \frac{(3x)^n}{n!} + C_1\sum_{n=0}^{\infty} \frac{(-3x)^n}{n!}$$

$$= C_0 e^{3x} + C_1 e^{-3x} \text{ where } C_0 + C_1 = a_0 \text{ and } C_0 - C_1 = \frac{a_1}{3}.$$

4. $y = C_0 e^{kx} + C_1 e^{-kx}$. Follow the solution to Exercise 3 with 9 replaced by k^2.

5. $y'' + 4y = 0$. Letting $y = \sum_{n=0}^{\infty} a_n x^n$,

$$y'' + 4y = \sum_{n=0}^{\infty} n(n-1)a_n x^{n-2} + 4\sum_{n=0}^{\infty} a_n x^n = \sum_{n=-2}^{\infty} (n+2)(n+1)a_{n+2}x^n + \sum_{n=0}^{\infty} 4a_n x^n = 0$$

$$(n+2)(n+1)a_{n+2} = -4a_n$$

$$a_{n+2} = \frac{-4a_n}{(n+2)(n+1)}$$

$$a_0 = a_0 \qquad\qquad\qquad a_1 = a_1$$

$$a_2 = \frac{-4a_0}{2} \qquad\qquad\qquad a_3 = \frac{-4a_1}{3 \cdot 2}$$

$$a_4 = \frac{-4a_2}{4 \cdot 3} = \frac{(-4)^2}{4!}a_0 \qquad\qquad a_5 = \frac{-4a_3}{5 \cdot 4} = \frac{(-4)^2 a_1}{5!}$$

$$\vdots \qquad\qquad\qquad\qquad \vdots$$

$$a_{2n} = \frac{(-1)^n 4^n}{(2n)!}a_0 \qquad\qquad a_{2n+1} = \frac{(-1)^n 4^n}{(2n+1)!}a_1$$

$$y = \sum_{n=0}^{\infty} \frac{(-1)^n 4^n a_0}{(2n)!}x^{2n} + \sum_{n=0}^{\infty} \frac{(-1)^n 4^n a_1}{(2n+1)!}x^{2n+1} = a_0\sum_{n=0}^{\infty} \frac{(-1)^n (2x)^{2n}}{(2n)!} + \frac{a_1}{4}\sum_{n=0}^{\infty} \frac{(-1)^n (2x)^{2n+1}}{(2n+1)!}$$

$$= C_0 \cos 2x + C_1 \sin 2x.$$

6. $y = C_0 \cos kx + C_1 \sin kx$. Follow the solution to Exercise 5 with 4 replaced by k^2.

7. $y' + 3xy = 0$. Letting $y = \sum\limits_{n=0}^{\infty} a_n x^n$,

$$y' + 3xy = \sum_{n=0}^{\infty} n a_n x^{n-1} + \sum_{n=0}^{\infty} 3 a_n x^{n+1} = 0$$

$$\sum_{n=-2}^{\infty} (n+2) a_{n+2} x^{n+1} = \sum_{n=0}^{\infty} -3 a_n x^{n+1}$$

$$a_{n+2} = \frac{-3 a_n}{n+2}$$

$a_0 = a_0$ $\qquad\qquad\qquad\qquad$ $a_1 = a_1$

$a_2 = -\dfrac{3 a_0}{2}$ $\qquad\qquad\qquad$ $a_3 = -\dfrac{3 a_1}{3}$

$a_4 = -\dfrac{3}{4}\left(-\dfrac{3 a_0}{2}\right) = \dfrac{3^2}{2^3} a_0$ $\qquad$ $a_5 = -\dfrac{3}{5}\left(-\dfrac{3 a_1}{3}\right) = \dfrac{3^2 a_1}{3 \cdot 5}$

$a_6 = -\dfrac{3}{6}\left(\dfrac{3^2}{2^3} a_0\right) = -\dfrac{3^3 a_0}{2^3(3 \cdot 2)}$ $\qquad$ $a_7 = -\dfrac{3}{7}\left(\dfrac{3^2 a_1}{3 \cdot 5}\right) = -\dfrac{3^3 a_1}{3 \cdot 5 \cdot 7}$

$a_8 = -\dfrac{3}{8}\left(-\dfrac{3^3 a_0}{2^3(3 \cdot 2)}\right) = \dfrac{3^4 a_0}{2^4(4 \cdot 3 \cdot 2)}$ $\qquad$ $a_9 = -\dfrac{3}{9}\left(-\dfrac{3^3 a_1}{3 \cdot 5 \cdot 7}\right) = \dfrac{3^4 a_1}{3 \cdot 5 \cdot 7 \cdot 9}$

$$y = a_0 \sum_{n=0}^{\infty} \frac{(-3)^n x^{2n}}{2^n n!} + a_1 \sum_{n=0}^{\infty} \frac{(-3)^n x^{2n+1}}{1 \cdot 3 \cdot 5 \cdot 7 \cdots (2n+1)}.$$

8. $y' - 2xy = 0$. Letting $y = \sum\limits_{n=0}^{\infty} a_n x^n$,

$$y' - 2xy = \sum_{n=0}^{\infty} n a_n x^{n-1} - \sum_{n=0}^{\infty} 2 a_n x^{n+1} = 0$$

$$\sum_{n=-2}^{\infty} (n+2) a_{n+2} x^{n+1} = \sum_{n=0}^{\infty} 2 a_n x^{n+1}$$

$$a_{n+2} = \frac{2 a_n}{n+2}$$

$a_0 = a_0$ $\qquad\qquad\qquad\qquad$ $a_1 = a_1$

$a_2 = \dfrac{2 a_0}{2} = a_0$ $\qquad\qquad\qquad$ $a_3 = \dfrac{2 a_1}{3}$

$a_4 = \dfrac{2}{4}\left(\dfrac{2 a_0}{2}\right) = \dfrac{2^2 a_0}{2^2 \cdot 2} = \dfrac{a_0}{2}$ $\qquad$ $a_5 = \dfrac{2}{5}\left(\dfrac{2 a_1}{3}\right) = \dfrac{2^2 a_1}{3 \cdot 5}$

$a_6 = \dfrac{2}{6}\left(\dfrac{2^2 a_0}{2^2 \cdot 2}\right) = \dfrac{2^3 a_0}{2^3 3 \cdot 2} = \dfrac{a_0}{3!}$ $\qquad$ $a_7 = \dfrac{2}{7}\left(\dfrac{2^2 a_1}{3 \cdot 5}\right) = \dfrac{2^3 a_1}{3 \cdot 5 \cdot 7}$

$a_8 = \dfrac{2}{8}\left(\dfrac{a_0}{3!}\right) = \dfrac{a_0}{4!}$ $\qquad\qquad$ $a_9 = \dfrac{2}{9}\left(\dfrac{2^3 a_1}{3 \cdot 5 \cdot 7}\right) = \dfrac{2^4 a_1}{3 \cdot 5 \cdot 7 \cdot 9}$

$$y = a_0 \sum_{n=0}^{\infty} \frac{x^{2n}}{n!} + a_1 \sum_{n=0}^{\infty} \frac{2^n x^{2n+1}}{1 \cdot 3 \cdot 5 \cdot 7 \cdots (2n+1)}.$$

9. $y'' - xy' = 0$. Letting $y = \sum_{n=0}^{\infty} a_n x^n$,

$$y'' - xy' = \sum_{n=0}^{\infty} n(n-1)a_n x^{n-2} - x\sum_{n=0}^{\infty} na_n x^{n-1} = 0$$

$$\sum_{n=0}^{\infty} n(n-1)a_n x^{n-2} = \sum_{n=0}^{\infty} na_n x^n$$

$$\sum_{n=-2}^{\infty} (n+2)(n+1)a_{n+2} x^n = \sum_{n=0}^{\infty} na_n x^n$$

$$a_{n+2} = \frac{na_n}{(n+2)(n+1)}$$

$a_0 = a_0$ $\qquad\qquad\qquad a_1 = a_1$

$a_2 = 0$ $\qquad\qquad\qquad a_3 = \dfrac{a_1}{3 \cdot 2}$

There are no even powered terms.

$$a_5 = \frac{3a_3}{5 \cdot 4} = \frac{3a_1}{5!}$$

$$a_7 = \frac{5a_5}{7 \cdot 6} = \frac{5 \cdot 3a_1}{7!}$$

$$y = a_1\sum_{n=0}^{\infty} \frac{1 \cdot 3 \cdot 5 \cdot 7 \cdots (2n-1)x^{2n+1}}{(2n+1)!} = a_1\sum_{n=0}^{\infty} \frac{(2n)!x^{2n+1}}{2^n n!(2n+1)!} = a_1\sum_{n=0}^{\infty} \frac{x^{2n+1}}{2^n n!(2n+1)}.$$

10. $y'' - xy' - y = 0$. Letting $y = \sum_{n=0}^{\infty} a_n x^n$,

$$y'' - xy' - y = \sum_{n=0}^{\infty} n(n-1)a_n x^{n-2} - x\sum_{n=0}^{\infty} na_n x^{n-1} - \sum_{n=0}^{\infty} a_n x^n = 0$$

$$\sum_{n=-2}^{\infty} (n+2)(n+1)a_{n+2} x^n = \sum_{n=0}^{\infty} (n+1)a_n x^n$$

$$a_{n+2} = \frac{a_n}{n+2}$$

$a_0 = a_0$ $\qquad\qquad\qquad a_1 = a_1$

$a_2 = \dfrac{a_0}{2}$ $\qquad\qquad\qquad a_3 = \dfrac{a_1}{3}$

$a_4 = \dfrac{a_2}{4} = \dfrac{a_0}{8} = \dfrac{a_0}{2^2 2!}$ $\qquad a_5 = \dfrac{a_3}{5} = \dfrac{a_1}{3 \cdot 5}$

$a_6 = \dfrac{a_4}{6} = \dfrac{a_0}{2^3 3!}$ $\qquad\qquad a_7 = \dfrac{a_5}{7} = \dfrac{a_1}{3 \cdot 5 \cdot 7}$

$a_8 = \dfrac{a^6}{8} = \dfrac{a_0}{2^4 4!}$ $\qquad\qquad a_9 = \dfrac{a_7}{9} = \dfrac{a_1}{3 \cdot 5 \cdot 7 \cdot 9}$

$$y = a_0\sum_{n=0}^{\infty} \frac{x^{2n}}{2^n n!} + a_1\sum_{n=0}^{\infty} \frac{x^{2n+1}}{1 \cdot 3 \cdot 5 \cdot 7 \cdots (2n+1)}.$$

11. $(x^2 + 4)y'' + y = 0.$ Letting $y = \displaystyle\sum_{n=0}^{\infty} a_n x^n,$

$$(x^2 + 4)y'' + y = \sum_{n=0}^{\infty} n(n-1)a_n x^n + 4\sum_{n=0}^{\infty} n(n-1)a_n x^{n-2} + \sum_{n=0}^{\infty} a_n x^n$$

$$= \sum_{n=0}^{\infty} (n^2 - n + 1)a_n x^n + \sum_{n=-2}^{\infty} 4(n+2)(n+1)a_{n+2} x^n = 0$$

$$a_{n+2} = \frac{-(n^2 - n + 1)a_n}{4(n+2)(n+1)}$$

$$a_0 = a_0 \qquad\qquad a_1 = a_1$$

$$a_2 = \frac{-a_0}{4 \cdot 2 \cdot 1} \qquad\qquad a_3 = \frac{-a_1}{4(3)(2)} = \frac{-a_1}{24}$$

$$a_4 = \frac{-3a_2}{4(4)(3)} = \frac{a_0}{128} \qquad a_5 = \frac{-7a_3}{4(5)(4)} = \frac{7a_1}{1920}$$

$$y = a_0 \left(1 - \frac{x^2}{8} + \frac{x^4}{128} - \cdots\right) + a_1 \left(x - \frac{x^3}{24} + \frac{7x^5}{1920} - \cdots\right).$$

12. $y'' + x^2 y = 0.$ Letting $y = \displaystyle\sum_{n=0}^{\infty} a_n x^n,$

$$y'' + x^2 y = \sum_{n=0}^{\infty} n(n-1)a_n x^{n-2} + \sum_{n=0}^{\infty} a_n x^{n+2} = 0$$

$$\sum_{n=-4}^{\infty} (n+4)(n+3)a_{n+4} x^{n+2} = -\sum_{n=0}^{\infty} a_n x^{n+2}$$

$$a_{n+4} = \frac{-a_n}{(n+4)(n+3)}.$$

Also, $y = a_0 + a_1 x + a_2 x^2 + a_3 x^3 + \cdots + a_n x^n + \cdots$

$$y'' = 2a_2 + 3 \cdot 2a_3 x + \cdots + n(n-1)a_n x^{n-2} + \cdots$$

$$y'' + x^2 y = 2a_2 + 3 \cdot 2a_3 x + (a_0 + 4 \cdot 3a_4)x^2 + (a_1 + 5 \cdot 4a_5)x^3 + \cdots = 0.$$

$$2a_2 = 0, \quad 6a_3 = 0, \quad 12a_4 + a_0 = 0, \quad 20a_5 + a_1 = 0$$

Thus, $a_2 = 0$ and $a_3 = 0 \Rightarrow a_6 = 0, \quad a_7 = 0, \quad a_{10} = 0,$ and $a_{11} = 0.$ Therefore, $a_{4n+2} = 0$ and $a_{4n+3} = 0.$

$$a_0 = a_0 \qquad\qquad\qquad a_1 = a_1$$

$$a_4 = -\frac{a_0}{4 \cdot 3} \qquad\qquad\qquad a_5 = -\frac{a_1}{5 \cdot 4}$$

$$a_8 = -\frac{a_4}{8 \cdot 7} = \frac{a_0}{8 \cdot 7 \cdot 4 \cdot 3} \qquad a_9 = -\frac{a_5}{9 \cdot 8} = \frac{a_1}{9 \cdot 8 \cdot 5 \cdot 4}$$

$$a_{12} = -\frac{a_8}{12 \cdot 11} = -\frac{a_0}{12 \cdot 11 \cdot 8 \cdot 7 \cdot 4 \cdot 3} \qquad a_{13} = \frac{a_9}{13 \cdot 12} = \frac{a_1}{13 \cdot 12 \cdot 9 \cdot 8 \cdot 5 \cdot 4}$$

$$y'' + x^2 y = a_0 \left(1 - \frac{x^4}{4 \cdot 3} + \frac{x^8}{8 \cdot 7 \cdot 4 \cdot 3} - \frac{x^{12}}{12 \cdot 11 \cdot 8 \cdot 7 \cdot 4 \cdot 3} + \cdots\right)$$

$$+ a_1 \left(x - \frac{x^5}{5 \cdot 4} + \frac{x^7}{9 \cdot 8 \cdot 5 \cdot 4} - \frac{x^9}{13 \cdot 12 \cdot 9 \cdot 8 \cdot 5 \cdot 4} + \cdots\right)$$

13. $y' + (2x - 1)y = 0$, $\quad y(0) = 2$

$$y' = (1 - 2x)y \qquad\qquad y'(0) = 2$$
$$y'' = (1 - 2x)y' - 2y \qquad\qquad y''(0) = -2$$
$$y''' = (1 - 2x)y'' - 4y' \qquad\qquad y'''(0) = -10$$
$$y^{(4)} = (1 - 2x)y''' - 6y'' \qquad y^{(4)}(0) = 2$$
$$\vdots \qquad\qquad\qquad\qquad \vdots$$

$$y(x) = 2 + \frac{2}{1!}x - \frac{2}{2!}x^2 - \frac{10}{3!}x^3 + \frac{2}{4!}x^4 + \cdots$$

Using the first five terms of the series, $y\left(\dfrac{1}{2}\right) = \dfrac{163}{64} \approx 2.547$.

14. $y' - 2xy = 0$, $\quad y(0) = 1$

$$y' = 2xy \qquad\qquad y'(0) = 0$$
$$y'' = 2(xy' + y) \qquad\qquad y''(0) = 2$$
$$y''' = 2(xy'' + 2y') \qquad\qquad y'''(0) = 0$$
$$y^{(4)} = 2(xy''' + 3y'') \qquad y^{(4)}(0) = 12$$
$$y^{(5)} = 2(xy^{(4)} + 4y''') \qquad y^{(5)}(0) = 0$$
$$y^{(6)} = 2(xy^{(5)} + 5y^{(4)}) \qquad y^{(6)}(0) = 120$$
$$\vdots \qquad\qquad\qquad\qquad \vdots$$

$$y(x) = 1 + \frac{2}{2!}x^2 + \frac{12}{4!}x^4 + \frac{120}{6!}x^6 + \cdots = 1 + x^2 + \frac{1}{2}x^4 + \frac{1}{6}x^6 + \cdots$$

Using the first four terms of the series, $y(1) = \dfrac{8}{3} \approx 2.667$.

15. $y'' - 2xy = 0$, $\quad y(0) = 1$, $\quad y'(0) = -3$

$$y'' = 2xy \qquad\qquad y''(0) = 0$$
$$y''' = 2(xy' + y) \qquad\qquad y'''(0) = 2$$
$$y^{(4)} = 2(xy'' + 2y') \qquad y^{(4)}(0) = -12$$
$$y^{(5)} = 2(xy''' + 3y'') \qquad y^{(5)}(0) = 0$$
$$y^{(6)} = 2(xy^{(4)} + 4y''') \qquad y^{(6)}(0) = 16$$
$$y^{(7)} = 2(xy^{(5)} + 5y^{(4)}) \qquad y^{(7)}(0) = -120$$
$$\vdots \qquad\qquad\qquad\qquad \vdots$$

$$y(x) = 1 - \frac{3}{1!}x + \frac{2}{3!}x^3 - \frac{12}{4!}x^4 + \frac{16}{6!}x^6 - \frac{120}{7!}x^7 + \cdots$$

Using the first six terms of the series, $y\left(\dfrac{1}{4}\right) \approx 0.253$.

16. $y'' - 2xy' + y = 0,$ $y(0) = 1,$ $y'(0) = 2$

$$y'' = 2xy' - y \qquad\qquad y''(0) = -1$$

$$y''' = 2xy'' + y' \qquad\qquad y'''(0) = 2$$

$$y^{(4)} = 2xy''' + 3y'' \qquad\quad y^{(4)}(0) = -3$$

$$y^{(5)} = 2xy^{(4)} + 5y''' \qquad y^{(5)}(0) = 10$$

$$y^{(6)} = 2xy^{(5)} + 7y^{(4)} \qquad y^{(6)}(0) = -21$$

$$y^{(7)} = 2xy^{(6)} + 9y^{(5)} \qquad y^{(7)}(0) = 90$$

$$\vdots \qquad\qquad\qquad\qquad \vdots$$

$$y(x) = 1 + \frac{2}{1!}x - \frac{1}{2!}x^2 + \frac{2}{3!}x^3 - \frac{3}{4!}x^4 + \frac{10}{5!}x^5 - \frac{21}{6!}x^6 + \frac{90}{7!}x^7 - \cdots$$

Using the first eight terms of the series, $y\left(\dfrac{1}{2}\right) \approx 1.911.$

17. $f(x) = e^x,$ $f'(x) = e^x,$ $y' - y = 0.$ Assume $y = \displaystyle\sum_{n=0}^{\infty} a_n x^n,$ then

$$y' = \sum_{n=0}^{\infty} n a_n x^{n-1}$$

$$\sum_{n=0}^{\infty} n a_n x^{n-1} = \sum_{n=0}^{\infty} a_n x^n$$

$$\sum_{n=-1}^{\infty} (n+1) a_{n+1} x^n = \sum_{n=0}^{\infty} a_n x^n$$

$$a_{n+1} = \frac{a_n}{n+1}, \quad n \geq 0.$$

$n = 0,$ $a_1 = a_0$

$n = 1,$ $a_2 = \dfrac{a_1}{2} = \dfrac{a_0}{2}$

$n = 2,$ $a_3 = \dfrac{a_2}{3} = \dfrac{a_0}{2(3)}$

$n = 3,$ $a_4 = \dfrac{a_3}{4} = \dfrac{a_0}{2(3)(4)}$

$n = 4,$ $a_5 = \dfrac{a_4}{5} = \dfrac{a_0}{2(3)(4)(5)}$

$$\vdots$$

$$a_{n+1} = \frac{a_0}{(n+1)!} \Rightarrow a_n = \frac{a_0}{n!}$$

$y = a_0 \displaystyle\sum_{n=0}^{\infty} \dfrac{x^n}{n!}$ which converges on $(-\infty, \infty).$ When $a_0 = 1,$ we have the Maclaurin Series for $f(x) = e^x.$

18. $f(x) = \cos x$, $f'(x) = -\sin x$, $f''(x) = -\cos x$, $y'' + y = 0$. Assume $y = \displaystyle\sum_{n=0}^{\infty} a_n x^n$, then

$$y'' = \sum_{n=0}^{\infty} n(n-1)a_n x^{n-2}$$

$$\sum_{n=0}^{\infty} n(n-1)a_n x^{n-2} + \sum_{n=0}^{\infty} a_n x^n = 0$$

$$\sum_{n=-2}^{\infty} (n+2)(n+1)a_{n+2} x^n = -\sum_{n=0}^{\infty} a_n x^n$$

$$a_{n+2} = -\frac{a_n}{(n+1)(n+2)}, \quad n \geq 0.$$

$$a_0 = a_0 \qquad\qquad a_1 = a_1$$

$$a_2 = -\frac{a_0}{(1)(2)} \qquad\qquad a_3 = -\frac{a_1}{(2)(3)}$$

$$a_4 = -\frac{a_2}{(3)(4)} = \frac{a_0}{4!} \qquad\qquad a_5 = -\frac{a_3}{(4)(5)} = \frac{a_1}{5!}$$

$$\vdots \qquad\qquad\qquad\qquad \vdots$$

$$a_{2n} = \frac{(-1)^n a_0}{(2n)!} \qquad\qquad a_{2n+1} = \frac{(-1)^n a_1}{(2n+1)!}$$

$$y = a_0 \sum_{n=0}^{\infty} \frac{(-1)^n x^{2n}}{(2n)!} + a_1 \sum_{n=0}^{\infty} \frac{(-1)^n x^{2n+1}}{(2n+1)!} \text{ which converges on } (-\infty, \infty).$$

When $a_0 = 1$ and $a_1 = 0$, we have the Maclaurin Series for $f(x) = \cos x$.

19.

$$f(x) = \arctan x$$

$$f'(x) = \frac{1}{1 + x^2}$$

$$f''(x) = \frac{-2x}{(1 + x^2)^2}$$

$$y'' = \frac{-2x}{1 + x^2} y'$$

$$(1 + x^2)y'' + 2xy' = 0$$

Assume $y = \sum_{n=0}^{\infty} a_n x^n$, then

$$y' = \sum_{n=0}^{\infty} n a_n x^{n-1}$$

$$y'' = \sum_{n=0}^{\infty} n(n-1) a_n x^{n-2}$$

$$(1 + x^2)y'' + 2xy' = \sum_{n=0}^{\infty} n(n-1) a_n x^{n-2} + \sum_{n=0}^{\infty} n(n-1) a_n x^n + \sum_{n=0}^{\infty} 2n a_n x^n = 0$$

$$\sum_{n=0}^{\infty} n(n-1) a_n x^{n-2} = -\sum_{n=0}^{\infty} n(n-1) a_n x^n - \sum_{n=0}^{\infty} 2n a_n x^n$$

$$\sum_{n=-2}^{\infty} (n+2)(n+1) a_{n+2} x^n = -\sum_{n=0}^{\infty} n(n+1) a_n x^n$$

$$(n+2)(n+1) a_{n+2} = -n(n+1) a_n$$

$$a_{n+2} = -\frac{n}{n+2} a_n, \quad n \geq 0.$$

$n = 0 \Rightarrow a_2 = 0 \Rightarrow$ all the even power terms have a coefficient of 0

$n = 1, \quad a_3 = -\frac{1}{3} a_1$

$n = 3, \quad a_5 = -\frac{3}{5} a_3 = \frac{1}{5} a_1$

$n = 5, \quad a_7 = -\frac{5}{7} a_5 = -\frac{1}{7} a_1$

$n = 7, \quad a_9 = -\frac{7}{9} a_7 = \frac{1}{9} a_1$

$$\vdots$$

$$a_{2n+1} = \frac{(-1)^n a_1}{2n+1}$$

$$y = a_1 \sum_{n=0}^{\infty} \frac{(-1)^n x^{2n+1}}{2n+1} \quad \text{which converges on } (-1, 1).$$

When $a_1 = 1$, we have the Maclaurin Series for $f(x) = \arctan x$.

20.

$$f(x) = \arcsin x$$

$$f'(x) = \frac{1}{\sqrt{1-x^2}}$$

$$f''(x) = \frac{x}{(1-x^2)^{3/2}}$$

$$y'' = \frac{1}{\sqrt{1-x^2}} \cdot \frac{x}{1-x^2} = \frac{x}{1-x^2}y'$$

$$(1-x^2)y'' - xy' = 0$$

Assume $y = \sum\limits_{n=0}^{\infty} a_n x^n$, then

$$\sum_{n=0}^{\infty} a_n n(n-1)x^{n-2} - \sum_{n=0}^{\infty} a_n n(n-1)x^n - \sum_{n=0}^{\infty} a_n n x^n = 0$$

$$\sum_{n=-2}^{\infty} (n+2)(n+1)a_{n+2}x^n = \sum_{n=0}^{\infty} n^2 a_n x^n$$

$$a_{n+2} = \frac{n^2}{(n+1)(n+2)}a_n, \quad n \geq 0.$$

$n = 0 \Rightarrow a_2 = 0 \Rightarrow$ all the even power terms have a coefficient of 0

$$a_1 = a_1$$

$$n = 1, \quad a_3 = \frac{1}{(2)(3)}a_1$$

$$n = 3, \quad a_5 = \frac{9}{(4)(5)}a_3 = \frac{9}{(2)(3)(4)(5)}a_1 = \frac{3}{(2)(4)(5)}a_1$$

$$n = 5, \quad a_7 = \frac{25}{(6)(7)}a_5 = \frac{(9)(25)}{(2)(3)(4)(5)(6)(7)}a_1 = \frac{(3)(5)}{(2)(4)(6)(7)}a_1$$

$$n = 7, \quad a_9 = \frac{49}{(8)(9)}a_7 = \frac{(9)(25)(49)}{(2)(3)(4)(5)(6)(7)(8)(9)}a_1 = \frac{(3)(5)(7)}{(2)(4)(6)(8)(9)}a_1$$

$$n = 9, \quad a_{11} = \frac{81}{(10)(11)}a_9 = \frac{(9)(25)(49)(81)}{(2)(3)(4)(5)(6)(7)(8)(9)(10)(11)}a_1 = \frac{(3)(5)(7)(9)}{(2)(4)(6)(8)(10)(11)}a_1$$

$$\vdots$$

$$a_{2n+1} = \frac{(2n)!}{(2^n n!)^2(2n+1)}a_1$$

$$y = a_1\sum_{n=0}^{\infty} \frac{(2n)!}{(2^n n!)^2(2n+1)}x^{2n+1} \text{ which converges on } (-1, 1).$$

When $a_1 = 1$, we have the Maclaurin Series for $f(x) = \arcsin x$.

Chapter 17   Review Exercises

1. $\dfrac{dy}{dx} - \dfrac{y}{x} = 2 + \sqrt{x}$

Integrating factor: $e^{-\int (1/x)\,dx} = e^{-\ln|x|} = \dfrac{1}{x}$

$y\left(\dfrac{1}{x}\right) = \displaystyle\int \dfrac{1}{x}(2 + \sqrt{x})\,dx = \ln x^2 + 2\sqrt{x} + C$

$\qquad y = x\ln x^2 + 2x^{3/2} + Cx$

2. $y' + xy = 2y$

$\dfrac{dy}{dx} = -(x - 2)y$

$\displaystyle\int \dfrac{1}{y}\,dy = \int -(x - 2)\,dx$

$\qquad \ln y = -\dfrac{1}{2}(x - 2)^2 + \ln C$

$\qquad y = Ce^{-(x-2)^2/2}$

3. $\dfrac{dy}{dx} - \dfrac{2y}{x} = \dfrac{1}{x}\dfrac{dy}{dx}$

$(x - 1)\dfrac{dy}{dx} = 2y$

$\displaystyle\int \dfrac{1}{y}\,dy = \int \dfrac{2}{x - 1}\,dx$

$\qquad \ln|y| = \ln(x - 1)^2 + \ln C$

$\qquad y = C(x - 1)^2$

4. $\dfrac{dy}{dx} - 3x^2 y = e^{x^3}$

Integrating factor: $e^{\int -3x^2\,dx} = e^{-x^3}$

$ye^{-x^3} = \displaystyle\int dx$

$ye^{-x^3} = x + C$

$\qquad y = (x + C)e^{x^3}$

5. $\dfrac{dy}{dx} - \dfrac{y}{x} = \dfrac{x}{y}$, Bernoulli

$n = -1, \quad P = -\dfrac{1}{x}, \quad Q = x,$

$e^{\int (-2/x)\,dx} = e^{\ln x^{-2}} = x^{-2}$

$y^2 x^{-2} = \displaystyle\int 2(x)x^{-2}\,dx = \ln x^2 + C$

$\qquad y^2 = x^2 \ln x^2 + Cx^2$

6. $y' - \dfrac{3y}{x^2} = \dfrac{1}{x^2}$

Integrating factor: $e^{\int -(3/x^2)\,dx} = e^{3/x}$

$ye^{3/x} = \displaystyle\int \dfrac{1}{x^2}e^{3/x}\,dx = -\dfrac{1}{3}e^{3/x} + C$

$\qquad y = -\dfrac{1}{3} + Ce^{-3/x}$

7. $y' - 2y = e^x$

Integrating factor: $e^{\int -2\,dx} = e^{-2x}$

$ye^{-2x} = \displaystyle\int e^{-2x}e^x\,dx + C = -e^{-x} + C$

$\qquad y = Ce^{2x} - e^x$

8. $y' + \dfrac{2y}{x} = -x^9 y^5$, Bernoulli

$n = 5, \quad P = \dfrac{2}{x}, \quad Q = -x^9,$

$e^{\int -4(2/x)\,dx} = e^{\ln x^{-8}} = x^{-8}$

$y^{-4}x^{-8} = \displaystyle\int -4(-x^9)(x^{-8})\,dx + C$

$\qquad\qquad = 2x^2 + C$

$x^8 y^4(2x^2 + C) = 1$

9. $(10x + 8y + 2)\, dx + (8x + 5y + 2)\, dy = 0$

Exact: $\dfrac{\partial M}{\partial y} = 8 = \dfrac{\partial N}{\partial x}$

$$U(x,\ y) = \int (10x + 8y + 2)\, dx = 5x^2 + 8xy + 2x + f(y)$$

$$U_y(x,\ y) = 8x + f'(y) = 8x + 5y + 2$$

$$f'(y) = 5y + 2$$

$$f(y) = \frac{5}{2}y^2 + 2y + C_1$$

$$U(x,\ y) = 5x^2 + 8xy + 2x + \frac{5}{2}y^2 + 2y + C_1$$

$$5x^2 + 8xy + 2x + \frac{5}{2}y^2 + 2y = C$$

10. $\quad (y + x^3 + xy^2)\, dx - x\, dy = 0$

$$y\, dx - x\, dy + x(x^2 + y^2)\, dx = 0$$

$$\frac{y\, dx - x\, dy}{x^2 + y^2} + x\, dx = 0$$

$$\arctan \frac{x}{y} + \frac{1}{2}x^2 = C_1$$

$$2 \arctan \frac{x}{y} + x^2 = C$$

11. $(2x - 2y^3 + y)\, dx + (x - 6xy^2)\, dy = 0$

Exact: $\dfrac{\partial M}{\partial y} = -6y^2 + 1 = \dfrac{\partial N}{\partial x}$

$$U(x,\ y) = \int (2x - 2y^3 + y)\, dx$$

$$= x^2 - 2xy^3 + xy + f(y)$$

$$U_y(x,\ y) = -6xy^2 + x + f'(y) = x - 6xy^2$$

$$f'(y) = 0$$

$$f(y) = C_1$$

$$U(x,\ y) = x^2 - 2xy^3 + xy + C_1$$

$$x^2 - 2xy^3 + xy = C$$

12. $3x^2y^2\, dx + (2x^3y + x^3y^4)\, dy = 0$

$$3x^2y^2\, dx + x^3(2y + y^4)\, dy = 0$$

$$\int \frac{3}{x}\, dx + \int \left(\frac{2}{y} + y^2\right)\, dy = 0$$

$$\ln x^3 + \ln y^2 + \frac{1}{3}y^3 = C_1$$

$$3 \ln x^3 y^2 + y^3 = C$$

13. $\quad x\, dy = (x + y + 2)\, dx$

$$\frac{dy}{dx} - \left(\frac{1}{x}\right) y = \frac{x + 2}{x}$$

Integrating factor: $e^{\int -(1/x)\, dx} = e^{-\ln |x|} = \dfrac{1}{x}$

$$y\left(\frac{1}{x}\right) = \int \frac{x + 2}{x^2}\, dx = \ln |x| - \frac{2}{x} + C$$

$$y = x \ln |x| - 2 + Cx$$

14. $ye^{xy}\, dx + xe^{xy}\, dy = 0$

$$y\, dx + x\, dy = 0$$

$$xy = C$$

15. $\ln(1+y)\,dx + \left(\dfrac{1}{1+y}\right) dy = 0$

$$\int dx + \int \frac{1}{(1+y)\ln(1+y)}\,dy = C_1$$

$$x + \ln|\ln(1+y)| = C_1$$

$$\ln|\ln(1+y)| = C_1 - x$$

$$\ln|1+y| = e^{C_1 - x} = Ce^{-x}$$

16. $(2x + y - 3)\,dx + (x - 3y + 1)\,dy = 0$

Exact: $\dfrac{\partial M}{\partial y} = 1 = \dfrac{\partial N}{\partial x}$

$$U(x,\ y) = \int (2x + y - 3)\,dx = x^2 + xy - 3x + f(y)$$

$$U_y(x,\ y) = x + f'(y) = x - 3y + 1$$

$$f'(y) = -3y + 1$$

$$f(y) = -\frac{3}{2}y^2 + y + C_1$$

$$U(x,\ y) = x^2 + xy - 3x - \frac{3}{2}y^2 + y + C_1$$

$$2x^2 + 2xy - 6x - 3y^2 + 2y = C$$

17. $dy = (y\tan x + 2e^x)\,dx$

$$\frac{dy}{dx} - (\tan x)y = 2e^x$$

Integrating factor: $e^{\int -\tan x\,dx} = e^{\ln|\cos x|} = \cos x$

$$y\cos x = \int 2e^x \cos x\,dx = e^x(\cos x + \sin x) + C$$

$$y = e^x(1 + \tan x) + C\sec x$$

18. $y\,dx - (x + \sqrt{xy})\,dy = 0$

Integrating factor: $\dfrac{1}{2x^{1/2}y^{3/2}}$

$$\frac{y\,dx - x\,dy}{2x^{1/2}y^{3/2}} - \frac{1}{2y}\,dy = 0$$

$$\sqrt{x/y} - \ln\sqrt{y} = C$$

19. $(x - y - 5)\,dx - (x + 3y - 2)\,dy = 0$

Exact: $\dfrac{\partial M}{\partial y} = -1 = \dfrac{\partial N}{\partial x}$

$$U(x,\ y) = \int (x - y - 5)\,dx = \frac{1}{2}x^2 - xy - 5x + f(y)$$

$$U_y(x,\ y) = -x + f'(y) = -x - 3y + 2$$

$$f'(y) = -3y + 2$$

$$f(y) = -\frac{3}{2}y^2 + 2y + C_1$$

$$U(x,\ y) = \frac{1}{2}x^2 - xy - 5x - \frac{3}{2}y^2 + 2y + C_1$$

$$x^2 - 2xy - 10x - 3y^2 + 4y = C$$

20.
$$y' = x^2 y^2 - 9x^2$$

$$\int \frac{1}{y^2 - 9}\, dy = \int x^2\, dx$$

$$\frac{1}{6} \ln \left| \frac{y-3}{y+3} \right| = \frac{1}{3} x^3 + C_1$$

$$\ln \left| \frac{y-3}{y+3} \right| = 2x^3 + C$$

21.
$$2xy' - y = x^3 - x$$

$$y' - \left(\frac{1}{2x} \right) y = \frac{1}{2}(x^2 - 1)$$

Integrating factor: $e^{\int -(1/2x)\, dx} = e^{\ln x^{-1/2}} = x^{-1/2}$

$$yx^{-1/2} = \int \frac{1}{2} x^{-1/2} (x^2 - 1)\, dx$$

$$= \frac{1}{5} x^{5/2} - x^{1/2} + C$$

$$y = \frac{1}{5} x^3 - x + C\sqrt{x}$$

22.
$$\frac{dy}{dx} = 2x\sqrt{1 - y^2}$$

$$\int \frac{1}{\sqrt{1 - y^2}}\, dy = \int 2x\, dx$$

$$\arcsin y = x^2 + C$$

$$y = \sin(x^2 + C)$$

23.
$$x + yy' = \sqrt{x^2 + y^2}$$

$$\int \frac{x\, dx + y\, dy}{\sqrt{x^2 + y^2}} = \int dx$$

$$\sqrt{x^2 + y^2} = x + C$$

$$x^2 + y^2 = x^2 + 2Cx + C^2$$

$$y^2 = 2Cx + C^2$$

24.
$$xy' + y = \sin x$$

$$y' + \left(\frac{1}{x} \right) y = \frac{1}{x} \sin x$$

Integrating factor: $e^{\int (1/x)\, dx} = e^{\ln x} = x$

$$yx = \int \sin x\, dx = -\cos x + C$$

$$yx + \cos x = C$$

25.
$$yy' + y^2 = 1 + x^2$$

$$y' + y = \frac{1}{y}(1 + x^2), \quad \text{Bernoulli}$$

$$n = -1, \ P = 1, \ Q = 1 + x^2, \ e^{\int 2\, dx} = e^{2x}$$

$$y^2 e^{2x} = \int 2(1 + x^2) e^{2x}\, dx$$

$$= \left(x^2 - x + \frac{3}{2} \right) e^{2x} + C$$

$$y^2 = x^2 - x + \frac{3}{2} + Ce^{-2x}$$

26.
$$2x\, dx + 2y\, dy = (x^2 + y^2)\, dx$$

$$\int \frac{2x\, dx + 2y\, dy}{x^2 + y^2} = \int dx$$

$$\ln(x^2 + y^2) = x + C$$

$$\ln(x^2 + y^2) - x = C$$

27.

$$(1 + x^2)\, dy = (1 + y^2)\, dx$$

$$\int \frac{1}{1 + y^2}\, dy = \int \frac{1}{1 + x^2}\, dx$$

$$\arctan y - \arctan x = C_1$$

$$\tan(\arctan y - \arctan x) = \tan C_1$$

$$\frac{y - x}{1 + xy} = C$$

28. $\quad y' = \dfrac{x^4 + 3x^2 y^2 + y^4}{x^3 y}$

$$x^3 y\, dy = (x^4 + 3x^2 y^2 + y^4)\, dx$$

Homogeneous

$$y = vx, \quad dy = v\, dx + x\, dv$$

$$vx^4(v\, dx + x\, dv) = (x^4 + 3v^2 x^4 + v^4 x^4)\, dx$$

$$v^2\, dx + vx\, dv = (1 + 3v^2 + v^4)\, dx$$

$$\int \frac{v}{(1 + v^2)^2}\, dv = \int \frac{1}{x}\, dx$$

$$\frac{-1}{2(1 + v^2)} = \ln |x| + C_1$$

$$\frac{-x^2}{x^2 + y^2} = \ln x^2 + C$$

29. $y' - \left(\dfrac{a}{x}\right) y = bx^3$

Integrating factor: $e^{-\int (a/x)\, dx} = e^{-a \ln x} = x^{-a}$

$$yx^{-a} = \int bx^3 (x^{-a})\, dx = \frac{b}{4 - a} x^{4 - a} + C$$

$$y = \frac{bx^4}{4 - a} + Cx^a$$

30. $\quad y' = y + 2x(y - e^x)$

$$y' - (1 + 2x)y = -2xe^x$$

Integrating factor: $e^{\int -(1 + 2x)\, dx} = e^{-(x + x^2)}$

$$ye^{-(x + x^2)} = \int e^{-(x + x^2)}(-2xe^x)\, dx = e^{-x^2} + C$$

$$y = e^x(1 + Ce^{x^2})$$

31. $y'' + y = x^3 + x$

$\quad m^2 + 1 = 0$ when $m = -i,\ i.$

$$y_h = C_1 \cos x + C_2 \sin x$$

$$y_p = A_0 + A_1 x + A_2 x^2 + A_3 x^3$$

$$y_p' = A_1 + 2A_2 x + 3A_3 x^2$$

$$y_p'' = 2A_2 + 6A_3 x$$

$$y_p'' + y_p = (A_0 + 2A_2) + (A_1 + 6A_3)x + A_2 x^2 + A_3 x^3$$

$$= x^3 + x$$

$$A_0 = 0, \quad A_1 = -5, \quad A_2 = 0, \quad A_3 = 1$$

$$y = C_1 \cos x + C_2 \sin x - 5x + x^3$$

32. $y'' + 2y = e^{2x} + x$

$\quad m^2 + 2 = 0$ when $m = -\sqrt{2}\,i,\ \sqrt{2}\,i.$

$$y_h = C_1 \cos \sqrt{2}\, x + C_2 \sin \sqrt{2}\, x$$

$$y_p = Ae^{2x} + B_0 + B_1 x$$

$$y_p' = 2Ae^{2x} + B_1$$

$$y_p'' = 4Ae^{2x}$$

$$y_p'' + 2y_p = 6Ae^{2x} + 2B_0 + 2B_1 x = e^{2x} + x$$

$$A = \tfrac{1}{6}, \quad B_0 = 0, \quad B_1 = \tfrac{1}{2}$$

$$y = C_1 \cos \sqrt{2}x + C_2 \sin \sqrt{2}x + \tfrac{1}{6}e^{2x} + \tfrac{1}{2}x$$

33. $y'' + y = 2\cos x$

$m^2 + 1 = 0$ when $m = -i,\ i.$

$$y_h = C_1 \cos x + C_2 \sin x$$

$$y_p = Ax\cos x + Bx\sin x$$

$$y_p{}' = (Bx + A)\cos x + (B - Ax)\sin x$$

$$y_p{}'' = (2B - Ax)\cos x + (-Bx - 2A)\sin x$$

$$y_p{}'' + y_p = 2B\cos x - 2A\sin x = 2\cos x$$

$$A = 0,\quad B = 1$$

$$y = C_1\cos x + (C_2 + x)\sin x$$

34. $y'' + 5y' + 4y = x^2 + \sin 2x$

$m^2 + 5m + 4 = 0$ when $m = -1,\ -4.$

$$y_h = C_1 e^{-x} + C_2 e^{-4x}$$

$$y_p = A_0 + A_1 x + A_2 x^2 + B_0 \sin 2x + B_1 \cos 2x$$

$$y_p{}' = A_1 + 2A_2 x + 2B_0 \cos 2x - 2B_1 \sin 2x$$

$$y_p{}'' = 2A_2 - 4B_0 \sin 2x - 4B_1 \cos 2x$$

$$y_p{}'' + 5y_p{}' + 4y_p = (4A_0 + 5A_1 + 2A_2) + (4A_1 + 10A_2)x + 4A_2 x^2 - 10B_1 \sin 2x + 10B_0 \cos 2x = x^2 + \sin 2x$$

$$A_0 = \tfrac{21}{32},\quad A_1 = -\tfrac{5}{8},\quad A_2 = \tfrac{1}{4},\quad B_0 = 0,\quad B_1 = -\tfrac{1}{10}$$

$$y = C_1 e^{-x} + C_2 e^{-4x} + \tfrac{21}{32} - \tfrac{5}{8}x + \tfrac{1}{4}x^2 - \tfrac{1}{10}\cos 2x$$

35. $y'' - 2y' + y = 2xe^x$

$m^2 - 2m + 1 = 0$ when $m = 1,\ 1.$

$$y_h = (C_1 + C_2 x)e^x$$

$$y_p = (v_1 + v_2 x)e^x$$

$$v_1{}'e^x + v_2{}'xe^x = 0$$

$$v_1{}'e^x + v_2{}'(x+1)e^x = 2xe^x$$

$$v_1{}' = -2x^2$$

$$v_1 = \int -2x^2\,dx = -\tfrac{2}{3}x^3$$

$$v_2{}' = 2x$$

$$v_2 = \int 2x\,dx = x^2$$

$$y = \left(C_1 + C_2 x + \tfrac{1}{3}x^3\right)e^x$$

36. $y'' + 2y' + y = \dfrac{1}{x^2 e^x}$

$m^2 + 2m + 1 = 0$ when $m = -1,\ -1.$

$$y_h = (C_1 + C_2 x)e^{-x}$$

$$y_p = (v_1 + v_2 x)e^{-x}$$

$$v_1{}'e^{-x} + v_2{}'(xe^{-x}) = 0$$

$$v_1{}'(-e^{-x}) + v_2{}'(-x+1)e^{-x} = \dfrac{1}{e^x x^2}$$

$$v_1{}' = -\dfrac{1}{x}$$

$$v_1 = \int -\dfrac{1}{x}\,dx = -\ln|x|$$

$$v_2{}' = \dfrac{1}{x^2}$$

$$v_2 = \int \dfrac{1}{x^2}\,dx = -\dfrac{1}{x}$$

$$y = (C_1 + C_2 x - \ln|x| - 1)e^{-x}$$

37.
$$(x - C)^2 + y^2 = C^2$$
$$x^2 - 2Cx + C^2 + y^2 = C^2$$
$$\frac{x^2 + y^2}{x} = 2C$$
$$\frac{x(2x + 2yy') - (x^2 + y^2)}{x^2} = 0$$
$$2x^2 + 2xyy' - x^2 - y^2 = 0$$
$$y' = \frac{y^2 - x^2}{2xy}$$

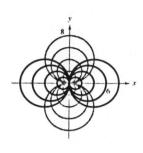

The negative reciprocal of y' is the slope of the orthogonal trajectories.
$$\frac{dy}{dx} = \frac{2xy}{x^2 - y^2}$$
$$2xy\,dx + (y^2 - x^2)\,dy = 0$$

Homogeneous

$$x = vy, \quad dx = v\,dy + y\,dv$$
$$2vy^2(v\,dy + y\,dv) + (y^2 - v^2y^2)\,dy = 0$$
$$\int \frac{2v}{1 + v^2}\,dv + \int \frac{1}{y}\,dy = 0$$
$$\ln(1 + v^2) + \ln|y| = \ln K_1$$
$$y^2 + x^2 = K_1 y$$
$$x^2 + (y - K)^2 = K^2$$

38. $y - 2x = C$
$$y = 2x + C$$
$$y' = 2$$

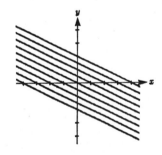

The negative reciprocal of y' is the slope of the orthogonal trajectories.
$$\frac{dy}{dx} = -\frac{1}{2}$$
$$y = -\frac{1}{2}x + K$$
$$x + 2y = K$$

39. $(x-4)y' + y = 0$. Letting $y = \sum_{n=0}^{\infty} a_n x^n$,

$$xy' - 4y' + y = \sum_{n=0}^{\infty} n a_n x^n - 4\sum_{n=0}^{\infty} n a_n x^{n-1} + \sum_{n=0}^{\infty} a_n x^n = \sum_{n=0}^{\infty}(n+1)a_n x^n - \sum_{n=0}^{\infty} 4n a_n x^{n-1}$$

$$= \sum_{n=0}^{\infty}(n+1)a_n x^n - \sum_{n=-1}^{\infty} 4(n+1)a_{n+1} x^n = 0$$

$$(n+1)a_n = 4(n+1)a_{n+1}$$

$$a_{n+1} = \frac{1}{4}a_n$$

$$a_0 = a_0, \quad a_1 = \frac{1}{4}a_0, \quad a_2 = \frac{1}{4}a_1 = \frac{1}{4^2}a_0, \quad \cdots a_n = \frac{1}{4^n}a_0$$

$$y = a_0 \sum_{n=0}^{\infty} \frac{x^n}{4^n}.$$

40. $y'' + 3xy' - 3y = 0$. Letting $y = \sum_{n=0}^{\infty} a_n x^n$,

$$y'' + 3xy' - 3y = \sum_{n=0}^{\infty} n(n-1)a_n x^{n-2} + 3x\sum_{n=0}^{\infty} n a_n x^{n-1} - 3\sum_{n=0}^{\infty} a_n x^n = 0$$

$$\sum_{n=-2}^{\infty}(n+2)(n+1)a_{n+2}x^n = \sum_{n=0}^{\infty}(3-3n)a_n x^n$$

$$a_{n+2} = \frac{3(1-n)a_n}{(n+2)(n+1)}$$

$a_0 = a_0$ $a_1 = a_1$

$a_2 = \frac{3}{2\cdot 1}a_0$ $a_3 = 0$

There are no odd-powered terms for $n > 1$.

$$a_4 = -\frac{3}{4\cdot 3}\left(\frac{3}{2\cdot 1}a_0\right) = -\frac{3(3)a_0}{4!}$$

$$a_6 = -\frac{3(3)}{6\cdot 5}\left(-\frac{3(3)a_0}{4!}\right) = \frac{3^3(3)a_0}{6!}$$

$$a_8 = -\frac{3(5)}{8\cdot 7}\left(\frac{3^3(3)a_0}{6!}\right) = -\frac{3^4(5\cdot 3)a_0}{8!}$$

$$a_{10} = -\frac{3(7)}{10\cdot 9}\left(-\frac{3^4(5\cdot 3)a_0}{8!}\right) = \frac{3^5(7\cdot 5\cdot 3)a_0}{10!}$$

$$y = a_0 + \frac{3}{2}a_0 x^2 + a_0\sum_{n=2}^{\infty}\frac{(-1)^{n+1}3^n[3\cdot 5\cdot 7\cdots(2n-3)]}{(2n)!}x^{2n}$$